T5-CQA-885

TEACHING AND LEARNING HIGH SCHOOL MATHEMATICS

CHARLENE E. BECKMANN
Grand Valley State University

DENISSE R. THOMPSON
University of South Florida

RHETA N. RUBENSTEIN
University of Michigan–Dearborn

WILEY
John Wiley & Sons, Inc.

VICE PRESIDENT AND PUBLISHER Laurie Rosatone

ASSOCIATE EDITOR Jennifer Brady

MARKETING MANAGER Sarah Davis

SENIOR PRODUCTION EDITOR Trish McFadden

MEDIA MANAGER Melissa Edwards

DESIGN DIRECTOR Harry Nolan

SENIOR DESIGNER Madelyn Lesure

PRODUCTION MANAGEMENT SERVICES Elm Street Publishing Services

This book was set in 10/12pt JansonText-Roman by Thomson Digital and printed and bound by Courier Kendallville. The cover was printed by Courier Kendallville.

This book is printed on acid free paper. ∞

Copyright © 2010 John Wiley & Sons, Inc. All rights reserved. No part of this publication may be reproduced, stored in a retrieval system ortransmitted in any form or by any means, electronic, mechanical, photocopying, recording, scanning, or otherwise, except as permitted under Sections 107 or 108 of the 1976 United States Copyright Act, without either the prior written permission of the Publisher, or authorization through payment of the appropriate per-copy fee to the Copyright Clearance Center, Inc., 222 Rosewood Drive, Danvers, MA 01923, (978)750-8400, fax (978)750-4470 or on the web at www.copyright.com. Requests to the Publisher for permission should be addressed to the Permissions Department, John Wiley & Sons, Inc., 111 River Street, Hoboken, NJ 07030-5774, (201)748-6011, fax (201)748-6008, or online at http://www.wiley.com/go/permissions.

To order books or for customer service please, call 1-800-CALL WILEY (225-5945).

ISBN-13 978-0470-45450-3
ISBN-10 0-470-45450-4

Printed in the United States of America

10 9 8 7 6 5 4 3 2 1

Dedicated to our families, our teachers, and our students,
who have nurtured our love of learning.
May the legacy continue.

Table of Contents

Annotated Table of Contents

360–364). Reston, VA: National Council of Teachers of Mathematics, 2000.

- "A Truth Table on the Island of Truthtellers and Liars" by Christopher Baltus. *Mathematics Teacher*, 94 (December 2001): 730–732.

1.4 Learning from Students: Valid and Invalid Arguments **50**

This lesson addresses two additional techniques for cooperative learning that help students rely on their peers while being individually accountable for their learning. The mathematical focus is valid forms of argument.

READINGS

- "A Unified Framework for Proof and Disproof" by Susanna S. Epp. *Mathematics Teacher*, 91 (November 1998): 708–713.
- "Proofs That Students Can Do" by Robert O. Stanton. *Mathematics Teacher*, 99 (March 2006): 478–482.

1.5 Summarizing Classroom Observations and Listening to Students **62**

This lesson provides an opportunity for teachers to share experiences from the observations and interviews with students assigned earlier in this unit. In particular, the observations were structured to focus on the nature of questions asked in the classroom and the expectations that teachers have of students in classes at different ability levels. The interviews provided an opportunity to explore students' thinking as they reasoned through problems.

Synthesizing Unit One **67**

The unit synthesis highlights the main concepts in the unit and contains a series of questions designed to help teachers connect the unit themes with the issues in the professional readings.

Unit One Investigation: Carpet Square Mazes **69**

Students need opportunities to engage in extended problems that require more than a couple of days to solve. Each unit of this textbook contains an investigation related to the content of the unit to give teachers an opportunity to engage in the types of extended problems they may want to consider using with their students. In the Unit One Investigation, teachers extend the work begun in the team builder to mazes of larger dimensions.

UNIT TWO: HIGH SCHOOL STUDENTS AND HOW THEY LEARN 71

Mathematics Strands: Geometry and Measurement

The pedagogical focus of this unit is on how high school students learn mathematics, highlighting the need for teachers to create classroom environments in which students are actively engaged in making sense of mathematics. Instructional approaches to assist students in learning mathematics, such as concrete materials, technology, multiple perspectives, and the use of contexts, are discussed within the mathematical content areas of geometry and measurement, with aspects of proof embedded throughout the unit. The van Hiele model of the development of geometric thought provides a framework for exploring how students' knowledge of geometry develops. An emphasis on questioning techniques and analyzing students' thinking, begun in Unit One, helps build prerequisite knowledge needed for lesson planning in Units Three and Four.

Unit Two Team Builder: Transformed Snowflakes **74**

As the unit begins, teachers engage in a new team builder to form teams for study during the unit. This team builder requires members to write their names and use various symmetries to transform their names to create a snowflake. Team members then design a team logo using their names and horizontal, vertical, or rotational symmetries.

Preparing to Observe Mathematics Classrooms: Focus on Learning **75**

The observation connected with this unit focuses on the Learning Principle from Principles and Standards for School Mathematics. *In addition, teachers use a general reading about issues related to mathematics learning to guide their observations of mathematics classrooms.*

READINGS

- "Learning Principle" from *Principles and Standards for School Mathematics* (pp. 20–21). Reston, VA: National Council of Teachers of Mathematics, 2000.
- "Introduction" by M. Suzanne Donovan and John D. Bransford. In *How Students Learn: Mathematics in the Classroom*, edited by M. Suzanne Donovan and John D. Bransford (pp. 1–28). Washington, DC: National Academies Press, 2005.

Listening to Students Reason about Geometry **77**

In this interview, students explore geometric figures formed by joining consecutive midpoints of sides of polygons. Students attempt to justify their conclusions and teachers interpret their responses in light of the van Hiele levels of geometric thought.

READING

- "Linking Theory and Practice in Teaching Geometry" by Randall E. Groth. *Mathematics Teacher*, 99 (August 2005): 27–30.

2.1 Understanding Geometry Learning: Coordinate Geometry **80**

The van Hiele model of geometric thought is introduced as a framework for considering how students' geometric knowledge develops. The framework is studied in the context of coordinate geometry, exploring such properties as distance, slope, parallelism, and midpoints on a plane, and then using these properties to make conjectures about geometric figures and then prove them.

READINGS

- "Geometry Standard for Grades 9–12" from *Principles and Standards for School Mathematics* (pp. 308–318). Reston, VA: National Council of Teachers of Mathematics, 2000.
- "Linking Theory and Practice in Teaching Geometry" by Randall E. Groth. *Mathematics Teacher*, 99 (August 2005): 27–30.

2.2 Building Conceptual Understanding: Congruence and Similarity **90**

When students build conceptual understanding rather than just procedural knowledge, they typically have a deeper understanding of the concepts and are flexible in their approaches to problems. This lesson explores activities that focus on conceptual understanding through examples related to congruence and similarity.

READINGS

- "The Human Body's Built-In Range Finder: The Thumb Method of Indirect Distance Measurement" by Michael Wong. *Mathematics Teacher*, 99 (May 2006): 622–626.
- "From Classroom Discussions to Group Discourse" by Azita Manouchehri and Dennis St. John. *Mathematics Teacher*, 99 (April 2006): 544–551.

simulating an experiment to obtain six specific players from repeated random draws of these six players.

Preparing to Observe Mathematics Classrooms: Focus on Teaching **294**

The observation connected with this unit focuses on the Teaching Principle from Principles and Standards for School Mathematics.

READING

- "Teaching Principle" from *Principles and Standards for School Mathematics* (pp. 16–19). Reston, VA: National Council of Teachers of Mathematics, 2000.

Listening to Students Reason about Data Analysis and Probability **296**

In this interview, high school students reason about probability through tasks on a student page similar to those used in research studies. As in the previous units, teachers consider possible questions to ask of students depending on their expected level of response.

READING

- "Research on Students' Understandings of Probability" by J. Michael Shaughnessy. In *A Research Companion to Principles and Standards for School Mathematics*, edited by Jeremy Kilpatrick, W. Gary Martin, and Deborah Schifter (pp. 216–226). Reston, VA: National Council of Teachers of Mathematics, 2003.

4.1 Planning a Lesson Launch and Explore: Data Analysis **299**

This lesson focuses on lesson planning, particularly the Launch and Explore phases, using the Thinking Through a Lesson Protocol developed by Smith, Bill, and Hughes. Data analysis, specifically the importance of variation, is the content focus for the lesson.

READINGS

- "Data Analysis and Probability" from *Principles and Standards for School Mathematics* (pp. 48–51). Reston, VA: National Council of Teachers of Mathematics, 2000.
- "How Faithful Is Old Faithful? Statistical Thinking: A Story of Variation and Prediction" by J. Michael Shaughnessy and Maxine Pfannkuch. *Mathematics Teacher*, 95 (April 2002): 252–259.
- "Integrating Statistics into a Course on Functions" by James E. Schultz and Rheta N. Rubenstein. *Mathematics Teacher*, 83 (November 1990): 612–617.

4.2 Planning a Lesson Share and Summarize: Probability **313**

This lesson continues work on lesson planning, specifically considering elements needed for an effective summary of the lesson. Teachers work on the Share and Summarize portions of lesson planning within the context of several problems involving probability.

READINGS

- "Orchestrating Discussions" by Margaret S. Smith, Elizabeth K. Hughes, Randi A. Engle, Mary K. Stein. *Mathematics Teaching in the Middle School,* 14 (May 2009): 548–556.
- "Data Analysis and Probability for Grades 9–12" *Principles and Standards for School Mathematics* (pp. 324–333). Reston, VA: National Council of Teachers of Mathematics, 2000.
- "Good Things Always Come in Threes: Three Cards, Three Prisoners, and Three Doors" by Laurie H. Rubel. *Mathematics Teacher*, 99 (February 2006): 401–405.

READINGS

- "Assessment Principle" from *Principles and Standards for School Mathematics* (pp. 22–24). Reston, VA: National Council of Teachers of Mathematics, 2000.
- "Improving Classroom Tests as a Means of Improving Assessment" by Denisse R. Thompson, Charlene E. Beckmann, and Sharon L. Senk. *Mathematics Teacher*, 90 (January 1997): 58–64.

Listening to Students Reason about Precalculus **373**

In this interview, teachers first explore the assessment tool on the website for the National Assessment of Educational Progress (NAEP). Teachers identify their own items for use with high school students and compare the performance of their students to the performance of students in a national sample.

5.1 Daily Assessments: Limits **374**

This lesson focuses on strategies for assigning homework and managing grading. Teachers work on one of several clusters to explore a range of issues related to homework. The mathematics focus of this lesson is on understanding limits in terms of behavior of rational functions at specific points.

READING

- "Assessing Students' Understanding of Functions in a Graphing Calculator Environment" by Charlene E. Beckmann, Sharon L. Senk, and Denisse R. Thompson. *School Science and Mathematics*, 99 (December 1999): 451–456.

5.2 Rubrics: Rates of Change **385**

In this lesson, teachers have an opportunity to consider general and item-specific rubrics and to apply a rubric to score sample student responses related to rates of change. In the questions, teachers develop rubrics for specific tasks.

READINGS

- "Using Rubrics in High School Mathematics Courses" by Denisse R. Thompson and Sharon L. Senk. *Mathematics Teacher*, 91 (December 1998): 786–793.
- "EMRF: Everyday Rubric Grading" by Rodney Y. Stutzman and Kimberly H. Race. *Mathematics Teacher*, 97 (January 2004): 34–39.

5.3 Designing and Aligning Tests with Instruction: Derivatives **396**

Developing tests that align with instruction is an important part of the job of mathematics teachers. This lesson has teachers critique a draft of a test for a sample unit introducing derivatives.

READING

- "Improving Classroom Tests as a Means of Improving Assessment" by Denisse R. Thompson, Charlene E. Beckmann, and Sharon L. Senk. *Mathematics Teacher*, 90 (January 1997): 58–64.

5.4 Alternative Assessments: Accumulations **404**

There are many issues teachers need to consider in relation to assessment, including concepts best assessed through strategies other than in-class tests. This lesson explores a variety of alternative assessment strategies. In addition, teachers have an opportunity to engage with an informal introduction to the integral.

READINGS

- "A, E, I, O, U, and Always Y: A Simple Technique for Improving Communication and Assessment in the Mathematics Classroom" by Lorna Thomas Vazquez. *Mathematics Teacher*, 102 (August 2008): 16–23.

READINGS

- "A Vision for School Mathematics" from *Principles and Standards for School Mathematics* (pp. 3–8). Reston, VA: National Council of Teachers of Mathematics, 2000.
- "Participation in Career-long Professional Growth," Standard 5 of Standards for the Education and Continued Professional Growth of Teachers of Mathematics. In *Mathematics Teaching Today*, edited by Tami S. Martin (pp. 157–170). Reston, VA: National Council of Teachers of Mathematics, 2007.

6.4 Summarizing Interviews on Educational Issues 445

This lesson provides an opportunity for teachers to share experiences from the interviews made in conjunction with this unit.

READING

- "The Role of Mathematics Instruction in Building a Just and Diverse Democracy" by Deborah Loewenberg Ball, Imani Masters Gaffney, and Hyman Bass. *The Mathematics Educator*, 15 (2005): 2–6.

Synthesizing Unit Six 448

The unit synthesis highlights the main concepts in the unit and contains a series of questions designed to help teachers connect the unit themes with the issues in the professional readings.

Appendix: Unit One 449

Blank Question Response Support (QRS) Guide 450
Stages of Questioning 451
Attribute Pieces for Colorful Shapes 452
Question Response Support (QRS) Guide for Colorful Shapes 453
Question Response Support (QRS) Guide for How Many Oranges? 454

Appendix: Unit Two 455

van Hiele Levels of Geometric Thought 456
Levels of Demand 457
Link Sheet 458
Frayer Model 459
Circular Geoboard 460
Square Geoboard 461
Cards for What Am I? 462
QRS Guide for What Am I? 463
QRS Guide for Exploring the Diagonals of Quadrilaterals 465
Sample Student Responses for Sprinkling around Obstacles 467
Sample Student Responses for Sprinkling around Obstacles when Using *The Geometer's Sketchpad*™ 470
QRS Guide for Sprinkling around Obstacles 472
Euclidean Geometry Postulates 473

Preface and Acknowledgments

Work on this text began with *Teaching and Learning Middle Grades Mathematics* (*TLMGM*). As we were writing the middle grades methods book, we knew we would want a high school content/methods textbook in the same inquiry-based format to help build teachers' pedagogical and content knowledge in a nurturing classroom environment that promotes equity. Thus, *Teaching and Learning High School Mathematics* (*TLHSM*) was born. The book arose from the same precepts as *TLMGM*:

- all students deserve excellent teachers;
- reasoning about concepts and problem solving is fundamental to mathematics learning;
- teaching is complex and involves numerous knowledge bases and decisions before, during, and after classroom activities;
- knowledge of teaching (methods) and knowledge of mathematics (content) must be integrated to be used effectively in practice; and
- teachers, like other learners, figure things out for themselves by working on problems.

With the support of peers, experiences with high school teachers and learners, insightful guidance of instructors, and high-quality professional resources, we believe teachers will become reflective practitioners and career-long learners.

In *Teaching and Learning High School Mathematics*, we built on the format we developed for *TLMGM*, using the lesson design format of Launch, Explore, and Share and Summarize (Lappan et al., 1998), with each lesson having a pedagogical focus developed through important high school mathematics. In the intervening years since writing the middle grades text, we have added to our repertoire and included elements that give teachers more help with developing independent, confident mathematics learners. Of particular note are Stages of Questioning (Goldin, 1998) and Question Response Support (QRS) Guides. These elements are designed to give teachers early and continual experience in planning lesson segments. These elements help teachers anticipate what and how students are learning so they can prepare questions in advance of class time that will give students just enough help to move forward without preempting their thinking. We have also included observations of high school mathematics classrooms to add another level of realism to the work teachers will do in their careers.

As in its predecessor, sample student pages appropriate for use in high school classrooms are included both in the text and in electronic form. Teachers can use them as-is or adapt them for their own students. Because we often teach as we are taught, we have integrated effective teaching approaches, such as cooperative learning strategies and the use of technological and physical tools in inquiry-based learning. Teachers learn about appropriate pedagogy through examples based on student pages, student work on problems, etc. More details about features and approaches appear in the following *Welcome to This Book!*

We are indebted to many colleagues for support and assistance through piloting various versions of this text and providing feedback:

Richard Austin, University of South Florida, Tampa, Florida

Marshall Lassak, Eastern Illinois University, Charleston, Illinois

Michael McDaniel, Aquinas College, Grand Rapids, Michigan

Michael Shelly, University of Michigan–Dearborn, Dearborn, Michigan

Liliana Toader, Siena Heights University, Adrian, Michigan

Rebecca Walker, Grand Valley State University, Allendale, Michigan

We are grateful to the many students at these institutions who willingly used pilot materials and provided suggestions for improvement. In particular, Grand Valley State University (GVSU) students, who used several drafts with Charlene Beckmann, were exceptional partners from whom we learned a great deal. Special thanks to Tim Steenwyk, a former GVSU student who used one of the initial versions of the text and who wrote and gave us permission to include problem 10 on p. 220 of the text. We also thank Stephen Blair for sharing his students' work on the problem about the intersection of a circle and a square in Lesson 2.6.

We are also indebted to several colleagues who reviewed and commented on all or part of early and later drafts of the manuscript:

Stephen Blair, Eastern Michigan University, Ypsilanti, Michigan

William Blubaugh, University of Northern Colorado, Greeley, Colorado

John Gabrosek, Grand Valley State University, Allendale, Michigan

Mike Hall, Arkansas State University, Jonesboro, Arkansas

Maria Mitchell, Central Connecticut State University, New Britain, Connecticut

Jeff Shamatha, Northern Arizona University, Flagstaff, Arizona

Wendy Hageman Smith, Longwood University, Farmville, Virginia

Thomas Snabb, University of Michigan–Dearborn, Dearborn, Michigan

Jan Yow, University of South Carolina, Columbia, South Carolina

Rose Mary Zbiek, Pennsylvania State University, State College, Pennsylvania

We appreciate the work of Myrna Jacobs from the National Council of Teachers of Mathematics, who negotiated the acquisition of permissions to use resources from the Council. We also appreciate the support and encouragement of our publishers, those from the former Key College Publishing who supported and encouraged our work from initial planning through submittal of the first full manuscript, and those at John Wiley & Sons, Inc. who nurtured the project through a final round of reviews and revision and through the publishing process. Persons especially supportive of our work have been Richard Bonacci, Casey Fitzsimons, Mike Simpson, and Steve Rasmussen from the former Key College Publishing; Jen Brady, Trish McFadden, and Laurie Rosatone from Wiley & Sons; and Debbie Meyer from Elm Street Publishing Services. Each of these persons has helped shape the final product through their suggestions, support, and encouragement.

Finally, work that is undertaken for as long as this textbook required does not occur without a great deal of support from the families of the authors. Special thanks to all of these special people in our lives, whose support, love, encouragement, and patience is most appreciated: Dave, Valen, Jackson, and Ben Beckmann; Melanie, Geoff, Gavin, and Grady Alm; Colleen, Scott, Aidan, and Tegan Mouw; Carleigh Van Allen; Donald and Sara Jo Thompson; and Howard Rubenstein.

Welcome to This Book!

"We live in a time of extraordinary and accelerating change. . . . In this changing world, those who understand and can do mathematics will have significantly enhanced opportunities and options for shaping their future. Mathematical competence opens doors to productive futures."
(*Principles and Standards for School Mathematics*, 2000, pp. 4–5)

The above statement, from the National Council of Teachers of Mathematics (NCTM), is a reminder to all mathematics teachers that success in mathematics is critically important to students' future educational and career options. Too many high school students find themselves struggling with abstract concepts and sometimes curtail engagement in further mathematics courses, likely impacting their college and career options. Thus, high school mathematics teachers have the responsibility to help students recognize the importance of mathematics while also designing instruction that makes mathematics accessible and meaningful to all students.

Ball and Bass (2000), as well as other mathematics educators, have recognized that mathematics teachers not only need to know mathematics content and mathematics pedagogy (i.e., teaching strategies) but they also need to know how these ideas are integrated. This *mathematical knowledge for teaching* is the knowledge that teachers of mathematics need and it differs from the knowledge that research or applied mathematicians must know. This textbook is designed to provide teachers opportunities to expand upon their mathematical knowledge for teaching.

Teaching and Learning High School Mathematics is likely different from many other textbooks you have used. It integrates both content and pedagogy to help you build your own understanding of teaching. The textbook is designed to help you develop "deep conceptual understanding of fundamental mathematics" (Ma, 1999) to enable you to approach mathematics from multiple perspectives with many tools. Such flexibility in teaching is essential if teachers are to help *all* students become mathematically proficient.

Throughout this textbook, you are encouraged to work in cooperative teams. There are two reasons for this work: (1) to help you develop a mathematics learning community and build a professional network that will become a valuable resource during your professional career, and (2) to help you consider how to encourage such learning environments in your own classroom.

Lesson planning is another pervasive element throughout this textbook. To help teachers plan for effective student-centered lessons, the **Question Response Support (QRS) Guide** is introduced in Lesson 1.1 and used throughout the remainder of the lessons. The **QRS Guide** is a tool in which teachers may record tasks or questions (**Q**) for students, expected and observed student responses (**R**), and teacher support (**S**) in the form of additional "just enough" questions to support students in their progress on the task. In each unit, teachers expand their repertoire of teaching and learning elements and strategies and incorporate these elements as they plan additional lesson segments. In Unit Four, lesson planning is formally introduced as teachers put together elements from previous units into complete, cohesive lesson plans.

This textbook includes *lessons organized in units*; there are six units in the course. Most units have a specific mathematics strand through which the pedagogical focus of the unit is unveiled. The mathematical focus is geared toward mathematics that high school teachers are likely to teach. Within each lesson, professional readings are suggested that correlate with the pedagogy or mathematics of the lesson. Topics from middle school mathematics taught in lower-level high school courses generally are not addressed in this textbook, as they are covered extensively in *Teaching and Learning Middle Grades Mathematics* (Rubenstein, Beckmann, & Thompson, 2004).

Each unit begins with a **Brief Overview** of the major focus of the unit and some historical perspective on the mathematical content of the unit. The course itself begins with a personal survey about mathematics education—*Where Do I Stand?*—to help you reflect on and continue to develop your personal teaching philosophy.

Also, each unit begins with a **Team Builder** activity. Teams might change as each new unit begins, so a team builder is incorporated into each unit to facilitate the formation of a cohesive team. These are also included as examples of activities in which high school students could be engaged when new teams are formed in classes you will teach.

The **Unit Investigations** in the first five units build from the Unit Team Builder and provide an opportunity for teachers to engage in a challenging mathematical problem over the course of the unit. (The Unit Investigation is found at the end of each unit.) Such extended investigations model those that are important for high school students to complete and relate to the mathematics strand of the unit. Investigations often illustrate how the mathematics of a particular unit connects to other disciplines or other strands.

The first five units also begin with **Preparing to Observe Mathematics Classrooms**; the tasks and related readings are designed to provide guidance for novice teachers during observations in classrooms of practicing teachers. Veteran teachers might use the observation guidelines to reflect on specific issues within their own classrooms that they want to investigate. Over the course, the observations focus on the six principles that undergird *Principles and Standards for School Mathematics* from the NCTM (2000). The final lesson of each unit provides an opportunity for discussion and sharing of the observations.

In addition, the first five units begin with **Preparing to Listen to Students,** a feature that provides an opportunity for teachers to listen to students as they work with a specific task to gauge how students think about the mathematics of the task. Such conversations with students often are a source of valuable insights that can influence the design of instruction. In the final unit, listening to students is replaced with **Preparing to Listen to Educational Partners** to ascertain how different stakeholders in the education of high school students interact within the larger educational system.

The **Unit Synthesis** gives users of this textbook an opportunity to synthesize ideas and issues from the unit's lessons and relate those to issues raised in the professional readings.

The lessons comprising each unit typically use a three-pronged approach: **Launch**, **Explore**, and **Share and Summarize**, adapted from the *Connected Mathematics Project* (Lappan et al., 1998) and used extensively in *Teaching and Learning Middle Grades Mathematics* (Rubenstein et al., 2004). In the **Launch** phase, mathematical or pedagogical problems are introduced to provide a starting point for discussion. During the **Explore** phase, teachers investigate problems in guided activities in pairs or small groups. In the **Share and Summarize** phase, important mathematical and pedagogical ideas are shared with the entire class. Thus, the lesson activities help you construct understanding of how mathematical concepts can be introduced, studied, and assessed in high school classrooms. Although neither high school nor college classrooms need to follow this format rigidly, this format is readily adaptable to inquiry, open-ended, and direct instruction aspects of a lesson. Hence, this format provides a flexible organizational frame for the lessons.

Questions in the lessons are of two types. **Deepening Mathematical Understanding (DMU)** questions encourage you to explore mathematics concepts in depth, often revisiting content you have previously studied but in new ways.

Questions in the **Developing Mathematical Pedagogy (DMP)** section enable you to enhance and refine teaching approaches by analyzing, modifying, or extending activities; by anticipating or analyzing students' responses; and by creating your own activities or lessons. The varied examples of student work allow you to gain insight into students' thinking to identify possible misconceptions and to consider how you would address those misconceptions individually or in class. The student work samples come from the authors' experiences with high school students.

Some questions are identified with a fieldwork icon. These questions encourage you to engage with one or more high school students in order to connect ideas from the textbook with the practical realities of working with students. Other questions are identified with a portfolio icon. The materials created in these questions are appropriate for inclusion in a professional portfolio to demonstrate your skills and professional knowledge as a mathematics teacher. Some questions are identified with a lesson planning icon. These questions focus on important aspects of creating effective lessons; they are also good candidates to include in a professional portfolio. Several questions are also identified with a World Wide Web icon (WWW). These questions reference resources available on the course website, http://www.wiley.com/college.

Throughout the textbook, in both lessons and questions, you have an opportunity to work with **sample student pages** that model the types of tasks and activities appropriate for use with high school students. These pages often have been compressed to include multiple concepts; you are encouraged to adapt, extend, or provide additional support when you use them with your own students. To facilitate such adaptation, the student pages are available in Word format on a password-protected course website http://www.wiley.com/college, and available to all purchasers of the textbook. You are encouraged to download the student pages and create a master file that can be revisited when necessary to adapt and revise the activities to meet varied student needs as your career progresses. Other resources for use in your classroom are also available on the course website. Website resources needed for particular lessons are identified prior to the **Launch** of each lesson.

Professional readings are suggested for each lesson; questions related to these readings are incorporated within the **Developing Mathematical Pedagogy** questions. The readings promote understanding of important ideas in mathematics education; the accompanying questions guide your reading of these primary sources. These readings provide exposure to the thinking of others and the opportunity to gain knowledge about other resources likely to be beneficial throughout your career. Most of the readings are from the *Mathematics Teacher*, one of the school-based journals published by the NCTM. Articles published after January 1999 are available online to members of the NCTM who select *Mathematics Teacher* as the journal to accompany their membership. You are strongly encouraged to become a member of the NCTM and to access these articles online as well as other articles that have the potential to enhance mathematics instruction. Some older articles of particular importance are available on the password-protected course website; in addition, the NCTM has graciously permitted the inclusion on the course website of articles written by one of this textbook's authors. Specific readings available on the course website are also identified with WWW.

The activities, readings, sample student pages, and questions are designed to help you construct a classroom that enables *all* students to be successful in mathematics. Equity in mathematics teaching requires teachers to plan instruction so that all students maximize their potential to learn—regardless of race, ethnicity, gender, socioeconomic status, prior mathematical experience, English language proficiency, or learning disabilities. Although meeting the varied needs of students in most classrooms today often is a challenge, it is vital for a democratic society.

Whether you are preparing for a teaching career or are a practicing teacher working to enhance your skills, the authors hope you enjoy your experience with this textbook. Your comments are welcomed.

COURSE INTRODUCTION

This is an exciting and challenging time to be teaching mathematics. Recent knowledge about learning mathematics has encouraged teachers to create engaging classroom environments. Students today explore mathematical concepts with materials and technology, work cooperatively in small groups, discuss and debate ideas, and focus on ways in which mathematics is used in the real world. The increased use of calculators that graph a variety of functions and perform symbolic algebra has caused educators to reevaluate the mathematics curriculum. Some topics are less important because of technology, while others are more important and accessible. In addition, today's high school mathematics curriculum must not only prepare students for college but also for the world of work, in which technology plays a major role. Further, in order for students to be prepared for college, the high school curriculum needs to prepare them for advanced study in mathematics, such as calculus, as well as for study in fields where discrete mathematics and statistics are increasingly important.

In 1989, the National Council of Teachers of Mathematics (NCTM)—the world's largest organization of mathematics teachers—challenged mathematics teachers to better prepare students for mathematics in the 21st century by enhancing their expectations for all students and by considering changes in content and in their teaching approach. In its *Curriculum and Evaluation Standards for School Mathematics* (1989), the NCTM outlined a set of standards for mathematics teaching and learning. Then, in 1991, it published *Professional Standards for Teaching Mathematics,* which provided more specific information related to issues of mathematics teaching, including the use of worthwhile tasks, teachers' and students' roles in discourse, and tools for enhancing discourse and for creating supportive learning environments. In 1995, the NCTM published its third document in the series, *Assessment Standards for School Mathematics*, which addressed issues related to assessing the curriculum envisioned in the 1989 *Standards*.

After nearly a decade of leading the reform movement in mathematics education, the NCTM updated its standards documents in 2000 with the publication of *Principles and Standards for School Mathematics* (*PSSM*). This document outlines content and process standards for four grade bands: preK–2, 3–5, 6–8, and 9–12. It incorporates all three previous documents and reflects changes based on what was learned from implementation of the earlier standards. In 2007, the NCTM published *Mathematics Teaching Today*, an update of standards for teaching. These documents and any future revisions will undoubtedly guide much of your career in teaching mathematics.

Throughout this text, you will explore issues of curriculum and teaching for the high school grades influenced by *PSSM*. Instructional programs are expected to enable all students to handle content in five content strands: number and operations, algebra, geometry, measurement, and data analysis and probability. In high school, algebra and geometry are expected to receive the greatest attention; although number and measurement have a place in the high school curriculum, their relative emphasis is much less than in earlier grades. Data analysis and probability should receive about the same attention as they received in the middle grades, about 10–12 percent of the curriculum.

A brief overview of the content and process standards that guide instruction for all programs from prekindergarten through grade 12 is provided here. Throughout this book, you will become more familiar with specifics within each standard and, in particular, the details for the high school curriculum.

- *Number and Operations*. Understand numbers, number systems, number relationships, and meanings of operations; and be able to compute fluently and estimate.
- *Algebra*. Understand patterns and functions; represent and analyze situations using variables; use mathematical models; and analyze change in different contexts.
- *Geometry*. Analyze properties of two- and three-dimensional shapes; develop arguments about relationships; use coordinates and other systems for locating points; apply transformations and use symmetry; and use visualization and spatial reasoning to solve problems.
- *Measurement*. Understand measurable attributes of objects, including systems and processes of measurement; and apply techniques and formulas to determine measures.
- *Data Analysis and Probability*. Pose questions that can be addressed with data and collect and analyze data to answer questions; use statistical methods to analyze data; develop and evaluate inferences and predictions for data; and understand and use concepts of probability.

In addition to the five content strands, there are five process strands that apply to all of the content strands and to all of the grade levels:

- *Problem Solving*. Build new knowledge from problem solving; solve problems using a variety of strategies; and monitor and reflect on the processes of problem solving.
- *Reasoning and Proof*. Recognize reasoning as fundamental to mathematics; make and explore conjectures; develop and evaluate mathematical arguments; and select and use different types of reasoning and methods of proof.
- *Communication*. Organize and clarify thinking through communication; communicate coherently and clearly with good mathematical language; and analyze the thinking of others.
- *Connections*. Find and use connections among mathematical ideas; see mathematical ideas in relation to one another; and recognize and apply mathematics to real contexts.
- *Representation*. Create and use tables, graphs, diagrams, symbols, or other representations to organize, record, and communicate; select, apply, and

TEACHING AND LEARNING HIGH SCHOOL MATHEMATICS © 2010 John Wiley & Sons, Inc.

translate among different representations; and use representations to model and interpret.

The recommendations outlined in *PSSM* provide a vision of a challenging curriculum. Effective teachers will need to know the mathematics outlined in the content and process strands thoroughly and deeply if they are to guide high school students toward a solid understanding of mathematics that will serve them well in this century.

Throughout this course, you will have an opportunity to investigate the content and processes outlined in *PSSM*. In addition, you will have a chance to consider effective ways to teach mathematics. As you embark on a career in teaching high school mathematics, your challenge is to create a learning environment in which all students find that mathematics makes sense and that they can be successful.

Mathematics Education: Where Do I Stand?

Welcome to *Teaching and Learning High School Mathematics*! One purpose of this text is to help you develop or refine your teaching philosophy. This introductory activity provides an opportunity to determine where you stand on a number of issues related to teaching and learning mathematics.

Just as it is important for high school teachers to work collaboratively to encourage communication in their mathematics classrooms, it is important for you to develop a learning community as you work through this text. Getting to know your peers and how they perceive mathematics teaching and learning is an important part of building a learning community. In addition, your peers are your colleagues and are an invaluable resource for you during your teaching career. Thus, this first activity starts with a survey designed to help you get to know your peers.

In this activity, you will:

- Learn about your colleagues.
- Share and compare where you and your colleagues stand on a number of survey items related to mathematics teaching and learning.

Materials

- Chart paper
- Markers

Website Resources

- *Where Do I Stand?* survey

LAUNCH

Many people have very strong views about mathematics, with those views ranging from love to hate. Think about what drew you to mathematics. Why did you decide to become a high school mathematics teacher?

EXPLORE

In the mathematics education community and the general public, differences of opinion exist about the balance between learning skills and learning concepts and about expectations for high school students in mathematics. Use the following survey to determine where you stand on the issues and how your views compare to those of your peers.

TEACHING AND LEARNING HIGH SCHOOL MATHEMATICS © 2010 John Wiley & Sons, Inc.

Where Do I Stand?

For statements 1–21, indicate the extent to which you agree or disagree with the statement. Be prepared to share your thinking. Please save your responses because you will revisit your responses throughout the text to learn how your views evolve.

Key

SD strongly disagree
D disagree
N neutral
A agree
SA strongly agree

1. Mathematics is a collection of abstract ideas. SD D N A SA
2. To be of value, mathematics must be applicable to one or more real-world situations. SD D N A SA
3. There is always a procedure or algorithm to solve a mathematics problem. SD D N A SA
4. High school students should be expected to complete proofs. SD D N A SA
5. Symbolic manipulation is the most important aspect of algebra. SD D N A SA
6. Students need to learn skills and procedures before they can learn the related concepts. SD D N A SA
7. Requiring all students to complete a course in algebra is unrealistic because some students can't understand algebra. SD D N A SA
8. Problem solving can only be introduced after students have learned all the necessary skills. SD D N A SA
9. Graphing calculators are useful tools to help students understand important algebraic concepts. SD D N A SA
10. Computer algebra systems should not be allowed at the high school level. SD D N A SA
11. Students should be expected to explain their mathematical ideas and understandings to their peers. SD D N A SA
12. Graphing calculators should only be used after students have demonstrated the ability to complete algebraic skills by hand. SD D N A SA
13. High school students often need to work with concrete materials to understand new ideas. SD D N A SA
14. Reasoning and proof should occur only in geometry. SD D N A SA
15. Mathematics teachers need to spend class time preparing their students to take high-stakes tests. SD D N A SA
16. Students who do poorly in mathematics should be placed in remedial classes. SD D N A SA

TEACHING AND LEARNING HIGH SCHOOL MATHEMATICS © 2010 John Wiley & Sons, Inc.

17. It is important for mathematics teachers to plan lessons in consultation with other mathematics teachers.	SD	D	N	A	SA
18. Cooperative learning is a way for slow students to succeed without learning.	SD	D	N	A	SA
19. Teachers should use their textbook and not deviate from it.	SD	D	N	A	SA
20. High school students should have opportunities to work on problems or projects for extended periods of time.	SD	D	N	A	SA
21. Learning mathematics is mostly memorizing formulas.	SD	D	N	A	SA

22. How do you best learn mathematical procedures? Mathematical concepts?

23. What are the qualities of the best mathematics teacher you have had?

24. What do you anticipate as one of the major challenges of teaching high school students?

SHARE AND SUMMARIZE

1. As a class, determine the number or percent of the class that responded in a given way to each of the survey items. For the purpose of aggregating the data, group together *strongly agree* and *agree* and group together *strongly disagree* and *disagree*.

2. Look at the class results. For which items is there strong consensus among your peers? For which items is there variability in responses? For which items did the response patterns surprise you?

3. As a class, share the responses to item 23 about the qualities of the best mathematics teacher you have had.

- **a.** Determine a way to categorize groups of responses.
- **b.** Write a brief paragraph that identifies the themes among the responses to the item.
- **c.** When this item was given to high school students in four different schools, responses were analyzed and grouped into four major categories: pedagogical qualities, relationships with students, personal qualities, and characteristics related to classroom management. How do these categories compare to the categories you used in question 3a?
- **d.** Regroup the responses of your peers to item 23 using the categories in question 3c. Based on this new grouping, how, if at all, would you change your paragraph describing the responses?

4. In terms of teaching high school students, summarize the major challenges identified by you and your peers. Which responses surprised you?

5. Based on the responses of you and your peers, complete the following metaphor: Being a mathematics teacher is like being a _____ because _____.

TEACHING AND LEARNING HIGH SCHOOL MATHEMATICS © 2010 John Wiley & Sons, Inc.

UNIT *one*

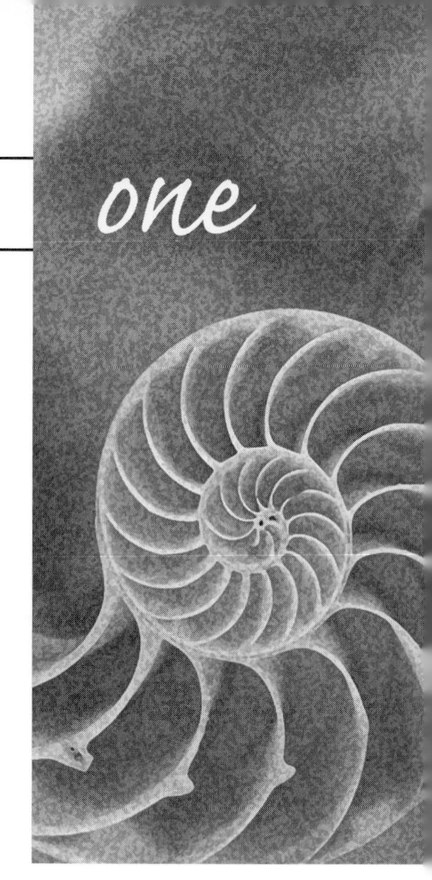

ENCOURAGING COMMUNICATION IN MATHEMATICS CLASSROOMS

Mathematics Strand: Logic and Reasoning

Today's high school students live in a world of stimulation—video games, music, the Internet, and more. Mathematics teachers need to take advantage of the energy and inquisitiveness typical of adolescents. Students need to be engaged socially and intellectually in order to be motivated to study mathematics, to maintain their interest in this study, and to succeed in mathematics learning. Communication is a major ingredient for success.

Equity is the first principle of the *Principles and Standards for School Mathematics* (*PSSM*) of the National Council of Teachers of Mathematics (NCTM, 2000). Schools must provide high expectations and strong support for all students. Teaching for equity involves many dimensions, as addressed in this text. One aspect of equity includes students sharing their different ways of thinking, learning to respect and understand one another, and appreciating each others' perspectives. Research on the use of cooperative learning suggests that such a format can play a significant role in building positive human relations and learning (Slavin, 1996). This unit introduces the use of cooperative groups, which is then extended throughout the text. Students in cooperative groups work together on mathematical tasks, conduct experiments to collect data or test solution strategies, investigate and generalize patterns, and explain their understanding. In such classrooms, students regularly engage in discussions and debate ideas with their classmates.

Unlike educational settings of the past in which the teacher was a giver of information and students were receivers, today's classrooms are "learning communities" where teachers and students learn from each other. In such environments, students are encouraged to communicate their ideas, conjectures, and discoveries in a supportive setting where their views are valued and respected. Finding ways to build such communities and encourage mathematical communication is one of the goals of this unit. To promote this, a variety of cooperative learning strategies are incorporated throughout.

Cooperative learning includes the following features:

- *Face-to-face interaction.* Group members sit and work facing one another, not just side by side.
- *Positive interdependence.* Challenging tasks are selected and orchestrated so that students must rely on each other's thinking in positive ways.
- *Individual accountability.* All students are expected to show that they have achieved instructional objectives.
- *Building social skills.* Group participants are expected to reflect regularly on and evaluate behaviors that support or hinder their cooperation with each other.

When working cooperatively, some general ground rules are necessary. At times, students need to document how well they and their group members satisfy one or more of the following general rules.

Cooperative Learning Ground Rules

1. Work together in a group of two, three, or four students.
2. Cooperate with everyone in your group.
   - **a.** Give praise liberally.
   - **b.** Do not use put-downs.
   - **c.** Share written and other materials.
3. Listen carefully to others.
   - **a.** Try, whenever possible, to build on each other's ideas.
   - **b.** Get all ideas out in the open. (There's no such thing as a bad idea!)
   - **c.** Ask for clarification if needed.
4. Achieve a group solution for each problem.
5. Make sure everyone understands the solution before the group proceeds.
6. Share the leadership of the group.
7. Make sure everyone participates and no one dominates.
8. Proceed at a pace comfortable for your group.

(Davidson, 1990)

In addition to creating a learning community in which students can thrive, teachers also need to be concerned with the mathematical content and processes that this community is exploring. Teachers need an integrated view of instruction, learning, and assessment as all three aspects inform each other. For example, as teachers design instruction, they consider their learning objectives. As instruction takes place, teachers constantly monitor students' learning and conduct informal assessments of understanding, modifying their questions and adjusting instruction to address misconceptions or misunderstandings that arise. This cycle repeats throughout the year. Through formal assessments, such as tests, teachers can determine what has been learned and then may revise future instruction to address areas or concepts that students still have not mastered.

Reasoning, a hallmark of mathematics and the mathematical focus of this unit, is one of the five process standards in the *PSSM* and is essential in helping students make sense of mathematics. High school students should be expected to provide explanations and justifications. Further, these experiences must become more

TEACHING AND LEARNING HIGH SCHOOL MATHEMATICS © 2010 John Wiley & Sons, Inc.

sophisticated as students build a repertoire of reasoning strategies and with the development of formal methods of proof.

Logic and reasoning have a long history. Early mathematicians engaged in reasoning by asking, "*Why?*" "*How do you know?*" or "*Will this always work?*" As early as 600 B.C., Greek mathematics began to change from satisfaction with finding an answer to providing a rationale for why an answer was correct (Katz, 1998). The city-state political environment in early Greece required individuals to know the rules and laws by which they were governed; thus, they had to develop skills in argument and debate. These skills carried over to mathematics, and the Greeks are credited with providing ideas related to mathematical proof.

Thales of Miletus (c. 624–547 B.C.), the first Greek mathematician to "prove" results, is credited with justifying that the base angles of an isosceles triangle are congruent and that vertical angles are congruent. Later, Pythagoras (c. 572–497 B.C.) and his followers provided proofs about numbers (e.g., the sum of an even number and an odd number is odd). Pythagoras is perhaps most well-known for the theorem that bears his name about the lengths of the sides of a right triangle. These early examples of proof laid the groundwork for further development of Greek mathematics and indicated that proof was required in areas of mathematics other than geometry.

Aristotle (c. 384–322 B.C.) is credited with codifying the rules of logic through syllogisms. A *syllogism* is a set of true statements that naturally lead to other true statements. Aristotle recognized that an assertion cannot be both true and false and that an assertion had to be *either* true or false. He also recognized that some truths must be accepted without argument, leading to ideas of postulates and axioms; *postulates* are specific to a discipline and *axioms* are common to all areas.

The forms of argument used today were first analyzed by Chrysippus (280–206 B.C.), who identified four basic rules of inference. If p, q, and r represent statements, then these four argument forms are as follows:

Modus Ponens
(Latin for "Method of Affirming")
If p, then q.
p.
Therefore, q.

Modus Tollens
(Latin for "Method of Denial")
If p, then q.
not q.
Therefore, not p.

Hypothetical Syllogism
(Transitive Reasoning)
If p, then q.
If q, then r.
Therefore, if p, then r.

Disjunctive Syllogism
(Reasoning with "Or")
p or q.
not p.
Therefore, q.

You may recognize these rules of inference from previous mathematics or logic courses.

Perhaps the most important mathematical text by the Greeks was Euclid's *Elements*, believed to have been written about 300 B.C. This work, written in 13 books, has been translated into many languages and has been in print since printing began; only the Bible has appeared in more editions. In the *Elements*, Euclid organized what was known about Greek mathematics and provided "a model of how 'pure mathematics' should be written, with well-thought-out axioms, precise definitions, carefully stated theorems, and logically coherent proofs" (Katz, 1998, p. 58). Even today, high school geometry courses are still based on Euclid's *Elements*.

Understanding the history of mathematics helps both students and teachers recognize the universality of the process of studying mathematics. In all generations, those working in the discipline have raised questions, looked for patterns, pursued conjectures, tried multiple approaches, persisted, and strived to find a logical foundation for what must be true. In an effort to bring to light the struggles and accomplishments of this human endeavor, each unit of this book includes some mathematics history.

Although much more could be said about the early history of reasoning, it is clear that ideas related to reasoning and proof have engaged the imagination, interest, and

TEACHING AND LEARNING HIGH SCHOOL MATHEMATICS © 2010 John Wiley & Sons, Inc.

time of mathematicians over the centuries. Neither this unit nor this text will address all aspects of reasoning and proof. Still, it is essential that reasoning and proof play an important part in school mathematics, particularly at the high school level. Thus, reasoning is a recurring theme throughout this text.

As you explore reasoning in this unit, consider how it provides an avenue for students to develop and communicate their emerging understandings of mathematical concepts and processes. Throughout this unit, you are encouraged to develop classrooms in which students communicate and justify their thinking as they work together to build a learning community so all students make sense of mathematics.

Unit One Team Builder: Carpet Square Maze

To encourage all students to engage in learning mathematics and to grapple with the challenges encountered in understanding mathematics, it is necessary to build a nurturing community where students feel safe in presenting their ideas. Each time groups are formed or reformed, it is helpful to take time for group members to get to know each other. In each unit of this text, a team builder is tied to the mathematics explored in the unit investigation and throughout the unit. The icon indicates that a student page, template, or table is available on the course website.

Carpet Square Maze is designed to encourage individuals to work together to find a predetermined and initially unknown route through a maze. Individuals may complete the team builder in teams of four; if the class is small, they may complete it with the whole class.

1. Before you begin, introduce yourself to your team members in a way that will help them remember your name. For example, you might use alliteration to help others recall your name. "I'm Aidan and I like acting." "I'm Gavin and I like the Florida Gators."
2. With your team members, complete the activity on the student page, **Carpet Square Maze**.

Materials

- 25 carpet squares or one 5 by 5 grid marked on the classroom floor per team

Website Resources

- ***Carpet Square Maze*** student page

For further study of carpet square mazes, see the unit investigation on pages 69–70 at the end of the unit.

TEACHING AND LEARNING HIGH SCHOOL MATHEMATICS © 2010 John Wiley & Sons, Inc.

Carpet Square Maze

Start				
				End

Materials

- 25 carpet squares or one 5 by 5 grid marked on the classroom floor per team

1. Identify a team member to be the Monitor. The Monitor should record a route through the maze from **Start** to **End**. All moves are down, to the right, or diagonal (down and right) from one square to another square that shares at least one vertex. The Monitor should not tell the other team members the route.
2. Team members work together to find the Monitor's predetermined route through the maze. Begin at **Start** in the upper left corner. Exit at **End** through the lower right corner. If a team member steps on a square on the route, he/she continues. When a team member steps on a square not on the route, the monitor signals an incorrect step and another team member must restart. On every turn, team members must begin the route at the square marked **Start**.
3. Once the route is determined, each team member must take the same route without making a mistake. If anyone makes a mistake, the whole team must start over, going through the maze until all team members complete the maze correctly. As they do so, they introduce one member of their team for each carpet square on which they step. For example, if Aidan is first and has three other team members, he steps onto the first square and says, "I'm Aidan and I like acting." When he steps to the second square in the maze, he says, "This is Gavin and he likes the Gators." Stepping onto the third square, he says, "This is Alex and he likes apples." And finally, when he steps to the fourth square, he introduces the Monitor by saying, "This is David; he likes drawing dogs."
4. For a 3 by 3 square, determine the following:
   - **a.** How many routes are possible if no diagonal moves are used? Explain how you know your answer is correct.
   - **b.** How many routes are possible if diagonal moves are allowed? Explain how you know your answer is correct.
   - **c.** If you found your answers to problems 4a and 4b in only one way, determine another approach.

TEACHING AND LEARNING HIGH SCHOOL MATHEMATICS © 2010 John Wiley & Sons, Inc.

Preparing to Observe Mathematics Classrooms: Focus on Equity

Each unit in this text includes an opportunity for you to observe high school mathematics classrooms and possibly interview high school teachers. In this first observation, you will focus on the types of questions asked in mathematics classrooms. By analyzing these questions, you will consider how teachers communicate their expectations to students, how those expectations are related to issues of equity, and how those expectations influence different opportunities students have to learn mathematics.

In *Principles and Standards for School Mathematics* (*PSSM*), equity is the first of six overarching principles for school mathematics. When teachers teach for equity, they have high expectations for all students and provide the support students need to meet those expectations. The equity principle is characterized by three components:

- Teachers need to have high expectations of all students and provide them worthwhile mathematical opportunities.
- Teachers need to accommodate differences among students to help all be able to learn mathematics.
- Schools need to provide sufficient and appropriate resources and support for all classrooms. (NCTM, 2000, pp. 12–14)

High schools typically offer several levels of mathematics courses ranging from pre-algebra to calculus. How do mathematics expectations compare for students at these different levels? The purpose of this observation is to compare and contrast expectations for students in classes at different ability levels, to consider how different expectations determine students' opportunities to learn mathematics, and how differences in opportunities raise questions about equity.

Website Resources

- Observation One Template

1. *Prepare for the Observation.* In preparation for observing high school classes, complete the following:
   a. Read "The Equity Principle" from *PSSM* (pp. 12–14) and the Equity in Mathematics Education position statement from the National Council of Teachers of Mathematics (nctm.org). Summarize these two readings. Based on these readings, describe student and teacher behaviors and actions you would expect to see in a mathematics classroom focused on equity for all students.
   b. Read "Unveiling Student Understanding: The Role of Questioning in Instruction" by Azita Manouchehri and Douglas A. Lapp (*Mathematics Teacher*, 96 (November 2003): 562–566). Think about mathematics classes in which you have been a student, either at the high school or college level. Compare the types of questions used in your classes with those advocated in the articles.
   c. Think about teacher practices in classes in which you have been a student. What practices supported equity? What practices, if any, did not support an equitable environment? Explain.
2. *Conduct the Observation.* Locate a high school that will permit you to observe several classes. Observe at least three full class periods. As much as possible, observe classes at various ability levels or observe different instructors who teach the same level class. If you are only able to observe one teacher teaching classes at

TEACHING AND LEARNING HIGH SCHOOL MATHEMATICS © 2010 John Wiley & Sons, Inc.

one level, work with a partner to compare observation responses to look for similarities and differences across course levels.

   a. As you observe, use the Observation One Template (on page 14) to record the questions asked in the class. Adjust the template as needed. Listen for questions teachers ask of students, that students ask of teachers, and that students ask of each other. Write down the questions asked during the main period of instruction (i.e., after any warm-ups and before students begin working on homework). Use a checkmark in the appropriate column to note who asks the question, teacher or student. (If students in the class are working in groups, observe one group. Record the questions students ask of each other.)

   b. Try to estimate the length of wait time, that is, the amount of time students have to answer a question before another is asked, or before someone responds to the original question. Also try to estimate the time the teacher waits after a student response to let everyone think before speaking again.

3. *Analyze the Observation Data.* After the observation, categorize the questions by type, *open* or *closed*, as described by Manouchehri and Lapp (2003).

   a. For each observed class, find the percentage of closed and open questions.

   b. How do these percentages compare for classes at different ability levels?

   c. What do these percentages suggest about teachers' expectations of students?

   d. What differences, if any, did you observe in wait time for classes at different ability levels? To what extent did all the students in the class have an opportunity to think about a question before a response was given?

4. *Critique the Questions.* Review the questions asked by the teacher.

   a. What percentage required students to provide a justification of their response?

   b. Give examples of good questions that needed justification. How is asking for justifications related to equity?

   c. What good open questions did you hear? How do open questions relate to equity?

5. *Interview the Teacher.* If possible, spend 5–10 minutes talking with one of the teachers you observed about his or her expectations for students in classes at different levels in terms of instruction, communication, and overall level of work.

6. *Summarize the Observations.* Write a summary of your observations. Start with a brief description of the school environment and the level of the classes you observed; include some brief demographic information about the class (e.g., balance between males and females, ethnic mix, and so on). Be sure to describe the intended mathematical content of each observed class. Justify any conclusions with specific examples from your observations.

   a. Summarize your responses to questions 1–4.

   b. Based on the teacher's comments in question 5, summarize the similarities and differences in expectations for students at different ability levels or in different courses.

Be prepared to discuss your observations and conclusions in the last lesson of this unit.

TEACHING AND LEARNING HIGH SCHOOL MATHEMATICS © 2010 John Wiley & Sons, Inc.

Observation One Template

Elapsed Time	Questioner		Questions	Question Type	
	Teacher	Student		Open	Closed
0–5					
5–10					
10–15					
15–20					
20–25					
25–30					
30–35					
35–40					
40–45					
45–50					
50–55					
55–60					

TEACHING AND LEARNING HIGH SCHOOL MATHEMATICS © 2010 John Wiley & Sons, Inc.

LISTENING TO STUDENTS REASON ABOUT MATHEMATICS

Each unit in this text includes one or more tasks designed to be completed with one or more high school students. The aim of these tasks is to provide an opportunity for teachers to talk with students and listen to their explanations as they engage with mathematics. Listening to students often provides insight into student thinking— a valuable resource when designing effective instruction.

How do students think about approaching a problem for which they have not learned a solution strategy? The tasks for this interview focus on reasoning and the thinking processes students use as they reason.

Materials

- Set of small attribute pieces (16 pieces, 4 shapes ■ ▲ ● ♦, 4 colors, no piece the same color and shape as any other piece)

Website Resources

- **QRS Guide**
- **Colorful Shapes** student page
- Stages of Questioning
- **Attribute Pieces for Colorful Shapes**
- **Listening to Students Reason about Proof** student page
- **QRS Guide for Colorful Shapes**
- "I Would Consider the Following to Be a Proof . . . " by Walter Dodge, Kathleen Goto, and Philip Mallinson (*Mathematics Teacher*, 91 (November 1998): 652–653)

1. Read **Colorful Shapes** on the accompanying student page.
   a. Solve the puzzle on your own. Record the strategies you used in the Expected Students' Responses column of the **Question Response Support (QRS) Guide** in the Unit One Appendix on page 450.
   b. Solve the puzzle another way. Record your strategies in the **QRS Guide**.
   c. Try to predict how students might solve the puzzle. Add these strategies to the **QRS Guide**. Consider three kinds of student responses: *misconceived*, *partial*, and *satisfactory*.
   d. Use the Stages of Questioning (see Lesson 1.1 on pages 19–20) to plan appropriate guidance to students without giving too much help. Record your questions in the Teacher Support column of the **QRS Guide**.
   e. In the partially completed **QRS Guide for Colorful Shapes** (in the Unit One Appendix, page 453), review the student responses and the sample teacher support questions for each strategy. Compare those in the sample **QRS Guide** to the expected student responses and teacher support you generated.
2. Give **Colorful Shapes** to one or more high school students. (To use the puzzle, you might copy the pieces from the **Attribute Pieces for Colorful Shapes** template on heavy paper or cardstock and cut them out. Alternatively, you might use commercial attribute pieces. A copy of the template is located in the Unit One Appendix.) As students solve the puzzle, ask them to "think out loud." Record their strategies in the **QRS Guide**, including useful steps, inappropriate steps, and any other difficulties along the way. What seemed easy for students? What was

TEACHING AND LEARNING HIGH SCHOOL MATHEMATICS © 2010 John Wiley & Sons, Inc.

difficult? In what ways did students pay attention to the conditions of the problem from the beginning?

3. Refer to the student page, **Listening to Students Reason about Proof**.
   a. Choose one task from the set. Solve it yourself.
   b. Repeat items 1a–1d for the task you chose, recording your responses in the **QRS Guide**.
   c. Read the article, "I Would Consider the Following to Be a Proof . . . " by Walter Dodge, Kathleen Goto, and Philip Mallinson (*Mathematics Teacher*, 91 (November 1998): 652–653). The authors suggest a proof they would accept from students at a particular course level for each problem on the **Listening to Students Reason about Proof** student page. For the task you chose in item 3a, compare the response in the article to the sample response you provided. Add responses to the **QRS Guide**.
   d. Give the task you chose in item 3a to one or more high school students for whom the task presents an appropriate challenge. As students solve the task, ask them to "think out loud." Record their strategies in the **QRS Guide**, including useful steps, inappropriate steps, and any other difficulties along the way. What seemed easy for students? What was difficult? Record student responses in the **QRS Guide**.
   e. Was the students' solution a "proof" you would be willing to accept? Explain.

Be prepared to share your insights into how students reasoned about **Colorful Shapes** and the problem you selected from **Listening to Students Reason about Proof** during the last lesson of this unit.

Colorful Shapes

Use the 4 by 4 grid below and one set of small attribute pieces. Place each piece in the grid so that each color and shape appears only once in any row, column, or main diagonal.

1. Solve the puzzle. Keep track of your thoughts as you do so.
2. What pieces did you place first? Why?
3. Describe how you solved the puzzle.

Materials

- Set of small attribute pieces (16 pieces, 4 shapes ■ ▲ ● ◆, 4 colors, no piece the same color and shape as any other piece)

TEACHING AND LEARNING HIGH SCHOOL MATHEMATICS © 2010 John Wiley & Sons, Inc.

Listening to Students Reason about Proof

Pre-Algebra

1. Find an equation that gives the sum of consecutive whole numbers from 1 to n. Verify that your equation is correct.

First-year Algebra

2. Prove that $(a+b)^2 = a^2 + 2ab + b^2$.

Geometry

3. What is the sum of the exterior angles of a convex polygon? How do you know?

Second-Year Algebra

4. Roll two normal six-sided die. Find the product of the numbers showing on the die. What is the probability that the product is greater than 14? How do you know?

Trigonometry

5. Prove that $\arctan(1) + \arctan(2) + \arctan(3) = 180^\circ$.

Calculus

6. Use a flat sheet of paper with one side length a. The other side length b is allowed to vary. Cut squares with side dimension x from the flat sheet and fold up the sides to make a box (see figure at right). Given a value of b, there is a value of x for which the box has maximum volume. Choose x so that it maximizes the volume of the box. Prove that

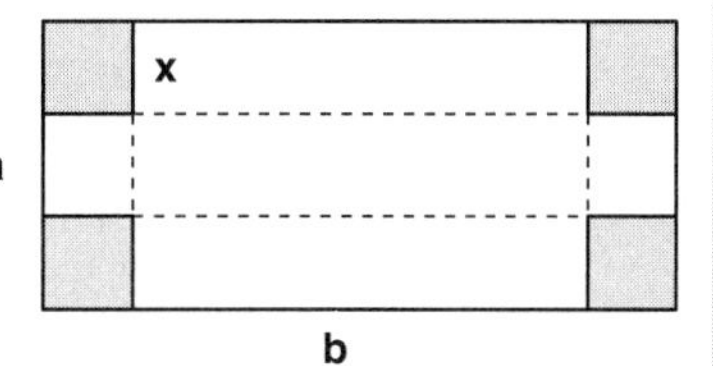

$$\lim_{b \to \infty} x = \frac{a}{4}.$$

Note: These tasks are adapted from the article, "I Would Consider the Following to Be a Proof . . . " by Walter Dodge, Kathleen Goto, and Philip Mallinson (*Mathematics Teacher,* 91 (November 1998): 652–653).

1.1 Developing Questioning Strategies: Conjecturing and Reasoning

Being able to reason about mathematics is a critical component of making sense of mathematics. Throughout their schooling, students should have many opportunities to build their conjecturing and reasoning skills. By high school, students should be able to determine patterns, conjecture rules or statements that describe these patterns, and decide under what conditions the patterns arise. They should also use their reasoning skills to understand and create mathematical proofs, including proofs or disproofs of their conjectures. In order to engage all students in reasoning and proof, teachers need to understand what supports and forwards that learning.

TEACHING AND LEARNING HIGH SCHOOL MATHEMATICS © 2010 John Wiley & Sons, Inc.

Learning is enhanced by engaging as many senses as possible. Katz and Rubin (1999) identify six senses: sight, sound, touch, taste, smell, and emotion. In mathematics classes, it is rare that taste and smell enter into learning. It is undeniable, however, that emotion is often a mitigating factor in students' learning. Math anxiety and all of its repercussions are well documented, though often ignored. Students need to engage their emotions when learning; they need to care about what they are learning, see a need for it, or enjoy it. The other senses, especially sound, are engaged when students work together. Learning theory indicates that students learn by checking their reality with others. To promote an atmosphere where reasoning is key, students should be encouraged to listen to each other's thoughts about mathematics problems and to share their thoughts with their peers. They need to be encouraged to be reflective problem solvers, making reasonable guesses or conjectures from patterns that seem to be arising in a problem and keeping track of the processes they try and their thoughts along the way.

To help teachers encourage collaborative work among students, several cooperative learning strategies are presented in this unit. Each lesson introduces a strategy by engaging you in the process. In this lesson, you are encouraged to solve problems in pairs. Paired groupings are easiest to orchestrate in a classroom in which students are first learning to work together. For this lesson, the cooperative learning strategy, *Think Pair Share* (McTighe and Lyman, 1988), is recommended for solving the problems posed. In *Think Pair Share*, both partners think about the problem posed and work quietly until time is called. Then partners work on the problem together, sharing their results.

Think Pair Share

1. Think quietly about the problem posed until "think" time ends. Write whatever occurs to you.
2. Pair with your partner and discuss the problem until you both agree on at least one way to arrive at a solution. Carry out your plan together.
3. When requested to do so, share your response with the larger group.

(McTighe & Lyman, 1988)

Asking good questions, of yourself or others, is important when solving a problem. Asking appropriate questions helps learners reflect on problem-solving efforts and describe the reasoning used to make sense of a problem. Teachers want to support students while at the same time helping them become as independent as possible. Consequently, it is important to ask questions that are "just enough" to move a student forward without pre-empting his or her opportunity to think. A valuable guide for this sort of questioning is found in the writings of Gerald A. Goldin (1998). The *Stages of Questioning* inspired by his work will be used throughout this text; it is a vital part of the lesson planning process. Starting in this lesson, you will be asked to think about your work and the work of others as you solve problems. You will also be asked to consider what questions to ask of persons in various stages of their problem-solving process. All of this work is essential for writing and implementing effective student-centered lesson plans.

TEACHING AND LEARNING HIGH SCHOOL MATHEMATICS © 2010 John Wiley & Sons, Inc.

Stages of Questioning

Questions are tools teachers can use to understand and support students' mathematical thinking. Teachers need to allow students the opportunity to think independently or in groups, asking questions only when students are not able to proceed. Questions should be phrased to be "just enough" to help students progress; they often are rhetorical and intended to help students think about alternatives. It is advisable to allow students the opportunity to pursue their own solution processes before asking them to consider other routes. Guidelines are provided below.

Stage 1: Asking Non-Directive Questions

Students need time to understand the problem and get started; the teacher should intervene as little as possible. Non-directive questions, such as the following, are used only when it is evident that a student has trouble getting started.

- What do you think the problem is asking you to do?
- What are you thinking? Please explain.

Stage 2: Suggesting General Problem-Solving Strategies

At times, a student is able to explain the problem but still struggles to begin. Questions at this stage may help a student think of ways to get started on the problem by giving him or her a range of strategies to consider. The teacher's tone of voice indicates that these strategies might or might not work. Teachers should be as non-directive as possible.

- Might drawing a picture help?
- Might a table help to solve this problem?
- Might any materials be helpful to model the problem?
- Have you solved similar problems that might help?
- Can you work backwards or simplify the problem?
- What must happen in the problem?
- What cannot happen in the problem?

Stage 3: Suggesting Specific Problem-Solving Strategies

At times, students understand the problem and have initiated a strategy, but are not able to use the work they have generated to bring the problem to closure. They need suggestions for more specific problem-solving strategies through questioning. Once again, the teacher should be as non-directive as possible, asking questions that suggest possibilities that might or might not help them move forward.

- What patterns do you see in the numbers? in the table? in the pictures?
- Is there a rule that describes the relationship in the table of values you created?
- Are the graph and the story related?

Stage 4: Asking for Metacognition

At this stage, the teacher helps the student think metacognitively about the problem, that is, to think about their thinking.

- How did you think about the problem?

(continued)

TEACHING AND LEARNING HIGH SCHOOL MATHEMATICS © 2010 John Wiley & Sons, Inc.

(*continued*)

- How would you convince someone who has never seen the problem that your approach is correct?
- How is this problem like others you have solved? How is it different?

The teacher's goal is to help students generate a coherent verbal explanation of the problem with representations as appropriate. Teachers might have a list of strategies posted in the classroom that students are reminded to consult when solving problems. In time, students begin to think about the list automatically and to ask themselves and each other such questions.

Table inspired by Goldin, Gerald A. "Observing Mathematical Problem Solving through Task-Based Interviews." In Anne R. Teppo (Ed.), *Qualitative Research Methods in Mathematics Education* (pp. 40–62). Reston, VA: National Council of Teachers of Mathematics, 1998.

In most lessons in this text, there are both mathematical and pedagogical objectives. In this lesson, you will:

- Begin to employ reasoning strategies as you solve an open-ended problem.
- Use the Stages of Questioning to aid yourself and your partner in reasoning through the problem.
- Experience the *Think Pair Share* cooperative learning strategy as one step toward building a learning community.

Materials

- graph paper
- two-color counters, about 50 per pair of students

Website Resources

- Cooperative Learning Powerpoint
- Stages of Questioning
- *How Many Oranges?* student page
- **QRS Guide**
- **QRS Guide for How Many Oranges?**

LAUNCH

Reasoning is important in all aspects of daily life. Choose one of the following questions and discuss your answer with your partner.

1. Think about a time when you had to solve an important problem. What was the problem? How did you decide to attack the problem? Which conjectures gave you some insight into what to try? How did you solve it? How did you use reasoning?
2. Recall a time when you did not have all of the materials needed to solve a problem that you were intent on solving. What did you do? How did you solve the problem without the necessary equipment? How did you use conjecturing and reasoning to help solve the problem?

EXPLORE

Open-ended problems provide opportunities for individuals to engage in mathematical reasoning. An open-ended problem can be solved in many different

TEACHING AND LEARNING HIGH SCHOOL MATHEMATICS © 2010 John Wiley & Sons, Inc.

ways. **How Many Oranges?** (below) is an open-ended problem that is possible to solve through multiple strategies.

Throughout this text, you will use the **Question Response Support (QRS) Guide** as a lesson-planning tool (see page 450 for the general template). As your first example, see page 454 in the Unit One Appendix for the **QRS Guide for How Many Oranges?** In the far left column of the **QRS Guide** in this sample, you will find the question (**Q**) to be asked of students. In the center column, record expected student responses (**R**), including at least one *misconceived* response, one *partial* response, and one *satisfactory* response. In the far right column, record teacher support (**S**) in the form of questions you would ask students for each of the responses recorded in the center column.

3. a. Work individually on the first three problems on the **How Many Oranges?** student page. In the **QRS Guide**, record approaches you try as you solve the problem. Record partial or misconceived as well as satisfactory responses.

   b. When time is called, work with a partner to solve the problems. In the Teacher Support column of the **QRS Guide**, write any questions that you ask yourself or your partner as you continue to work on the problem.

How Many Oranges?

Materials

- graph paper
- two color counters, about 50 per pair of students

Melanie, Colleen, and Ben picked oranges together. After picking a large number of oranges, they stopped to take a break and fell asleep under a tree. As they slept, the weather turned very hot. Melanie woke before the others. Being too warm, she decided to take her third of the oranges and head home. She counted the oranges, and found that the number was divisible by 3 if she took one of the oranges first. Melanie took one orange and then took one-third of the remaining oranges. She left quietly.

Ben awoke a short time later, noticed that Colleen was sleeping, and presumed Melanie had gone for a walk. He decided to head home because he wasn't feeling well. Not realizing that anyone had left, he counted the oranges and found that the number of oranges was two more than a number that could be easily divided by 3. So he took the two extra oranges and then one-third of what was left. He left without waking Colleen and without waiting for Melanie to return.

Colleen finally awoke and wondered where her friends had gone. She assumed they were both busy somewhere nearby. Before she left, she counted the oranges and was surprised at how small the number was. She found that the number of oranges was one more than a multiple of three, so she took the extra orange and one third of those remaining.

1. What was the total number of oranges the friends picked?
2. How many oranges were left behind when Colleen departed?
3. Is more than one answer possible for the total number of oranges? If so, what other numbers are possible? Explain.
4. Based on the values you found in problem 3, determine a reasonable solution to the problem. Explain your answer.
5. Is it possible that five oranges remained after Colleen took her "share"? Explain.
6. Could there have been six oranges left after Colleen departed? Explain.
7. Could there have been 46 oranges initially? Explain.
8. Find a pattern that gives all the possible numbers of oranges that the friends picked. Explain your pattern.

TEACHING AND LEARNING HIGH SCHOOL MATHEMATICS © 2010 John Wiley & Sons, Inc.

4. With another pair, share your approaches, conjectures, and solutions to **How Many Oranges?**
   a. How did conjecturing arise as you worked on **How Many Oranges?**
   b. How were your approaches or conjectures similar to your partner's? How did they differ?
   c. Explain any differences between your solutions and your partner's solutions. Record additional solutions in the **QRS Guide**.
5. Place yourself in the role of teacher. Use the Stages of Questioning to critique the questions you asked yourself and each other as you worked through the problem.
   a. Do these questions naturally increment the level of help? Explain.
   b. Did the questions provide "just enough" help? If not, revise any questions that give too much help and those that are not helpful enough.
   c. Anticipate student solutions for **How Many Oranges?** Record these in the **QRS Guide**. Solutions might be satisfactory, partial, or misconceived. For each, what questions would you ask students to help them move forward, clarify their work, or extend their understanding?
   d. Categorize your questions from items 5b and 5c according to the Stages of Questioning.

SHARE

Completing **QRS Guides**, a first step in lesson planning, is a complex task. It requires you to anticipate a variety of ways that students might respond to a problem and then to consider how you might support students who respond as expected. Your responses and those of your peers provide a good base of possible student responses. To this end, complete the following as a full class.

6. Share one of your responses to **Explore** questions 4–5 as requested by your instructor. The response might be correct, partial, or misconceived, your own solution, or one you expect of students. (Note: If possible, record the publicly shared work with a digital camera and copy the work into a **QRS Guide**.) Add any new solutions from the class to your **QRS Guide for How Many Oranges?**
7. Analyze each of the new solutions without help from the persons who wrote them. What questions would you like to ask the persons who wrote these solutions? Add the questions to the **QRS Guide**.

SHARE AND SUMMARIZE

8. For each of the publicly shared solutions to **How Many Oranges?**:
   a. Share the questions you would like to ask the author of the solution.
   b. Categorize the questions offered by your peers according to the Stages of Questioning. Which questions provide "just enough" help?
   c. Identify any questions that either give too much help or questions that do not provide enough help. Why do you think so? Suggest revisions.
9. Comment on how beneficial you believe the **QRS Guide** and the Stages of Questioning can be to teachers in mathematics classrooms. Justify your position.
10. What reasoning did you use as you solved **How Many Oranges?** How did you use conjecturing in your solution process?
11. Use metacognitive strategies to consider your thinking in solving the problem. Which strategy was easiest? How did you keep track of the approaches you tried?
12. Comment on how the cooperative learning strategy *Think Pair Share* supported student learning in this lesson.

TEACHING AND LEARNING HIGH SCHOOL MATHEMATICS © 2010 John Wiley & Sons, Inc.

DEEPENING MATHEMATICAL UNDERSTANDING

Throughout this text you will be asked to respond to items called "Deepening Mathematical Understanding" (DMU). These items are intended to help you strengthen and extend your mathematical background.

1. *Pictorial vs. Algebraic Solutions*. Revisit the student page **How Many Oranges?**
   a. If you originally solved the problem with a picture, attempt an algebraic solution. If you originally solved the problem algebraically, attempt a pictorial solution.
   b. Compare the pictorial solution with the algebraic solution. Discuss advantages and disadvantages of each solution approach.
2. *Diophantine Equations*. Diophantine equations, named for the ancient mathematician Diophantus who lived around the third century A.D., have special characteristics.
   a. Search the Web or other resources to learn about Diophantine equations.
   b. How do these equations relate to the problem in **How Many Oranges?**
   c. For **How Many Oranges?** write one of the variables as a function of the other. How is the slope of this equation related to the difference between the values of consecutive solutions?
   d. Create or find another problem whose solution is a Diophantine equation.
3. *Card Arrangement*. Make a deck of cards, creating one card for each of the numbers 1 through 10. Arrange the top card face up so the number 1 is showing. Move the next card to the bottom of the pile. Turn over the third card to display the number 2. Continue this process, alternating between turning up a card and moving the next card to the bottom of the pile. The cards are arranged properly if they are revealed in the order of 1 through 10.
   a. Solve the puzzle.
   b. Describe your thought process as you worked toward a solution.
   c. Does the same solution process work for any deck with an even number of cards?
   d. Does the same solution process work for any deck with any number of cards? Explain any differences if the number of cards is odd.
4. *Russell's Paradox*. Bertrand Russell (1872–1970) is credited with publishing several paradoxes in set theory. One of these paradoxes is called the barber paradox: *In a particular town, a barber cuts the hair of all those and only those who do not cut their own hair. Does the barber cut his own hair?* Explain why this statement is called a paradox.
5. *True and False with Mom and Dad*. In *Math Curse* by Jon Scieska and Lane Smith (Viking Books, 1995), there is a part of the story when a mother says, "What your father says is false" and the father responds, "What your mother says is true." Analyze this pair of statements. Compare and contrast the *Math Curse* problem with Russell's Paradox in problem 4.

DEVELOPING MATHEMATICAL PEDAGOGY

Throughout this text you will be asked to respond to items called "Developing Mathematical Pedagogy" (DMP). These items are intended to help you apply your evolving teaching concepts and skills. You will sometimes find an icon next to specific problems. The icon indicates that the problem involves working with students. The icon means the item is especially well-suited for inclusion in a professional portfolio. The icon indicates that the problem is related to lesson planning; these items are also good candidates for a professional portfolio. The icon indentifies resources found on the course website.

TEACHING AND LEARNING HIGH SCHOOL MATHEMATICS © 2010 John Wiley & Sons, Inc.

Reflecting on Our Own Learning

6. Consider your own experiences with reasoning and proof.
   a. In which mathematics courses, other than geometry, were you expected to justify your thinking (more than "show your work") or construct proofs of mathematical concepts and relationships?
   b. What about reasoning and writing proofs has been easy for you? What has been difficult?
   c. As a teacher, what concerns do you have about helping your students develop their abilities in reasoning and proof?

Preparing for Instruction

7. George Pólya (1887–1985), a renowned mathematician and teacher, identified four basic steps in problem solving, as outlined in the following table.

Pólya's Problem-Solving Steps

1. Understand the problem. What is given? What do you need to find?
2. Devise a plan. Find ways to get started; here are some suggestions.

- Make the problem simpler
- Use diagrams
- Collect data systematically
- Act it out
- Use cases
- Use organized lists to find a pattern
- Use variables
- Work backwards
- Simplify the conditions
- Build a model
- Solve a related problem
- Use graphs
- Do another activity and return to the problem later

3. Implement a plan. Follow through with one or more of the plans in the previous step. If necessary, devise and use a different plan.
4. Reflect. After solving the problem, think about questions like the following:
   - What have you learned?
   - What larger generalizations can you make?
   - What insights have you gained into the problem-solving process?
   - What related problems can you solve?
   - Is there an easier way to solve this problem?

Adapted from Pólya (1957)

   a. Explain how Pólya's problem-solving steps relate to the Stages of Questioning.
   b. For which of these strategies does conjecturing play a role? Explain.

8. Consider **DMU** problem 3, the Card Arrangement activity. Suppose you observe a student working on the problem. Anticipate ways the student might work on the problem and create four questions to support the student, one for each of the Stages of Questioning.

TEACHING AND LEARNING HIGH SCHOOL MATHEMATICS © 2010 John Wiley & Sons, Inc.

Analyzing Student Thinking

9. Extend the **Question Response Support (QRS) Guide** you began when initially working on the How Many Oranges? problem (refer to page 454 in the Unit One Appendix for the **QRS Guide** for this student page).

   a. Recall the different approaches to How Many Oranges? shared by your peers. Record these in the **QRS Guide** in a way that will help you remember the approaches.

   b. For the approaches listed in problem 9a, list any questions you would ask the learner in the Teacher Support column of the **QRS Guide**.

   c. Some student responses on How Many Oranges? are shown below. For student work different from the approaches taken by your peers, record the approach in a succinct but memorable way in the **QRS Guide**. Provide questions you would ask students who gave these responses in the Teacher Support column of the **QRS Guide**.

   d. How did students use conjecturing in the sample responses below?

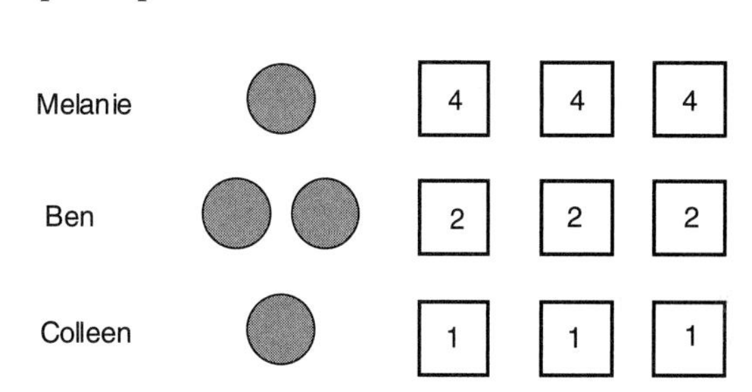

Adele: *I thought about what Melanie must have found . . . 3 equal portions of oranges plus one additional orange. Imagine that Melanie puts the oranges in three boxes with one extra orange; Melanie left behind two boxes. Ben took 2 oranges and then had 3 equal groups or boxes of oranges. The total of Ben's oranges has to be the same as the two groups Melanie left behind. Ben also left behind 2 boxes. Colleen took 1 orange, then had 3 equal groups or boxes. She also left behind 2 boxes. At first, I just had empty boxes but then I tried numbers to see which ones would work. This one in the picture works, so Melanie took 5 oranges, Ben took 4 oranges, and Colleen took 2 oranges. There were a total of 13 oranges and 2 oranges got left behind.*

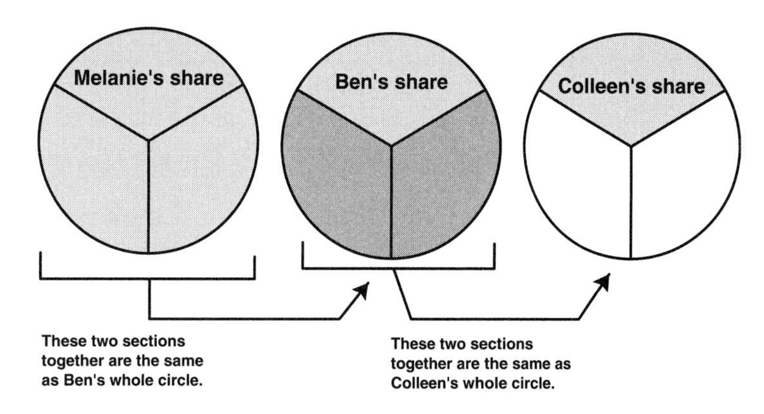

Celeste: *The first circle shows the total number of oranges. Melanie's share includes 1 more orange than each of the other sections. The second circle represents what Ben saw when he woke. Ben's share includes 2 more oranges than each of the other sections. The third circle shows the oranges Colleen found. Colleen's share includes the orange she took out. The algebra shows what happens at each step.*

Melanie left behind: $\frac{2}{3}(T-1)$

Ben left behind: $\frac{2}{3}[\frac{2}{3}(T-1)-2]$

Colleen left behind: $L = \frac{2}{3}\left\{\frac{2}{3}[\frac{2}{3}(T-1)-2]-1\right\} = \frac{8}{27}T - \frac{50}{27}$

Solving for T, $T = \frac{27}{8}L + \frac{50}{8}$.

Beth: *I started with the total number of oranges and tried out different numbers to see which ones would work. The total number of oranges has to be a multiple of 3 plus 1 more orange, so I started with 4 and added 3 to get the next total number of oranges and*

TEACHING AND LEARNING HIGH SCHOOL MATHEMATICS © 2010 John Wiley & Sons, Inc.

so on. 13 was the first number that worked. 40 was the second number that worked. After that, I tried 67 because $40 - 13 = 27$, so I added 27 to 40 and 67 worked! Then I started adding 27 to see if I could find any more numbers that worked—and they did. So, to find the total number of oranges, $T = 27n + 13$, where n is a non-negative integer. I also noticed a pattern in the leftover oranges, $L = 8n + 2$.

Total Oranges	Melanie Took	Melanie Left	Ben Took	Ben Left	Colleen Took	Colleen Left
4	1 + 1 = 2	2				
7	1 + 2 = 3	4				
10	1 + 3 = 4	6				
13	1 + 4 = 5	8	2 + 2 = 4	4	1 + 1 = 2	2
16	1 + 5 = 6	10				
19	1 + 6 = 7	12				
22	1 + 7 = 8	14	2 + 4 = 6	8		
25	1 + 8 = 9	16				
28	1 + 9 = 10	18				
31	1 + 10 = 11	20	2 + 6 = 8	12		
34	1 + 11 = 12	22				
37	1 + 12 = 13	24				
40	1 + 13 = 14	26	2 + 8 = 10	16	1 + 5 = 6	10
67	1 + 22 = 23	44	2 + 14 = 16	28	1 + 9 = 10	18
94	1 + 31 = 32	62	2 + 20 = 22	40	1 + 13 = 14	26
121	1 + 40 = 41	80	2 + 26 = 28	52	1 + 17 = 18	34

Using Internet Resources

10. Sudoku puzzles have become very popular all around the world. Several websites offer free puzzles to print out or complete online. Find a Sudoku website and learn the rules.

a. Solve a puzzle and discuss several reasoning strategies that could be used in solving Sudoku puzzles.

b. What mathematical benefits could be derived from solving Sudoku puzzles?

c. Refer to **Colorful Shapes** from **Listening to Students Reason about Mathematics** (page 16). How is **Colorful Shapes** related to Sudoku puzzles?

11. The Math Forum website (http://mathforum.org/, downloaded May 8, 2009) offers several puzzles and games to engage students in mathematical reasoning.

TEACHING AND LEARNING HIGH SCHOOL MATHEMATICS © 2010 John Wiley & Sons, Inc.

a. Find two puzzles or games you think will encourage students to develop and use their reasoning skills.

b. Share what you like about the puzzles or games. Describe how they support the development of reasoning.

Analyzing Instructional Activities

12. Refer to **Colorful Shapes** from **Listening to Students Reason about Mathematics** on page 16.

a. Find some way to prove that the puzzle can be solved. How many unique solutions exist (disregarding placement of pieces by specific color and shape)? Explain.

b. Suppose you simplified the problem to a 3 by 3 grid to be completed with 3 colors and 3 shapes.

i. Can the puzzle still be completed? Justify your answer.

ii. Can the rules be modified slightly so that the puzzle can be completed? Justify your answer.

c. **Colorful Shapes** might appeal to a kinesthetic learner who would benefit from moving the pieces around. The same puzzle might also appeal to a visual learner who can easily distinguish between the attributes of each puzzle piece. Characterize the problems and puzzles in this lesson (including **DMU** and **DMP**) by the types of learning styles (kinesthetic, visual, auditory, reading/writing) students might employ to solve them.

d. When might you use a puzzle like **Colorful Shapes** in your classroom?

e. How does the puzzle help learners get an intuitive feel for deduction (drawing conclusions that *must* follow from given conditions)?

Making Historical Connections

The introductions to the first five units contain brief remarks about the history of mathematics related to the mathematical focus of the unit. The first lesson of each unit contains questions referring to these historical connections.

13. There have been many famous theorems in mathematics that have intrigued mathematicians for centuries. For instance, Fermat's last theorem was only proved in the 1990s, the Four-Color Theorem was proved in the 20th century using computers, and Goldbach's Conjecture is still unproved.

a. Explore a bit about the history of one of these problems.

b. Some famous problems find their way into pop culture through plays, poems, or songs. According to Wikipedia, Fermat's last theorem is referenced in numerous fictional works. Find out more about at least one of these.

c. How might you consider making historical connections in your mathematics classroom?

Teaching for Equity

Given the diverse student population in many schools throughout the country, teachers need to be aware of cultural and learning differences that may influence students' success in the mathematics classroom. Every unit has several readings or activities under the heading, ***Teaching for Equity****, to provide teachers with some insights into cultural, ethnic, language, socioeconomic, or learning differences that may be important to consider to help all students be successful with mathematics.*

14. Teaching for equity means that all students—regardless of gender, ethnic, or cultural background; geographical location; or socioeconomic status—are provided opportunities to learn advanced mathematics that will develop their full

TEACHING AND LEARNING HIGH SCHOOL MATHEMATICS © 2010 John Wiley & Sons, Inc.

potential and keep their mathematics and career options open. Go to NCTM's website (nctm.org). Locate the section on position statements (under "About NCTM") and find the January 2008 position statement on *Equity in Mathematics Education*.

a. Read the position statement. Summarize the rationale for teachers to develop a culture of equity in their mathematics classrooms.

b. Identify at least two strategies mathematics teachers might use to create an equitable learning environment.

c. Search for a mathematician, current or past, who is female or from an ethnic minority. Learn something about this mathematician's history and accomplishments. How might you integrate information about this mathematician into the high school mathematics curriculum?

d. One concern among mathematics educators is the achievement disparities that exist among white, African American, Hispanic/Latino, Asian, and Native American students. Use the Internet to research mathematics achievement and/or high school graduation rates in your state or local school district by ethnic group. What differences, if any, exist in your location? What commitments can you make to address these differences in your own classroom?

e. The 1997 yearbook from NCTM, *Multicultural and Gender Equity in the Mathematics Classroom: The Gift of Diversity*, edited by Janet Trentacosta and Margaret J. Kenney, contains a series of chapters on various aspects of teaching for equity. The first chapter by Lucille Croom, "Mathematics for All Students: Access, Excellence, and Equity" (pp. 1–9), provides a rationale for teaching from a perspective that focuses on equity, including gender and cultural diversity. Locate the yearbook and read this chapter. Compare the suggestions in this chapter to those you developed in problems 14b and 14d.

Reflecting on Professional Reading

15. Read the general "Reasoning and Proof" standard (pp. 56–59) and "Reasoning and Proof Standard for Grades 9–12" (pp. 342–346) in *Principles and Standards for School Mathematics*.

a. Identify the four aspects of the Reasoning and Proof standard that should permeate the mathematics curriculum for all students in grades preK–12.

b. The Reasoning and Proof standard for grades 9–12 identifies several vignettes or examples that describe the role reasoning can play in the mathematics curriculum at these grades. How have these examples influenced your thinking about the role that reasoning and proof will play in your classroom?

16. Read the general "Problem Solving" standard (pp. 52–55) and "Problem Solving Standard for Grades 9–12" (pp. 334–341) from *Principles and Standards for School Mathematics*.

a. Mathematics homework historically has involved solving "problems." But the *PSSM* view on what constitutes a problem is different from what traditionally has been expected for homework. Explain.

b. What are some of Pólya's problem-solving strategies? (Be sure to include some that were not already mentioned in this book.)

c. What are some of the dispositions that play a role in problem solving?

d. What are the dual roles of problem solving in grades 9–12 as seen by the *PSSM* authors?

Problems 17–19 refer to "Types of Students' Justifications" by Larry Sowder and Guershon Harel (*Mathematics Teacher*, 91 (November 1998): 670–675). (See the course website for this article.)

TEACHING AND LEARNING HIGH SCHOOL MATHEMATICS © 2010 John Wiley & Sons, Inc.

17. Read "Types of Students' Justifications" by Larry Sowder and Guershon Harel.

   a. Consider your solution to the **How Many Oranges?** problem. According to the scheme suggested by Sowder and Harel, how would you classify your reasoning or proof for this problem?

   b. Refer to the student work provided in **DMP** problem 9. Determine the types of reasoning or proof strategies each student used to solve the problem.

   c. What questions might a teacher ask to help a student realize that the reasoning they provided is not complete? Add such questions to your **QRS Guide for How Many Oranges?** if you have not already done so.

18. a. Suppose a student in an algebra class attempts to justify the algebraic relationship $(x + 3)^2 = x^2 + 6x + 9$ with a graph done on a graphing calculator. How would you classify this "proof" according to the scheme suggested by Sowder and Harel?

   b. A student disproves a conjecture by providing a counterexample to show that the conjecture is not true in general. How would you classify this "proof" according to the scheme suggested by Sowder and Harel?

   c. Students are working in a small group on a conjecture stated by one of the group members. The students are not sure whether or not the conjecture is reasonable, so they try several examples and find that the conjecture is true for all of the examples. What are likely some of your challenges as a teacher regarding this group's progress toward the development of reasoning and proof skills?

19. a. Inductive reasoning refers to making a generalization based on many specific cases. For example, students measure the three angles of many triangles and conclude that the sum of the angle measures of all triangles is 180°. Where is inductive reasoning in the Sowder and Harel scheme?

   b. Deductive reasoning refers to making a convincing general argument that some assertion is true by using logic and what has already been established as true (axioms and theorems). Where is deductive reasoning in the Sowder and Harel scheme?

1.2 Exploring Mathematical Concepts Cooperatively: Reasoning with Conditional Statements

As indicated earlier in this unit, today's mathematics classrooms should be characterized by less teacher telling and by more student inquiry. With appropriately designed tasks and activities, students can explore mathematical concepts individually or with their peers. The teacher's role in this process is crucial. Not only must he or she use tasks that help students uncover the mathematics, but the teacher must also orchestrate discourse so that students summarize the concepts learned through the task. In this lesson, these processes are modeled through activities that focus on understanding important ideas related to conditional (*if–then*) statements.

Inference permeates daily life. For example, a person getting ready to leave for school might think, "If I leave home at 7:45 A.M., I run into all kinds of traffic. If I leave home at 7:30 A.M., the traffic is lighter." A mother asks her young children what they want for supper; when they can't decide, she says, "I guess I'll fix spaghetti." The children are always able to make a decision after that. So the mother likely thinks, "If I offer to make the kids spaghetti for supper, then they will make a decision on what they would like instead." In mathematics, implication statements must have some truth value, either true or false. In daily life, inferences are not always so determinate.

TEACHING AND LEARNING HIGH SCHOOL MATHEMATICS © 2010 John Wiley & Sons, Inc.

For a sentence to be a mathematical statement, it must be possible to determine whether or not the statement is true. A **conditional, implication,** or **if–then statement** is one in which a condition or **hypothesis** (also referred to as *antecedent*) precedes a **conclusion** (also referred to as *consequent*). In the example, "If I leave home at 7:45 A.M., I run into all kinds of traffic," the hypothesis is "I leave home at 7:45," and the conclusion is "I run into all kinds of traffic." In this text, the term **dictionary** is used to refer to the translation of statements into symbols. In the traffic example, one can represent the hypothesis and conclusion as follows:

p: I leave home at 7:45.

q: I run into all kinds of traffic.

Then, the conditional statement can be symbolized as: *If p then q*, *p implies q*, or $p \rightarrow q$. In symbolic form, the underlying logic is sometimes easier to analyze.

When is the statement "If I leave home at 7:45, then I run into all kinds of traffic" true? When is the statement false? Clearly, if one leaves at 7:45 and runs into traffic, the statement must be true. What happens if one leaves at 7:45 and does not run into traffic? Then the statement is clearly false. If one does not leave home at 7:45, then whether one hits traffic or not is immaterial because the hypothesis is not true. In this case, the conditional statement is considered to be true. Thus, a conditional statement is false only when the hypothesis is true and the conclusion is false.

Subtle differences exist in the way implication statements are made and understood. A statement can sometimes be rewritten using p and q to obtain a related statement equivalent in terms of its truth value. At other times, a rewritten statement using p and q no longer has the same truth value as the original conditional. Using common sense with conditional statements involving well-known facts can help you determine the equivalence or nonequivalence of a conditional and its related statements.

You will work on the tasks in this lesson in pairs using the cooperative learning strategy, *Pairs Check* (Kagan, 1992), an adaptation of *Think Pair Share*. However, with *Pairs Check*, although both partners think about the problem, they take turns listening to and monitoring the work of each other.

Pairs Check

1. Student A solves a problem or carries out a procedure, demonstrating or explaining the process employed and why it makes sense.
2. Student B watches and listens carefully, asking questions at any point when the explanation or process is unclear or in error.
3. Student A corrects the work, if necessary, until both students are satisfied with their solution.
4. Students switch roles after a predetermined number of problems.

(Kagan, 1992)

In this lesson, you will:

- Translate English sentences to conditional statements in *if–then* form.
- Create Venn diagrams to model English sentences and conditional statements.
- Write the converse, inverse, and contrapositive of a conditional statement.
- Determine the equivalence of two or more conditional statements, assuming the original conditional statement is true.
- Experience an inquiry lesson and its summary discourse.
- Experience the cooperative learning strategy, *Pairs Check*, to continue building a learning community.

TEACHING AND LEARNING HIGH SCHOOL MATHEMATICS © 2010 John Wiley & Sons, Inc.

Materials

- 10 coins or markers

Website Resources

- Cooperative Learning Powerpoint
- Stages of Questioning

LAUNCH

To determine if one's reasoning is correct or not, it often helps to write implications as *if-then* statements and to visualize the statements whenever possible. Venn diagrams provide one way to represent conditional statements visually. The visual representation can help to determine the truth or falsity of conditional statements that have subtle differences.

Work with your partner on items 1–3 below. Use *Pairs Check* on item 1, taking turns as you determine the meaning of each of the Venn diagrams. While listening to and monitoring your partner, use the Stages of Questioning on pages 19–20 from Lesson 1.1 as a guide to ask good questions if your partner gets stuck.

1. Consider the following Venn diagrams. Recall that the Great Lakes refer to Lakes Huron, Ontario, Michigan, Erie, and Superior. Describe what each of the Venn diagrams represents. How do you know?

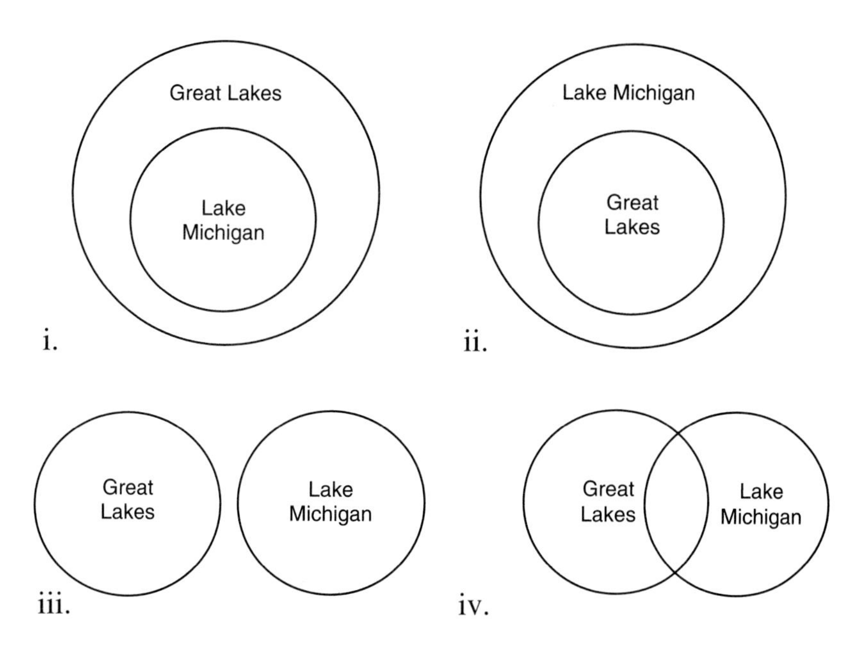

2. Which Venn diagram represents the statement, *Lake Michigan is one of the Great Lakes*?
3. Translate the above statement into *if–then* form.

EXPLORE

Several statements are related to conditional statements, and there are many ways to write them. Now you will explore these related statements.

4. Which of the following is logically equivalent to the conditional statement you wrote in **Launch** item 3? Explain in terms of your use of the English language and in terms of the Venn diagram you chose in **Launch** item 2.

TEACHING AND LEARNING HIGH SCHOOL MATHEMATICS © 2010 John Wiley & Sons, Inc.

a. If the lake is not one of the Great Lakes, then the lake is not Lake Michigan.

b. If the lake is one of the Great Lakes, then the lake is Lake Michigan.

c. If the lake is not Lake Michigan, then the lake is not one of the Great Lakes.

5. To symbolize each of the statements in items 3 and 4, use the following dictionary for the statements:

p: The lake is Lake Michigan.

q: The lake is one of the Great Lakes.

Identify the statements from items 3 and 4 associated with each of the following.

a. If p then q: the original conditional statement, symbolized as $p \rightarrow q$.

b. If q then p: the **converse** of the original conditional, symbolized as $q \rightarrow p$.

c. If not p then not q: the **inverse** of the original conditional, symbolized as $\sim p \rightarrow \sim q$. (The tilde symbol $\sim$ represents *not* or *negation*.)

d. If not q then not p: the **contrapositive** of the original conditional, symbolized $\sim q \rightarrow \sim p$.

6. Consider statements i–viii below. Then, use *Pairs Check* to complete items 6a–6d; members of each pair should alternate between working on the problem and monitoring their partner.

i. Numbers less than 10 must be less than 20.

ii. Odd integers greater than 11 are prime.

iii. All ducks are birds.

iv. All triangles are polygons.

v. None of the defenders of the Alamo lived.

vi. No woman has been president of the United States of America.

vii. All equilateral triangles are isosceles triangles.

viii. Vertical angles are congruent to each other.

For each statement in i–viii:

a. Create a Venn diagram to represent the statement.

b. Write the statement as a conditional statement in *if–then* form.

c. Create a dictionary to use when symbolizing the conditional statement.

d. Complete the following table and determine whether or not each statement is true or false, assuming the original conditional is true.

Statement	Symbolic Form	English Sentence	T or F?
Conditional	$p \rightarrow q$		
Converse			
Inverse			
Contrapositive			

SHARE AND SUMMARIZE

7. Share your pair's responses with the responses of another pair. Resolve any differences.

TEACHING AND LEARNING HIGH SCHOOL MATHEMATICS © 2010 John Wiley & Sons, Inc.

8. a. When the original conditional statement is true, which of the following statements is always true?
    i. Converse
    ii. Inverse
    iii. Contrapositive
  b. What do your results in item 8a tell you about the logical equivalence of the two statements?
  c. How do Venn diagrams help explain which statements are logically equivalent?
9. What statement is logically equivalent to the converse of a conditional?
10. In this lesson, an abstract concept (conditionals and their related statements) is introduced through a visual approach and common sense English statements.
  a. Discuss the pros and cons of such an approach.
  b. How do Venn diagrams help you think about conditional statements and their related statements?
  c. How did the questions in the **Explore** help you decide which statements are logically equivalent?
  d. What are the "big" ideas of the lesson?
11. Throughout the lesson, you worked with a partner to aid each other in making sense of conditional statements.
  a. Discuss pros and cons of using *Pairs Check* when working on "new" concepts that arise from common experience.
  b. In what other ways might you use *Pairs Check* in your classroom?
  c. In this lesson, pairs checked their work with another pair. How does this approach support students in gaining a deeper understanding of the big ideas of the lesson?

DEEPENING MATHEMATICAL UNDERSTANDING

1. *Conditionals: True or False?* Consider the conditional statement, *If Kathryn works on Wednesday, then Sara babysits David.*
  a. Provide a situation for which the conditional is true.
  b. Provide a situation for which the conditional is false.
  c. Write the contrapositive of the conditional. Provide a situation for which the contrapositive is false.
2. *More Work with Conditionals.*
  a. Write each of the following as a conditional statement in *if–then* form.
    i. It's raining, so I'll take an umbrella.
    ii. All squares are rhombi.
    iii. All high school students need to take algebra.
    iv. The mountains are beautiful when they are not covered with fog.
    v. All equilateral triangles are regular polygons.
    vi. An apple a day keeps the doctor away.
    vii. A penny saved is a penny earned.
  b. Choose two of the conditionals in problem 2a, one that you know to be true and one that is subjective. Write the converse, inverse, and contrapositive for each statement you chose.
  c. Assuming the original conditional statement is true, which of the converse, inverse, and/or contrapositive is also true?

TEACHING AND LEARNING HIGH SCHOOL MATHEMATICS © 2010 John Wiley & Sons, Inc.

d. In real life, inferences are not always true. Which of the statements in problem 2a might have questionable truthfulness? Comment on the precision of the English language as compared to the precision of mathematical language.

e. Write the converse, inverse, and contrapositive for the statement(s) you chose in problem 2d. Is the truthfulness of these statements consistent with your decision in problem 2c?

Problems 3–6 refer to the **Moving Markers** student page (below). **Moving Markers** is an adaptation of a puzzle that was part of a traveling exhibit sponsored by the National Science Foundation.

Moving Markers

Materials

- 10 coins or markers

	1	2	3	4
A				
B				
C				
D				

Place markers on the shaded cells of the table. Move two of the markers to cells that are currently empty so that all rows and columns have an even number of markers in them.

1. Is it helpful to move the marker in B3 to B4? Why or why not?
2. Solve the puzzle and write your moves as *if–then* statements, such as the bulleted statements that follow.
   - *For the first move, if the marker in B3 is moved to B4, then there are an odd number of markers in row B.*
   - *For the first move, if the marker in B3 is moved to B4, then there are an odd number of markers in column 4.*
3. Consider the statement: *After the first move, if there are an odd number of markers in column 4, then the marker in B3 was moved to B4.*
   Is the statement true or false? Explain. How does this statement compare to the second conditional statement in problem 2?
4. Write additional *if–then* statements in which the hypothesis is true but the conclusion is false. Determine in each case if the conditional statement is true or false. Explain your decision.

3. *Conditionals and Moving Markers.* Complete the student page, **Moving Markers**.

a. Use the following dictionary to symbolize the conditional statement from problem 2 on the **Moving Markers** student page.

For the first move, if the marker in B3 is moved to B4, then there are an odd number of markers in column 4.

p: The marker in B3 is moved to B4.

q: There are an odd number of markers in column 4.

TEACHING AND LEARNING HIGH SCHOOL MATHEMATICS © 2010 John Wiley & Sons, Inc.

b. Write the conditional statement that arises from $q \rightarrow p$. What is the name of this reversed conditional statement?

c. For the first move in solving the puzzle, determine the truth or falsity of the statements $p \rightarrow q$ and $q \rightarrow p$.

4. *Moving Markers Revisited.* Before making his first move in solving this puzzle, a high school student noticed the following move and wrote:

If I move the marker from B2 to A2, then column 2 still has an even number of markers in it.

a. Determine if this student's conditional statement is true for the **Moving Markers** puzzle. Explain.

b. Write the converse. How does it compare to the original conditional statement? Is it true or false? Explain your thinking.

c. Write the inverse. How does it compare with the original conditional statement? Is it true or false? Explain your thinking.

d. Write the contrapositive. How does it compare with the original conditional statement? Is it true or false? Explain your thinking.

5. *Puzzles and Conditionals.*

a. How does reasoning arise in solving the **Moving Markers** puzzle?

b. How did the **Moving Markers** puzzle help you think about conditional statements and their inverses, converses, and contrapositives?

6. *The Converse of Conditional Statements.*

a. Find an example of a true conditional statement for which the converse is also always true.

b. Find an example of a true conditional statement for which the converse might be true but is not always true.

c. What can you tell about the converse when you know the conditional statement is true? Is it false? Why or why not?

7. *Moving More Markers.* At most, 36 markers can be placed on a 6 by 6 square grid.

a. Arrange 14 markers on the grid so that each row and column has an even number of markers.

b. Comment on your reasoning as you place the markers. Write several of the moves you made as conditional *if–then* statements.

c. Compare and contrast this problem with the one presented on the **Moving Markers** student page. Which would you consider more challenging? Why do you think so?

8. *Toy Storage.* Tori enjoys arranging and rearranging her miniature toys on a wall display case that has 25 small openings as in the figure. She has 14 toys to arrange.

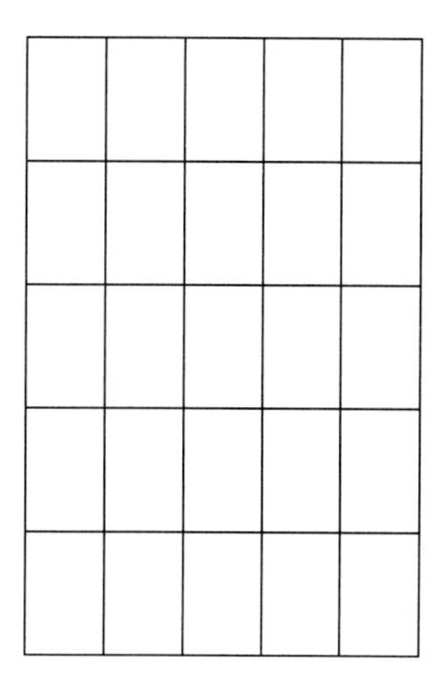

a. How can she arrange all 14 toys so that every row and column has an even number of toys in it?

b. Can she arrange the toys so that no row or column is empty? Why or why not?

c. Can she arrange the toys so that all rows and columns have an odd number of toys in them? Why or why not?

d. If the display case had 36 small openings in a square arrangement, could Tori accomplish the task in problem 8c?

e. For each of the problems in 8a–8d, show an arrangement that works—if one exists.

9. *Universal Statements.* A **universal statement** is a statement of the form *For all x in set S, p(x), where p(x) is a property that may or may not be true for elements in set S.* Determine whether each of the following universal statements is true or false. If

TEACHING AND LEARNING HIGH SCHOOL MATHEMATICS © 2010 John Wiley & Sons, Inc.

false, find an example that shows the statement is false. An example for which x is in set S but $p(x)$ is false is called a **counterexample** to the statement.

a. For all squares q, q is a rhombus.

b. For all positive real numbers x, $x^3 > 1$.

c. For all rectangles r, r is a square.

d. For all triangles t in the plane, the sum of the angle measures of t is 180°.

e. For all data sets d, d has a mode.

10. *Rewriting Universal Statements*. Rewrite each of the universal statements in *if–then* form. Determine if each statement is true or false.

a. All certified high school mathematics teachers have completed a course in calculus.

b. All hurricanes in the Atlantic have female names.

c. All snakes are reptiles.

d. All rectangles are parallelograms.

e. Vertical angles have equal measures.

f. The equation of a linear function can be written in the form $y = mx + b$.

11. *Existential Statements*. An **existential statement** is a statement of the form *There exists an element x in set S such that p(x), where p(x) is a property that may or may not be true for elements in set S.*

a. Determine whether each of the following existential statements is true or false. If true, provide an example to show the statement true; if false, explain why the statement is false.

i. There exists a candy bar that has nuts in it.

ii. There exists a quadrilateral that has no right angles.

iii. There exists a quadrilateral with all sides congruent but with no right angles.

iv. There exists a bear that is not a mammal.

b. Consider the statement(s) in problem 11a that is(are) false. Show that you can write a true *if–then* statement that disproves the existential statement.

12. *If and Only If Statements*. In mathematics, there are many statements of the form *p if and only if q*. Such statements assert *If p, then q* and, at the same time, assert the converse *If q, then p*. Rewrite each of the following *if and only if* statements as two *if–then* statements.

a. A satellite is in geosynchronous orbit if and only if it flies at 24,000 miles above the earth's surface.

b. Devin has a fever if and only if his temperature is above 98.6°.

c. A number is even if and only if it can be written in the form $2k$, where k is an integer.

d. $\sqrt{3x + 1} = x - 3$ if and only if $x = 8$.

e. A quadratic function has a maximum value if and only if its leading coefficient is negative.

13. *Variants of If–Then*. There are several variants of *if–then* statements. For each of the following, rewrite the statement in *if–then* form. Convince yourself that each variant makes sense.

a. The form ***p* only if *q*** is logically equivalent to *if p then q*.

i. A teacher is certified to teach mathematics in a state only if the teacher satisfies the state's certification requirements for mathematics at the time of certification.

ii. A triangle is equilateral only if it is isosceles.

b. The form ***p* is a sufficient condition for *q*** is logically equivalent to *if p then q*.

i. Eating's one's vegetables is a sufficient condition for having dessert.

TEACHING AND LEARNING HIGH SCHOOL MATHEMATICS © 2010 John Wiley & Sons, Inc.

ii. Having two congruent bisecting diagonals is a sufficient condition for being a rectangle.

c. The form **p is a necessary condition for q** is logically equivalent to *if q then p*.

i. Being a rhombus is a necessary condition to being a square.

ii. Being male is a necessary condition for being a father.

DEVELOPING MATHEMATICAL PEDAGOGY

Analyzing Students' Thinking

14. In a classroom, the teacher wrote the following statement on the board: *If Cassie finished third at the track meet, then she threw the shot put at least 50 feet.* In each case, evaluate the student's reasoning to determine whether or not it makes sense. What questions might you ask to help students whose reasoning is incorrect? (Consult the Stages of Questioning from Lesson 1.1 to help you.)

a. ***Sam:*** *Cassie did not throw the shot put at least 50 feet, so she did not finish third at the track meet.*

b. ***Marsha:*** *Cassie threw the shot put 60 feet, so she must have finished first.*

c. ***Bonnie:*** *Cassie did not finish third at the meet, so she did not throw the shot put at least 50 feet.*

Preparing for Instruction

15. Some teachers believe they have found "just the right case" for helping students with implications when they use the conditional, *If a figure is a parallelogram, then the figure is a quadrilateral.* It translates to $p \rightarrow q$ with easy translations for the two letters. The relationship is well understood.

a. Use the format of the Venn diagrams in the Great Lakes problem from the **Explore** in this lesson. Draw a Venn diagram using the parallelogram/quadrilateral context.

b. Using Venn diagrams, design a task or series of questions that help students realize that when $p \rightarrow q$ is true, so is its contrapositive.

c. Using Venn diagrams, design questions or a task that help students see that when $p \rightarrow q$ is true, its converse is not clearly true or false.

d. Create another *if–then* statement that uses geography or geometry or some other context that you think will make conditionals more accessible to students.

Teaching for Equity

16. Some educators believe that one piece of generic advice they can give to any parent of a student is to encourage the student to have a study partner, a variant of cooperative learning—a focus of this unit. What do you think might be some potential benefits to students if they have both out-of-class as well as in-class study partners?

Working with Students

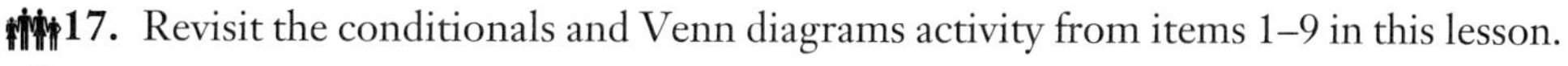

17. Revisit the conditionals and Venn diagrams activity from items 1–9 in this lesson.

a. Complete a **QRS Guide** that includes the problems students will work on in column 1, your solutions and possible student solutions in column 2, and teacher

TEACHING AND LEARNING HIGH SCHOOL MATHEMATICS © 2010 John Wiley & Sons, Inc.

support in terms of questions in column 3. To complete column 3, consult the Stages of Questioning in Lesson 1.1.

b. Try the activity with one or more high school students. Record any student solutions that you did not anticipate on the **QRS Guide**.

c. How did high school students make sense of the equivalence or nonequivalence of a conditional and its related converse, inverse, and contrapositive?

d. How did Venn diagrams help students see how some conclusions must follow from certain given conditions?

Reflecting on Professional Reading

18. Read "Communication Standard for Grades 9–12" (pp. 348–352) in *Principles and Standards for School Mathematics*.

a. According to this standard, what role does communication play in reasoning and proof?

b. Consider the three writing examples on p. 352 in the reading. Develop a writing assignment for students that would be appropriate with this lesson.

19. Read "From Exploration to Generalization: An Introduction to Necessary and Sufficient Conditions" by Martin V. Bonsangue and Gerald E. Gannon (*Mathematics Teacher*, 96 (May 2003): 366–371).

a. Using the technique described in the article, express 40 and 62 as the sum of consecutive natural numbers.

b. In the conclusion, the authors cite *PSSM* and the need for students to learn mathematics with understanding and to build new knowledge from previous knowledge. How did students' work on the initial problem influence their ability to attach understanding to the proof?

c. In both the article and this lesson, English sentences were used to provide a context in considering the truth of a conditional as well as its converse, inverse, and contrapositive. To what extent did you find the use of English sentences helpful? What difficulties, if any, resulted from using English sentences to make sense of conditionals?

d. The article provides several sequences of teacher and class dialogue. How did the discourse help students come to grips with the meanings of "necessary" and "sufficient"? In what ways would lack of teacher and student discourse make the task more difficult?

1.3 Using Representations to Investigate Mathematics: Reasoning with Conjunctions, Disjunctions, and Negations

Mathematics teachers have long used pictures, graphical displays, and symbolic expressions to represent mathematics. In today's classrooms, representations of mathematics also include contexts (story lines), verbal expressions, spreadsheets, charts, physical materials, diagrams, and more. Further, the focus includes student-generated representations, not just ones offered by teachers or textbooks. Inviting students to generate representations that make sense to them honors and respects their different ways of learning and provides teachers a way to understand and build on student thinking. Both of these benefits promote equity with strong achievement by all students.

Representations are an essential ingredient in the mathematics classroom. They help students understand concepts by focusing the user on important mathematical structures. In addition, representations help students make connections between

TEACHING AND LEARNING HIGH SCHOOL MATHEMATICS © 2010 John Wiley & Sons, Inc.

concepts, for example, showing with graphs how geometry relates to algebra. In this lesson, we continue with the use of visual displays, specifically Venn diagrams, to help students make sense of logical statements involving conjunctions (and), disjunctions (or), and negations (not). These concepts often seem abstract for students; a visual representation of the concept helps make meanings and relationships more concrete. In addition, translation to symbols helps one focus on the underlying mathematical structure of arguments.

To continue developing a classroom community, the cooperative learning strategies, *Think Pair Share* (McTighe and Lyman, 1988) and *Pairs Check* (Kagan, 1992), are extended to help students transition from working in pairs to working in groups of four. Students are arranged in groups of four but pair off to complete a task or two. After each pair has found a solution or gotten as far as they can in their work on a task, the pairs return to their group of four. Pairs then share their ideas with the small group. The classroom teacher monitors the amount of time for each phase of paired work and small group work.

In this lesson, you will:

- Compare the mathematical and everyday meanings of the conjunction *and* as well as the disjunction *or*.
- Use visual representations, specifically Venn diagrams, to show relationships between statements that use *and* or *or*.
- Use symbolic representations to represent statements and objects in order to analyze mathematical arguments.
- Extend the cooperative learning strategies, *Think Pair Share* and *Pairs Check*, to discussion in small groups of four.

Website Resources

- **Multiples** student page
- **Baseball and Soccer** student page

LAUNCH

1. As a class, create a large Venn diagram. The diagram should resemble the figure shown. Write your name in the region that best describes you.

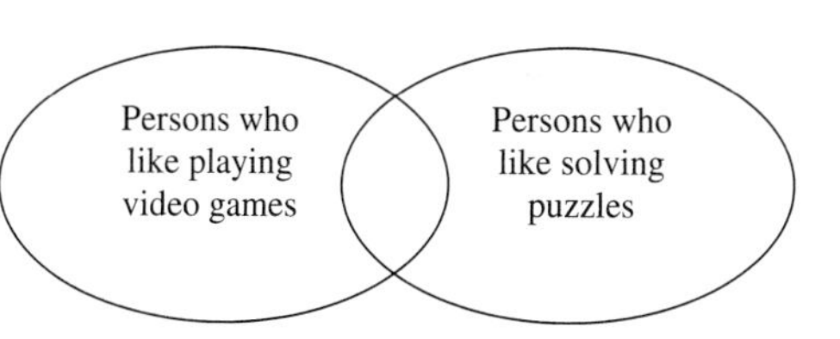

The symbol $\wedge$ represents the conjunction *and*. The symbol $\vee$ represents the disjunction *or*. The symbol $\sim$ represents the *negation* of the statement following it. The dictionary provided symbolizes the statements relevant to the Venn diagram.

G: Persons who like playing video games

S: Persons who like solving puzzles

2. For each of the following, list the names of persons who fit the category described. Draw a Venn diagram and shade the regions that visually represent such persons. Use the dictionary provided above for G and S to represent symbolically the category described.
   - **a.** Persons who like playing video games and solving puzzles
   - **b.** Persons who like playing video games or who like solving puzzles
   - **c.** Persons who either like playing video games or like solving puzzles, but not both
   - **d.** Persons who do not like playing video games
   - **e.** Persons who do not like solving puzzles

TEACHING AND LEARNING HIGH SCHOOL MATHEMATICS © 2010 John Wiley & Sons, Inc.

3. What is meant by each of the following questions asked by waiters and waitresses in restaurants?
   a. "Cream and sugar?"
   b. "Fries or onion rings?"
   c. "Coffee or dessert?"
4. a. Describe differences in the use of *inclusive or* (juice or coffee, or both) and *exclusive or* (regular or diet, but not both).
   b. Identify the use of *inclusive or* in items 2 and 3.
   c. Identify the use of *exclusive or* in items 2 and 3.
   d. Discuss differences between English language and mathematical uses of *and* and *or*.

EXPLORE

Venn diagrams are effective visual tools to represent conjunctions and disjunctions. Student pages **Multiples** (page 41) and **Baseball and Soccer** (page 42) provide two different scenarios in which conjunctions, disjunctions, and the negations of these are used.

5. Use the cooperative learning strategy *Pairs Check* to complete one of the student pages. Each pair of students in a group of four should complete a different student page. Be ready to share your results with the other pair of students in your group of four.

TEACHING AND LEARNING HIGH SCHOOL MATHEMATICS © 2010 John Wiley & Sons, Inc.

Multiples

Consider the following Venn diagram. One circle represents the positive multiples of 2. The other circle represents the positive multiples of 9. The contents of the entire rectangle represent the set of counting numbers.

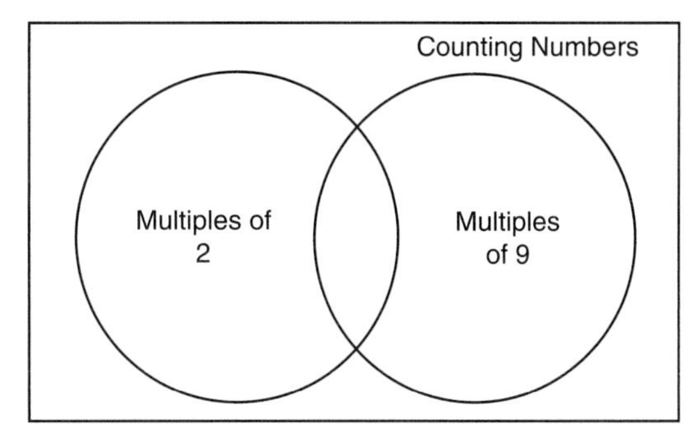

1. Sort the counting numbers 1–40 into the appropriate regions of the Venn diagram. How would you describe the numbers in the intersection of the circles?

2. What would you find in the region outside of the circles?
Draw Venn diagrams to illustrate the solution for each of the following:

3. Shade the regions that represent *Multiples of 2 and multiples of 9.*

4. Shade the regions that represent *Multiples of 2 or multiples of 9.*

5. Statements describing numbers in each region can be represented symbolically by creating a dictionary of statements:

t: The number is a multiple of 2.

n: The number is a multiple of 9.

c: The number is a counting number.

The symbols $\wedge$ and $\vee$ represent the **conjunction** *and* and the **disjunction** *or*, respectively. (Note that the And symbol, $\wedge$, resembles a capital A without the crossbar.)

a. Distinguish between the meanings of items 3 and 4.

b. Represent each of the statements in items 3 and 4 using symbols.

6. Shade the regions that represent *Numbers that are not multiples of both 2 and 9.*

7. Shade the regions that represent *Numbers that are not multiples of 2 or are not multiples of 9.*

8. The statement in item 6 is a **negation** and can be symbolized as $\sim(t \wedge n)$, in which $\sim$ symbolizes *not.*

a. Symbolize the statement in item 7.

b. How do the two diagrams in items 6 and 7 compare? What does this tell you about how to find the negation of a conjunction?

9. Consider a statement symbolized as $\sim(t \vee n)$.

a. Write an English statement for $\sim(t \vee n)$.

b. Draw a Venn diagram for $\sim(t \vee n)$.

c. In order for $\sim(t \vee n)$ to be true, what has to be true for $t \vee n$? Use ordinary English to help you determine conditions under which $\sim(t \vee n)$ is true.

d. Draw a diagram for $\sim t \wedge \sim n$.

e. What conjecture would you make about the negation of a disjunction based on your Venn diagrams in items 9b and 9d?

TEACHING AND LEARNING HIGH SCHOOL MATHEMATICS © 2010 John Wiley & Sons, Inc.

Baseball and Soccer

The figures in the diagram represent students who play baseball, soccer, other sports, or no sports at all.

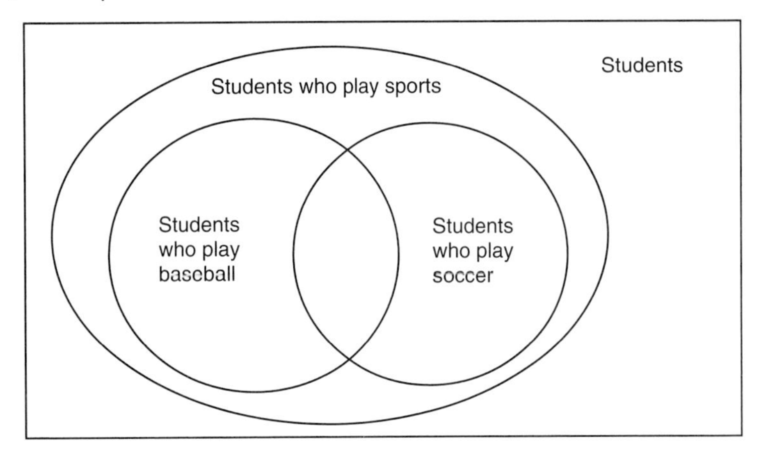

1. Categorize the students based on the sports they play. Write their names in the appropriate region(s).
   - **a.** Turan plays only baseball.
   - **b.** Caro plays both baseball and soccer.
   - **c.** Alexia plays only soccer.
   - **d.** Kris plays only football.
   - **e.** Pat does not play any sports.
   - **f.** Sweda will play baseball, soccer, or both.
2. The various regions can be represented symbolically by creating a dictionary of statements:

   b: A student plays baseball.

   s: A student plays soccer.

   p: A student plays sports.

   The symbols $\wedge$ and $\vee$ represent the **conjunction** *and* and the **disjunction** *or*, respectively. (Note that the And symbol, $\wedge$, resembles a capital A without the crossbar.)
   - **a.** Draw Venn diagrams and shade the region(s) corresponding to $b \wedge s$.
   - **b.** Draw Venn diagrams and shade the region(s) corresponding to $b \vee s$.
3. A person does not play baseball is symbolized $\sim b$ and read *not b*. (This statement is called a **negation**.) Draw a Venn diagram for $\sim b$.

   Use the original Venn diagram to identify appropriate regions for each of the statements in items 4 and 5. For parts a and b, draw a Venn diagram and shade the appropriate regions, then translate the sentences into mathematical symbols.
4. **a.** A person does not play both baseball and soccer.
   - **b.** A person does not play baseball or a person does not play soccer.
   - **c.** Compare your responses in items 4a and 4b.
   - **d.** What can you say about the negation of a conjunction?
5. **a.** A person does not play either of the sports baseball or soccer.
   - **b.** A person does not play baseball and does not play soccer.
   - **c.** Compare your responses in items 5a and 5b.
   - **d.** What can you say about the negation of a disjunction?

TEACHING AND LEARNING HIGH SCHOOL MATHEMATICS © 2010 John Wiley & Sons, Inc.

SHARE

Discuss questions 6–8 in your small group of four.

6. What is the underlying mathematics in these two student pages? In what ways were the activities on the two student pages similar? In what ways were they different?
7. To what extent do you think the context of the student pages would influence students' engagement with the content? Explain.
8. How do you think the use of Venn diagrams would influence students' ability to understand conjunctions, disjunctions, or the negations of the two?

SUMMARIZE

Discuss items 9–14 with the other members of your class.

9. In the activities of this lesson, the conjunction *and* and disjunction *or* are not defined. Instead, students who complete these activities build intuition about the differences in the uses of *and* and *or* in mathematics versus their uses in English.
   a. What are the differences between mathematical and English uses of *or?*
   b. What are the differences between mathematical and English uses of *and?*
   c. How will you make certain that students are able to use the mathematical terms *and* and *or* correctly while assisting them in minimizing any interference from their use in everyday language?
10. Problems 6–9 from the student page **Multiples** and problems 4 and 5 from the student page **Baseball and Soccer** help students develop an understanding of DeMorgan's Laws that state how to negate a conjunction or a disjunction. From what you learned in these problems, state DeMorgan's Laws.
11. Students often make errors when evaluating the negations of conjunctions and disjunctions. Consider the following sample student response:

    To find $\sim(t \wedge n)$, I would negate each one and get $\sim t \wedge \sim n$. To find $\sim (b \vee c)$, I would get $\sim b \vee \sim c$.

   a. What property of the real numbers is this student applying inappropriately?
   b. How might you help the student develop understandings of a conjunction and its negation or a disjunction and its negation?
12. a. Consider items 9–11. Which of these questions would you ask of students? Why?
   b. How do the questions help students focus on the important mathematical ideas of the lesson?
   c. What additional important mathematical ideas should students take from these student pages?
   d. How will you help students focus on what they learned through these student pages in a class summary?
13. Discuss the potential of representations of concepts to help students make sense of mathematics.
14. How is student learning enhanced by having students work in pairs for part of the time and having students discuss that work in groups of four?

DEEPENING MATHEMATICAL UNDERSTANDING

1. *Inclusive or* versus *Exclusive or*.
   a. Find a real-life example other than those in the lesson of *inclusive or*.
   b. Find a real-life example other than those in the lesson for *exclusive or*.

TEACHING AND LEARNING HIGH SCHOOL MATHEMATICS © 2010 John Wiley & Sons, Inc.

c. Construct a Venn diagram to represent an example that uses the *exclusive or*. How does it compare to a diagram for *inclusive or* for the same example?

2. *Combining Statements with Conjunctions and Disjunctions.*

a. For each of the following, write an equivalent expression. Use Venn diagrams to prove at least one of the equivalent relationships is true.

i. $p \wedge (q \wedge r)$
ii. $p \vee (q \vee r)$
iii. $p \wedge (q \vee r)$
iv. $(p \vee q) \wedge (p \vee r)$

b. Of which properties of arithmetic are you reminded in problem 2a?

3. *Conditionals and Negations.* Consider the statement: *If a person plays basketball, then the person is tall.*

a. Under what conditions would this conditional statement be false?

b. If the conditional is false, then its negation is true and vice versa. Use your answer in problem 3a to draw a conclusion about the negation of a conditional statement.

c. Create your own conditional statement and write its negation.

4. *Revisiting Baseball and Soccer.* Consider the statement: *If Marc plays baseball, then he plays soccer.*

a. Under what conditions would this conditional be false?

b. If the conditional is true, then its negation is false. Write the conditions you cited as a compound sentence in English. Then use the dictionary from the student page to write the conditions symbolically (i.e., $\sim (b \rightarrow p) =$________).

c. Use your results in problems 4a and 4b to draw a conclusion about the negation of a conditional statement.

5. *Truth Tables.* Another way to visualize statements is to use truth tables, in which the various combinations of *T* (True) and *F* (False) are written and then applied to determine the truth value of a compound statement.

a. Complete the table to show the truth values of a conjunction, disjunction, and conditional.

p	q	$p \wedge q$	$p \vee q$	$p \rightarrow q$	$p \wedge \sim q$
T	T				
T	F				
F	T				
F	F				

b. Explain how the truth table provides a tool to understand the negation of a conditional.

c. Build a truth table to determine the truth value of the statement. $p \wedge (q \vee r)$. (Hint: Think about the number of rows needed to show all the possibilities of True and False for three different simple statements p, q, and r.)

6. *Pandora's Boxes.* Solve the puzzles on the student page, **Pandora's Boxes** (pages 45–46).

a. The dictionary provided symbolizes the following statements relevant to **Pandora's Boxes.** S_N means, "The statement in box *N* is true." *N* can be box *A*, *B*, or *C*. Complete the table.

Statement	**Symbolic Notation**	**Negation of Symbolic Notation**	**English Statement of Negation**
Statement *N* is true.	S_N		
Box *N* contains Happiness.	H_N		
Box *N* contains Evil.	V_N		
Box *N* is Empty.	E_N		

TEACHING AND LEARNING HIGH SCHOOL MATHEMATICS © 2010 John Wiley & Sons, Inc.

b. Use the dictionary in problem 6a and the symbols for conjunction, disjunction, or negation as needed to symbolize the statements on each of Pandora's boxes. Then symbolize your solution to each of the **Pandora's Boxes** puzzles.

c. Comment on the ease or difficulty of translating words to symbols. How do symbols help or hinder finding a solution to these puzzles?

d. How were the solutions to these puzzles dependent on the mathematical uses of the terms *and* and *or*?

e. For puzzle 3 from **Pandora's Boxes**, can you determine what is in each box? If not, what additional clue(s) would you need to do so?

Pandora's Boxes

Pandora, in her curiosity, is said to have opened a forbidden box and let the four evils—disease, sorrow, vice, and crime—out into the world. What if she had been presented with more than one box and clues as to their contents? Solve the following puzzles to assist Pandora in letting happiness into the world while keeping evil contained!

1. The gods presented Pandora three boxes because the gods knew Pandora was inquisitive and would not be satisfied with leaving the boxes alone. She was also given the following clues:

i. One of the boxes contains evil, one contains happiness, and one is empty.

ii. The labels on all of the boxes are false.

Determine the contents of each box.

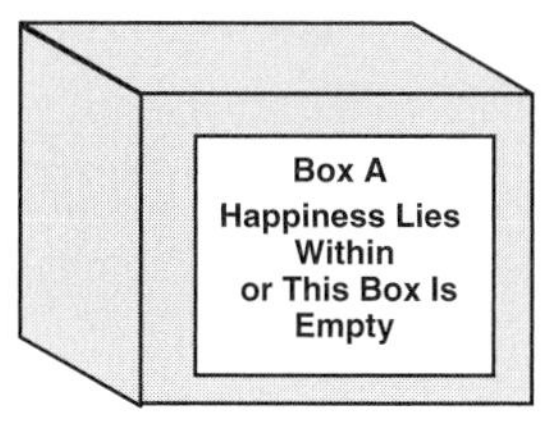

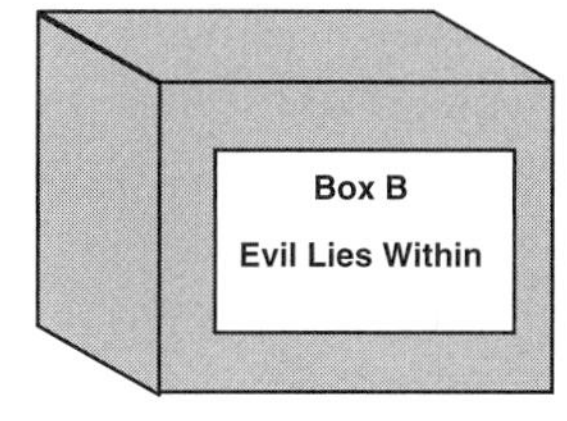

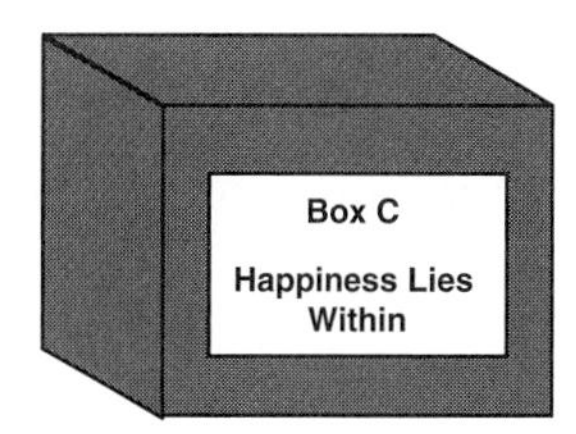

2. Pandora proved herself to be an able puzzle solver. To test her further, the gods presented the following set of boxes to Pandora. In this test, Pandora only had two boxes from which to choose. This time the clues were different; Pandora needed to be especially careful of what she assumed to be true!

i. Both boxes are labeled with statements whose truth value is the same (i.e., both boxes are labeled with true statements or both are labeled with false statements).

ii. A box contains either evil or happiness but not both and neither box is empty.

Determine the contents of each box.

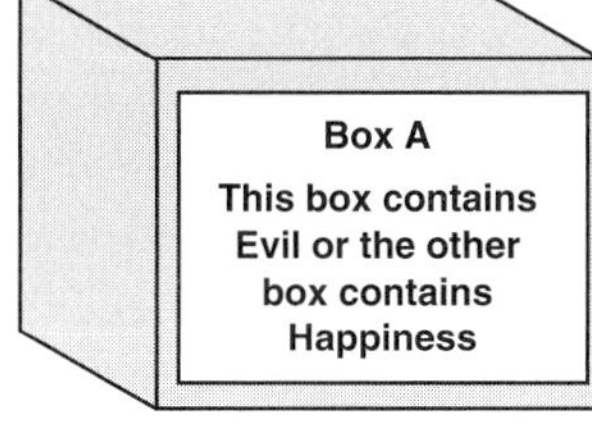

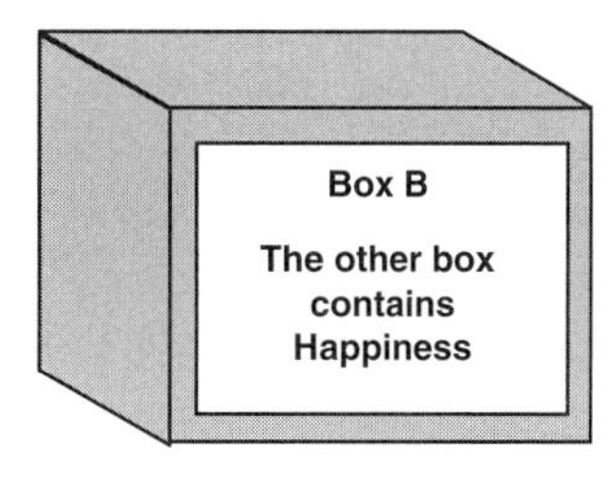

(*continued*)

TEACHING AND LEARNING HIGH SCHOOL MATHEMATICS © 2010 John Wiley & Sons, Inc.

(*continued*)

3. Pandora was not fooled, so the gods presented her with yet another set of boxes and clues. Help Pandora find happiness and keep evil from being let out into the world.

i. One of the boxes contains evil, one contains happiness, and one is empty.

ii. The box containing happiness is labeled with a true statement.

iii. The box containing evil is labeled with a false statement.

iv. The truth or falsity of the statement on the empty box is unknown.

Which box should Pandora open?

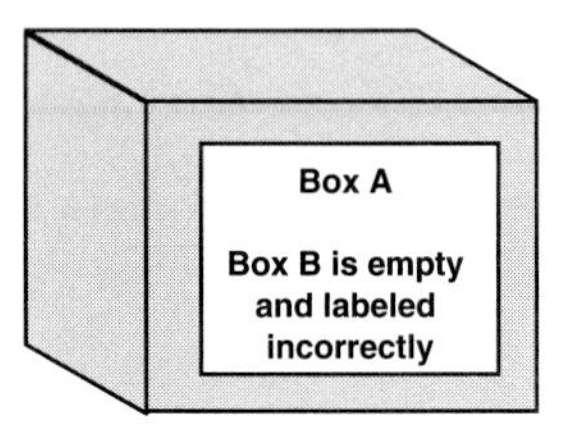

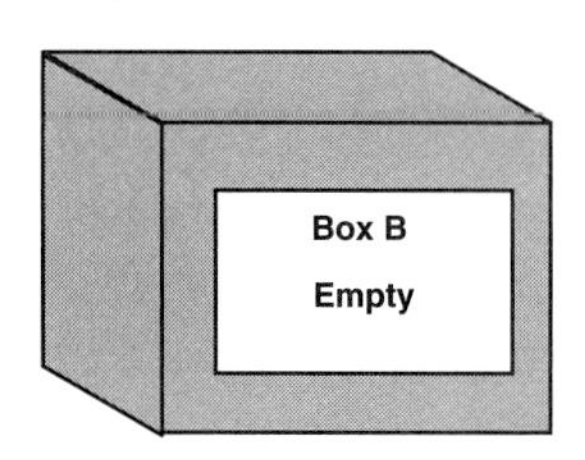

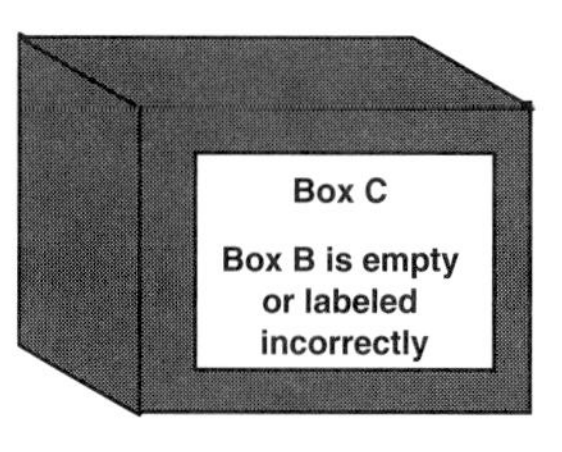

DEVELOPING MATHEMATICAL PEDAGOGY

Reflecting on Our Own Learning

7. In this lesson, you investigated conjunctions, disjunctions, and their negations with contexts, Venn diagrams, and symbols.
   a. Which approach did you prefer? Why?
   b. Which approach(es) do you believe are likely to appeal to high school students? Explain.
   c. Share how you might combine truth tables and Venn diagrams.

Analyzing Students' Thinking

8. Revisit problem 3 on the student page, Pandora's Boxes, in **DMU** problem 6. While working on the student page, high school students responded in the following ways:

   Joe: *Box C contains Happiness. The message on C is true no matter what, so Happiness is in the box with the true label.*

   Wade: *Box C contains Happiness. The message on C is always true. The message on Box A is always false. Can't tell about Box B, so Box C has Happiness in it.*

   Warren: *Pandora should open Box C because it contains Happiness. The message on Box A is always false because Box B can't both be labeled incorrectly and be empty. The message on C is always true because if B isn't empty it is labeled incorrectly, so the second half of the statement is true and if B is empty, the first half is true. With OR, only one or the other must be true for the whole thing to be true. I can't tell if Box B is labeled correctly or incorrectly, so its truth value is unknown. I know Happiness isn't in Box B because that would make its label false and the box containing Happiness has a true statement on it.*

TEACHING AND LEARNING HIGH SCHOOL MATHEMATICS © 2010 John Wiley & Sons, Inc.

a. Which student(s) are correct? Explain why you think so.

b. Which student(s) explain their reasoning to your satisfaction? Explain.

c. What questions would you be prepared to ask each student to help them make their responses more complete? (Revisit Lesson 1.1 for suggestions from the Stages of Questioning.) Complete a **QRS Guide** to indicate the problem students are solving in the left-hand column, student work in the center column, and teacher support in the right-hand column.

d. How does **Pandora's Boxes** help students engage with deduction, that is, drawing necessary conclusions from specific given conditions?

9. Students were working on the following problem. Some student responses appear below.

Kim is organizing her trophy case. She has one trophy each for Basketball, Volleyball, Figure Skating, and Gymnastics. How many ways might she arrange them on a single shelf?

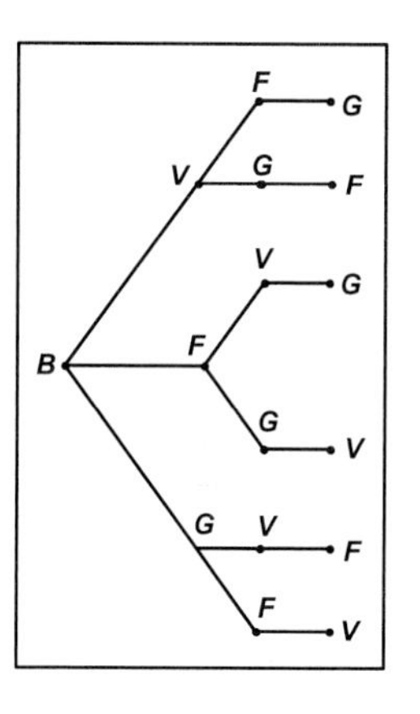

Ira: *I started to make a tree diagram: I figured there were four choices for her first trophy on the left. I only showed what happens for the Basketball trophy going first, but the rest of the trophies would have a similar tree diagram. In each of those cases, there were three others she could put next. That makes 12 possibilities. Then I just thought she could double this by taking one of the two remaining. The last one is whatever is left, so there are 24 possibilities.*

Callie: *I figured there were 4 choices for the first and 4 choices for the second. That makes 16. Then there are 4 choices for the third, so $16 \times 4 = 64$. Finally there are 4 choices for the last for $64 \times 4 = 256$. It seems like a lot.*

Mary: *Once she chooses one of 4 for the first location, she has three left to choose for the second spot. So that makes 12. Then she has 2 choices for the third spot, for 24. There is no choice at the end.*

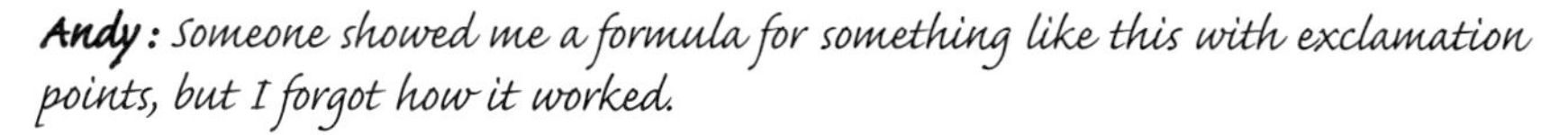

Andy: *Someone showed me a formula for something like this with exclamation points, but I forgot how it worked.*

a. Characterize each student's use of reasoning.

b. How does the tree diagram involve the logic in this unit?

c. Complete a **QRS Guide** to indicate the problem students are solving in the left hand column, student work in the center column, and teacher support in the right hand column. Use the Stages of Questioning to write some questions you would pose for each student.

d. What does Andy's comment reveal about student learning?

Preparing for Instruction

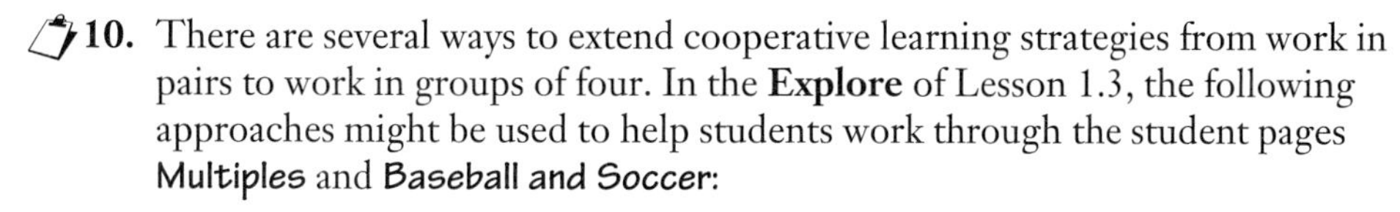

10. There are several ways to extend cooperative learning strategies from work in pairs to work in groups of four. In the **Explore** of Lesson 1.3, the following approaches might be used to help students work through the student pages **Multiples** and **Baseball and Soccer**:

i. Each pair of students completes both student pages and then discusses their results with another pair of students.

ii. Each pair of students chooses and completes one of the student pages and then discusses their results with another pair of students who might have completed the same page.

iii. In a group of four, one pair of students works on the student page **Multiples**; the other pair works on the student page **Baseball and Soccer**. When both

TEACHING AND LEARNING HIGH SCHOOL MATHEMATICS © 2010 John Wiley & Sons, Inc.

pairs have completed their assigned tasks, they form a group of four to share their findings.

iv. In a group of four, each pair of students works on the same student page, discusses it in the group of four, and then prepares to share their work with the class. Other groups of four are assigned the other student page to discuss and share with the class.

a. In planning for lessons, it is important for teachers to think about how to structure student learning opportunities. Which cooperative learning strategy might give students the best opportunity to learn this material? Why do you think so?

b. For each suggested approach, determine a purpose for which the approach would be preferred. If there are any approaches that you do not think appropriate, state why.

11. Thinking about contexts while preparing to teach addresses students' motivation to learn and connects what they are learning to what they already know; both of these are important concerns while planning lessons. Find at least five real-world contexts appropriate for use with high school students in which *and* or *or* is used. Are the uses of *or* inclusive or exclusive?

12. For an example of the use of conjunctions and disjunctions in electrical circuits, locate the following article: "Demonstrating Boolean Logic Using Simple Electrical Circuits" by Kevin W. McElhaney (*Mathematics Teacher*, 97 (February 2004): 126–134).

a. Work through the student activity pages that accompany the article. Evaluate the activities. What did you like? What might you change?

b. The article includes truth tables as a method to understand the functioning of the electrical circuits. How useful were the truth tables in making sense of the circuits and their actions? Do you believe this context would be of interest to high school students? Why or why not?

c. Compare and contrast the use of electrical circuits with the use of Venn diagrams to help students make sense of conjunctions and disjunctions. What are the advantages and disadvantages of each representation?

d. The context in this article is one instance of integrating mathematics with other disciplines, such as physics or engineering. What are the advantages of this sort of integration? What are some of the challenges?

13. Graphing often entails logic with *and* or *or*. Consider the zoo admissions context. Admission to The Kangaroo Zoo is as follows:

Regular visitor	\$12
Age 65 or older	\$ 9
Under age 5	free

a. Produce a graph showing the ages of those who pay \$12.

b. How does this graph involve logic?

c. Why might teachers want to use a context like this one when helping students explore the meanings of logical connectors?

d. Create another context that uses *and* or *or* to produce a number line.

14. Consider graphs for the following.

i. $x > 3$ or $x > 5$ **ii.** $x > 3$ and $x > 5$ **iii.** $|x - 3| < 5$ **iv.** $|x - 3| > 5$

a. Produce a graph for each sentence.

b. How do the concepts of *and* and *or* fit in these graphing tasks?

c. What difficulties do you anticipate students might have producing some of these graphs?

d. For one sentence that some students might find difficult to graph, suggest questions you could pose to support them.

e. Select one of the sentences and create a context (word problem) for it.

TEACHING AND LEARNING HIGH SCHOOL MATHEMATICS © 2010 John Wiley & Sons, Inc.

15. Tables and databases are other representations that may involve conjunctions, disjunctions, and negations. Consider the database on football players:

Player	Abe	Ben	Cal	Dan	Ed	Fin	Gil
Weight (kg)	86	108	82	104	82	90	75
Height (cm)	179	190	180	188	180	173	178

a. Write a conjunction, a disjunction, and a negation problem for this set of data.

b. Create another set of data you think would be of interest to high school students and repeat problem 15a.

16. Geometry often provides contexts for conjunctions, disjunctions, and negations.

a. Sketch triangles that are right and isosceles.

b. Sketch triangles that are right or isosceles.

c. Identify two other contexts in geometry that could be used to engage students with conjunctions, disjunctions, and negations. Write two items students could solve for each context.

17. a. Develop a student page using your own context that deals only with conjunctions and their negations.

b. Try your student page with a high school student. What was easy for the student? What misconceptions did the student have? What surprised you about the student's interaction with the activity?

c. Depending on the level of your student, some incremental help might be useful on the student page by providing questions according to the Stages of Questioning. Where might you add some hints or additional support for students?

Using Technological Tools

18. Many graphing calculators have features that enable the user to input a statement using a logical connector such as *and*, *or*, *not*. Determine how this feature works on your graphing calculator.

Teaching for Equity

19. Jones (2004) identifies a culturally responsive teacher as one who knows the subject matter, who listens to students to assess their thinking, who uses cultural knowledge to help students connect mathematics to their lives, who respects students' cultural beliefs, and who is reflective. (For more details, read "Promoting Equity in Mathematics Education through Effective Culturally Responsive Teaching" by Joan Cohen Jones. In *Perspectives on the Teaching of Mathematics*, edited by Rheta N. Rubenstein and George W. Bright (pp. 141–150). Reston, VA: National Council of Teachers of Mathematics, 2004.)

a. What aspects of a culturally responsive teacher have you noticed in your classroom observations?

b. The article raises the issue of *challenge* as a necessary part of an equitable environment in terms of the tasks asked of students and the actions of teachers. How might *challenge* be applied in the mathematics classroom?

c. How could preparation of support questions, such as those included in a **QRS Guide**, help teachers prepare to create a classroom environment in which students are appropriately challenged?

d. Which of the characteristics of a culturally responsive teacher do you hope to put into practice in your first year of teaching?

TEACHING AND LEARNING HIGH SCHOOL MATHEMATICS © 2010 John Wiley & Sons, Inc.

20. Some educators believe that using and encouraging students to use a variety of representations in mathematics courses is a practice that supports equity, that is, more students learning more mathematics successfully. Give your thoughts on this idea.

Reflecting on Professional Reading

21. Read the general "Representation" standard (pp. 67–71) and the "Representation Standard for Grades 9–12" (pp. 360–364) from *Principles and Standards for School Mathematics*.
    a. How do the representations discussed in this lesson connect with the recommendations in *PSSM* for this standard?
    b. The general portion of this standard discusses the use of abstraction in mathematics. How do representations, such as the Venn diagrams used in this lesson, help students understand abstract symbolizations of mathematics?
22. Read "A Truth Table on the Island of Truthtellers and Liars" by Christopher Baltus (*Mathematics Teacher*, 94 (December 2001): 730–732).
    a. Truth tables are another representation used to determine the truth values for conjunctions, disjunctions, and conditionals. Compare the use of truth tables to Venn diagrams.
    b. Compare and contrast the riddle in the article with the puzzles you solved throughout this unit.
    c. Create your own riddle that is in the same vein as the one in the article. Have one of your peers or friends try to solve your riddle. Revise it as needed.

    d. Revisit one or more of the **Pandora's Boxes** puzzles in this lesson. Try to solve the puzzle(s) using truth tables.

1.4 Learning from Students: Valid and Invalid Arguments

When teachers ask students to share their thinking and then listen carefully to what students are saying, they gain information valuable for teaching. For example, teachers can learn what students do or do not understand and what strategies students are using or considering. This information places the teacher in a stronger position to help students develop their understanding by building on what they already know. Consequently, tasks that draw out student thinking—and teaching that attends to that thinking—are an important avenue to support excellence and equity. As illustrated in this unit, the use of cooperative learning strategies encourages public sharing. When students work in pairs or small groups, they share their reasoning more readily than in whole-class settings. In such an environment, teachers have an opportunity to listen and learn about what their students are thinking as they solve problems. As an added benefit, use of cooperative learning strategies helps continue the goal of building a learning community.

In Unit One, the mathematical focus has been on reasoning and logic. In Lesson 1.1, you solved the **How Many Oranges?** problem using guiding questions to help find a solution. In Lesson 1.2, you learned about rewriting English sentences into conditional statements and determining the truth value of related statements based on the truth value of the original conditional statement. In Lesson 1.3, you learned about the mathematical meanings of the conjunction *and*, the disjunction *or*, and the negation of these. In this lesson, you will put these ideas together to construct valid and invalid arguments using direct and indirect reasoning.

TEACHING AND LEARNING HIGH SCHOOL MATHEMATICS © 2010 John Wiley & Sons, Inc.

To understand direct reasoning, consider the following examples. Parents sometimes get their children to eat their meals by promising a treat when they're finished. For example, a parent may say, "Valen, if you eat your dinner, then you can have dessert." Valen eats his dinner. So Valen gets dessert. Another form of direct reasoning includes stringing together several conditional statements to arrive at a valid conclusion. "Valen, if you eat your vegetables, then you will finish your dinner. If you finish your dinner, then you can have dessert." Valen eats his vegetables. So Valen gets dessert.

In the first example, Valen satisfied the condition or hypothesis of the conditional statement, so the conclusion occurred. In the second example, Valen satisfied the hypothesis of the first conditional statement whose conclusion satisfied the hypothesis of the second conditional statement; so, the conclusion of the second conditional statement occurred. Direct reasoning encompasses two basic types of arguments:

Modus Ponens

(*Latin for "Method of Affirming"*)
If p then q.
p
Therefore q.

Transitive Reasoning

(*Hypothetical Syllogism*)
If p then q.
If q then r.
Therefore: If p then r.

To understand indirect reasoning, revisit the example above: "Valen, if you eat your dinner, then you may have dessert." Valen did not get dessert. So it must be that Valen did not eat his dinner. One form of indirect reasoning (*modus tollens*) recognizes a basic relationship between the truth values of *If p then q* and *If not q then not p*. The basic indirect argument is:

Modus Tollens

(*Latin for "Method of Denial"*)

If p then q.

$\sim q$

Therefore $\sim p$.

Another form of indirect reasoning uses contradiction. In Proof by Contradiction, one assumes that a statement to be proved is false. Reasoning is used until a contradiction is reached. The contradiction demonstrates that assuming falsity was wrong. Therefore, the given initial statement must be true.

Proof by Contradiction

Given statement to prove: If p then q.

Assume $\sim (p \rightarrow q) = p \wedge \sim q$.

Reason until a contradiction is obtained.

Then, $p \wedge \sim q$ is false, so its negation, $\sim (p \wedge \sim q) = \sim (\sim (p \rightarrow q)) = p \rightarrow q$ is true.

Therefore $p \rightarrow q$.

Building on the pair cooperative learning strategies of *Think Pair Share* and *Pairs Check*, students' repertoires can be expanded to include strategies in which three or four students work together. In this lesson, you will work on the **Explore** problems using two complementary small group strategies, Co-Op and Random Reporting. In *Co-Op*, each team of students is assigned a different task. Teams share their results with the rest of the class once all teams have completed their assigned tasks. *Random Reporting* is a cooperative learning strategy in which a team member is randomly chosen to present the team's findings. *Random Reporting* helps ensure that no one student does all the work for the team and that all students are responsible for their own learning. *Co-Op* combined with *Random Reporting* encourages students to be interdependent, that is, to work together and share their understanding, while also being individually accountable for their learning. The rules for both cooperative learning strategies follow.

TEACHING AND LEARNING HIGH SCHOOL MATHEMATICS © 2010 John Wiley & Sons, Inc.

Co-Op

1. Choose a topic from among those given.
2. Work in your group of three or four to further each member's understanding and produce a group product according to the activity guidelines.
3. As a group, share your product or experience with the whole class.

(Davidson, 1990)

Random Reporting

1. Work on the assigned problem together in your group. Make sure everyone understands the problem, process, and solution.
2. Remember your assigned number.
3. Be prepared to report for your group when called on by number.

(Adaptation of *Numbered Heads Together* by Kagan, 1992)

To make best use of cooperative learning strategies that encourage groups of three or four students working together, the tasks assigned should be more difficult and require more minds working on them than those assigned for partner work.

In this lesson, you will:

- Employ what you have learned in previous lessons to analyze a problem logically.
- Use valid arguments of direct (including transitive) or indirect reasoning to solve problems and analyze solutions.
- Consider the benefits of listening carefully to students to gauge understanding and provide support.
- Engage in problem solving using the cooperative learning strategies of *Co-Op* and *Random Reporting*.

Website Resources

- Cooperative Learning Powerpoint
- **Pandora's Puzzler** student page
- "Types of Students' Justifications," by Larry Sowder and Guershon Harel (*Mathematics Teacher*, *91*, (November 1998): 670–675)

LAUNCH

1. **a.** Work in a small group of four to solve **Pandora's Puzzler**.

   b. Write out your solution to the problem using conditional statements or use the dictionary from Lesson 1.3 (**DMU** problem 6, page 44) to symbolize your solutions to the puzzles.

Pandora's Puzzler

According to Greek mythology, Pandora opened a box that let all the evils of mankind into the world. Rather than tempt Pandora with a single box and dare her not to open it, the gods took pity on her and gave her the following challenge. Help Pandora find happiness and keep evil from being let out into the world.

Pandora was presented three boxes and clues to their contents.

i. One of the boxes contains evil, one contains happiness, and one is empty.

ii. Only the box containing happiness is labeled with a true statement.

Suppose the three boxes were labeled as below. Determine the contents of each box.

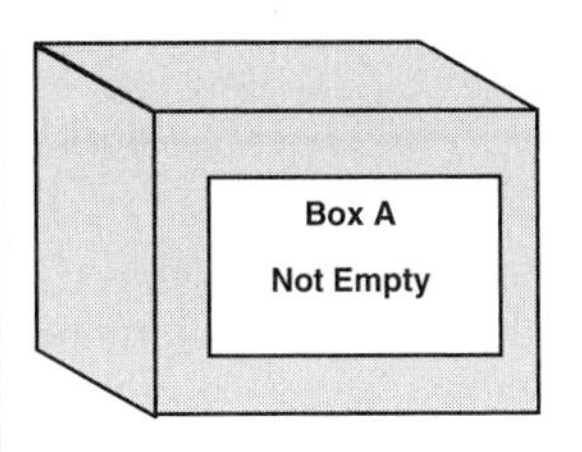

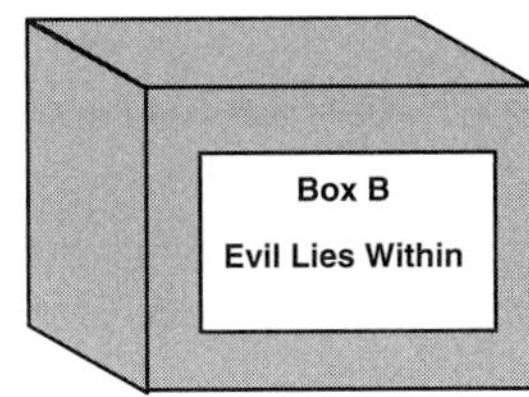

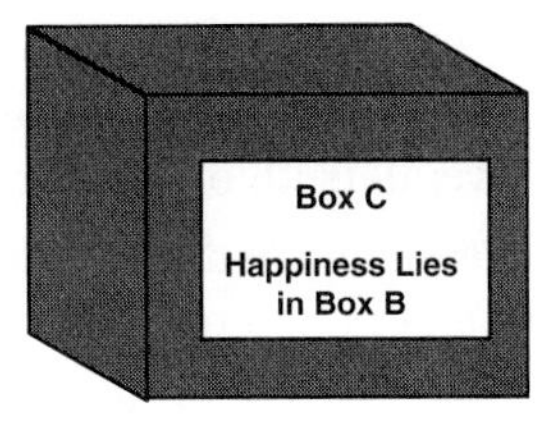

2. Which type(s) of reasoning did you use in your solution? If possible, match your arguments to the types identified in the introduction to this lesson. Give examples of each type of reasoning you used.
3. Is the statement on Box B true or false? Prove your conjecture. Determine what type(s) of reasoning you used to arrive at your conclusion.
4. Think about this lesson's theme of listening to students. How did listening to one another enable you to better work toward a solution to the puzzle? Give one specific example.

EXPLORE

Work in your *Co-Op* teams on the following problems. *Random Reporting* will be used when teams share their work with the class.

5. Consider the following statement: *For all integers m, n, and p, if m is a factor of n and m is a factor of p, then m is a factor of* $n + p$. Three sample student proofs are provided. Analyze the responses.
   a. Which proofs use appropriate reasoning?
   b. Identify any errors in reasoning.
   c. Identify uses of direct or indirect reasoning (including proof by contradiction).
   d. Classify the types of reasoning used by students according to the article, "Types of Students' Justifications," by Larry Sowder and Guershon Harel (*Mathematics Teacher*, 91, (November 1998): 670–675) from Lesson 1.1.

Clay: *Because m is a factor of n, I can write* $n = mk$ *for some integer k. I can also write* $p = mk$. *Then* $n + p = mk + mk = 2mk$. *So,* $n + p = m(2k)$. *This means that m is a factor of* $n + p$.

Michelle: *Because m is a factor of n, I can use the definition of factor to write* $n = mk$ *where k is an integer. Because m is a factor of p, I can also write* $p = mr$ *where r is an integer. I have to use different letters k and r because the integers might not be the same. Then,* $n + p = mk + mr = m(k + r)$. *But* $k + r$ *is an integer because the*

TEACHING AND LEARNING HIGH SCHOOL MATHEMATICS © 2010 John Wiley & Sons, Inc.

sum of any two integers is always another integer. If I let $s = k + r$, then, $n + p = m(k + r) = ms$ where s is an integer. So, by the definition of factor, m is a factor of $n + p$.

Sherry: *If $m = 3$, $n = 12$, and $p = 15$. Then m is a factor of n and m is a factor of p. But $n + p = 12 + 15 = 27$ and m is a factor of 27. So, the statement is true.*

Counterexamples typically are used to disprove a statement or conjecture a student might make. Some research indicates many students do not realize that a single counterexample is sufficient to disprove a claim. Rather, they view the counterexample as an exception to the rule.

6. Consider the following statement: *If m is a factor of $n + p$, then m is a factor of n and m is a factor of p.*

- **a.** Find a set of three numbers, m, n, and p, that make the statement true.
- **b.** Find a set of three numbers, m, n, and p, that make the statement false.
- **c.** Based on your responses to items 6a and 6b, what can you conclude about the overall truth of the given statement?

7. As mentioned in the introduction, proof by contradiction is a valid form of reasoning. Use a proof by contradiction to prove each of the following statements:

- **a.** There is no largest integer.
- **b.** There is no largest prime number. (Hint: Create a number that is one more than the product of all the known prime numbers. Use this result to generate a contradiction.)
- **c.** A triangle can have at most one right angle.
- **d.** A quadrilateral cannot have four obtuse angles.
- **e.** If n^2 is odd, then n is odd.
- **f.** If 3 is a factor of a^2, then 3 is a factor of a.

SHARE AND SUMMARIZE

Use the cooperative learning strategy, *Random Reporting*, to share your team's findings. Your reporting number will be assigned by your instructor.

8. Share your team's thinking about the student responses in item 5.

- **a.** How does paying close attention to students' thinking enhance your ability to help them understand?
- **b.** Think about the Stages of Questioning. What questions might you use with Clay and Sherry to help them address their misconceptions in thinking?
- **c.** How does paying attention to students' thinking promote equity?

9. Share your proof to one of the statements in item 7. Ensure that at least one team provides a sample proof for each statement.

- **a.** What similarities and differences do you notice about the proofs by contradiction?
- **b.** Was one of the statements harder to prove than the others? If so, which one? What made this statement more difficult to prove?

10. You used the cooperative learning strategy, *Random Reporting*, to share what your team did to complete the problems in the **Explore** section. Discuss benefits and difficulties with using *Random Reporting* to report team findings.

DEEPENING MATHEMATICAL UNDERSTANDING

1. *Proof by Contradiction*. Consider the following theorem known as the Prime Factor Theorem: *Every integer greater than 2 is either prime or has a prime factor.*

TEACHING AND LEARNING HIGH SCHOOL MATHEMATICS © 2010 John Wiley & Sons, Inc.

a. Create a dictionary and symbolize the theorem as a conditional.

b. Write the negation of the statement in problem 1a. This is the beginning assumption. (Hint: Recall from Lesson 1.3 that the negation of a conditional is not another conditional.)

c. Show that a number satisfying the statement in problem 1b can be factored into the product of at least two numbers where neither number is prime. Why is this true?

d. Reason from your work in problem 1c to obtain a contradiction, leading to the truth of the original statement.

2. *Irrational Numbers and Proof by Contradiction*. Proof by contradiction can be used to show that $\sqrt{2}$ (or any nonperfect square) is an irrational number. Follow the steps below to complete a proof.

a. With what assumption should you begin the proof?

b. In what form can any rational number be represented?

c. Use your results from problems 2a and 2b to express $\sqrt{2}$ as a rational number using integers x and y. Square both sides and simplify to remove fractions. Is x^2 even or odd? What does this tell you about x, y^2, and y?

d. Show that simplifying leads to a contradiction.

3. *Defining Quadrilaterals*. In geometry, students play a game, *What Am I?*, by choosing the fewest possible clues to determine a quadrilateral. Their team members use these clues to guess the quadrilateral.

a. A student is trying to get his team members to guess the word *square*. He uses the two clues below. Are these clues adequate? Why or why not?

A quadrilateral with at least one right angle.
A quadrilateral with two pairs of parallel sides.

b. If the clues in problem 3a are not sufficient, what additional clue(s) need to be given? Give additional clue(s) to ensure that the only possible quadrilateral described by all the clues together is a square. Make certain not to use more clues than necessary.

c. Repeat problem 3b using a different set of additional clues.

4. *Solving Equations*. Consider the following equation, $ax + b = 3x + 7$, where a and b are parameters that can assume any values. Think of each side of the equation as a separate line, $y = ax + b$ and $y = 3x + 7$. For each of the following, determine a pair of values to replace a and b so that

a. The equation has one solution.

b. The equation has no solution.

c. The equation has infinitely many solutions.

d. Explain how you know your choice works in each case.

e. Repeat problems 4a–4d for the following equation: $a(x + b) = 3x + 7$, where a and b are again parameters that can assume any values.

5. *Revisiting* Modus Tollens.

a. Using what you know about the equivalence of a conditional and its contrapositive, show how *Modus Tollens* can be derived from *Modus Ponens*.

b. Use a Venn diagram to demonstrate the validity of *Modus Tollens*. For example, show $p \rightarrow q$, then locate where *not* q must fall to show why *not* p must be the case.

6. *Disjunctive Syllogism*. Another valid argument form is $p \vee q$, $\sim p$, therefore q.

a. Explain why this argument form is valid.

b. Use a Venn diagram to show why this argument form is valid.

TEACHING AND LEARNING HIGH SCHOOL MATHEMATICS © 2010 John Wiley & Sons, Inc.

c. Consider the following dictionary of statements:

S: David likes to swim.

G: David likes to play video games.

David likes to swim or he likes to play video games. He does not like to play video games.

i. Use the dictionary to symbolize the argument.
ii. What can you conclude?

d. Construct your own English statements for p and q. Apply the disjunctive syllogism form to these statements. How does the English help make sense of the argument form?

7. *Inverse Errors*. Consider the following argument offered by a student.

If a lake is Lake Michigan, then it is one of the Great Lakes.
Lake Superior is not Lake Michigan.
Therefore, Lake Superior is not a Great Lake.

a. Create a dictionary for the statements in the response.

b. Use the dictionary to symbolize the student's reasoning.

c. Use a Venn diagram to explain why this argument is invalid. (Hint: Show the ambiguity by locating where Lake Superior may be in the diagram.)

d. This student's error often is called an *inverse error*. Why is this label appropriate for this type of error?

8. *Converse Error*. Consider the following argument offered by another student.

If a person is a gymnast, then the person goes to sporting meets.
Jason goes to sporting meets.
Therefore, Jason is a gymnast.

a. Create a dictionary for the statements in the response.

b. Use the dictionary to symbolize the student's reasoning.

c. Use a Venn diagram to explain why this argument is invalid.

d. This student's error often is called a *converse error*. Why is this label appropriate for this type of error?

e. High school students often think, $x = 5$ implies $x^2 = 25$ so if $x^2 = 25$, then $x = 5$. What is the nature of their error?

9. *Knightly Knews*. Solve the problems on the student page, **Knightly Knews** (page 57).

a. Write out your solution to the problem using conditional statements.

b. Create a dictionary to help symbolize your solution to the puzzle. (Hint: Let G_n: Gold knight is in position n.)

c. Which types of reasoning did you use in your solution? Cite examples of each.

d. Is your solution unique? If not, what seating guidelines might you add to be certain only one arrangement of knights is possible?

e. Suppose White is positioned next to the knight on King Arthur's right side. What are the logical consequences of this seating arrangement? Determine which types of reasoning you used to arrive at your conclusion.

 © 2010 John Wiley & Sons, Inc.

Knightly Knews

King Arthur and Guinevere are having a dinner party and have invited 10 of the most important knights of the realm to feast at the round table. Guinevere will sit immediately to her husband's left. She is sensitive to color, so she will not allow knights whose colors clash to sit next to each other; she and Arthur will be the only persons to wear Red to the dinner. Arthur wants the dinner to be peaceful but also productive, helping his knights to form more friendly alliances. The cook has asked that knights with similar food preferences be seated together. So far, Arthur and Guinevere have compiled the following seating guidelines. Each knight is identified by color.

1. Colors that cannot be seated together: Brown with Black, Red with Orange, Yellow with Gold, Purple with Brown.
2. These persons will be seated together with no one between. Food preferences of knights who sent in their choices (others will have to eat what is served!):

   Beef eaters: Arthur and Guinevere, Orange, and Blue

   Vegetarians: Green and Silver

   Chicken choosers: Gold, Yellow, and Brown.
3. Arthur thinks that Yellow is attracted to Guinevere so he wants Yellow seated opposite him in order to keep an eye on him.
4. White, Yellow, and Orange already are in a strong alliance, so they will be distributed at equal distances around the table.
5. Black and Blue make each other black and blue if seated next to each other, so they will be seated directly opposite each other.

Your job is to determine seating arrangements for the guests and make suggestions to Arthur so that the guests may find the one and only seat reserved for them. What suggestions will you give Arthur? What news (knews!) will Arthur provide his guests? Give reasons for each seating recommendation. Be creative.

10. *The Ring of the Lords.* **The Ring of the Lords** puzzle arises from a quote in *The Lord of the Rings* (J.R.R. Tolkien, 1954, frontispiece) that describes the creation and distribution of rings of power.
    - **a.** Solve **The Ring of the Lords** puzzle. Write down questions, strategies, and thoughts while solving the puzzle.
    - **b.** Determine any uses of direct or indirect reasoning or reasoning by contradiction in your solution to the puzzle in problem 10a.
    - **c.** Is the solution unique? If so, how do you know? If not, what edict(s) must be added to make the solution unique?
    - **d.** What seating arrangements are possible if edict (iv) is removed? Explain.

TEACHING AND LEARNING HIGH SCHOOL MATHEMATICS © 2010 John Wiley & Sons, Inc.

The Ring of the Lords

In the beginning, when the Dark Lord and the leaders of the Elven kingdoms, dwarf lands, and mankind agreed to share the power of Middle Earth, they met to receive their rings. To make certain all were equal, the Dark Lord and all leaders sat at a round table. There were three Elven Kings, seven Dwarf-lords, nine Mortal Men, and the Dark Lord. The rings were distributed, one to each leader.

The leaders of Middle Earth were seated around the table according to the edicts:

i. Dwarf-lords feel not Dark Lord's influence; Dwarves shall sit next to the Dark Lord on both right and left.

ii. Leaders shall disperse so that no two of the same race are neighbors at the table.

iii. Beware of the Dark Lord's power. Consecutive Elves and the Dark Lord must be separated by different numbers of leaders.

iv. Three of the separation numbers (in edict iii) must be consecutive and all must be greater than one.

1. Determine where the Dark Lord, the leaders of the Elven kingdoms, the Dwarf-lords, and Mortal Men can sit relative to each other. Explain how you know you are right.

2. **a.** Is the seating arrangement you found in problem 1 the only arrangement possible? Why or why not?

b. If not, what additional rule(s) should be included to ensure a unique seating arrangement (by race)?

3. What seatings are possible if edict iv reads: "Three of the separation numbers (in edict iii) must be consecutive"? (Note: There is no requirement that the separation numbers be used in a particular order around the table.)

Axiomatic Systems. Problems 11–12 address the relationships between the problems in this lesson and axiomatic systems. With definitions and axiomatic systems, one strives to give necessary and sufficient information to define an object or a mathematical system completely. When not enough conditions are presented to completely define the system (or definition), the system (or definition) is **incomplete**. A **redundant** system (or definition) provides more conditions than necessary. An internally contradictory system is considered **inconsistent**.

11. *Knightly Knews Axiomatic System.* Revisit the **Knightly Knews** student page.

a. Are the conditions presented sufficient to provide a full seating chart for all of the knights without ambiguity?

b. Is the system described by the seating guidelines in **Knightly Knews** consistent? Why or why not?

c. If the system is not fully defined, what additional guidelines will complete the system?

d. If the system is redundant, what seating guidelines are unnecessary in solving the puzzle?

e. Students suggested each of the following conditions with the intent that each individual condition completes the system. Determine if the student succeeded or if the rule makes the system redundant or inconsistent.

TEACHING AND LEARNING HIGH SCHOOL MATHEMATICS © 2010 John Wiley & Sons, Inc.

i. *White, green, and silver are vegetarians.*

ii. *Purple is vain and likes to see his reflection in Silver's armor. He asked to have Silver seated directly across the table.*

iii. *Purple announced when he arrived that he is a vegetarian so needs to be seated with them.*

12. *The Ring of the Lords Axiomatic System.*

a. Are the conditions presented sufficient to provide a full seating chart for all of the leaders without ambiguity?

b. Is the system described by the edicts in **The Ring of the Lords** consistent? Why or why not?

c. If the system is not fully defined, what additional guidelines will complete the system?

d. If the system is redundant, which edicts are unnecessary in solving the puzzle?

13. *Playing with Palindromes.* A palindrome is a number or word whose digits or letters are in the same order when read from left to right as from right to left. For example, 373, racecar, and 45654 are palindromes.

a. Investigate the claim: All 4-digit palindromes are divisible by 11.

b. Is 11 a factor of all palindromes? Explain.

c. Tegan claims that 11 is a factor of any palindrome of even length. Is she right? How do you know?

d. Which type(s) of reasoning did you use to solve each of problems 13a–13c?

DEVELOPING MATHEMATICAL PEDAGOGY

Analyzing Students' Thinking

14. For each of the following student proofs, evaluate the validity of the student's argument. Justify your response by referring to the valid argument forms. What question(s) would you ask each student whose understanding is not complete? Consult the Stages of Questioning in Lesson 1.1 for help.

a. *If Michaele likes red, then she buys red shoes. Michaele does not like red, so she does not buy red shoes.*

b. *Jose takes French or Spanish. He takes French. Therefore, he does not take Spanish.*

c. *If Sharon likes cats, then she has cats as pets. If Sharon has cats as pets, then she does not have mice. Sharon has mice. So, Sharon does not like cats.*

Analyzing Instructional Materials

15. Sherry's response in problem 5 of the **Explore**, attempting a proof by example, is typical of responses by many high school students. Curriculum materials may contribute to this misconception by providing examples that use particular theorems rather than providing proofs of them. Find one or more high school textbooks other than geometry. Look through several sections of the text. Are theorems proved or just illustrated with specific examples?

16. One common format often used in geometry textbooks is the two-column proof. Consider the following proof in this format.

Given: In $\Delta\ ABC$, $\overline{DE} \parallel \overline{BC}$

Prove: $\Delta ADE \sim \Delta ABC$

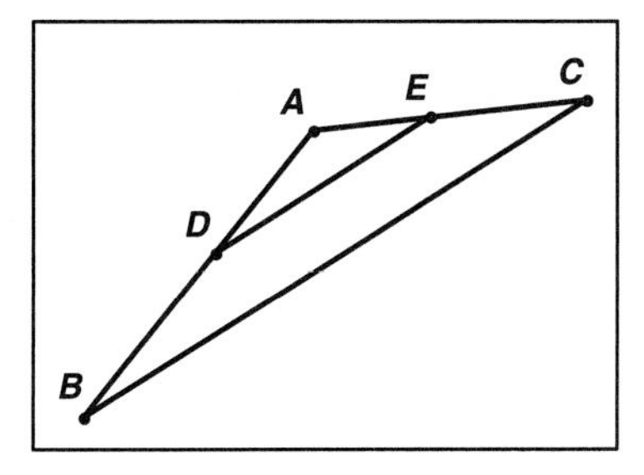

TEACHING AND LEARNING HIGH SCHOOL MATHEMATICS © 2010 John Wiley & Sons, Inc.

Proof:

Statements	Reasons
1. $\overline{DE} \parallel \overline{BC}$	1. Given
2. $\angle ADE \cong \angle ABC$	2. If parallel lines are intersected by a transversal, then corresponding angles are congruent.
3. $\angle A \cong \angle A$	3. Reflexive Property of Congruence
4. $\Delta ADE \sim \Delta ABC$	4. If two angles of one triangle are congruent to two angles of another triangle, then the two triangles are similar.

a. The two-column proof uses a direct argument (actually twice), but the location of the three elements in the argument masks its presence. Find the two instances of direct argument. (Hint: Find a conditional, an instance of the hypothesis, and an instance of the conclusion.)

b. The reasons in a geometry proof are the critical "glue" that link one or more statements (antecedents) to a conclusion (consequent). How might you help students see this relationship?

Preparing for Instruction

17. You used the cooperative learning strategy, *Co-Op*, to "divide and conquer" throughout this lesson so that no group had to solve too many problems. Discuss benefits and difficulties with using *Co-Op* to solve problems whose mathematical content is the same but varies in some other way (context, choice of parameters, etc).

18. a. Solve **Pandora's Puzzler, Too**. Keep track of your reasoning. Identify the types of reasoning you used to solve this puzzle.

b. Refer to the Stages of Questioning in Lesson 1.1. Identify questions at different levels you might use with students solving **Pandora's Puzzler, Too**. Record the puzzle, student work, and teacher support in a **QRS Guide**.

Pandora's Puzzler, Too

According to Greek mythology, Pandora opened a box that let all the evils of mankind into the world. Rather than tempt Pandora with a single box and dare her not to open it, the gods took pity on her and gave her the following challenge. Help Pandora find happiness and keep evil from being let out into the world.

Pandora was presented three boxes and clues to their contents.

i. One of the boxes contains evil, one contains happiness, and one is empty.

ii. Only the box containing happiness is labeled with a true statement.

Suppose these were the three boxes presented to Pandora. Determine the contents of each box.

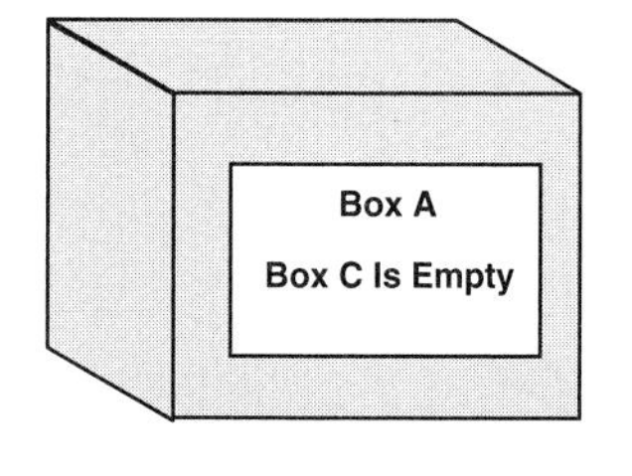

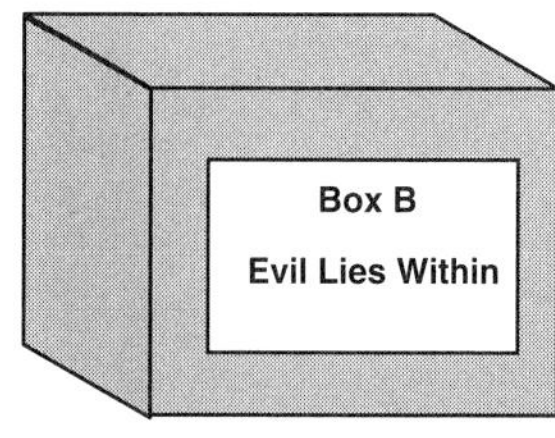

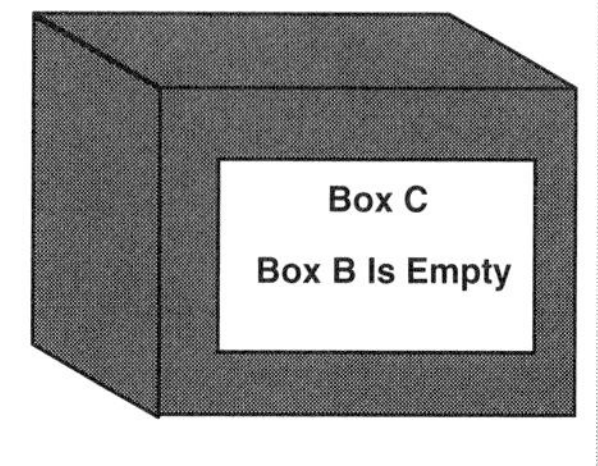

TEACHING AND LEARNING HIGH SCHOOL MATHEMATICS © 2010 John Wiley & Sons, Inc.

19. Return to one of the puzzles in the lesson or **DMU**.
    a. Complete a **QRS Guide** for the puzzle as you work through the problem. Anticipate how students will solve the puzzle and include possible student work in the **QRS Guide**. Include teacher support questions for the anticipated work.
    b. Try the puzzle with a student. Ask the student to think aloud while working on the puzzle and to share any diagrams or notes used. Listen carefully to the student's thinking. Use information you glean to pose one or two questions like those from the Stages of Questioning. Report the student's solution and teacher support questions in the **QRS Guide**.
    c. How difficult was it to ask questions that provide "just enough" guidance without being overly leading?
    d. What would you do differently next time to pose more effective questions?

Using Technological Tools

20. Students who have access to graphing calculators can use those tools to test conjectures, such as $(x + 5)^2 = x^2 + 25$. Which student arguments would you consider valid? Explain.
    a. ***Ruby:*** *To test whether $(x + 5)^2 = x^2 + 25$ is true, I graphed $y = (x + 5)^2$ and $y = x^2 + 25$ on the same axes. When $x = 0$, the two graphs intersect and the equation is true. For all other values of x, the graphs do not coincide. So, I have many counterexamples to prove that the statement is false.*
    b. ***Jenn:*** *I typed $(x + 5)^2 = x^2 + 25$ into my calculator and got 0. So the statement is always false.*
    c. ***Ron:*** *I typed $(x + 5)^2 = x^2 + 25$ into my calculator and got 1. So the statement is always true.*
    d. ***Sam:*** *I typed $y = (x + 5)^2$ and $y = x^2 + 25$ into the Y= menu. I looked at the table of values. When $x = 7$, I get 144 and 74, so the statement is not true.*

Reflecting on Professional Reading

21. Read "A Unified Framework for Proof and Disproof" by Susanna S. Epp (*Mathematics Teacher*, 91 (November 1998): 708–713). In this article, Epp discusses different types of proofs.
    a. In your own words, explain how Epp describes a direct proof, a disproof by counterexample, and a proof by contradiction.
    b. Some students do not believe one counterexample is enough to disprove a statement. Rather, they think one case is an exception and the theorem generally is true. What suggestions does Epp's article offer teachers as they try to help the student?
    c. Identify at least one strategy Epp recommends for expanding the role of proof at the high school level.
    d. On page 712 of the article, Epp provides several problems to support proof at the high school level. Choose one, solve it, and pose questions at each of the Stages of Questioning that could be used with students working on the problem.
22. Read "Proofs That Students Can Do" by Robert O. Stanton (*Mathematics Teacher*, 99 (March 2006): 478–482).
    a. Stanton describes a forward–backward method to address proofs. Provide your personal reaction to this approach.

TEACHING AND LEARNING HIGH SCHOOL MATHEMATICS © 2010 John Wiley & Sons, Inc.

b. What might be advantages, if any, to the forward–backward method? What disadvantages, if any, do you anticipate?

c. Use the forward–backward approach on problem 1 in the article (page 482).

d. Use the forward–backward approach to prove one of problems 5–9 in the article (page 482).

e. Use the forward–backward approach to prove one of problems 13–15 in the article (page 482).

Preparing to Discuss Observations

23. In preparation for Lesson 1.5, review your notes from your observations in high school mathematics classes. Be prepared to discuss your findings with your peers.

a. Which teacher practices supported equity? Which practices, if any, did not support an equitable environment?

b. For each observed class, find the percentage of closed and open questions. How do these percentages compare for classes at different ability levels? How do open questions relate to equity?

c. What differences, if any, did you observe in wait time for classes at different ability levels? To what extent did all the students in the class have an opportunity to think about a question before a response was given?

d. Review the questions asked by the teacher. Of these questions, what percentage required students to provide a justification of their response? How is asking for justifications related to equity?

e. Summarize any interviews you had with the teacher about his or her expectations for students in classes at different levels in terms of instruction, communication, and overall level of work.

Preparing to Discuss Listening to Students

24. In preparation for Lesson 1.5, review your notes from your interviews with students.

a. Summarize the strategies students used to solve **Colorful Shapes**. What seemed to be easy for students? What was difficult? In what ways did students pay attention to the conditions of the problem from the beginning? Add any student solutions to the **QRS Guide for Colorful Shapes** from the Unit One Appendix.

b. Summarize how students approached the proof task you selected from the set of tasks on **Listening to Students Reason about Proof** (adapted from the article by Dodge, Goto, & Mallinson, 1998). Were your students' solutions "proofs" you would be willing to accept? Add any student solutions to the **QRS Guide for Listening to Students Reason about Proof**.

1.5 Summarizing Classroom Observations and Listening to Students

At the beginning of this unit, you might have been required to observe in mathematics classrooms, focusing on the types of questions asked and the frequency with which students were expected to justify their thinking. The nature of the questions and discourse in a classroom influence expectations teachers have of students, and those expectations influence students' opportunities to engage in higher-level thinking in mathematics to the best of their ability. Taken together, all of these influence the extent to which the mathematics opportunities for students at the school level are equitable.

TEACHING AND LEARNING HIGH SCHOOL MATHEMATICS © 2010 John Wiley & Sons, Inc.

You might also have worked with one or more high school students as they reasoned through a puzzle and a nonroutine mathematics problem, recording students' strategies and any difficulties they faced. Listening to students on a regular basis as they think through mathematics problems gives teachers insights into their understanding of mathematics and can be used to design instruction more effectively.

In this lesson, you will:

- Compare the results of your observations with those of your peers.
- Share students' reasoning as they worked on tasks.
- Reflect on what your observations suggest about equity in mathematics experiences and learning opportunities for students.
- Reflect on the extent to which current high school mathematics classes encourage reasoning among students.

LAUNCH

1. Think about your own experiences in mathematics classes, either at the high school or college level.

a. To what extent did the instructor attempt to create a learning community among the individuals in the class?

b. What actions or words encouraged you to achieve at high levels?

c. What actions or words discouraged you or made you want to give up?

EXPLORE RESPONSES TO OBSERVING MATHEMATICS CLASSROOMS

2. With the members of your group, share your overall impressions about the classes you observed. What did you like? What concerned you? What surprised you? Justify your responses in each case.

3. If you had a chance to interview the teacher, share the teacher's responses about his or her expectations for students in classes at different levels in terms of instruction, communication, and overall level of work.

SHARE

Stand on the Line is an activity that may be used to address human relations issues in classrooms, one step toward building equitable learning communities. The activity was used in *Freedom Writers*, a film that showed a teacher in a school where students belonged to gangs, distrusted those not in their gang, said inflammatory things to one another, refused to work in groups, and displayed other behaviors that prevented learning from taking place. The teacher used masking tape to create a line down the center of the room and told students to "stand on the line" as she stated various life experiences they might have had:

- Stand on the line if you know someone who ruined their life by using drugs.
- Stand on the line if you know someone who was killed by gang violence.
- Stand on the line if you know someone in prison.
- Stand on the line if you know someone who has lost a job.

As students stood for each item, she asked them to think about how each event made them feel. Many students from different gangs stood on the line for each experience the teacher mentioned. The activity helped them realize they had a great deal in common, including many painful events. The teacher used this experience to break down animosities and begin to build a safe, nurturing, learning community.

TEACHING AND LEARNING HIGH SCHOOL MATHEMATICS © 2010 John Wiley & Sons, Inc.

A modified version of *Stand on the Line* is used to help you get a physical observable measure of some aspects of equity in the classrooms you observed. Identify a location in your classroom for a line. For each of the items you were to observe, stand on the line as requested.

4. Stand on the line . . . if at least 5 percent of the questions asked by the teacher you observed were open questions. Stay on the line . . . if at least 10 percent of the questions asked were open. Stay on the line . . . if at least 15 percent of the questions asked were open. Stay on the line . . . if at least 20 percent of the questions asked were open. What good open questions did you hear? How do open questions relate to equity?

5. Stand on the line . . . if teachers asked students to justify their thinking at least 5 percent of the time. Stay . . . on the line if teachers asked students to justify their thinking at least 10 percent of the time. Stay on the line . . . if teachers asked students to justify their thinking at least 15 percent of the time. Give examples of good questions that needed justification. How is asking for justification related to equity?

6. Stand on the line . . . if you observed wait times of at least 2 seconds. Stay on the line . . . if you observed wait times of at least 3 seconds. Stay on the line . . . if you observed wait times of at least 5 seconds. Stay on the line . . . if you observed wait times of 10 seconds or longer. How is wait time related to equity?

7. Revisit the "Equity Principle" from *PSSM* and the "Equity in Mathematics Education" position statement from NCTM. Discuss the extent to which you believe this principle is evident in the classes you observed. If equity does not seem evident, what suggestions would you have for school mathematics departments to achieve greater equity?

EXPLORE RESPONSES TO LISTENING TO STUDENTS

8. Share the strategies students used to solve **Colorful Shapes**. Among your classmates, compare the questions you prepared to address students' potential difficulties as they worked through the problem.

9. Compare the approaches students used to solve tasks from **Listening to Students Reason about Proof** (adapted from the article by Dodge, Goto, & Mallinson 1998). Were your students' solutions "proofs" you would be willing to accept?

10. What issues arose when trying to encourage students to explain their thinking? What surprised you about students' engagement with the problems?

11. During the interview, how useful did you find your pre-interview work of solving the problem, anticipating student approaches, and preparing teacher support?

SHARE AND SUMMARIZE

12. Share significant insights you gleaned from the interview with students.

13. Based on your observations and interviews—and those of your peers—what appear to be some of the challenges you will face as teachers who want to create an equitable learning community in which there is a high expectation for reasoning from all students?

14. How does the use of the **QRS Guide** help you with initial lesson planning?

DEVELOPING MATHEMATICAL PEDAGOGY

Preparing for Instruction

1. Consider the following problems. Determine which are open and which are closed. For problems you believe are closed, rewrite them in a way to make them more open.

TEACHING AND LEARNING HIGH SCHOOL MATHEMATICS © 2010 John Wiley & Sons, Inc.

a. Solve $4x + 3 = 2(x - 7)$. Explain your solution process.

b. Write a real-world problem for which $8x + 2y = 15$ is a solution.

c. Compare and contrast the graph of $y = x^2 + 8x + 16$ with the graph of $y = (x - 4)^2$.

d. In a triangle, two of the sides have lengths of 5 inches and 12 inches and the third side has a length of 13 inches. Show that the triangle is a right triangle.

Debating Issues

2. Suppose you made a commitment to use cooperative learning in your classroom, but some of your colleagues have concerns about its use during instruction.

   a. What are the points you would use to argue for its use in teaching and learning mathematics?

   b. What are the main concerns you expect your colleagues to raise about the use of cooperative learning?

Teaching for Equity

3. Teachers often devise strategies to ensure every student in the class has as much chance to be called on as every other student. To ensure all students are asked to respond and not just those who quickly raise their hands, some teachers put students' names on small tags and place them in a bag. Then, the teacher randomly draws from the bag when calling on students.

   a. Discuss the pros and cons of such an approach.

   b. Create at least one other strategy that can be used to ensure all students in a class are thinking about the questions asked and taking responsibility for their own learning.

   c. How can you be certain that all students can answer the questions asked without having to ask every student every question?

4. Many teachers struggle with "wait time." The silence between the time a question is asked and the time a response is permitted can seem very long, even when it is only a few seconds. Devise at least one strategy you could use to ensure you give all students time to think about a question before you permit a response to be shared.

5. How are the issues raised in problems 3 and 4—equal access to participation and sufficient wait time—related to building a classroom learning environment that promotes equity for all?

6. Following are comments that different teachers have been heard making in class:

   - "That is a novel approach. Explain to the rest of the class why that works."
   - "You should have been paying attention. I can't repeat this example 30 times."
   - "Sean just had a nice explanation for that problem. Sean, would you repeat that explanation for Rosa?"
   - "This is a really easy concept. I don't know why you're having such trouble with it."
   - "You should already know this. If you don't, you are probably going to have difficulty with this class."
   - "Take two minutes and work with the person beside you to analyze the problem. Be prepared to share your thoughts with the rest of the class when we come back together."

   a. Which comments encourage students while also building a learning community?

TEACHING AND LEARNING HIGH SCHOOL MATHEMATICS © 2010 John Wiley & Sons, Inc.

b. Which comments likely cause students to shut down and disengage from the lesson? Why?

c. What are some ways teachers can monitor their comments to ensure their communications do not shut down student communication?

d. Reflect again on your observations.

i. What comments did you hear teachers make that might discourage students in a mathematics class, particularly students who are already struggling?

ii. What comments did you hear teachers make that seemed to encourage students in a mathematics class?

7. For the high school where you observed (or where you attended as a student), consider how you might use the activity *Stand on the Line*.

a. What are some life experiences like those described in the movie you might use with that population of students?

b. What are some issues related to mathematics learning you might use to build better human relations? Examples: Stand on the line if you are sometimes nervous in math class . . . if you are afraid to say you are good at math.

8. How do the following pedagogical processes introduced in this unit relate to building classroom equity?

a. Asking questions that provide "just enough" support or guidance

b. Using cooperative learning strategies

c. Asking students to develop representations that make sense to them

d. Paying close attention to students' verbal and written responses

TEACHING AND LEARNING HIGH SCHOOL MATHEMATICS © 2010 John Wiley & Sons, Inc.

Synthesizing Unit One

Mathematics Strand: Logic and Reasoning

It is important that students work on reasoning throughout high school and that their experiences with reasoning become increasingly more complex as they progress through the mathematics curriculum. Regardless of the level of the course (honors, regular, noncollege bound), students should be expected to justify their thinking and explain their reasoning. Equity requires high expectations for all students, not just those in a school's honor track. Although the level and sophistication of reasoning may vary—depending on mathematical content and a student's initial mathematical prowess—all students can be expected to justify their thinking.

The questions teachers ask set the stage for the learning that occurs. As Goldin (1998) suggests, teachers need to be prepared to ask students questions that will help them make progress on a problem. As students work on a problem, teachers need to provide incremental guidance as needed, moving from nondirective statements to suggestions of broad problem-solving strategies to specific strategies helpful for a given problem to helping students think metacognitively about their solution. These stages of questioning provide a means to support students in complex problem solving. They also provide teachers with an important tool for lesson planning to promote equity, challenging all students to engage in mathematics at their highest level.

Throughout this unit, you have had opportunities to analyze samples of student work. In analyzing such samples, you have undoubtedly realized that teachers gain a great deal of insight into students' thinking when they look at the justifications students provide, rather than just at a final answer. As you go through this text, you are encouraged to continue to reflect on the value of such analysis and how such information can help you learn what your students know so you can modify instruction accordingly.

In addition to focusing on reasoning, high school mathematics teachers need to consider the learning environments they create. Students should be encouraged to work collaboratively with their peers and to communicate about mathematics. Working cooperatively is a lifelong skill that employers expect. The cooperative learning strategies shared in this unit provide examples of ways teachers can structure their classrooms so students have an opportunity to practice important life skills that simultaneously enhance their learning of mathematics.

It is time to reflect on your understanding of the lessons learned throughout this unit. The following items provide an overview of the unit. You should be able to respond meaningfully to each item, providing support for your answers with specifics drawn from the various lessons, discussions, readings, homework problems, observations in high school classes, or interviews with high school students.

Website Resources

- Stages of Questioning
- **QRS Guide**
- Cooperative Learning Powerpoint
- **Where Do I Stand?** survey

MATHEMATICAL LEARNINGS

1. Given a conditional, write the converse, inverse, and contrapositive. Identify which statements are equivalent in meaning. Translate *p only if q*, *p is a sufficient condition for q*, and *p is a necessary condition for q* to equivalent *if–then* statements.

TEACHING AND LEARNING HIGH SCHOOL MATHEMATICS © 2010 John Wiley & Sons, Inc.

2. Distinguish between universal and existential statements and identify the conditions under which each is true.
3. Construct Venn diagrams to illustrate a conditional, a conjunction, or a disjunction. Discuss the role that Venn diagrams play in providing a visual representation to help students make sense of these mathematical concepts.
4. Given a conjunction, disjunction, or conditional statement, identify the negation of the statement.
5. Recognize valid forms of direct and indirect argument. Given a problem requiring an argument or proof, determine which argument form is appropriate.

PEDAGOGICAL LEARNINGS

6. Identify some cooperative learning techniques that can be used in high school mathematics classrooms to create supportive learning environments. Compare and contrast the techniques outlined in this unit.
7. Given a specific problem, use the Stages of Questioning to outline possible questions that could be used with students to help them be successful with the problem—without providing more help than they need.
8. Discuss the use of the **QRS Guide** in recording problems students will solve, expected and observed student solutions, and teacher support (in light of the Stages of Questioning). Discuss the **QRS Guide** as an initial lesson planning tool. For what types of problems is this form helpful as a lesson planning tool?
9. Compare and contrast the Stages of Questioning with Pólya's problem-solving process.
10. Discuss the role communication should play in the high school mathematics classroom. How does communication contribute to the creation of a supportive learning environment?
11. Use a variety of representations to elucidate mathematical concepts or relationships.
12. In this unit, reasoning has been explored through the use of open-ended problems, puzzles, and real contexts. Discuss advantages and disadvantages of using these kinds of problems to engage students with the process of reasoning.
13. Explain how various strategies help promote classroom equity. Consider, for example, appropriate questioning, cooperative learning, having students create their own representations, attending to student thinking, and so on.
14. Revisit your responses to the survey, **Where Do I Stand?** (pages 4–5), for items 4, 11, 14, 16, 18, and 20. How have your responses changed, deepened, or grown?
15. Revisit your responses to item 24 on the **Where Do I Stand?** survey. Are there any additional challenges you expect to face as a high school mathematics teacher?

TEACHING AND LEARNING HIGH SCHOOL MATHEMATICS © 2010 John Wiley & Sons, Inc.

UNIT ONE INVESTIGATION: CARPET SQUARE MAZES

Many of the teaching devices used in this text can also be used with high school students. Open-ended investigations provide a rich problem to explore over the duration of the unit of study. This first investigation is related to the pattern searching and reasoning done in this unit.

The diagrams below represent carpet squares arranged to form 3 by 3, 4 by 4, and 5 by 5 mazes. The object of the **Carpet Square Maze** activity is to guess a predetermined route through the maze and then walk the maze perfectly. The only allowable moves are from one carpet square to another square that shares at least one vertex. Directions for moves are to the right, downward, or diagonally downward (left to right).

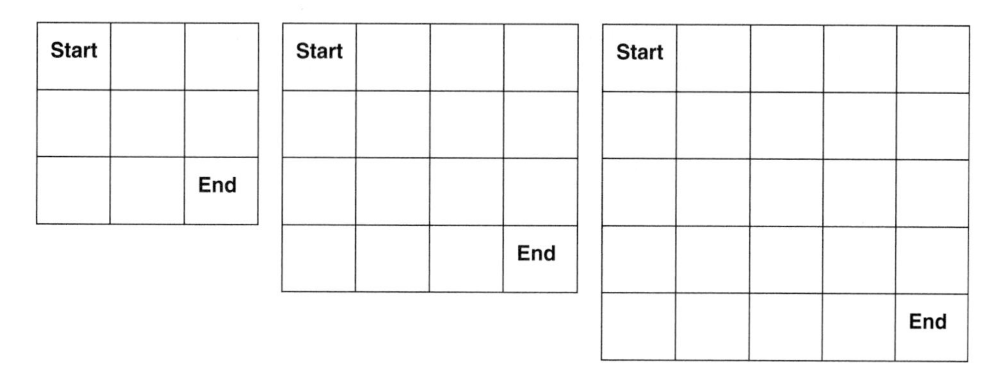

The following questions are designed to guide your investigation. Choose some that interest you and take them as far as you can. Explain your solutions in complete, coherent sentences. Include diagrams to clarify your arguments.

Website Resources

- **Carpet Square Maze** student page

1. What is the number of moves for the shortest route for each Carpet Square Maze? For a maze with dimensions n by n?
2. What is the number of moves for the longest route for each Carpet Square Maze? For a maze with dimensions n by n?
3. How many nondiagonal routes are possible in the Carpet Square Mazes with dimensions 3 by 3, 4 by 4, and 5 by 5? Find a pattern that can help you find the number of nondiagonal routes in Carpet Square Mazes of any square dimensions. Explain your pattern.
4. How many routes are possible if diagonal moves are allowed? Find a pattern that can help you find the number of routes, allowing for diagonal moves in Carpet Square Mazes of any square dimensions. Explain your pattern.
5. *Rectangular Mazes.* What if the Carpet Square Maze is not square but instead any rectangular arrangement, n by m? How do the patterns you found in problems 1–4 change?
6. *Race to the Corner.*

a. Play the following adaptation to **Carpet Square Maze**:
Player 1 places a marker in the **Start** square. Player 2 places his or her marker on a square that shares at least one vertex with player 1's last square. As with **Carpet Square Maze**, directions for moves are to the right, downward, or diagonally downward (left to right). Players alternate turns. The winner is the player whose marker lands on the **End** square.

TEACHING AND LEARNING HIGH SCHOOL MATHEMATICS © 2010 John Wiley & Sons, Inc.

b. Determine and explain strategies to ensure that you always win, whether you are player 1 or player 2.

c. Alter the playing board so that it is of any rectangular dimensions, n by m. How do your strategies for winning change when the board is not a square?

7. *Landing on All Squares.* Alter the **Carpet Square Maze** rules to allow moves to any square that shares at least one vertex in any direction. Begin at **Start** and end at **End.**

a. Is it possible to find a route through an n by n maze that lands on every square only once and uses no diagonal moves? If so, show your route and describe its generalizability to the n by n case. If not, write a convincing argument to explain why not. Be sure to consider square mazes of several dimensions.

b. Is it possible to find a route through the maze that lands on every square only once and uses at least one diagonal move? If so, show your route and describe its generalizability to the n by n case. If not, write a convincing argument to explain why not.

c. Which of the two types of routes is easier to find? Explain.

d. What changes occur if the maze is rectangular but not square? Try different dimensions.

8. For each of problems 1–7, discuss how reasoning is used to:

a. Find each solution,

b. Explain solution processes.

9. What connections do you see to other areas of mathematics?

10. While you were working through this investigation, questions likely came up for other avenues of exploration. Pose and answer some of your own questions related to mazes and maze routes.

TEACHING AND LEARNING HIGH SCHOOL MATHEMATICS © 2010 John Wiley & Sons, Inc.

UNIT two

HIGH SCHOOL STUDENTS AND HOW THEY LEARN

Mathematics Strands: Geometry and Measurement

High school students are moving from adolescence into young adulthood. They need to see that what they are learning today will be useful later in life. In *How Students Learn: Mathematics in the Classroom*, Donovan and Bransford (2005) describe four lenses that can be used to determine the effectiveness of the environments for teaching and learning:

- "The *community-centered lens* encourages a culture of questioning, respect, and risk taking." This overarching lens incorporates all the others. Developing a classroom culture in which questions are a normal occurrence and there is respect for everyone's views was the focus of Unit One of this text.
- "The *learner-centered lens* encourages attention to preconceptions, and begins instruction with what students think and know." A focus on the learner is the theme of Unit Two.
- "The *knowledge-centered lens* focuses on what is to be taught, why it is taught, and what mastery looks like." Focusing on content and why it is taught relates to the instructional decisions teachers make on a regular basis as they plan their lessons; this knowledge-centered lens focuses on curriculum and instruction. These are major goals and learning outcomes in Units Three and Four.
- "The *assessment-centered lens* emphasizes the need to provide frequent opportunities to make students' thinking and learning visible as a guide for both the teacher and the student in learning and instruction" (p. 13). A focus on assessment occurs in Unit Five.

Creating a learner-centered classroom requires that teachers attend to what students bring to the learning environment, that is, teachers need to assess what knowledge students already possess about a topic and use that knowledge as a starting point for learning. Vygotsky, a Russian psychologist, coined the term *zone of proximal development* to indicate those environments, or zones, in which learning can take place. If students are too far removed from what they already know, then they have nothing to help them build a bridge to new ideas; if they already know what they are to learn, they will disengage. Lauren Resnick (1991) has characterized the same concept via Venn diagrams similar to the following:

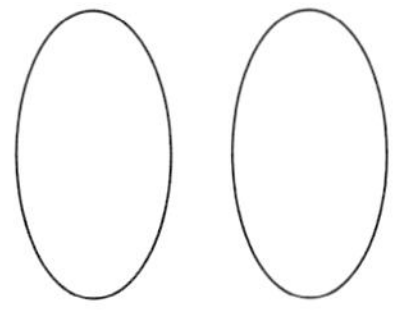

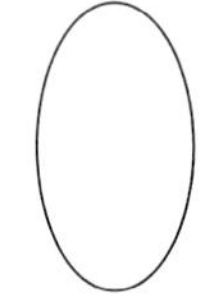

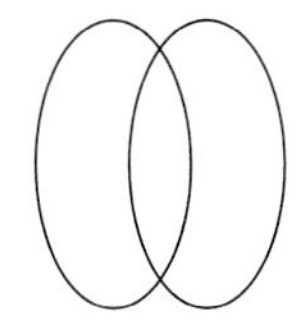

Prior knowledge and proposed new knowledge do not overlap. Learning is difficult.

Prior knowledge and proposed new knowledge are the same. There is nothing new to learn.

Prior knowledge and proposed new knowledge overlap. Learning is accessible.

Donovan and Bransford use the term "just-manageable difficulties" to describe appropriate learning environments, that is, environments "challenging enough to maintain engagement and yet not so challenging as to lead to discouragement. They [teachers] must find the strengths that will help students connect with the information being taught. Unless these connections are made explicitly, they often remain inert and so do not support subsequent learning" (p. 14). Connections to what students know can be triggered by the questions teachers ask. Teachers can use the Stages of Questioning introduced in Unit One to help students recall what they might already know about a topic and to provide just enough help to pose "just-manageable difficulties." In Lesson 2.2, you will be introduced to the cooperative learning strategy, *Roundtable*, which is especially helpful to activate students' prior knowledge.

Typical mathematics classes are composed of students of varying ability levels, all of whom are capable of achieving at high levels if provided with appropriate instruction. Using Piaget's classification, many teachers would identify students who function at the concrete operational stage, meaning they can put things in order, classify, conserve, and reverse operations. Other students in the class function at the formal operational level, indicating that they are able to manipulate symbols, isolate variables, and handle abstractions. Because most high school mathematics classes consist of students of varying abilities and prerequisite knowledge, the challenge for teachers is to create learning environments that have multiple points of entry for students. Centering instruction on tasks that are accessible to a wide range of students is a major step toward affording equity.

In Unit One, you explored the use of cooperative learning strategies to encourage communication in the classroom. These strategies take advantage of the social nature of high school students, and give teachers insight into their students' thinking processes. In this unit, you will broaden your repertoire of cooperative learning strategies as you consider additional instructional strategies useful for the mathematics classroom. Today's students have grown up in a stimulating environment; therefore, classrooms in which lecture is the only or the predominant form of instruction are not likely to stimulate many students. Although lecture is sometimes necessary, today's mathematics classrooms can be active environments in which students investigate and explore mathematics using concrete materials, technological tools, and writing to provide multiple representations of mathematical ideas and to communicate developing understandings. In this unit, you will continue to think about the questions you should ask your students and begin to incorporate those questioning techniques into aspects of planning lessons. Implemented instructional techniques are used to create an effective learning environment.

TEACHING AND LEARNING HIGH SCHOOL MATHEMATICS © 2010 John Wiley & Sons, Inc.

There is much about geometry and measurement, the mathematical foci of this unit, that high school students need to learn. The Geometry and Measurement Standards in *Principles and Standards for School Mathematics* (NCTM, 2000) include the following goals for students in instructional programs in grades 9–12:

- Analyze properties of objects in two- and three-dimensional space.
- Investigate relationships among classes of two- and three-dimensional geometric objects, including proposing and evaluating conjectures.
- Determine the validity of geometric arguments.
- Use coordinate systems (e.g., Cartesian, polar) to analyze geometric situations.
- Understand and represent transformations (e.g., translations, rotations) in the plane in various forms (e.g., coordinates, function notation, matrices).
- Determine units and scales appropriate for measurement problems.
- Understand and use measurement formulas for area and volume for two- and three-dimensional objects (pp. 308, 320).

Helping students develop proficiency with all of these topics throughout the high school years will require teachers in different grades and courses to work together to ensure that students have exposure to a coherent curriculum.

Much of the geometry that students are likely to study in high school has a long history. For instance, in ancient times, mathematicians knew how to find the areas of rectangles, triangles, and other common shapes. In addition, the Babylonians and the Chinese discovered geometric and measurement relationships for a circle, including how its area, circumference, and diameter are related. Even though these mathematicians often used 3 to approximate π, their results were close enough for practical purposes; for example, determining the volume of a dam meant they could determine how many workers were needed for its construction (Katz, 1998, pp. 20–23).

Ideas of proof in early Greek mathematics are related to attempts to solve three classic problems: squaring the circle (constructing a square with area equal to that of a circle); doubling the cube (constructing a length of a side of a cube with volume double that of the original); and trisecting an arbitrary angle (constructing an angle with the measure $\frac{1}{3}$ of any given angle)—all using only the basic construction tools of straightedge and compass. The various theorems developed by the Greeks and the body of knowledge handed down in Euclid's *Elements* provided the foundations for solutions to these problems (Katz, 1988, p. 51).

Over the centuries, mathematicians used geometry and related areas to solve problems of their day. For instance, the *Almagest*, compiled by Ptolemy in the second century, provided a compendium of knowledge about Greek astronomy with models for motions of the sun, moon, and planets. His work helped astronomers to create mathematical models to describe natural phenomena (Katz, 1988, p. 145). As travelers disseminated knowledge throughout the known world, mathematicians in many countries learned of the work of others and built on that knowledge to further their own work.

High school students should be aware of the rich history of mathematics and realize that mathematics is a living discipline with discoveries made on a regular basis. Students often think that all of what is known in geometry was handed down from the Greeks in the set of books known as Euclid's *Elements*. Euclid's geometry was based on four simple postulates and a rather complex fifth one that was shown by John Playfair to be equivalent to the following: *Given a line and a point not on the line, it is possible to draw exactly one line through the given point parallel to the line.* Over the ages, mathematicians studied Euclid's parallel postulate and tried to prove it from the other postulates. Their work led to the introduction of consistent geometric systems in which the parallel postulate is not true. That is, there are geometries in which for a given line and point there are no parallel lines (elliptical geometry) or many parallel lines (hyperbolic geometry). Carl Friedrich Gauss (1777–1855), Nicolai

TEACHING AND LEARNING HIGH SCHOOL MATHEMATICS © 2010 John Wiley & Sons, Inc.

Lobachevesky (1793–1856), and Janos Bolyai (1802–1860) used the assumption that led to hyperbolic geometry. Georg Bernhard Riemann (1826–1866) used the assumption that was the basis for elliptical geometry. By keeping their minds open but subjecting their ideas to rigorous proof, mathematicians changed mathematics in ways that could not have been envisioned by the ancients.

Information about how students learn mathematics supports teachers in focusing mathematics instruction so that all students develop the conceptual understanding and procedural fluency they need. This unit focuses on the use of different instructional approaches (e.g., using tools, different representations, or contexts) to help students work as mathematicians capable of discovering important concepts and justifying the validity of their discoveries. Becoming familiar with these approaches and how they influence students' learning is a prerequisite for teachers to be able to design appropriate instruction. During the observation connected with this unit, you will have an opportunity to see how teachers activate students' learning in an actual classroom environment. This information will likely be helpful as you continue to develop your ability to plan lesson segments and add new instructional strategies to your teaching repertoire. Planning and writing complete lesson plans will be addressed more formally in Units Three and Four.

Unit Two Team Builder: Transformed Snowflakes

In **Transformed Snowflakes**, each person is given a square piece of tracing paper on which to write their name before transforming it in various ways. Individuals are assigned new teammates and will name their group based on a transformation of the letters in their collective names as well as the symmetries of these letters.

Materials

- two pieces of patty paper (square tracing paper) per person

Website Resources

- **Transformed Snowflakes** student page

For further study of transformations, see the Unit Two Investigation at the end of the unit.

TEACHING AND LEARNING HIGH SCHOOL MATHEMATICS © 2010 John Wiley & Sons, Inc.

Transformed Snowflakes

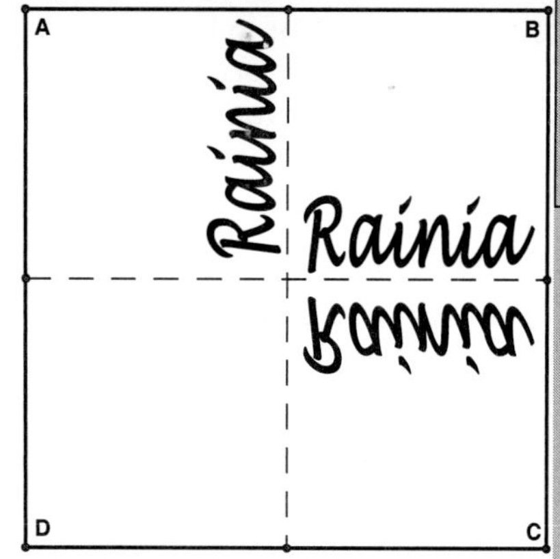

Materials

- two pieces of patty paper (square tracing paper) per person

1. Fold a square piece of tracing paper in half and then in half again across the fold lines shown so that vertices A, B, C, and D coincide.
2. Unfold the paper and impose coordinate axes with the origin at the intersection of the folds.
3. Write your name in large script along the positive *x*-axis in Quadrant I, using most of the space between the origin and the edge of the paper.
4. Figure out a way to fold the paper so that you can trace your name along the positive *x*-axis in Quadrant IV, along both sides of the negative *x*-axis (Quadrants II and III) and along both sides of the positive and negative *y*-axis. See the sample to help you get started.
5. Place the "snowflake" on colored paper and hang it or post it on a classroom bulletin board.
6. The alphabet and first ten digits are typed in Eurostile 18-point font below.
   a. List the letters and numbers with horizontal reflective symmetry.
   b. List the letters and numbers with vertical reflective symmetry.
   c. List the letters and numbers with rotational symmetry.

A B C D E F G H I J K L M N O P

Q R S T U V W X Y Z

0 1 2 3 4 5 6 7 8 9

7. Write a word that has
   a. Horizontal reflective symmetry.
   b. Vertical reflective symmetry.
   c. Rotational symmetry.
8. Create a team name or logo based on the letters in your collective first or last names. The end result must have at least one of the symmetries listed in items 7a–c.

Preparing to Observe Mathematics Classrooms: Focus on Learning

The Learning Principle from the *Principles and Standards for School Mathematics* is the focus for the observations in this unit. The Learning Principle states that students need to build understanding of concepts through experience and by connecting new knowledge to previous knowledge. Many educators would argue that conceptual understanding is at least as important in learning mathematics as procedural fluency and factual knowledge. That is, *how* students learn mathematics is at least as important as *what* mathematics students learn.

In *How Students Learn: Mathematics in the Classroom*, Donovan and Bransford (2005) identify three fundamental principles of learning that teachers of all subjects

TEACHING AND LEARNING HIGH SCHOOL MATHEMATICS © 2010 John Wiley & Sons, Inc.

should use: (1) Students must be engaged for learning to occur; (2) students must know facts and how they connect to concepts; and (3) students must learn to monitor their own progress in learning.

How do these principles influence the instruction and learning that occur in the mathematics classroom? How is students' prior knowledge used to introduce concepts? These questions are the focus of the observations for this unit.

1. *Prepare for the Observation.* In preparation for the observations, read the following selections:
   - "The Learning Principle" from *Principles and Standards for School Mathematics* (pp. 20–21).
   - Pages 1–21 from "Introduction" by M. Suzanne Donovan and John D. Bransford in M. Suzanne Donovan and John D. Bransford (Eds.), *How Students Learn: Mathematics in the Classroom*. Washington, DC: National Academies Press, 2005 (available online at http://books.nap.edu/openbook.php?record_id=11101&page=1)

   a. How does the reading from *How Students Learn* relate to the Learning Principle from *Principles and Standards*?

   b. Think about your own learning in mathematics. In what ways were you encouraged to connect factual knowledge or procedures to the underlying concepts? In what ways were you guided to think about your own learning?

   c. Based on your reading, what is a concept? What is a procedure?
2. *Conduct the Observations.* It is likely that you will want to observe at the same high school that you visited for the observations in Unit One; perhaps you have become acquainted with some of the teachers and students and are familiar with the school and classroom environments. If you do change classrooms, try to observe a class studying geometry. Observe at least three full class periods, preferably in the same class. If possible, begin your observations at the introduction of a new concept. If possible, try to observe for at least two consecutive days to investigate how lessons build on each other. Complete Observations 2.1 and 2.2.
3. *Observation 2.1: Learning Related to Students' Prior Knowledge or Experiences.* Base your responses on observations completed during two class periods. Make note of the mathematical intent of the lessons. During the observations, write down any examples used to introduce students to the concept studied, whether used by the teacher or referenced in the textbook. After the observations, consider the following questions:

   a. In what ways did the examples refer to students' prior experiences?

   b. How did the examples motivate students to engage in mathematical study?

   c. In what ways did a task provide accessibility to a wide range of students?

   d. How did the examples bridge everyday experiences with abstract mathematical concepts?

   e. In what ways did the examples and tasks focus on conceptual understanding, procedural fluency, or connections between the two?

   f. If no examples referred to students' prior experiences, choose one concept that arose in one of the classes. Determine and describe a real-life or fantasy context that could have been used to help students connect the concept to their prior experiences.
4. *Observation 2.2: Monitoring Students' Progress.* Begin the observation by noting the mathematical intent of the lesson.

   a. In what ways did the teacher make the students aware of the goals and/or learning outcomes of the lesson?

   b. What approaches did the teacher use to monitor students' understanding during the lesson?

TEACHING AND LEARNING HIGH SCHOOL MATHEMATICS © 2010 John Wiley & Sons, Inc.

c. What opportunities did students have to monitor their own understanding of the mathematical goals and learning outcomes of the lesson?

d. In what ways were students encouraged to think about their thinking, to look back to what they were learning? Give specific examples of such opportunities. If no such opportunities existed, identify one time in class when the teacher could have encouraged students to monitor their own understanding and thinking about the lesson.

5. *Summarize the Observations.* In summarizing your observations, start by providing a brief description of the level of the classes you observed. Also, be sure to indicate the mathematical intent of each of the lessons you observed.

   a. Summarize your comments about the readings and related questions from item 1.

   b. For each lesson, identify the mathematical goals and the tasks, questions, or activities that were used.

   c. Refer to the questions associated with each of Observations 2.1 and 2.2. From what you learned during the observations, summarize your responses to the questions. Justify any conclusions with specific instances from the readings or the classroom observations.

   d. Tell how the lessons you observed embodied the Learning Principle from *Principles and Standards* and the three principles of learning from *How Students Learn*.

Be prepared to discuss your observations and reflections during the last lesson of this unit.

Listening to Students Reason about Geometry

High school students should have had numerous opportunities to study topics in geometry, both in elementary school and in middle school. How robust is their understanding? As students plan to take a high school geometry course, it is helpful for their teachers to learn the depth of their geometry knowledge. The tasks for this interview provide an opportunity for you to engage students in thinking and reasoning about geometric figures.

Materials

- patty paper (optional)
- straightedge (optional)
- compass (optional)
- MIRA, a reflection tool (optional)

Website Resources

- **Connecting Midpoints of Sides of Polygons** student page
- **QRS Guide**
- Stages of Questioning

1. Refer to the **Connecting Midpoints of Sides of Polygons** student page.

   a. Work through problem 1, recording the problems, your solutions, and your conjectures on a **QRS Guide.** (A blank **QRS Guide** can be found in the Unit One Appendix.)

   b. How does the task support equity? What is it about the task that allows students at different levels to engage in it?

TEACHING AND LEARNING HIGH SCHOOL MATHEMATICS © 2010 John Wiley & Sons, Inc.

c. In what ways do you anticipate students will solve the problem? Add these expected responses to your **QRS Guide**. Anticipate student responses for each of the three levels: misconceived, partial, and satisfactory.

d. For each recorded solution, use the Stages of Questioning (see Lesson 1.1 and the Unit One Appendix) to provide appropriate guidance to students, without giving too much help. Record your questions in the Teacher Support column of your **QRS Guide**, immediately to the right of the expected response to which it pertains.

2. a. Repeat items 1a, 1c, and 1d for the second problem on the **Connecting Midpoints of Sides of Polygons** student page.

   b. How do you expect students' work on the triangle problem to influence their work on the quadrilateral problem?

3. Identify a pair of high school students who have not yet taken a high school course in geometry or are in the first few weeks of a geometry course.

   a. Provide each student a blank copy of the **Connecting Midpoints of Sides of Polygons** student page. Ask them to work through all of the problems, showing their work on the student page. Ask questions only as needed, using the Stages of Questioning and your anticipated teacher support from your **QRS Guide**.

   b. What, if anything, surprised you by their responses? What difficulties, if any, did they have that you had not anticipated?

   c. Record any new responses on the appropriate **QRS Guide**. Also include teacher support questions for any new responses.

4. Read "Linking Theory and Practice in Teaching Geometry" by Randall E. Groth (*Mathematics Teacher*, 99 (August 2005): 27–30).

   a. Suppose the students you interviewed entered your high school geometry course at the beginning of a school year. Based on their responses and your reading of the various van Hiele Levels of Geometric Thought (see Lesson 2.1), what issues or challenges would you expect to face in the classroom?

   b. To what extent did the coordinate grid for the first quadrilateral make it easier or harder for students to justify their conjectures?

Be prepared to share your insights into how students reasoned about these geometric tasks during the last lesson of this unit.

TEACHING AND LEARNING HIGH SCHOOL MATHEMATICS © 2010 John Wiley & Sons, Inc.

Connecting Midpoints of Sides of Polygons

Sometimes interesting relationships arise when you connect the midpoints of sides of different polygons.

1. Consider triangle PIE.

a. Find the midpoint of $\overline{PI}$ and label it T. Find the midpoint of $\overline{IE}$ and label it S. Draw $\overline{TS}$.

b. What conjectures can you make about $\overline{TS}$ or about triangle *TIS* in relation to triangle *PIE*?

2. For each quadrilateral in I–V,

a. Find the midpoint of each side. Label the midpoints consecutively as *W*, *X*, *Y*, and *Z*. Then connect the consecutive midpoints forming quadrilateral *WXYZ*.

b. Make a conjecture about the shape of *WXYZ*.

c. How can you justify your conjecture to someone who is not convinced about your conclusion?

i. *HELP* is a general quadrilateral.

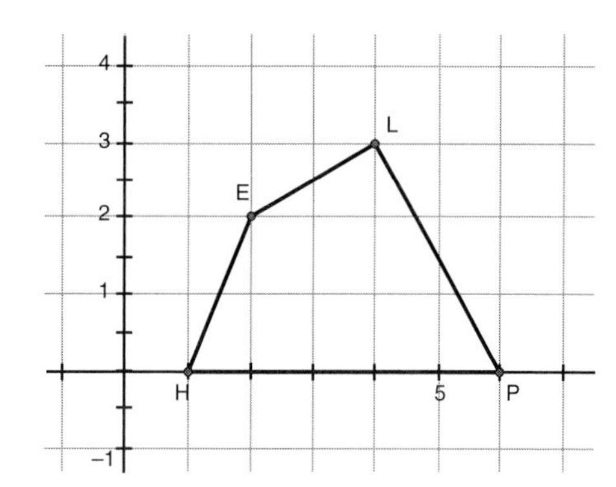

ii. Quadrilateral *BOAT* is a trapezoid with $\overline{OA} \parallel \overline{BT}$.

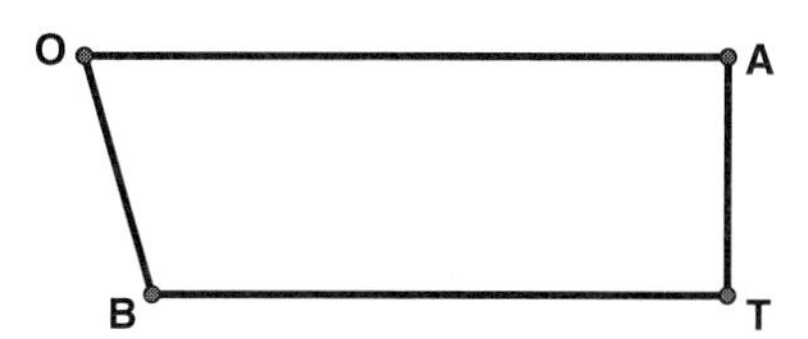

iii. Quadrilateral *FRED* is an isosceles trapezoid with $\overline{RE} \parallel \overline{FD}$.

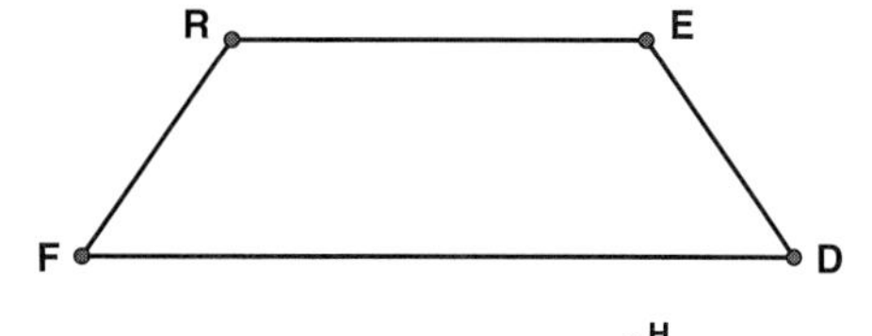

iv. Quadrilateral SHIP is a kite with $\overline{HS} \cong \overline{HI}$ and $\overline{IP} \cong \overline{SP}$.

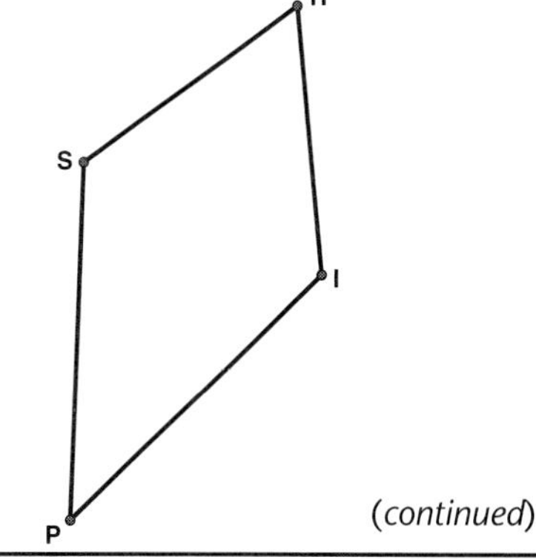

(*continued*)

TEACHING AND LEARNING HIGH SCHOOL MATHEMATICS © 2010 John Wiley & Sons, Inc.

v. Quadrilateral *BAKE* is rectangle.

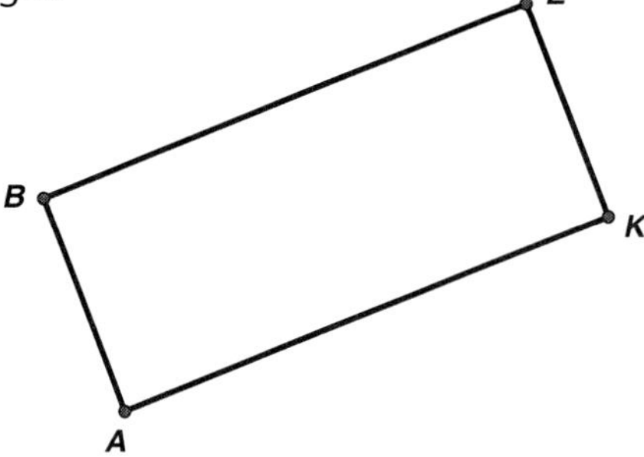

3. In what way does the shape of the original quadrilateral affect the quadrilateral formed by connecting the midpoints of each side? Explain.

2.1 Understanding Geometry Learning: Coordinate Geometry

Ask an adult about their experiences in learning geometry in school and you might hear, "I really liked geometry" or "I never understood geometry, but I was good at algebra." Geometry learning has a visual side that some students understand easily and others struggle to "see." Much is currently known about students' levels of geometric thinking. In the 1950s, two Dutch educators, Pierre van Hiele and Dina van Hiele-Geldof, introduced a theory about students' levels of geometric thinking based on experiences in Dutch secondary schools. Their work began to gain prominence in the United States in the 1970s. Additional research has validated their work in identifying levels of geometric understanding and shown that students' van Hiele levels are related to their success in high school geometry, especially with proof.

Throughout this unit, you will be exploring how students learn, particularly how they learn geometry. Many high school students need opportunities to explore topics concretely; they need experiences that support their learning and guide them from basic everyday experiences to more abstract concepts.

In this lesson, you will:

- Explore the van Hiele levels of geometric thought.
- Analyze geometric tasks with coordinates in relation to the van Hiele levels.
- Investigate mathematics concepts related to coordinate geometry on the plane, such as parallel lines, perpendicular lines, slope, midpoints, and distance.
- Construct and evaluate proofs about geometric figures using coordinates.

Website Resources

- van Hiele Levels of Geometric Thought
- **QRS Guide** (used twice)

LAUNCH

In order for learning to occur, experiences must be within each student's zone of proximal development as described by Vygotsky. The same is true for learning in geometry. In addition, the van Hieles' research indicates that the instructional level must be within a student's van Hiele level in order to be effective. When a student's van Hiele level is lower than the instructional level, understanding is unlikely to occur.

TEACHING AND LEARNING HIGH SCHOOL MATHEMATICS © 2010 John Wiley & Sons, Inc.

1. Consider the descriptions of the van Hiele model in the table *van Hiele Levels of Geometric Thought.* Based on your own experiences, which level best describes the instructional level of most high school geometry classes? Give your rationale.

van Hiele Levels of Geometric Thought

Level 1: Visualization

At this level, students view geometric objects in their entirety, rather than considering attributes. For instance, a figure has a given shape based on its physical appearance. Students are able to reproduce figures and learn vocabulary.

Level 2: Analysis

Students begin to consider the attributes and characteristics of figures. Through examples, students are able to make generalizations for classes of figures. However, students have difficulty with interrelationships between figures, do not understand definitions, and do not appreciate proof.

Level 3: Informal Deduction

Students are able to deduce properties of various figures, deal with issues related to class inclusion, and understand definitions. Although students can follow formal proofs, they have difficulty constructing proofs on their own.

Level 4: Deduction

Students understand geometric structure, recognize the importance of proof, and understand the roles of definitions, theorems, and axioms. Students are able to create their own proofs and recognize that proofs may be completed in more than one way. Students rely on models to make sense of geometric concepts and proofs.

Level 5: Rigor

Students are able to understand axiomatic systems and study non-Euclidean geometries. They no longer need to rely on a particular model to understand and create proofs.

Adapted from Crowley, Mary L. "The van Hiele Model of the Development of Geometric Thought." In Mary Montgomery Lindquist and Albert P. Shulte (Eds.), *Learning and Teaching Geometry, K–12* (pp. 1–16). Reston, VA: National Council of Teachers of Mathematics, 1987 and Groth, Randall E. "Linking Theory and Practice in Teaching Geometry." *Mathematics Teacher*, 99 (August 2005): 27–30.

EXPLORE

Early research on geometry (e.g., Senk, 1985; Usiskin, 1982) found that most high school students were at one of the first two van Hiele levels, even at the end of high school. But, most high school geometry courses were taught at the fourth level. This disconnect may explain why students find high school geometry difficult to understand. The goal for teachers is to provide students with experiences to help them progress to at least the level of deduction. The van Hieles focused their attention on the first four levels because these levels relate to school instruction.

2. Consider the following problem related to geometric concepts with coordinates. Plot quadrilateral *TOPS* with coordinates $T = (0, 3)$, $O = (-3, -1)$, $P = (1, -4)$, and $S = (4, 0)$. Then identify the quadrilateral with as specific a name as possible. Justify your classification.
   a. Individually, solve the task. Think about possible misconceived or partial responses students might make. Record these in a **QRS Guide** (see Unit One Appendix).

TEACHING AND LEARNING HIGH SCHOOL MATHEMATICS © 2010 John Wiley & Sons, Inc.

b. Consider the following student responses to the task. Work with your group members to determine at which of the four van Hiele levels you would classify each response. Provide a rationale for your classification.

i. *I plotted the four points. Then I found the slopes of all four sides:* $m_{\overline{TO}} = \frac{4}{3}$, $m_{\overline{OP}} = -\frac{3}{4}$, $m_{\overline{PS}} = \frac{4}{3}$, $m_{\overline{ST}} = -\frac{3}{4}$. *Opposite sides have the same slope, so opposite sides are parallel. This means the quadrilateral is a parallelogram.*

ii. *I plotted the four points. The quadrilateral looks like a square.*

iii. *I plotted the four points. Then I found the slopes of opposite sides:* $m_{\overline{TO}} = \frac{4}{3}$ *and* $m_{\overline{SP}} = \frac{4}{3}$. *Because these sides have the same slope, they are parallel. So TOPS is a parallelogram.*

iv. *I plotted the four points. I found the slopes of all four sides:* $m_{\overline{TO}} = \frac{4}{3}$, $m_{\overline{OP}} = -\frac{3}{4}$, $m_{\overline{PS}} = \frac{4}{3}$, $m_{\overline{ST}} = -\frac{3}{4}$. *So the quadrilateral is a parallelogram. I also noticed that the slopes of sides next to each other multiply to* -1, *so these sides are perpendicular. So the figure is a rectangle, which is a special kind of parallelogram.*

v. *I plotted the four points. Then I found that each side has a length of 5 units. So the figure is a rhombus.*

vi. *I plotted the four points. I found the slopes of all four sides:* $m_{\overline{TO}} = \frac{4}{3}$, $m_{\overline{OP}} = -\frac{3}{4}$, $m_{\overline{PS}} = \frac{4}{3}$, $m_{\overline{ST}} = -\frac{3}{4}$. *Opposite sides have the same slope so opposite sides are parallel. I also noticed that the slopes of adjacent sides multiply to* -1, *so these sides are perpendicular and make* 90° *angles. When I found the lengths of the sides, they all equal 5 units. So the figure is a parallelogram because opposite sides are parallel. The figure is a rectangle because the angles are all right angles. The figure is also a rhombus because all four sides have the same length. Because the quadrilateral is a rectangle and a rhombus, its most specific name is a square.*

vii. *I plotted the four points. Then I found the slopes of opposite sides:* $m_{\overline{TO}} = \frac{4}{3}$ *and* $m_{\overline{SP}} = \frac{4}{3}$. *Because these sides have the same slope, they are parallel. So TOPS is a trapezoid.*

viii. *I plotted the four points. The figure looks like a square. I found the lengths of all four sides to be 5 units. So my initial guess was correct and the quadrilateral is a square.*

SHARE

3. a. Compare your group's classification of the student responses in item 2 to those of other groups.

   b. What difficulties did you encounter in classifying these responses?

   c. As needed, include student approaches (misconceived, partial, or satisfactory) from item 2 in your **QRS Guide**.

4. It is likely that concepts related to slopes, parallel lines, perpendicular lines, and finding lengths would have been discussed over several lessons, perhaps separated over time. So, students may need an opportunity to review previously learned concepts in preparation for new learning. What are some strategies teachers might use to review the prerequisite concepts prior to needing them for the problem in item 2?

5. For one of the less successful responses in item 2, use the Stages of Questioning (see Unit One Appendix) to generate a series of questions you could use to help students move forward. Record these questions as Teacher Support in your **QRS Guide**.

TEACHING AND LEARNING HIGH SCHOOL MATHEMATICS © 2010 John Wiley & Sons, Inc.

EXPLORE MORE

Consider the following problems in a high school geometry unit on proofs with coordinates.

I. Triangle *TAN* has coordinates $T = (0, 8)$, $A = (10, 0)$, and $N = (-12, 0)$. Determine the coordinates of the midpoints of $\overline{TA}$ and $\overline{TN}$. What relationships exist between the segment formed by these midpoints and $\overline{AN}$? Prove your conjectures.

II. Quadrilateral *FROG* has coordinates $F = (1, 2)$, $R = (6, 4)$, $O = (3, 10)$, and $G = (-1, 6)$. Connect the midpoints of each segment in order, forming quadrilateral *LEAP*. What types of quadrilaterals are *FROG* and *LEAP*? Prove your conjectures.

III. Quadrilateral *MARE* has coordinates $M = (-3, 5)$, $A = (3, 5)$, $R = (7, 0)$, and $E = (-7, 0)$. What type of quadrilateral is *MARE*? What relationships exist between the diagonals? Prove your conjectures.

IV. Quadrilateral *BECK* has coordinates $B = (0, 0)$, $E = (3, 4)$, $C = (8, 4)$, and $K = (5, 0)$. What type of quadrilateral is *BECK*? What relationships exist between the diagonals? Prove your conjectures.

V. Quadrilateral *ARTS* has coordinates $A = (0, 0)$, $R = (5, 7)$, $T = (16, 9)$, and $S = (11, 2)$. What type of quadrilateral is *ARTS*? What relationships exist between the diagonals? Prove your conjectures.

VI. Quadrilateral *BEAR* has coordinates $B = (4, 1)$, $E = (8, 7)$, $A = (5, 9)$, and $R = (1, 3)$. What type of quadrilateral is *BEAR*? What relationships exist between the diagonals? Prove your conjectures.

As a class, select one of the problems I–VI so that each problem is selected by at least one group. Answer questions 6, 7, and 8 for your group's selected problem.

6. To be successful, which prerequisite skills or knowledge does a student need?

7. Students' prerequisite knowledge might be incomplete or rusty. Complete the following to determine which questions you might ask to help them "discover" a proof. Record your responses for items 7a and 7b in a **QRS Guide**.

- **a.** For each of the first three van Hiele levels of geometric thought, write a possible student response. Record these as expected student responses, and determine whether these should be considered as misconceived or partial. Justify your choices.
- **b.** For each of the possible student responses you wrote for item 7a, use the Stages of Questioning from Lesson 1.1 to determine appropriate questions to help students move beyond their current level of geometric thinking. Record these in the Teacher Support column of the **QRS Guide**.

8. The six problems that began this **Explore More** are all specific examples of quadrilaterals or triangles; that is, each problem is addressed with a specific polygon with numerical coordinates.

- **a.** What might be some advantages of having students work with a specific example before completing a proof for a generic case using variables?
- **b.** What might be some disadvantages of having students first complete a proof with a specific instance?
- **c.** Rewrite your selected problem for a generic case using variables. Which general coordinates are most appropriate to use in your problem? (Consider placing one vertex at the origin so one side of the quadrilateral coincides with the positive x-axis.)

SHARE AND SUMMARIZE

9. Share expected student responses and possible Teacher Support questions to address those responses for problems I–VI. What similarities and differences

TEACHING AND LEARNING HIGH SCHOOL MATHEMATICS © 2010 John Wiley & Sons, Inc.

exist in the types of questions asked? Add possible student responses and Teacher Support to the **QRS Guide** as needed.

10. Share your group's responses to item 8c with other groups.
11. What concerns do you have about helping your students learn to become successful with writing proofs?

DEEPENING MATHEMATICAL UNDERSTANDING

1. *Midpoint and Distance Formulas*. Several of problems I–VI in the **Explore More** required the use of either the midpoint formula or the distance formula. Rather than simply memorizing formulas, high school students should know how these formulas arise so they can reconstruct them if necessary.
   a. Given the ordered pairs (x_1, y_1) and (x_2, y_2). Provide a justification appropriate for use with high school students of the coordinates of the midpoint:

$$midpoint = \left(\frac{x_1 + x_2}{2}, \frac{y_1 + y_2}{2}\right).$$

   b. Given the ordered pairs (x_1, y_1) and (x_2, y_2). Provide a justification appropriate for use with high school students of the distance formula:

$$distance = \sqrt{(x_2 - x_1)^2 + (y_2 - y_1)^2}.$$

   c. Extend these formulas to three-dimensional space. That is, find the midpoint and the length of a segment joining (x_1, y_1, z_1) and (x_2, y_2, z_2).
2. *Perpendicular Segments and the Pythagorean Theorem*. Consider triangle *SET* with coordinates $S = (0, 0)$, $E = (4, 2)$, and $T = (-1, 2)$.
   a. Determine the slopes of each side of the triangle.
   b. Use the slopes to determine whether or not triangle *SET* is a right triangle.
   c. Find the lengths of each side of triangle *SET*. Do these side lengths satisfy the Pythagorean Theorem? Why or why not?
3. *Kites and Coordinates*. Given quadrilateral *BOAT* with coordinates $B = (1, 12)$, $O = (-2, -5)$, $A = (15, -8)$, and $T = (8, 2)$.
   a. What type of quadrilateral is *BOAT*?
   b. What relationships exist between the diagonals?
   c. Find the coordinates of the point at which the diagonals intersect.
   d. Prove your conjectures in problems 3a and 3b.
   e. For the type of quadrilateral you determined in problem 3a, determine general coordinates so the quadrilateral has all of the same properties as *BOAT*.
4. *Circles and Coordinates*. Consider the circle with center (4, 7) and radius 5.
   a. Show that the segment from the center to the midpoint of the chord with endpoints $A = (4, 12)$ and $B = (8, 10)$ is perpendicular to the chord.
   b. Consider another chord with endpoints $D = (-1, 7)$ and $E = (7, 11)$. Find the coordinates of the point where chords $\overline{AB}$ and $\overline{DE}$ intersect. Label this point *H*.
   c. Prove that for this circle, $AH \cdot HB = DH \cdot HE$.

TEACHING AND LEARNING HIGH SCHOOL MATHEMATICS © 2010 John Wiley & Sons, Inc.

d. Use *The Geometer's Sketchpad*™ or some other interactive geometry software to investigate in general the relationship in problem 4c between segments of intersecting chords.

e. Is the assertion in problem 4c true for all circles? If so, prove it.

5. *Coordinates and Quilts.* Quilt squares provide an opportunity to blend geometry and algebra. Complete the **Coordinatizing a Star Quilt Pattern** student page.

Coordinatizing a Star Quilt Pattern

Stars appear in many quilts, often symbolizing a guiding light. Use the quilt block shown to answer the following questions. The obtuse angle in quadrilateral *QRST* measures 113°. The point, *C*, at the center of the quilt block has coordinates (7, 7). The quilt pieces that comprise the star are congruent to triangle *CQR*, the four corner pieces are congruent to quadrilateral *QRST*, and the four remaining triangles near the border of the quilt block are congruent to triangle *NWQ*. In the children's book, *The Quilt-Block History of Pioneer Days* by Mary Cobb (1995), this quilt block is called the Lemoyne Star. The design is found in other sources, sometimes with other names.

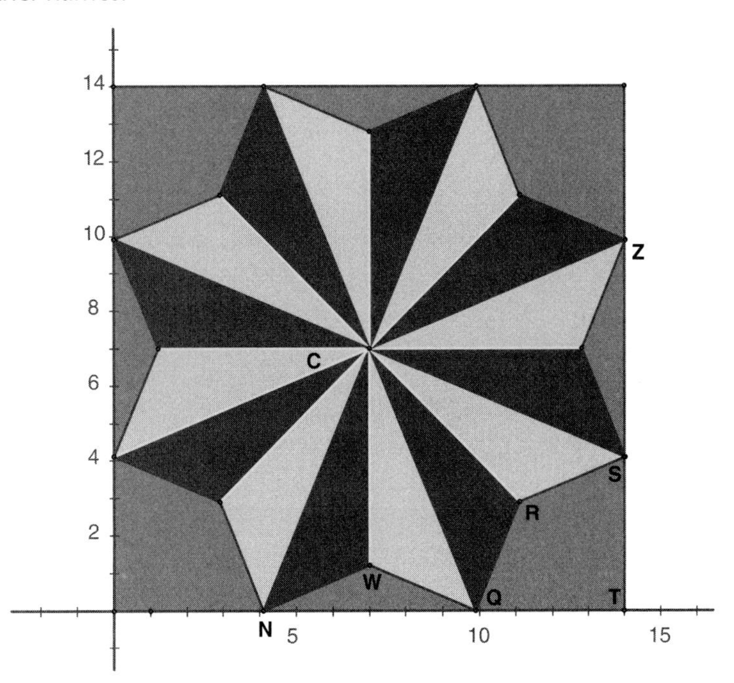

Use the given relationships to answer the following. Do not just attempt to read coordinates off the grid.

1. Find as many different triangles and quadrilaterals as possible; no two should be congruent to another. You may combine adjacent pieces in the quilt block (those that share edges). However, do not add line segments to the quilt design to create new figures. Trace each triangle and quadrilateral onto patty paper, showing the component quilt pieces that make up each. Categorize each triangle and quadrilateral as specifically as possible. Briefly explain how you know the shape belongs to the category you say it does.
   a. What types of triangles did you find? Label each triangle by its type.
   b. What types of quadrilaterals did you find? Label each quadrilateral by its type.

(*continued*)

TEACHING AND LEARNING HIGH SCHOOL MATHEMATICS © 2010 John Wiley & Sons, Inc.

2. Specifically describe at least four different symmetries in the quilt block.
3. The points (2, 5) and (5, 2) are on the edges of triangles forming the star. Identify these points.
   a. Use the fact that (5, 2) is on the edge of one of the triangles to find the slope of $\overline{CN}$.
   b. Use the result in item 3a to find an equation for $\overline{CN}$ and to find the coordinates of N.
   c. Use symmetry and your results from item 3b to find the coordinates of Q.
   d. Use symmetry to find the coordinates for S.
   e. Points N, W, R, and S are collinear (on the same line). Find an equation of line NS; then find the coordinates for W.
   f. Points C, R, and T are collinear. Find the coordinates for R.
4. Find the lengths of each side of triangle CQR.
5. Ribbon is to be used to outline the star on its outermost edge. Disregarding seam allowances and extra ribbon needed to go around corners, how much ribbon is needed to outline the star?
6. What is the measure of angle QCR? How do you know?
7. Shown below are three of the pieces from the quilt. Find the angle measures in each piece and label them in each figure. Explain how you found the measures of each angle.

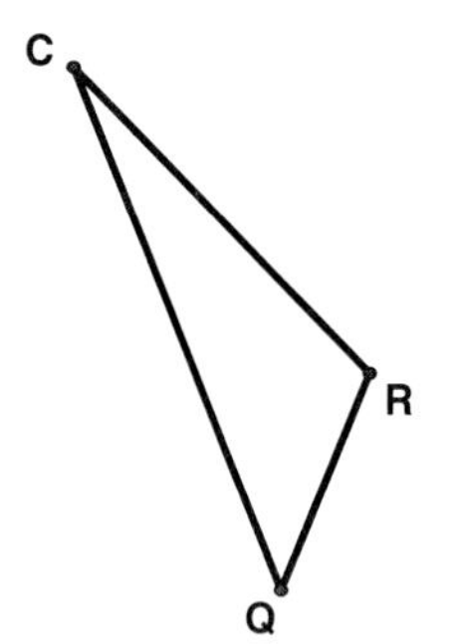

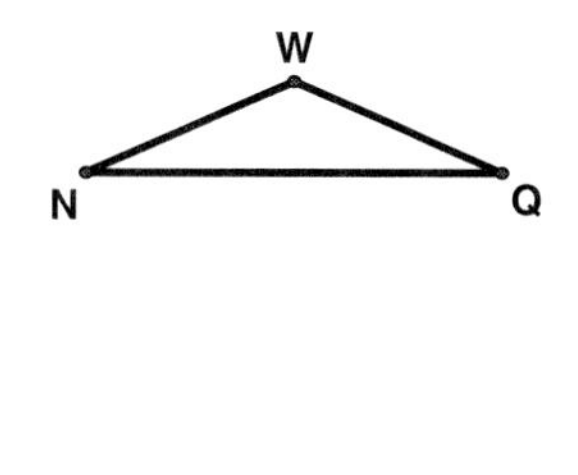

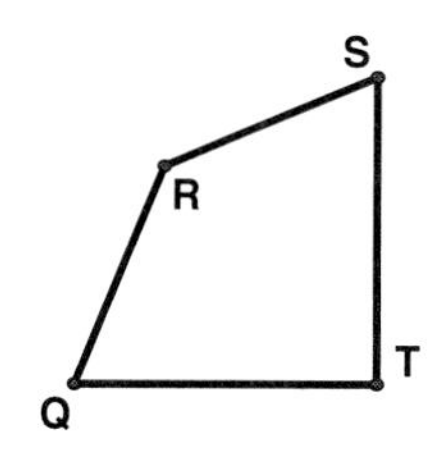

8. If the pieces congruent to quadrilateral $QRST$ and triangle NWQ are removed, can the star pattern be continued using only pieces congruent to triangle CQR? If so, show how you would continue the pattern using patty paper. If not, explain.

6. *Generalized Quadrilaterals.* Many of the problems in this lesson involved quadrilaterals with specific coordinates. So, properties you have proved are valid for only the quadrilateral whose coordinates were given. To prove a property in general, you need to use a generic quadrilateral; that is, you need to use variables for coordinates so that the quadrilateral is representative of all quadrilaterals in that class.
   a. For simplicity, generalized quadrilaterals often are placed with one side or one diagonal along either the x- or y-axis and one vertex or intersection of diagonals at the origin. Provide a rationale for why such placement can be made without loss of generality.
   b. Place each of the following quadrilaterals on a coordinate grid with generalized coordinates. Identify the coordinates you would use in each case.

TEACHING AND LEARNING HIGH SCHOOL MATHEMATICS © 2010 John Wiley & Sons, Inc.

i. trapezoid | ii. parallelogram | iii. rectangle
iv. rhombus | v. square | vi. kite
vii. isosceles trapezoid

c. Determine the smallest number of variables needed to coordinatize each of the quadrilaterals. Choose your variables so that each quadrilateral is as general as possible. For instance, the rhombus should not also be a square.

d. For each of the quadrilaterals from problem 6c, count the number of variables used. Arrange the quadrilaterals in a hierarchy with the most general quadrilateral at the top and more specific quadrilaterals below. Quadrilaterals that require the same number of variables should be shown on the same level of the hierarchy.

e. What relationships does the hierarchy help you see?

DEVELOPING MATHEMATICAL PEDAGOGY

Preparing for Instruction

7. When students are first learning to write proofs in a new content area, teachers often use several strategies to guide them.

   a. One strategy is to have students work on proofs in pairs or small groups. Comment on the benefits of cooperative work in writing proofs. What are some possible concerns?

   b. A second strategy is to give partially completed proofs. That is, a proof might be written, perhaps in two-column format, with missing statements, missing reasons, or steps with both a missing statement and a missing reason. Generate a partially completed proof for one of the problems I–VI from the **Explore More**. What are some considerations about which statements or reasons to omit?

   c. A third strategy is to write statements and reasons on index cards, with separate cards used for each statement and each reason. The cards are a "puzzle" that students must arrange in order to complete the proof. Generate a set of cards for the same problem you used in problem 7b.

   d. A fourth strategy is to write a completed proof. Sometimes the proof is valid and sometimes the proof is not valid or has incorrect statements or reasons. Students are expected to read the proof and critique it. Generate a valid proof and an invalid "proof" for your selected problem from 7b.

   e. Compare and contrast the four strategies (outlined in problems 7a–d) for helping students with proof writing. Which strategies would you consider using in your own classroom? How do your choices make proof accessible to all students? Explain your decision.

8. The figures in Problems I, III, and IV are oriented with a base along the x-axis. Comment on using figures with such an orientation. How does this orientation simplify a proof or make it more difficult for students?

9. Revisit the **Coordinatizing a Star Quilt Pattern** student page. Questions 3b, 3e, and 3f require multiple steps to complete. Choose one of these questions. Anticipate student responses at three levels: misconceived, partial, and satisfactory. For each expected response, determine which additional questions (re: Stages of Questioning) you would use to provide incremental help to guide struggling students toward a solution. Record this work in a **QRS Guide.**

10. When students use interactive geometry software to explore figures, they often make conjectures about figures and classes of figures. That is, they engage in **inductive reasoning,** the production of tentative generalizations. In contrast,

TEACHING AND LEARNING HIGH SCHOOL MATHEMATICS © 2010 John Wiley & Sons, Inc.

when given a particular statement to prove, they engage in **deductive reasoning**, the process of drawing logical conclusions and proving.

a. Identify examples of each type of reasoning in this lesson's activities.

b. What are some reasons for engaging students in both types of reasoning in the mathematics classroom?

c. What instructional issues arise with inductive reasoning?

11. Investigate your district and state mathematics standards.

a. Describe the mathematics requirements students must meet to graduate.

b. What expectations exist for students relative to geometry?

Teaching for Equity

12. Revisit Problems I–VI in the **Explore More**. The problems were open in that students were able to make and prove their own conjectures. A more closed version of Problem III is below:

Quadrilateral $MARE$ has coordinates $M = (-3, 5)$, $A = (3, 5)$, $R = (7, 0)$, and $E = (-7, 0)$. Prove that quadrilateral $MARE$ is an isosceles trapezoid and that its diagonals are congruent.

a. Comment on the pros and cons of having students work on an open problem versus a closed problem.

b. How might a lesson that begins with an open problem facilitate students' work with coordinate proofs in more specific situations?

c. A problem stated in an open fashion allows students of varying abilities to enter the problem at different levels and is a potentially important component of equity. Some students may explore only the most basic characteristics of the quadrilateral. Other students may work to find less obvious relationships. What are some strategies you can use to ensure the class benefits and learns from the investigations and discoveries of all students?

d. In what ways might the use of interactive geometry software (e.g., *The Geometer's Sketchpad*™ or geometry applications on a graphing calculator) facilitate exploration of this problem?

e. Teachers who teach using open problems also need to assess with such problems so that instruction and assessment are aligned. Yet, teachers are concerned with the time needed to give such an open problem on a classroom test. What are some ways teachers might give open problems in an assessment context that is fair and equitable to students?

13. Hilberg, Doherty, Dalton, Youpa, and Tharp (2002) identify the following seven standards for mathematics instruction. Five are designed to enhance mathematical success for a variety of at-risk students and two are designed specifically to address issues of mathematics education for American Indian students. The standards are:

- Teachers and students should work together to develop mathematics concepts.
- Students need to learn to use the language of mathematics.
- Teachers need to connect mathematics ideas to students' lives.
- Students, including those at risk, need to engage with cognitively challenging mathematics.
- Students need to be engaged in their learning through discourse.
- Students need to be engaged in activities in which they can direct their own learning, either individually or in small groups.
- Teachers need to use performance and observation as part of assessment.

TEACHING AND LEARNING HIGH SCHOOL MATHEMATICS © 2010 John Wiley & Sons, Inc.

a. In what ways would the instructional approaches in these standards help build an equitable learning environment for all students?
b. In what ways does communication about mathematics, either orally or in writing, enhance or hinder students' learning?
c. How should students' ethnicity play a role in the curriculum teachers use or in their expectations for students' learning of mathematics?
d. What standards do you see yourself working to implement in your own classroom?

Analyzing Instructional Materials

14. Find two or three current high school geometry textbooks. Investigate the lessons on proofs with coordinates. Compare and contrast the treatment of this topic in the different textbooks and as presented in this lesson.

15. In many school situations, high school students take geometry after one year of algebra. That is, geometry is sandwiched between Algebra I and Algebra II. Writing coordinate proofs provides an opportunity for students to work with concepts from algebra I that they will need in Algebra II.
    a. List some of the algebra concepts students use in their work with coordinate proofs.
    b. What are some of the instructional challenges you are likely to face related to students' use of these algebra concepts?
    c. Investigate lessons in geometry on coordinates. In what ways are algebraic concepts integrated with the geometry concepts?

Making Historical Connections

16. Consider the three classical construction problems of antiquity mentioned in the unit introduction: squaring the circle, doubling the cube, or trisecting an arbitrary angle.
    a. Research the history of one of the three problems.
    b. How might historical connections be used to motivate students' interest in geometry?

Communicating Mathematically

17. As students explore geometric figures and make conjectures, they need a way to keep track of their discoveries, including conjectures, some that they ultimately prove or disprove, and some that are open questions. One strategy for documenting student learning is a *Learning Log*, a running record in which students record their initial understandings about a concept and their updates as they learn more. When students maintain the log, they can correct their own misconceptions and add to their correct conceptions. In the end, students reflect on and write what they learned about the concept (Barton & Heidema, 2002, pp. 132–133).
    a. Consider the problem (I–VI) you investigated as part of the **Explore More**. What are some entries you might expect to include in a learning log related to this problem?
    b. Some teachers expect students to write in learning logs several times per week. They collect the logs weekly or biweekly. How does this strategy benefit students and teachers?
    c. Some teachers create a class learning log. Students take turns writing in the log on a daily basis. The log is accessible for review by any student in the class. How might such a strategy be used to support student learning?

TEACHING AND LEARNING HIGH SCHOOL MATHEMATICS © 2010 John Wiley & Sons, Inc.

Reflecting on Professional Readings

18. Read the "Geometry Standard for Grades 9–12" in *Principles and Standards for School Mathematics* (pp. 308–318).

a. What issues did the reading raise in terms of your future classroom?

b. Complete the proof of the problem in Figure 7.16 from the reading to show the medians of a triangle intersect.

c. Design a student page for the problem in Figure 7.16 to provide incremental help for students needing assistance.

d. Compare the geometry expectations outlined in *PSSM* with your own geometry experiences in high school or college.

19. Read "Linking Theory and Practice in Teaching Geometry" by Randall E. Groth in *Mathematics Teacher*, 99 (August 2005): 27–30. The article provides more detailed descriptions of the knowledge at each of the van Hiele levels than is provided in the table within this lesson.

a. In your own words, briefly indicate your understanding of each level of the model.

b. You have likely studied Piaget in courses in educational psychology. What are some similarities and differences between Piaget's model of cognitive development and the van Hiele model of geometric thought?

2.2 Building Conceptual Understanding: Congruence and Similarity

How does one come to understand mathematics? Is it through memorization, accurate replication, and careful use of teacher-demonstrated rules? Is it through making connections to related mathematical ideas, life experiences, or other content areas? Are these two types of learning related? Must one precede the other?

Consider the following scenario. Traveling to and from school each day, a person generally takes the same route. What happens when one unexpectedly comes across a delay, such as an accident or traffic detour? How does one manage? If a person is familiar with nearby roads, then another route likely springs to mind, moderating the delay. If a person is unfamiliar with any route beyond the one taken each day, then the options for avoiding the delay are more limited. One could drive blindly down other streets and hope the alternate route will not be too time-consuming. One could consult an available map or global positioning system (GPS) to find an alternate route. Having a single or limited number of routes to a destination is similar to learning that occurs only on the procedural level where the procedures are not connected to underlying concepts. Having a broader view of the possible routes to avoid a traffic delay illustrates the connected understanding that conceptual learning provides.

The **Launch** and **Explore** activities in this lesson are designed to help you understand more about conceptual understanding for triangle congruence relationships. Such activities provide the mathematical equivalent of finding new routes and determining which ones work and which ones don't.

To get students thinking about their prior experiences or to recall what they know about a mathematical concept, the brainstorming technique *Roundtable* (Kagan, 1992) can be useful.

TEACHING AND LEARNING HIGH SCHOOL MATHEMATICS © 2010 John Wiley & Sons, Inc.

Roundtable

1. Use only one sheet of paper for your group.
2. In turn, write down one response, say it out loud, and then hand the paper to the person on your left.
3. You may choose to skip your turn on one round only.
4. Continue passing the paper to the left until time is called.
5. Group members may ask for clarification—but may not critique responses.

(Kagan, 1992)

In this lesson, you will:

- Investigate mathematics concepts related to congruence and similarity, such as congruent triangles, their theorems, axioms and constructions, and similar triangles and trigonometry.
- Consider synthetic proofs related to congruence or similarity of geometric figures.
- Discuss relationships between procedural and conceptual understanding.
- Use the cooperative learning strategy *Roundtable*.

Materials

- several pieces of spaghetti per pair
- several sheets of patty paper per pair
- one roll of transparent tape per pair
- protractors and straight edges

Website Resources

- Cooperative Learning powerpoint
- Link Sheet
- Frayer Model

LAUNCH

To aid students in recalling what they know and recording their thoughts when learning mathematics conceptually, the Link Sheet and the Frayer Model are helpful tools. The Link Sheet (Shield & Swinson, 1996) is designed to help students organize their thoughts and consider multiple representations so they can employ more of their senses to understand information. The Frayer Model is similar to the Link Sheet in that students provide examples and characteristics of the concept. In addition, the Frayer Model encourages students to define the term and contrast examples with non-examples of the concept to develop criteria to distinguish between them.

1. In your small group, fold two pieces of paper into four quadrants. Use one piece of paper to prepare a Link Sheet and the other piece of paper to prepare a Frayer Model. Use the appropriate model and *Roundtable* to complete items 1a and 1b.
   a. Complete the Link Sheet for Congruent Triangles by finding an appropriate entry for each cell.

TEACHING AND LEARNING HIGH SCHOOL MATHEMATICS © 2010 John Wiley & Sons, Inc.

b. Fill in the Frayer Model for Similar Triangles by finding an appropriate entry for each cell:

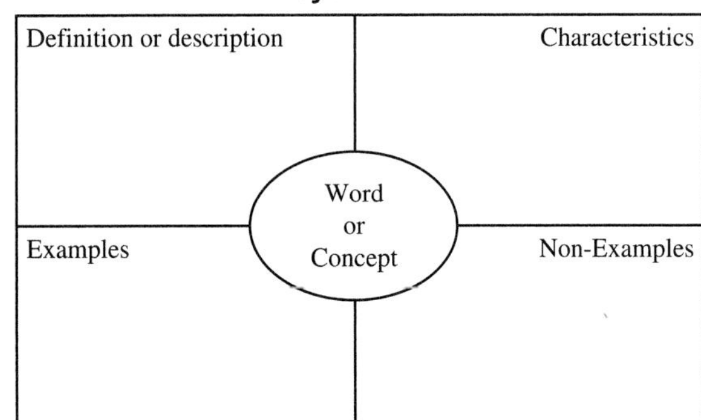

Link Sheet

Concept:	
Mathematics (symbolic) example	Everyday example
Diagram/picture/graph	My explanation

c. Which of these models do you prefer to use for the concept of congruent triangles? Which of these models do you prefer to use for the concept of similar triangles? Explain your choice(s).

d. Both models provide helpful guides for students when learning new vocabulary words or concepts. What would you include in a combined version for your own students? Why?

e. How does the *Roundtable* cooperative learning strategy support classroom equity?

EXPLORE

One general problem-solving strategy is to work backwards. This strategy can also help teachers gain insights into teaching. In this case, we begin with a rich problem that high school students should be able to solve. When teachers work on the problem themselves, they can get clues to the concepts and thought processes students will need to do comparable work.

2. In the drawing provided,

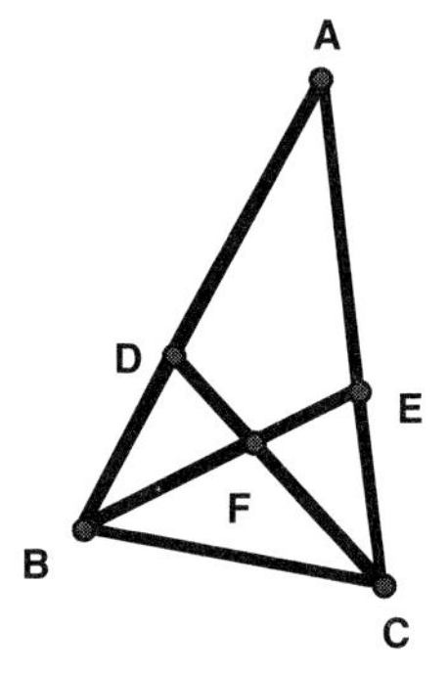

$\triangle ABC$ is isosceles with $AB = AC$.
$\overline{CD}$ bisects $\angle ACB$ with point D on $\overline{AB}$.
$\overline{BE}$ bisects $\angle ABC$ with point E on $\overline{AC}$.
$\overline{CD}$ intersects $\overline{BE}$ at point F.

Study the figure. Find as many pairs of congruent triangles as possible. How do you know the triangles are congruent? Outline a proof for at least two pairs of congruent triangles.

3. What concepts do students need to understand to do this problem?

SHARE

4. Share your responses to items 2 and 3. Note which theorems you used to prove triangles congruent.

TEACHING AND LEARNING HIGH SCHOOL MATHEMATICS © 2010 John Wiley & Sons, Inc.

EXPLORE MORE

Triangle congruence relationships are building blocks for many geometric concepts and proofs. High school students often learn them one at a time, though sometimes in comparison with each other. One major concept often overlooked in teaching and learning about triangle congruence is the concept of sufficiency, that is, identifying which conditions are just enough to ensure that two triangles are congruent. For example, SAS is just enough. (If two pairs of corresponding sides and the angles included between them are congruent, then the triangles are congruent.) SS is insufficient. (Knowing only that two pairs of corresponding sides are congruent does not guarantee that two triangles are congruent.)

5. In addition to SS, what other correspondences of congruent parts between two triangles are insufficient to prove that the two triangles are congruent?
6. **a.** What correspondences of congruent parts between two triangles are just enough to prove that the two triangles are congruent?
   b. Between pairs of right triangles, what correspondences are just enough to prove that the two triangles are congruent?
   c. Use tools provided (patty paper, protractors and straightedges, spaghetti, interactive geometry software) to make convincing cases for your claims in items 6a and 6b.
7. What correspondences of congruent parts are more than enough? Why?
8. In deciding whether two triangles are similar, a comparable analysis can be made.
   a. What major modification needs to be made in the analysis?
   b. What are some minimal sets of conditions that are just enough to prove two triangles are similar?

SHARE AND SUMMARIZE

9. Share your responses to items 5 through 8.
10. How do the tools help:
    a. To make a case for different triangle congruence theorems (just enough conditions)?
    b. To make the task accessible to a broad range of students?
11. The **Launch** of this lesson focused on communication activities related to congruence and similarity. How might differences in students' responses before and after a lesson inform teachers about the robustness of students' understanding of the concepts?
12. **a.** How was *Roundtable* useful in finding out what members of the group knew?
    b. Compare the value of small group sharing to whole class only sharing to encourage student participation. How might small group sharing impact equity, especially for English language learners?

DEEPENING MATHEMATICAL UNDERSTANDING

1. *Congruence and Similarity in the Real World.* List real-world examples that use congruent or similar objects. Describe the objects, their purpose, and why congruence or similarity of these objects is important.
2. *SSA and HL.* The correspondence SSA (two pairs of corresponding sides congruent and corresponding non-included angles congruent) is sometimes called "the ambiguous case."

TEACHING AND LEARNING HIGH SCHOOL MATHEMATICS © 2010 John Wiley & Sons, Inc.

a. Find cases where SSA works and cases where it produces two different triangles.

b. In two right triangles, is the relationship HL (corresponding hypotenuses and one pair of corresponding legs congruent) sufficient to prove that the triangles are congruent? Why or why not?

c. What is the relationship between HL and SSA?

3. *Medians, Altitudes, and Angle Bisectors of Isosceles Triangles.* Consider isosceles $\triangle ABC$ at right with $AB = CB$.

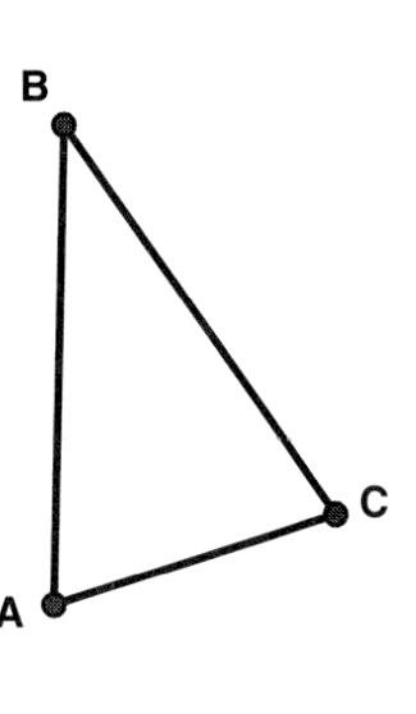

a. What is a median of a triangle? Construct one.

b. Construct the altitude of $\triangle ABC$ to $\overline{AC}$.

c. Construct the angle bisector for $\angle B$ in $\triangle ABC$.

d. What conclusions can you make about the relationships between the perpendicular bisector of $\overline{AC}$, the line segment from vertex B to the midpoint of $\overline{AC}$ (called the median), and the altitude from vertex B to $\overline{AC}$?

e. Prove one of the relationships you found in problem 3d.

4. *More on Medians, Altitudes, and Angle Bisectors of Triangles.*

a. Investigate relationships among the medians of a triangle. For which triangles do the medians have a special relationship? State the relationship and justify it.

b. Investigate relationships among the altitudes of a triangle. For which triangles do the altitudes have a special relationship? State the relationship and justify it.

c. Investigate relationships among the angle bisectors of a triangle. For which triangles do the angle bisectors have a special relationship? State the relationship and justify it.

d. Complete a coordinate proof of one of the results in problems 4a, 4b, or 4c. Choose the coordinates of A, B, and C wisely so that you need the fewest number of variables to label the vertices but so the triangle is general.

5. *More Angle Bisectors of Isosceles Triangles.* In the sketch at right,

$\triangle ABC$ is isosceles with $AB = AC$.
$\overrightarrow{CD}$ bisects $\angle ACB$ with point D on $\overline{AB}$.
$\overrightarrow{BE}$ bisects $\angle ABC$ with point E on $\overline{AC}$.
$\overrightarrow{BJ}$ bisects $\angle ABH$ with point J on $\overrightarrow{CD}$.
$\overrightarrow{CK}$ bisects $\angle ACG$ with K on $\overrightarrow{BE}$.
$\overrightarrow{CD}$ intersects $\overrightarrow{BE}$ at point F.

a. Study the figure. Determine all the relationships you can. Prove some of them. Use the questions below to get started (adapted from Becker, Jerry P. and Shigeru Shimada. *The Open-Ended Approach: A New Proposal for Teaching Mathematics.* Reston, VA: National Council of Teachers of Mathematics, 1997.)

b. Which pairs of line segments are congruent? How do you know?

c. What type of quadrilateral is $BCKJ$? Prove your result.

d. Determine any similar triangles created in the figure. How can you be sure they are similar?

6. *Similar Triangles and Fractals.* Draw $\triangle ABC$. Dissect $\triangle ABC$ into four congruent triangles that are similar to $\triangle ABC$. For the resulting congruent triangles, repeat the process two or three more times, each time dissecting all but the center

TEACHING AND LEARNING HIGH SCHOOL MATHEMATICS © 2010 John Wiley & Sons, Inc.

triangle of the newly constructed triangles. Determine the proportional relationship between the size of each new triangle and the original.

7. *The Golden Spiral and Golden Ratio*. The golden spiral gives rise to the famous golden ratio, phi. Conduct an Internet search to determine how to construct the golden spiral. What is the value of phi? In what other ways does phi arise?

8. *Spiral of Isosceles Right Triangles*. Construct a spiral of isosceles right triangles using these directions. To begin, construct an isosceles right triangle whose legs are 1 unit long. Construct a second isosceles right triangle using the hypotenuse of the first triangle as one of its legs. The triangles should have only the one side in common and should not overlap. Continue this process, each time creating an isosceles right triangle with legs the length of the hypotenuse of the previous right triangle, sharing a vertex with all of the previous triangles and none of the triangles overlapping.
   - **a.** What relationship exists between successive triangles?
   - **b.** Find the constant of proportionality between successive triangles.
   - **c.** Without overlapping, how many such triangles can be created? Explain.

9. *Similar Right Triangles*. Complete the **Similar Right Triangles** student page. What relationships are possible among the three right triangles in the figure provided in problem 4 on the student page?

Similar Right Triangles

Recall that two triangles are similar if all of their corresponding angles are congruent and all of their corresponding side lengths are in the same proportion. Determine the least number of conditions necessary for two right triangles to be similar by completing the following.

1. To determine if two triangles are similar, what is the least number of corresponding side lengths that must be in the same proportion? How do you know? Do not assume any relationships about the corresponding angles.
2. To determine if two *right* triangles are similar, what is the least number of corresponding side lengths that must be in the same proportion? How do you know? Do not assume any relationships about the remaining corresponding angles.
3. If one right triangle has an acute angle congruent to an acute angle in a second right triangle, will they be similar? Why or why not?
4. Consider the figure at right. Angle D is a right angle. Segments $\overline{BG}$, $\overline{CF}$, and $\overline{DE}$ are parallel. Points A, B, C, and D are collinear. Points A, G, F, and E are collinear.

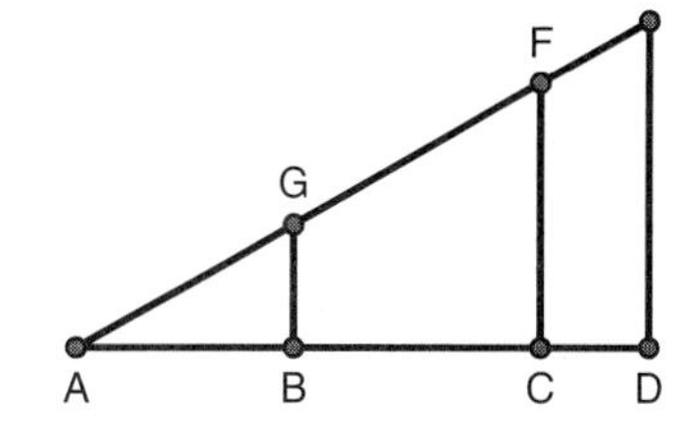

   - **a.** Determine all of the similar triangles in the figure.
   - **b.** How do you know the triangles you found are similar?
   - **c.** For each triangle found in problem 4a, carefully measure each side. Find the constant of proportionality for each pair of similar triangles.
5. Determine as many relationships as possible in the figure in problem 4 and write them as conditional statements in if–then form. The following categories might help you find relationships:
   - sides
   - angles
   - triangles
   - ratios

(continued)

TEACHING AND LEARNING HIGH SCHOOL MATHEMATICS © 2010 John Wiley & Sons, Inc.

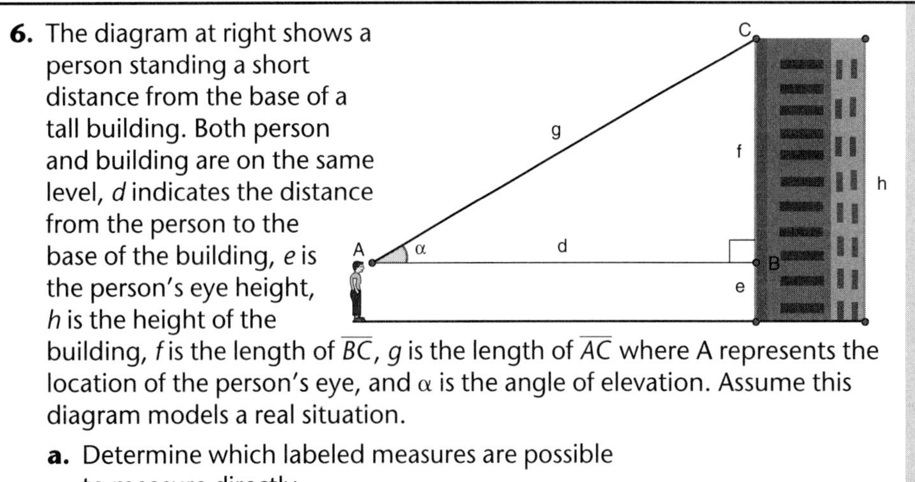

6. The diagram at right shows a person standing a short distance from the base of a tall building. Both person and building are on the same level, d indicates the distance from the person to the base of the building, e is the person's eye height, h is the height of the building, f is the length of $\overline{BC}$, g is the length of $\overline{AC}$ where A represents the location of the person's eye, and α is the angle of elevation. Assume this diagram models a real situation.

a. Determine which labeled measures are possible to measure directly.

b. Determine which labeled measurements can be found through calculation.

c. Explain your work.

7. The river banks are the same height on both sides of a straight stretch of river. A person is standing on the bank of a river at point P. A tall building stands across the river and down a short distance from where the person is standing. Angles at points A, B, C, D, and F are all right angles. (See the diagram at right.)

a. What measurements do you need to determine the width of the river?

b. What measurements are needed to find the height of the building?

c. Explain your work.

10. *Right Triangle Relationships.* Complete the **Exploring Relationships in Right Triangles** student page. It is possible to complete this student page by hand, but it is best to use an interactive geometry drawing tool for the necessary accuracy.

Exploring Relationships in Right Triangles

Materials

- an interactive geometry drawing tool

1. Construct a figure like the one shown in the diagram on p. 97. $\overline{AN}$, $\overline{BM}$, $\overline{CL}$, $\overline{DK}$, $\overline{EJ}$, and $\overline{FH}$ are perpendicular to $\overleftrightarrow{GN}$. Points B, C, D, E, and F are on $\overleftrightarrow{AG}$; points H, J, K, L, and M are on $\overleftrightarrow{GN}$.

2. a. Determine as many ratios as possible in the figure.

b. Which of the ratios are related?

c. Why do the relationships between the ratios make sense?

3. For each triangle in the figure, compare the length of the vertical side to the length of the corresponding hypotenuse.

(*continued*)

TEACHING AND LEARNING HIGH SCHOOL MATHEMATICS © 2010 John Wiley & Sons, Inc.

a. What do you notice?

b. Is your observation always true? If so, explain why. If not, find an example for which your answer to part 3a is not true.

c. Change the measure of $\angle G$. Does the relationship you described in problem 3a still hold? Why or why not?

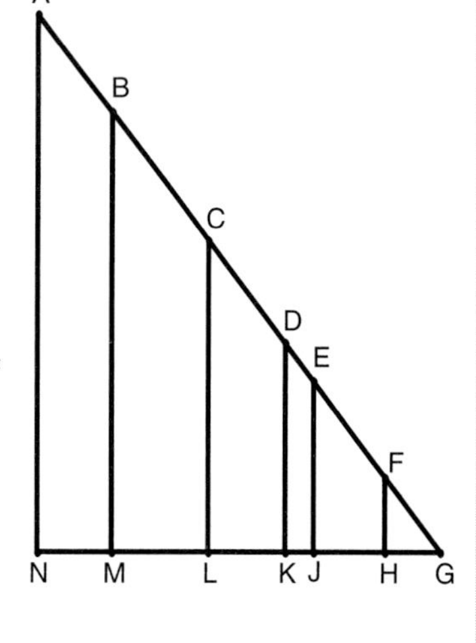

4. For each triangle in the figure, compare the length of the horizontal side to the length of the corresponding hypotenuse for each triangle.

a. What do you notice?

b. Is your observation always true? If so, explain why. If not, find an example for which your answer to part 4a is not true.

c. Change the measure of $\angle G$. Does the relationship you described in problem 4a still hold? Why or why not?

5. Choose one of the triangles. Let the length of the horizontal segment be a, the length of the vertical segment be v, and the length of the hypotenuse be h.

a. Complete the table for several different measures for $\angle G$ between 0 and 90 degrees.

$m\angle G$	0		30		45		60		90
a									
v									
h									
$\frac{a}{h}$									
$\frac{v}{h}$									

b. Plot the points $(m\angle G, \frac{a}{h})$ on a coordinate plane. Describe the shape of the graph.

c. Plot the points $(m\angle G, \frac{v}{h})$ on a coordinate plane. Describe the shape of the graph.

d. Compare the graphs you created in items 5b and 5c. Do these graphs seem to be related? Explain.

e. Would the results in items 5b and 5c be different if you started with a different triangle for item 5a? Why or why not?

6. a. Compare the graphs you created in problems 5b and 5c to the graphs of $y = \sin x$ and $y = \cos x$.

b. From your comparisons in problem 6a, define each of these functions.

11. *Similar Triangles and Trigonometry*. Trigonometry is based on the relationships between similar right triangles. In the top figure $\angle ABD$ and $\angle ACE$ are right angles.

TEACHING AND LEARNING HIGH SCHOOL MATHEMATICS © 2010 John Wiley & Sons, Inc.

a. Indicate the corresponding sides of the similar triangles in the figure.

b. From your work in trigonometry in mathematics classes, choose one of the right triangles from the top figure, label it, and state the sine, cosine, and tangent relationships.

c. Use the triangle you chose in problem 11b to explain why similar right triangles have the same sines, cosines, and tangents for the corresponding congruent acute angles.

d. In the lower figure at right, the segment of length t is tangent to the unit circle. The length t is the tangent of $\angle\theta$. Why?

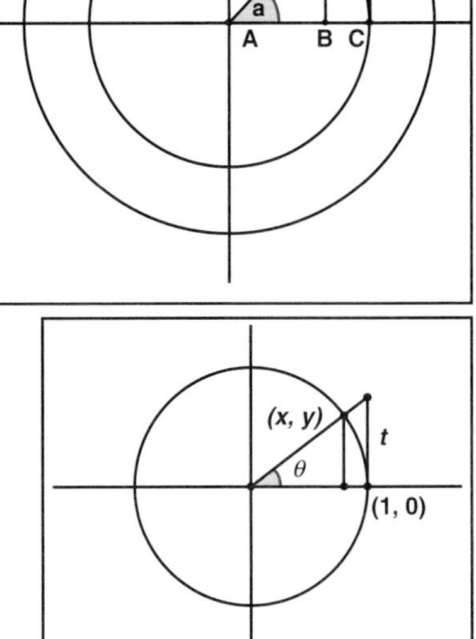

12. *Measuring Inaccessible Distances.*

a. Explain how similar triangles can be used to determine the height of a tall tree.

b. Milwaukee, Wisconsin and Muskegon, Michigan are on opposite shores of Lake Michigan, approximately 90 miles apart and close to the 45th parallel of latitude. The circumference of the earth on a great circle through Muskegon and Milwaukee is approximately 24,881 miles.

i. A person is standing on a 150-foot-tall sand dune in Muskegon. On a clear day, how far should this person be able to see to the horizon?

ii. What if the sand dune is 100 feet tall? 200 feet tall?

iii. How high above the lake must a person in Muskegon be in order to see across Lake Michigan to Milwaukee?

iv. On dark summer nights, when the conditions are right, people standing on sand dunes in Muskegon have claimed they can see the headlights of cars traveling along the highways of Milwaukee. Some even have photographs to "prove" their claim. Is this possible?

13. *Undoing Sines or Cosines.*

a. A carpenter builds model lighthouses with regular octagonal bases. He is interested in knowing how long to make each side of the base so that the finished lighthouse is approximately 2 feet in diameter measured from the midpoint of one side to the opposite parallel side. Find the length of a side of the base of the lighthouse.

b. Suppose the carpenter decided the base should be 2 feet from one "corner" to the opposite corner of the base of the lighthouse. How long is a side of the base of the lighthouse?

c. How big is the base of the lighthouse if the length of each side of the base is 1 foot?

DEVELOPING MATHEMATICAL PEDAGOGY

Communicating Mathematically

14. In the **Launch**, you used *Roundtable* to complete either the Link Sheet or a Frayer Model for triangle congruence or similarity. Consider the following adaptation.

TEACHING AND LEARNING HIGH SCHOOL MATHEMATICS © 2010 John Wiley & Sons, Inc.

Fold a sheet of paper to make three columns. Label one column **Congruence Concepts,** the second **Similarity Concepts**, and the third **Comparisons**. Pass this sheet of paper and use *Roundtable* in a group of four to generate responses to the following questions.

- What does it mean for two geometric figures to be congruent?
- What does it mean for two geometric figures to be similar?
- What is alike (A) or different (D) about these two concepts? (Use A or D in the third column to identify these.)

a. Which of the three models is useful when introducing triangle congruence?

b. Which of the three models is most useful when introducing triangle similarity?

c. Which of the three models is most useful when comparing triangle congruence and similarity?

d. Provide support for your choices in problems 14a–c.

Analyzing Instructional Materials

15. Study the student page **How Many Triangles?** How does this activity develop students' conceptual understanding of triangle congruence relationships?

How Many Triangles?

Materials
- several uncooked spaghetti noodles for each pair
- several sheets of patty paper for each pair
- one roll of transparent tape for each pair

1. What does it mean for two shapes to be congruent?

2. Suppose you know that $\triangle ABC$ is congruent to $\triangle DEF$. List all the possible congruence statements.

Although two congruent triangles have six different pairs of corresponding congruent parts, generally fewer than six pairs need to be checked for congruence. Often, knowing that fewer than six pairs of corresponding parts are congruent is enough to ensure that all the other parts are congruent, too. The question is: What pairs of congruences are just enough?

Example

Suppose a triangle has sides with lengths *S*1 and *S*3, and the angle between the sides measures *A*3.

- Trace an angle measuring *A*3 on patty paper.
- Then use the spaghetti to make *S*1 and *S*3 on the rays; measure carefully.
- Suppose another triangle has the same given information. Must it be congruent to the first triangle? Check with the materials.

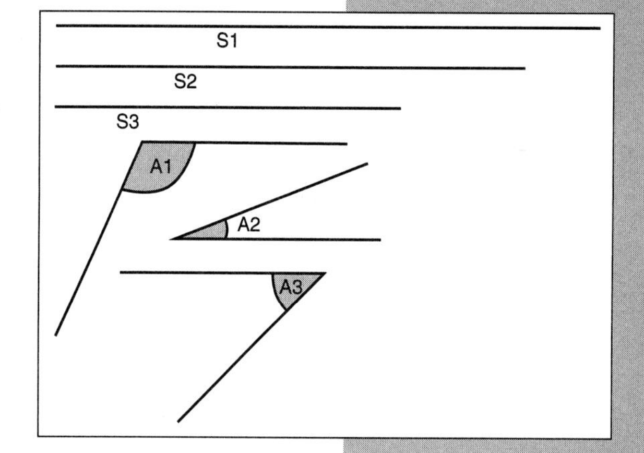

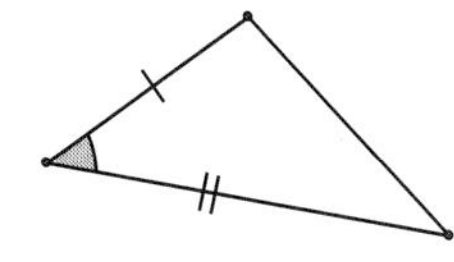

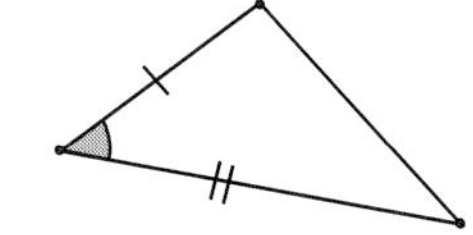

The Example relationship is called Side-Angle-Side (SAS) because the three given congruences relate two pairs of sides and the angle included between them. See figures above.

In this investigation, you will explore which pairs of congruences between two triangles guarantees that the triangles must be congruent. The exploration uses the lengths and angles indicated in the diagram. Use the spaghetti and patty paper to help with the exploration.

3. Refer to the given side lengths (*S*1, *S*2, *S*3) and the angle measures (*A*1, *A*2, *A*3). Break the spaghetti to the given lengths and use a

(*continued*)

TEACHING AND LEARNING HIGH SCHOOL MATHEMATICS © 2010 John Wiley & Sons, Inc.

straightedge to carefully trace each angle measure on a piece of patty paper. Extend the sides of each angle. For each row in the table, use the indicated side length(s) or angle measure(s); other side lengths or angle measures can vary. (You can tape spaghetti to your patty paper as necessary.) For each relationship:

a. In the second column, sketch two congruent triangles and show the congruence relationship being investigated.

b. Find as many noncongruent triangles as you can using the components listed in the third column and the relationship (in the order shown) listed in the first column.

c. Determine if the relationship in the first column is a congruence relationship. Explain your findings.

Relationship	Sketch	Components	Congruent?	Explanation
SSS		*S*1, *S*2, *S*3		
SSA		*S*1, *S*2, *A*2		
SAS		*S*1, *A*1, *S*2		
AAS		*A*1, *A*2, *S*1		
ASA		*A*1, *S*1, *A*2		
AAA		*A*1, *A*2, *A*3		
HL		*S*1, *S*3, Right Angle		

4. Test some of the relationships using other choices of angle measures or side lengths (beyond those in the table). Did your congruence decisions change? Why or why not?

5. Are there any other relationships between sides and angles that were not explored? Explain.

6. Which relationships, if any, are obvious from other relationships? When possible, state both relationships and explain how one is the direct result of the other.

16. Refer back to the student pages **Similar Right Triangles** and **Exploring Relationships in Right Triangles.**

a. Which major mathematical ideas are developed?

b. In what ways do the student pages develop students' conceptual understanding of the concepts?

c. In what ways do the student pages develop students' procedural understanding of the concepts?

d. Find several high school geometry books. Investigate how each text introduces concepts related to similar triangles and to the sine and cosine relationships in right triangles. Compare and contrast the development in these texts with the development of these two concepts on the student pages.

TEACHING AND LEARNING HIGH SCHOOL MATHEMATICS © 2010 John Wiley & Sons, Inc.

Preparing for Instruction

17. To ask appropriate *Roundtable* questions, teachers need to be aware of prerequisite knowledge necessary to engage students in learning mathematics.

a. Determine prerequisite knowledge needed to solve each of the following problems.

i. Prove whatever you can about the plane figure at right, given that:

$\overline{AB} \cong \overline{BC}$ and $\overline{DB} \cong \overline{EB}$

B is the intersection of $\overline{CD}$ and $\overline{AE}$.

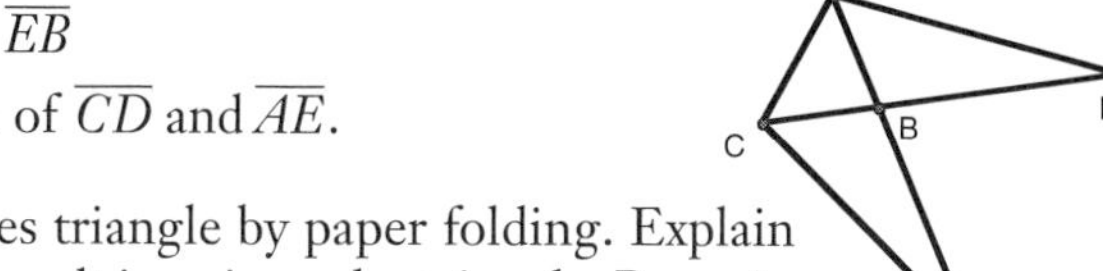

ii. Construct an isosceles triangle by paper folding. Explain how you know your result is an isosceles triangle. Prove it.

iii. How are the median, angle bisector, and altitude of an isosceles triangle related? Explain. Prove your conjecture(s).

b. What initial question(s) would you pose for a *Roundtable* activity in order to determine if students have sufficient prerequisite knowledge to solve each problem in item 17a?

c. Order the problems (i–iii) from least challenging to most challenging. Explain your choice.

d. Which of the problems would you expect students who have not studied triangle congruence to be able to answer? Explain.

e. Which of the problems would you expect students who have studied triangle congruence to find difficult?

i. Identify the difficulties you might expect students to have.

ii. What teacher support would you be prepared to give to make the problem(s) more accessible? (Consult the Stages of Questioning from Lesson 1.1.)

18. In some geometry texts, students learn the triangle congruence relationships one at a time and apply them to a limited collection of problems that use just the one relationship.

a. Once students have learned all the triangle congruence theorems, which *Roundtable* question(s) might you pose to remind students of what they have learned to prepare them to solve problems?

b. Once each group of students has generated lists through *Roundtable*, what should be done with the lists?

19. Teachers sometimes provide a real-life task to gauge what prerequisite knowledge students bring to a lesson. What might be an appropriate real-life task that students could consider before engaging in a lesson on congruent triangles?

20. Suppose **Explore** problem 2, **DMU** problem 5, and **DMP** problem 17a(i) are assigned to high school students to complete.

a. Discuss the pros and cons of presenting such open-ended problems to students.

b. Each problem and its various solutions involve several important mathematical ideas. Identify several for each problem.

c. Suppose some of your students complete the tasks before most of the rest of the class. What more would you want to ask of those students?

Analyzing Students' Thinking

21. In the figure at the right, H is the midpoint of both $\overline{GI}$ and $\overline{FA}$. Students were asked to prove $\overline{FG} \cong \overline{AI}$. Following are several sample responses.

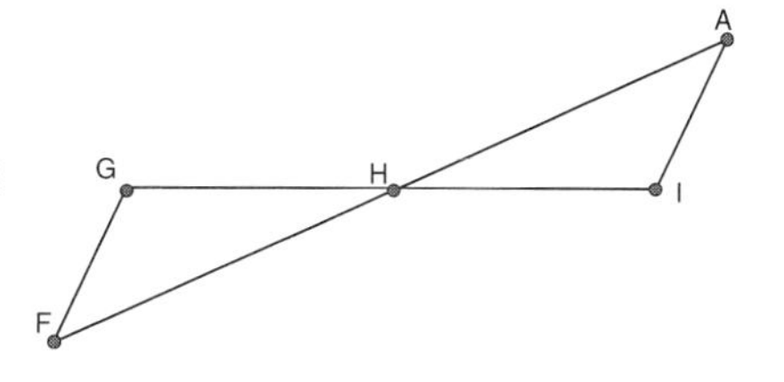

TEACHING AND LEARNING HIGH SCHOOL MATHEMATICS © 2010 John Wiley & Sons, Inc.

a. Critique each response to determine whether it is a valid proof.

b. Generate sample questions you might use to help any students whose proof contains misconceptions.

c. Complete a **QRS Guide** to record the problem, the student work on this problem, and the teacher support.

i. $\angle GHF \cong \angle IHA$ *because they are vertical angles.* $GH = IH$ *because H is the midpoint and cuts the length of* $\overline{GI}$ *in half. In the same way,* $FH = AH$ *because H is the midpoint of* $\overline{FA}$*. Then* $\triangle FGH \cong \triangle AIH$ *by SAS. So,* $\overline{FG} \cong \overline{AI}$ *because they are congruent parts of congruent figures.*

ii. *If I rotate* $\triangle AIH$ 180° *around point H, it lines up with* $\triangle FGH$*. So, this means* $FG = AI$*.*

iii. $\angle GHF \cong \angle IHA$ *because they are vertical angles.* $\angle G$ *and* $\angle I$ *are alternate interior angles so* $\angle G \cong \angle I$*. Then* $\triangle FGH \cong \triangle AIH$ *by ASA. So* $\overline{FG} \cong \overline{AI}$ *because they are congruent parts of congruent figures.*

Using Technological Tools

22. In the **How Many Triangles?** student page, the tools used to develop conceptual understanding are spaghetti, patty paper, and transparent tape.

a. What other tools could be used?

b. How would you modify the activities if you had computer tools available (e.g., *The Geometer's Sketchpad*™)?

Working with Students

23. Each of the articles "A Triangle Divided: Investigating Equal Areas" by Daniel Scher (*Mathematics Teacher*, 93 (October 2000): 608–611) and "The Angles of a Star" by Alan Lipp (*Mathematics Teacher*, 93 (September 2000): 512–516) contains an open problem relating to triangles. In the article by Scher, students subdivide any triangle into four triangles all with the same area. In the article by Lipp, students determine the sum of the angle measures at the vertices of a star with five points.

a. Investigate one of these problems on your own. Anticipate possible student responses to the problem. What questions might you prepare to help students as they work on this problem? Record expected responses and possible Teacher Support in a **QRS Guide**.

b. Try one or more of these problems with high school students. Record their responses in the **QRS Guide**. What surprised you about their engagement with the problem?

c. What are some advantages to using such open problems with high school students? What are some concerns?

d. Each problem can be explored with interactive geometry software. What are some issues and challenges of using technology to explore such problems?

Reflecting on Professional Reading

24. Read "The Human Body's Built-In Range Finder: The Thumb Method of Indirect Distance Measurement" by Michael Wong (*Mathematics Teacher*, 99 (May 2006): 622–626).

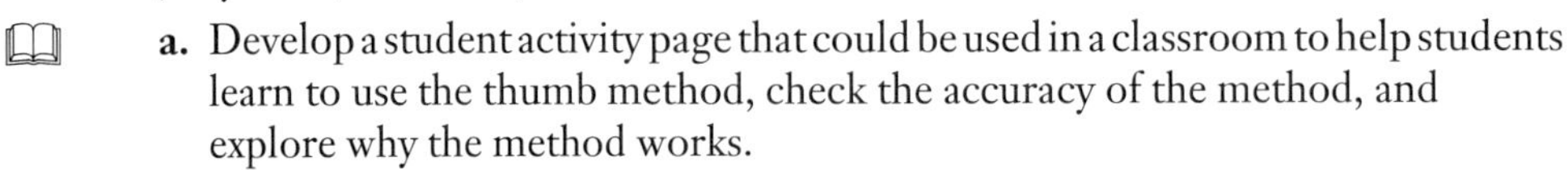

a. Develop a student activity page that could be used in a classroom to help students learn to use the thumb method, check the accuracy of the method, and explore why the method works.

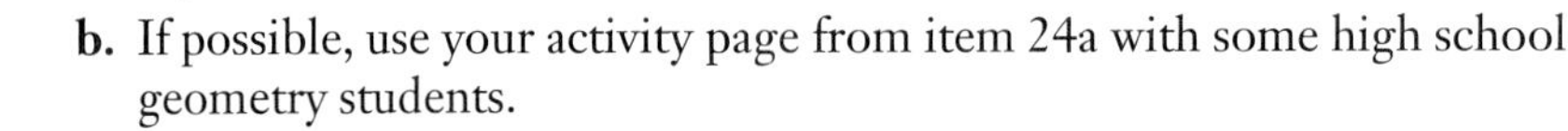

b. If possible, use your activity page from item 24a with some high school geometry students.

c. What are some advantages of using such an activity in a high school geometry classroom?

TEACHING AND LEARNING HIGH SCHOOL MATHEMATICS © 2010 John Wiley & Sons, Inc.

25. Read "From Classroom Discussions to Group Discourse" by Azita Manouchehri and Dennis St. John (*Mathematics Teacher*, 99 (April 2006): 544–551).

   a. According to the authors, discourse requires *participation*, *commitment*, *reciprocity*, *content*, and *purpose*. Define these terms in your own words.

   b. Summarize the differences between class discussion in a typical classroom and discourse within a learning community as distinguished by the authors. Which type of communication has existed in the classes you have visited or in your own classroom?

   c. What are issues and concerns you have about developing a classroom capable of supporting the type of discourse community evident in Lyle's class?

   d. Tell how at least one discourse requirement is relevant to building classroom equity.

2.3 Learning Mathematics through Multiple Perspectives: Quadrilaterals and Constructions

As mathematics teachers plan for student learning, they need to be cognizant of lessons drawn from learning theory about how adolescents learn. For instance, Dienes identifies four principles related to mathematics learning, two of which are pertinent here: *perceptual variability* and *constructivity*. Perceptual variability suggests that students' learning is enhanced when they have an opportunity to learn a concept from more than one perspective. Constructivity suggests that students need concrete experiences prior to studying abstract concepts; they need to make sense of mathematics through meaningful experiences. Ausubel, another learning theorist, focused on students' learning through *meaningful verbal discourse* in which students describe mathematics concepts in their own words.

Three specific stages of learning were identified as important by Jerome Bruner: the *enactive*, the *iconic*, and the *symbolic*. At the enactive stage, students work with concrete materials to explore concepts and begin to develop meaning. At the iconic stage, pictures and diagrams are used to represent the concepts being studied. At the symbolic stage, students are able to understand and use symbols and other abstract forms to represent concepts.

The principles raised by these theorists reinforce the messages of this text, namely that students need to make sense of mathematics, they need to develop conceptual understanding in addition to procedural fluency, and they need to engage in discourse with peers as they explore mathematics. In this lesson, learning at the enactive stage is supported through paper folding, using compass and straightedge, working with a MIRA, and using an interactive geometry drawing tool. Interpreting drawings produced during concrete activities or from explorations with an interactive geometry drawing tool provide experiences at the iconic stage of learning. Finally, proving conjectures in symbolic form requires the symbolic stage. All of this occurs in an environment in which students engage in meaningful verbal discourse.

For equitable classrooms, teachers need to approach a concept from many perspectives in order to help all students be successful and to offer alternative approaches to students who may struggle with an initial approach. However, teachers need to use multiple perspectives with sensitivity. Dienes cautions teachers not to introduce multiple embodiments of a concept at the same time. Also, physical materials are tools to help students make sense of mathematics but they are not a panacea; in planning lessons, teachers need to consider which tools are appropriate to help students build meaning.

In this lesson, you will:

- Compare and contrast the use of different construction tools and methods on students' understandings of quadrilaterals.

TEACHING AND LEARNING HIGH SCHOOL MATHEMATICS © 2010 John Wiley & Sons, Inc.

- Consider multiple perspectives from which quadrilaterals can be constructed and studied and analyze their impact on student learning.

Materials

- compass and straightedge
- patty paper
- MIRA
- interactive geometry drawing tool

LAUNCH

1. As a team, choose two categories of construction tools from the second column. Construct each of the geometric objects in the first column using each of the chosen tools. Keep track of your steps. Justify them.

Construct	Tools
Perpendicular lines	Compass and straightedge
Parallel lines	Patty paper
Midpoint of a segment	MIRA
Congruent segments	Interactive geometry drawing tool

Note: **DMU** problem 3 provides an introduction to *The Geometer's Sketchpad*™, if needed.

2. Share your constructions with at least one other group. Compare and contrast the constructions of the same geometric object completed with different tools.
   a. What do students need to know to complete each construction?
   b. For each construction tool, rank the construction of the four geometric objects in order of difficulty. What made some objects more difficult to construct than others?
   c. Which of the tools is easiest to learn to use? Which is most difficult? Why do you think so?
3. What different perspectives about each geometric object are obtained through use of the various construction tools? For example, when constructing parallel lines, what more do students have to think about to use a compass and straightedge than they need to consider to use patty paper or vice versa?

EXPLORE

Quadrilaterals are studied at some level beginning in very early grades and continuing through high school and college geometry. The goal of this long-range study of quadrilaterals is to deepen students' understanding, moving them further through the van Hiele levels of geometric thought with each immersion. For this lesson, assume that students have at least informal understandings of each of the quadrilaterals (trapezoids, isosceles trapezoids, rectangles, parallelograms, rhombi, kites, and squares) using their sides or angles.

4. For each of the quadrilaterals, complete the table to indicate some relationships among sides and some relationships among angles.

TEACHING AND LEARNING HIGH SCHOOL MATHEMATICS © 2010 John Wiley & Sons, Inc.

Quadrilateral	Relationships among Sides	Relationships among Angles
Square		
Rectangle		
Rhombus		
Parallelogram		
Kite		
Isosceles Trapezoid		
Trapezoid		

5. In a team of four, choose one of the quadrilaterals. Construct the quadrilateral using relationships among sides or angles. Each team member should complete one of the following. Keep track of which side or angle relationships you use in your construction. If a particular tool does not work for constructing your team's quadrilateral, describe why it is not useful.
   a. Construct the quadrilateral using compass and straightedge.
   b. Construct the quadrilateral using patty paper.
   c. Construct the quadrilateral using a MIRA.
   d. Construct the quadrilateral using interactive geometry software.
6. If possible, construct your group's quadrilateral using different relationships among its sides or angles than used in item 5. If only one method is possible with your construction tool, explain why that is the case.
7. a. Compare and contrast the relationships you used to construct each of the quadrilaterals with the particular tools chosen.
   b. Could you use the same construction techniques with one or more of the tools? Explain.
   c. What relationships in your quadrilateral are highlighted with each of the construction techniques used?
   d. Did you use other relationships besides sides or angles in constructing your quadrilateral? Explain.
   e. Were some construction tools easier than others to use to construct your quadrilateral? Explain.
   f. If you repeated the same construction techniques with different side or angle measures, could you construct a less specific quadrilateral? Why or why not?

A good definition places an object to be defined into a class of well-defined similar things and then states how it differs from other things in that class. Words used must be commonly understood, defined earlier, or left purposely undefined (such as point, line, and plane). In addition, the definition must include all information necessary to describe the idea—but no more. An example of a good definition for a rhombus is: *A rhombus is a quadrilateral with all sides congruent*. This definition indicates that a rhombus is in the set of quadrilaterals and distinguishes it from other quadrilaterals by indicating that the sides must all be the same length. An example of a definition that gives too much information is: *A rhombus is a quadrilateral with four congruent sides and opposite congruent angles*. Having opposite angles congruent can be proved from knowing the quadrilateral has four congruent sides.

TEACHING AND LEARNING HIGH SCHOOL MATHEMATICS © 2010 John Wiley & Sons, Inc.

The choice of definition is important and somewhat arbitrary. For example, a rhombus can be defined without any mention of the length of its sides. Consider the following definition: *A rhombus is a quadrilateral whose diagonals are perpendicular bisectors of each other*. Then the statement referring to congruent sides is a theorem resulting from the definition based on perpendicular bisecting diagonals.

8. **a.** For each of the constructions your group completed in items 5 and 6, use the steps in your construction to write a definition of the quadrilateral based on side or angle relationships.
   b. If you have written more than one definition for your quadrilateral, prove that one is a consequence of the other.
   c. From your work in items 8a and 8b, how do constructions of quadrilaterals lead to understanding definitions or theorems about quadrilaterals?

SHARE AND SUMMARIZE

9. Share your responses to items 5 through 8 with the class.
10. In his perceptual variability principle, Dienes suggests that students benefit from studying mathematics through a variety of materials in order to make mathematical constants explicit and to determine what aspects of a particular context do not apply to the underlying mathematics. He also recommends that teachers introduce unfamiliar tools one at a time.
    a. What different perspectives about each quadrilateral arise through the use of each tool?
    b. In what ways might teachers introduce the tools to students?
    c. What other uses do you see for each of the tools used in this lesson?
11. Bruner suggests three stages of learning: enactive, iconic, and symbolic.
    a. How does each stage unfold in the activities in this lesson?
    b. Which tools are useful in the enactive stage? Why?
    c. Symbolism in mathematics takes many forms. Often, we think of symbols as Arabic or Greek letters and numbers that we use in equations. What are the symbols of geometry? What might the symbolic stage look like in a geometry classroom?
12. Ausubel suggests that students learn mathematics through meaningful verbal learning or discourse. What role did meaningful verbal learning play as you worked through the activities in this lesson?
13. **a.** What issues do you anticipate might arise when using technology in your geometry classroom?
    b. What issues do you anticipate might arise when using compass and straightedge?
    c. What issues might arise when students complete the constructions using patty paper or a MIRA?

DEEPENING MATHEMATICAL UNDERSTANDING

1. *Quadrilaterals in Our World*. Look around you. What quadrilaterals do you see? How prevalent are quadrilaterals in objects you use or see every day? Why are quadrilaterals used for the objects in your daily life?
2. *Competing Definitions for Trapezoid*. Geometry books are likely to have one of two different definitions for trapezoid.
   I. *A trapezoid is a quadrilateral with exactly one pair of parallel sides.*
   II. *A trapezoid is a quadrilateral with at least one pair of parallel sides.*

TEACHING AND LEARNING HIGH SCHOOL MATHEMATICS © 2010 John Wiley & Sons, Inc.

a. These definitions are sometimes labeled *inclusive* and *exclusive*. Which trapezoid definition is which?

b. What other quadrilaterals are defined inclusively?

c. What are implications of an inclusive definition for how the trapezoid is related to other quadrilaterals?

d. Based on the properties of quadrilaterals you have studied, which definition seems consistent with the ways properties are inherited among the quadrilaterals?

3. *Basic Constructions with Interactive Geometry Software*. The student page **Basic Constructions with *The Geometer's Sketchpad*™** was designed to introduce students to *The Geometer's Sketchpad*™. It was written for version 4.0 on a PC (versus Macintosh). Some variations occur for different versions of the software. (Alternatively, you might use the introductory tours that come with the software.)

   a. Work through the student page, **Basic Constructions with *The Geometer's Sketchpad*™**. Keep track of cases in which you might prefer more or less direction.

   b. When using a dynamic geometry tool such as *The Geometer's Sketchpad*™, what constructions preserve congruence? Explain.

Basic Constructions with *The Geometer's Sketchpad*™

Materials

- *The Geometer's Sketchpad*™

Orientation to *The Geometer's Sketchpad*™ Screen

1. Drawing tools are found on the left side of the screen. Play with each tool to see what it does.

   a. **Arrow tool:** Select drawn objects. Point at an object you want to select. Click the left mouse button to select the object. To select more than one object, in turn, point the arrow tool at the object and then left click the mouse button to highlight each object in the order that you want to select them. To select the objects in one section of the screen, hold down the left mouse button when the arrow tool is active, trace a rectangular box around the objects you want to select, and then release the mouse button. All selected items will be highlighted.

   b. **Point tool:** Construct points on the screen or on objects already on the screen. Click on the point tool and then move the arrow to the location at which you would like a point. If the point is to be constructed on an already drawn object, move the arrow until the object is highlighted. Click the left mouse button to construct a point where desired.

   c. **Circle tool:** Construct circles. Position the center of the circle while holding down the left mouse button. Position a point on the circle and then release the mouse button to construct a circle with the desired center and radius length.

   d. **Line tool:** Construct lines, segments, or rays. Hold down the left mouse button when activating the line tool to select the type of line tool (segment, ray, or line). Position the first point while holding down the left mouse button. Position the second point and release the mouse button to construct the desired segment, ray, or line.

   e. **Labeling tool:** Label objects and create text boxes. Select the A button, the labeling tool. Point the hand at the object to be labeled. Left click. The object is automatically labeled. To change the label,

(*continued*)

TEACHING AND LEARNING HIGH SCHOOL MATHEMATICS © 2010 John Wiley & Sons, Inc.

double click on the label. A dialogue box appears, allowing you to change the label. To create a text box, double click at any blank place on the screen and begin typing.

2. Additional menus are found across the top of the screen. Familiarize yourself with the **Construct** and **Measure** menus.

a. Construct a segment with the line tool; notice that the line is highlighted until you click on some other location on the screen. While the line is highlighted, left click on the **Construct** menu. Notice that some menus are active (in black) and some are not active (in gray). Move the cursor to one of the construction items in the list. Notice the information displayed in the lower left corner of the screen. This information tells you what objects need to be selected before the tool can be used to perform the desired construction. This information is invaluable when you need a reminder in the future. Construct a midpoint to the segment.

b. Select the segment and midpoint from item 2a. Construct a line perpendicular to the segment through the point.

c. Construct and select two points. (If you hold down the Shift key on your keyboard while constructing the points, they will stay highlighted and are selected in the order you constructed them.) Left click on the **Construct** menu and construct a circle. Notice which point is the center and which point determines the length of the radius. Notice also how the information in the lower left hand corner of the screen changes at each step.

d. Select a segment. Left click on the **Measure** menu. Notice the active items in the **Measure** menu. Measure the length of the segment and determine its slope. What happened when you asked for the slope?

e. Left click on the **Measure** menu again. Determine what objects need to be selected in order to find an angle measure. Construct the objects, select the necessary objects. Notice the order in which the objects were selected. Measure the angle. Which point is the vertex of the angle?

Basic Constructions

Choose New Sketch in the **File** menu before beginning each of the constructions in items 3–7 below.

3. a. Construct an angle.

b. Use the arrow tool to select the angle and its vertex.

c. Use the **Edit** menu to copy and paste the angle. While the new angle is highlighted, use the arrow tool to drag it to a new position on the screen.

d. Use the **Measure** menu to determine the measure of both angles that now appear on the screen.

e. What relationships exist between the angles?

f. Now drag on one of the points or sides of one of the angles. What happens?

g. Will the technique in item 3c guarantee a pair of angles that stay congruent?

4. Construct a segment. Repeat items 3b–3f, replacing "angle" with "segment."

5. a. Construct $\angle ABC$. Label the points using the labeling tool.

b. Select points B and C in that order. Left click on the **Transform** menu and choose **Mark Vector**. (A vector from B to C is constructed.)

(*continued*)

TEACHING AND LEARNING HIGH SCHOOL MATHEMATICS © 2010 John Wiley & Sons, Inc.

c. Select $\angle ABC$. Left click on the **Transform** menu and choose **Translate**. Label the new angle $\angle ECF$. Select the labeling tool, double click on a point, and change the automatic label in the dialogue box that appears.

d. Measure both angles. Use the arrow tool to move any point or side of $\angle ABC$. How are $\angle ABC$ and $\angle ECF$ related?

6. a. Construct $\overline{AB}$. Construct $\overleftrightarrow{CD}$.

   b. Double click on $\overleftrightarrow{CD}$. This action marks the line as a line of reflection.

   c. Select $\overline{AB}$, including its endpoints. Left click on the **Transform** menu and choose **Reflect**. Note how the reflection points are labeled.

   d. Measure the preimage, $\overline{AB}$, and its image, $\overline{A'B'}$.

   e. Use the arrow tool to move any point on the preimage. How are the preimage and image related?

7. a. Construct an angle and a point Q not on the angle.

   b. Double click on point Q. This action marks the point as a center of rotation.

   c. Select the angle. Left click on the **Transform** menu and choose **Rotate**. Rotate the angle 70° counterclockwise. Note the direction associated with the sign of the rotation (+ or −) and how the rotated points are labeled.

   d. Measure both the preimage and image.

   e. Use the arrow tool to move any point on the preimage. How are the preimage and image related?

4. *More Constructions with Interactive Geometry Software*. Complete the student page Mystery Constructions. Justify your responses. Prove some of them.

Mystery Constructions

Using *The Geometer's Sketchpad*, follow the directions. Which geometric figures are constructed in each case?

1. a. Construct $\angle ABC$. Label the points using the labeling tool.

   b. Select points B and C in that order. Left click on the **Transform** menu and choose **Mark Vector**. A vector from B to C is constructed.

   c. Select $\angle ABC$. Left click on the **Transform** menu and choose **Translate**. Label the new angle $\angle ECF$.

   d. Select point C and $\overline{BC}$. Use the **Construct** menu to construct a perpendicular line to $\overline{BC}$ through C.

   e. **Reflect** $\angle ECF$ across the perpendicular line. Label the reflection of point E as point D. Construct $\overline{AD}$. Drag a point or segment to make $ABCD$ a convex quadrilateral.

   f. What type of quadrilateral is $ABCD$? How do you know? Use the arrow tool to drag one or more of the vertices to test whether your conjecture is true.

   g. What relationships between sides or angles were used to construct this quadrilateral?

2. a. Construct $\overline{AB}$. Label the points.

   b. Construct point C not on $\overline{AB}$.

(*continued*)

TEACHING AND LEARNING HIGH SCHOOL MATHEMATICS © 2010 John Wiley & Sons, Inc.

c. Construct a line through C parallel to $\overline{AB}$.

d. Construct a point on the new line and then construct and label convex quadrilateral $ABCD$.

e. Repeat 1f and 1g for this construction.

3. a. Construct a segment. Label it $\overline{AB}$.

b. Construct a circle whose center is at A with point B on the circle.

c. Construct a circle whose center is at B with point A on the circle.

d. Construct another segment, with one endpoint at A and the other endpoint at D on the circle constructed in item 3b.

e. Construct a circle whose center is at D with point A on the circle.

f. Construct the intersection of the circles with centers D and B. (Highlight both circles and construct their intersection using the **Construct** menu.) Label this point C.

g. Construct $\overline{CD}$ and $\overline{BC}$.

h. Repeat 1f and 1g for this construction.

4. a. Construct segment $\overline{AC}$. Find its midpoint M. Label the points.

b. Construct a circle with center M and radius $\overline{AM}$.

c. Construct a line through point M that does not coincide with $\overline{AC}$.

d. Construct the intersection points of the line with the circle. Label the points B and D.

e. Construct quadrilateral $ABCD$.

f. Repeat 1f and 1g for this construction.

5. *Line and Rotational Symmetry for Quadrilaterals.* A quadrilateral has **line symmetry** if it coincides with itself when reflected across a line. A quadrilateral has **rotational symmetry** of degree d about point A if the quadrilateral coincides with itself when it is rotated d degrees about point A. (Because all shapes have $360°$ rotational symmetry, this degree rotation is ignored.)

   a. For each of the quadrilaterals studied in this lesson (trapezoid, isosceles trapezoid, parallelogram, square, rectangle, kite, and rhombus), find all symmetry lines that exist.

   b. For each of the quadrilaterals studied in this lesson that has rotational symmetry, find the number of degrees of rotational symmetry.

6. *Diagonals of Quadrilaterals.* Complete the activities on the student page Exploring the Diagonals of Quadrilaterals.

TEACHING AND LEARNING HIGH SCHOOL MATHEMATICS © 2010 John Wiley & Sons, Inc.

Exploring the Diagonals of Quadrilaterals

Materials

- strips (paper or plastic) to use as diagonals, 2 short strips and 1 long strip per pair of students
- paper fasteners to connect the strips, 1 per pair of students

1. Experiment with the diagonals of quadrilaterals using the materials provided. Position the diagonals and then connect their endpoints to create a quadrilateral. Try at least three different arrangements for each diagonal relationship listed. Determine which quadrilaterals must always satisfy the diagonal relationships indicated for each cell of the table. List the quadrilaterals in the appropriate cells.

Relationships between Diagonals	Quadrilaterals whose Diagonals Might or Might Not Be Congruent	Quadrilaterals whose Diagonals Must Be Congruent
Diagonals bisect each other.		
Diagonals are perpendicular.		
Diagonals are perpendicular and one bisects the other.		
Diagonals are perpendicular bisectors of each other.		
Diagonals of *ABCD* intersect in point *E* such that $\frac{AE}{EC} = \frac{BE}{ED}$.		

2. Characterize the relationship between the diagonals for each of the following quadrilaterals. Write a definition for each of the quadrilaterals based on the relationships between their diagonals.

Quadrilateral	Definition
Trapezoid	
Isosceles Trapezoid	
Kite	
Parallelogram	
Rectangle	
Rhombus	
Square	

3. a. From the relationships between the diagonals of the quadrilaterals, which categories of quadrilaterals belong to other categories? Explain briefly.

b. Create a diagram showing a hierarchy of quadrilaterals.

TEACHING AND LEARNING HIGH SCHOOL MATHEMATICS © 2010 John Wiley & Sons, Inc.

Problems 7 through 10 are based in part on work by Craine and Rubenstein (1993), who suggest a hierarchy to synthesize and clarify relationships among quadrilaterals. Their hierarchy carefully uses rows and links to emphasize particular relationships. Here is the hierarchy, using names of quadrilaterals.

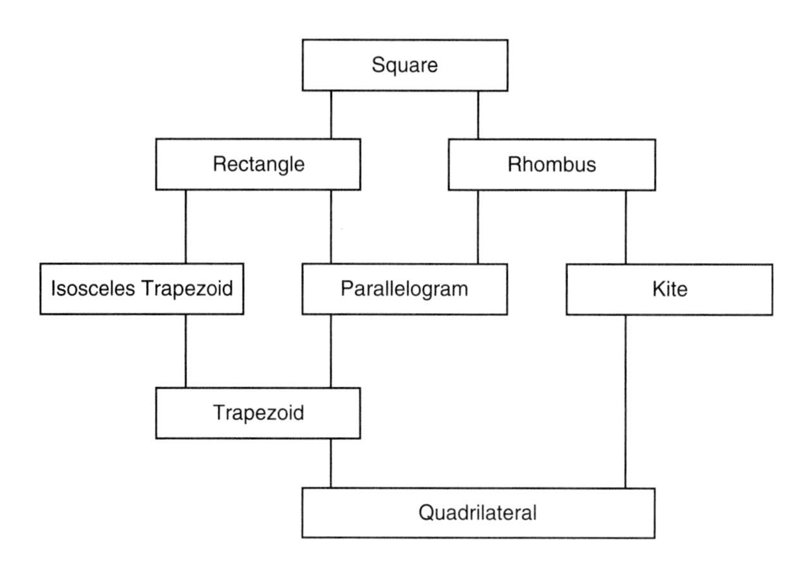

7. *A Quadrilateral Hierarchy.*

a. What definition of trapezoid does this hierarchy assume?

b. Redraw the hierarchy. Make it larger and include a figure that is a general example of each type. Make four copies of this image.

c. On one copy of the hierarchy, beside each figure write characteristics that define the figure. How are defining properties "inherited" from one generation to the next?

d. On one copy of your hierarchy, draw the diagonals for each figure. How are properties of diagonals inherited across generations?

e. On another copy of the hierarchy, show the symmetries of each figure. Explain how these are inherited.

f. Create a Venn diagram showing the interrelationships between the figures. Make sure there are no empty regions in your diagram. You will likely need to use shapes other than circles or ovals to create your Venn diagram.

8. *Coordinatizing with the Hierarchy.* Find a simple way to coordinatize each quadrilateral. For example, if possible, place one vertex at the origin and one side on the positive x-axis. Then determine the other coordinates using as few variables as possible. (You may want to refer back to your work in Lesson 2.1, **DMU** problem 6.) How do the rows of the hierarchy correlate with the number of variables needed to coordinatize each quadrilateral?

9. *Hierarchy Investigations.* Investigate other relationships suggested by the hierarchy. For example, there are dualities between angles and sides: *The kite has one pair of congruent angles, and the isosceles trapezoid has one pair of congruent sides.*

10. *Hierarchy Gap.* There seems to be a missing figure on the hierarchy that might fall between the kite and the general quadrilateral. What properties would such a figure have? (Revisit your solutions to problems 7–8.) What are some possible figures that would satisfy this location?

TEACHING AND LEARNING HIGH SCHOOL MATHEMATICS © 2010 John Wiley & Sons, Inc.

DEVELOPING MATHEMATICAL PEDAGOGY

Reflecting on Our Own Learning

11. Consider your own study of quadrilaterals in high school or college geometry courses.

a. Compare the manner in which you learned about constructing quadrilaterals and their properties with the approach taken in this lesson.

b. What are some pros and cons of an approach to learning in which students create their own understanding rather than the teacher sharing what is known about a topic?

c. Different tools may appeal to students of varying ability levels. How might the use of different tools play a part in creating an equitable learning environment?

Preparing for Instruction

12. Locate several current high school geometry textbooks. How is trapezoid defined in these books? (Refer to **DMU** problem 2 for two competing definitions.)

13. In **DMU** problem 2, two different definitions for a trapezoid were suggested. These definitions have different consequences for which theorems arise and how these theorems are linked. Suppose one uses the inclusive definition: *A trapezoid is a quadrilateral with at least one pair of parallel sides.*

a. Consider this definition: *An isosceles trapezoid is a trapezoid with congruent base angles.* For the inclusive definition of trapezoid:

i. Is a rectangle an isosceles trapezoid? Justify your answer.

ii. Is a parallelogram an isosceles trapezoid? Justify your answer.

b. Consider the following theorem: *An isosceles trapezoid has congruent diagonals.*

i. Prove this theorem.

ii. Assume an inclusive definition of trapezoid and the theorem proved in problem 13b.i. Is it necessary to prove that a rectangle has congruent diagonals? Why or why not?

iii. If you use the exclusive definition of trapezoid, is it necessary to prove that a rectangle has congruent diagonals? Why or why not?

c. In calculus, integrals often use trapezoids as a step to develop the concept of area under a curve. Under what conditions, if any, would these trapezoids be rectangles?

d. What are some advantages of an inclusive definition of trapezoid?

14. With a friend, complete the activities on the **What Am I?** student page.

a. Determine what prerequisite knowledge is needed for students to be successful.

b. Middle school students classify quadrilaterals by their sides and angles. They might also begin to investigate the diagonals of quadrilaterals. For what purpose might **What Am I?** be used with high school students?

c. Expected student responses and possible teacher support are provided for the **What Am I?** game. Complete the missing parts of the **QRS Guide for What Am I?**, incorporating expected student responses and teacher support into the guide. Keep the Stages of Questioning in mind as you plan teacher support for students' errors and misconceptions. (A partially completed **QRS Guide for What Am I?** is found in the Unit Two Appendix.)

TEACHING AND LEARNING HIGH SCHOOL MATHEMATICS © 2010 John Wiley & Sons, Inc.

What Am I?

Materials

- **Property Cards**, one deck per person
- **Quadrilateral Cards,** one deck per team

Game Objective

Determine each quadrilateral when a minimal set of property cards is shown, one at a time.

Preparation

Play as a team of 2–4 players. Shuffle and deal all of the Quadrilateral Cards so that each person has approximately the same number of cards. Quadrilaterals listed are: trapezoid, isosceles trapezoid, parallelogram, rectangle, rhombus, kite, and square. Each player should also have a full deck of Property Cards.

Choose a team leader for the first round. For each subsequent round, the new team leader is the person immediately to the left of the previous team leader.

Game Rules

- Choose a Quadrilateral Card from the cards dealt.
- Choose a minimal set of Property Cards from your deck that defines the quadrilateral chosen. Do not use too many properties and do not show others the properties you have chosen.
- The team leader presents his or her property cards one at a time, challenging team members to determine the quadrilateral using only the properties shown.
- Team members eliminate possibilities for the quadrilateral based on the properties shown by their team leader. Random guessing is discouraged. The person who determines the correct quadrilateral keeps the Quadrilateral Card.
- Play continues until all Quadrilateral Cards have been played, with students rotating the team leader in each round.
- The winner is the person who correctly determines the most quadrilaterals by the end of the game.

1. Play the game What Am I? until all of the Quadrilateral Cards have been used.

2. Use the results of the game to define each of the quadrilaterals based on relationships among their sides or angles. Record your work.

Quadrilateral	Definition
Trapezoid	
Isosceles Trapezoid	
Kite	
Parallelogram	
Rectangle	
Rhombus	
Square	

3. Share your definitions with another group. Note any differences.

TEACHING AND LEARNING HIGH SCHOOL MATHEMATICS © 2010 John Wiley & Sons, Inc.

Cards for What Am I?

Quadrilateral Cards

Trapezoid	Isosceles Trapezoid	Kite	Rhombus
Parallelogram	Rectangle	Square	

Property Cards

At least one pair of parallel sides	Two distinct pairs of adjacent congruent sides	Equilateral
Two pairs of parallel sides	At least one pair of consecutive supplementary angles	Equiangular
At least one pair of opposite congruent sides	At least one pair of opposite congruent angles	Congruent base angles
Two pairs of opposite congruent sides	Two pairs of opposite congruent angles	Two distinct pairs of consecutive supplementary angles

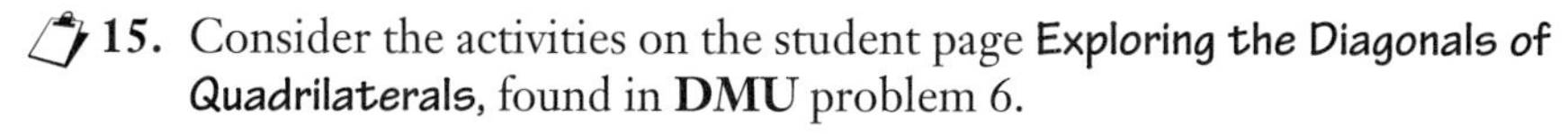

15. Consider the activities on the student page **Exploring the Diagonals of Quadrilaterals**, found in **DMU** problem 6.
   - **a.** Determine what prerequisite knowledge is needed for students to be successful.
   - **b.** Try to predict how high school students will respond to each activity.
   - **c.** Expected student responses and possible teacher support are provided for the first problem of the student page **Exploring the Diagonals of Quadrilaterals**. Complete the missing parts of a **QRS Guide**, incorporating expected student responses and teacher support into the template. Keep in mind the Stages of Questioning as you plan teacher support for students' errors and misconceptions. (A partially completed **QRS Guide for Exploring the Diagonals of Quadrilaterals** is found in the Unit Two Appendix.)

16. **a.** Complete the student page **Exploring Quadrilaterals and Their Definitions**.
   - **b.** Some of the concepts explored on this student page go beyond those studied by most high school students. Which concepts might be appropriate for all high school students? Which might be more appropriate for advanced students?
   - **c.** How might the design of the activity make the concepts accessible to a wide range of students?

   - **d.** Choose one quadrilateral. Use a blank copy of a **QRS Guide** in the Unit One Appendix to record expected student responses and possible teacher support as students work on the activity for the selected quadrilateral.

TEACHING AND LEARNING HIGH SCHOOL MATHEMATICS © 2010 John Wiley & Sons, Inc.

Exploring Quadrilaterals and Their Definitions

Materials

- *The Geometer's Sketchpad*™

1. Choose a number corresponding to a column in the table. Each team member must choose a different number. Indicate your choice.

Team member 1	Team member 2	Team member 3	Team member 4
rectangle	isosceles trapezoid	trapezoid	isosceles trapezoid
kite	square	rhombus	parallelogram

2. Focus on one of the quadrilaterals in the column you chose.

a. Using *The Geometer's Sketchpad*™ software, **construct** the quadrilateral from its **side definition** or **angle definition**. Save your constructions!

b. Move vertices or other parts of the quadrilateral to show that your construction allows you to form more specific quadrilaterals and will not allow less specific ones.

c. What other quadrilaterals can be formed from your construction?

d. Measure sides, angles, and diagonals. What relationships stay the same when vertices are moved?

e. Determine symmetries.

f. What properties does your quadrilateral seem to have?

3. Repeat problems 2a–2f to construct the same quadrilateral using its **diagonal definition**.

4. Repeat problems 2 and 3 for the second quadrilateral.

5. Compare your constructions with other persons who constructed the same quadrilaterals. If your construction was incorrect, reconstruct the quadrilateral.

Homework

6. For each of the quadrilaterals you investigated, use triangle congruence theorems and other Euclidean geometry axioms, theorems, and definitions to prove that your quadrilateral definitions are equivalent. Use definitions that do not rely on parallel lines.

a. Start with the side/angle definition and prove that you get the properties required in the diagonal definition.

b. Start with the diagonal definition and prove that you get the properties required in the side/angle definition.

c. Conclude that the definitions are equivalent or determine that the definitions do not result in the same quadrilaterals.

7. Repeat the investigation in problems 2–6 for the remaining quadrilaterals. Investigate sides, angles, diagonals, and symmetries. Compare your results with your team members.

TEACHING AND LEARNING HIGH SCHOOL MATHEMATICS © 2010 John Wiley & Sons, Inc.

17. Students live in a geometric world. To help them realize how often quadrilaterals (or other geometric figures) appear, it can be useful to have them go on a Scavenger Hunt.

a. During your visit to one of the schools for your observations, take a walk around the school. Make note of different quadrilaterals and their functions.

b. How would you engage students in a Quadrilateral Scavenger Hunt?

c. How might you modify the Quadrilateral Scavenger Hunt if you are not able to send your students exploring the school grounds?

Teaching for Equity

18. In high school classrooms, some students often finish an activity ahead of most of the class. It is helpful to plan extensions of activities for such students. One such extension is provided in the student page **Quadrilaterals on Spheres.**

a. Complete the extension problems.

b. What challenges do you expect students to face when completing these problems?

c. Do you think these challenges are beyond the reach of most high school students? Provide a rationale for your decision.

d. What more do students learn when completing the extension problems?

Quadrilaterals on Spheres

1. Repeat problems 2–4 of the student page **Exploring Quadrilaterals and their Definitions** to construct the same two quadrilaterals on a sphere. (Stretch a string as tight as you can along the surface of the sphere to make a line segment.)
2. For the quadrilaterals you constructed:
   a. When constructed on a sphere, do the resulting quadrilaterals have the same relationships between sides, angles, and diagonals as they do in the plane?
   b. Which of the properties of the quadrilateral still hold when you construct it on a sphere? Explain.
   c. Which properties of the quadrilateral no longer hold when you construct it on a sphere? Explain.
3. Which relationships seem easier to use to construct quadrilaterals on different surfaces (plane or sphere)? Why do you think this is true?

Materials

- small spherical balls at least the size of a baseball, one per team
- string, cut to different lengths
- compass, one per team

Analyzing Students' Thinking

19. Here are some student responses when writing quadrilateral definitions. How should the teacher support a student who gives each response? Add these responses and teacher support to the **QRS Guide for Exploring the Diagonals of Quadrilaterals** found in the Unit Two Appendix.

a. *A kite is a quadrilateral with at least one pair of opposite congruent angles.*

b. *A kite is a quadrilateral with two distinct pairs of congruent adjacent sides.*

c. *A rhombus is equilateral.*

d. *A parallelogram has two pairs of parallel sides.*

e. *A square is an equilateral equiangular quadrilateral.*

TEACHING AND LEARNING HIGH SCHOOL MATHEMATICS © 2010 John Wiley & Sons, Inc.

f. *A rectangle is equiangular.*

g. *A rectangle is a quadrilateral with 2 pairs of parallel sides and congruent base angles.*

20. One common "definition" for a rectangle that many students have from elementary school is a quadrilateral with four right angles and two long sides and two short sides. How does such a definition influence students' perception of the relationship between a square and a rectangle?

Communicating Mathematically

21. Word origins sometimes help students make sense of mathematics vocabulary. For example, students who know that *quad* means *four* have a start on understanding the meaning of the word *quadrilateral*. (More information about word origins can be found in "Word Histories: Melding Mathematics and Meanings" by Rheta N. Rubenstein and Randy K. Schwartz. (*Mathematics Teacher*, 93 (November 2000): 664–669).)

a. Research the word origins of familiar (e.g., pentagon, hexagon) or other geometry terms (e.g., perpendicular).

b. What strategies might teachers use to incorporate prefixes and suffixes into mathematics instruction on a regular basis?

Reflecting on Professional Readings

WWW **22.** Read "Learning Mathematics: Perspectives from Researchers" by Rheta N. Rubenstein found on the course website.

a. How have your experiences in this course or this lesson modeled some of the major ideas from research about learning mathematics?

b. What does research say about the use of materials in teaching mathematics? Cite specific researchers or theorists.

c. What did you learn from this reading that clarified an idea about students' learning from previous study, possibly in other education courses?

d. What did you learn from this reading that challenged or questioned some previous ideas about students' learning? What are you now thinking about that idea?

23. Read "Using Conjectures to Teach Students the Role of Proof" by Rhonda L. Cox (*Mathematics Teacher*, 97 (January 2004): 48–52).

a. Reflect on your own ability to write proofs. What have been some of the challenges and rewards for you in developing your proof-writing ability?

b. Consult at least three geometry textbooks. Scan the books to gain a sense of how they develop students' ability to write proofs. Summarize your findings.

c. How does Cox use conjecturing (inductive reasoning) to motivate proof (deductive reasoning)?

d. Summarize the benefits and drawbacks of using a unit as described by Cox.

2.4 Using Physical Tools and Technology: Circles

Students today bring to school a wide range of experiences and ability levels. Although some high school students are capable of thinking abstractly about mathematical concepts, many students benefit from experiences with concrete materials or visual aids to provide a bridge to abstract ideas. Mathematics, like science, is effectively learned when students work with a variety of hands-on materials or tools to experiment, make observations, and then make conjectures.

TEACHING AND LEARNING HIGH SCHOOL MATHEMATICS © 2010 John Wiley & Sons, Inc.

Tools—such as geoboards, MIRAs, patty paper, compasses, and straightedges—enable students to use their visual and kinesthetic senses while they explore important geometric concepts. Interactive geometry drawing software on computers and graphing calculators is also an important tool to explore significant geometric concepts; such software enables students to use their visual and numeric senses to make connections among geometric ideas. Technology offers opportunities to experiment with mathematical ideas in ways not possible in the past. With technology, what was once the goal of a problem—to construct a geometric object—is now just its beginning. Students are able to construct the object, drag points or segments, and observe what stays constant and what changes in the construction. From their observations, they may conjecture relationships and then determine some way to prove that what they are seeing on the screen is always true.

Teachers are rightly concerned about management issues when using concrete materials, such as geoboards, or technological tools, such as interactive geometry drawing software. When teachers use such tools, students often need a few minutes of free time to explore so that their need to play with the tools is satisfied and they are ready to focus on their mathematical use. Issues of safety, health, and appropriate behavior must also be addressed; for example, eating reused uncooked spaghetti or shooting rubber bands should be discouraged, as should using technological tools to instant message friends, read email, or surf the Web.

A second issue that arises during explorations with physical and technological tools is how to make efficient use of the tools to explore many aspects of a concept. Rather than have each student investigate every aspect of a particular concept, it is sometimes helpful to share the learning among class members. One way to do this is through the cooperative learning strategy *Jigsaw* (Davidson, 1990). Each member of a home team becomes an expert on a single part of a task and then shares what he or she has learned with the other members of his or her team.

Jigsaw

1. Each member of your regular or *home* team is assigned a topic on which to become an expert.
2. Students from *home* teams who are assigned the same topic meet in an *expert* group to discuss and master the topic and plan how to teach it to their *home* group members.
3. All students return to their *home* teams and teach what they have learned to their team members.
4. Each individual is accountable for learning all of the material.

(Davidson, 1990)

In the case below, each team member chooses a student page to study and joins an expert group whose role is to become experts on that student page. Once each expert group has completed its assigned tasks, home teams reconvene and team members each share what they have learned. Often, home teams complete a summary activity that incorporates all the major ideas studied in expert groups. When using *Jigsaw* with a large class, the instructor may need to split expert groups so that three to four students work together in each group.

In this lesson, you will:

- Investigate the geometry of circles.
- Reflect on benefits and management issues related to the use of physical and technological tools to explore geometric concepts.

TEACHING AND LEARNING HIGH SCHOOL MATHEMATICS © 2010 John Wiley & Sons, Inc.

- Experience the cooperative learning technique *Jigsaw* and reflect on its benefits.

Materials

- circular geoboards and rubber bands or circular geoboard templates, one per pair (see the Unit Two Appendix)
- compass and straightedge, one per person
- protractor or goniometer (angle measuring tool), one per pair
- flat, circular objects from a half-inch to three inches in diameter (e.g., coins, bingo or poker chips, buttons, washers), 5 objects of different diameters per group
- MIRA, one per group
- patty paper, several sheets per group
- interactive geometry drawing software on a computer or graphing calculator, one per group

Website Resources

- Cooperative Learning powerpoint
- **Angles, Arcs, and Chords in Circles** student page
- **Special Triangles on a Geoboard** student page
- **Finding Tangent Circles** student page
- **Circles and Tangent Lines** student page
- Circular Geoboard Template
- **Sprinkling around Obstacles** student page
- **QRS Guide**

LAUNCH

Though circles are important geometric shapes, too often their study is given little attention.

1. Think of several circular or partially circular objects. Why do you think circles are used for these objects?
2. Complete the student page **Angles, Arcs, and Chords in Circles** to remind you of some of the relationships you might have learned about circles. Be prepared to share your responses with the class.

TEACHING AND LEARNING HIGH SCHOOL MATHEMATICS © 2010 John Wiley & Sons, Inc.

Angles, Arcs, and Chords in Circles

Use the circular geoboard template (provided at right) or a circular geoboard to complete the following.

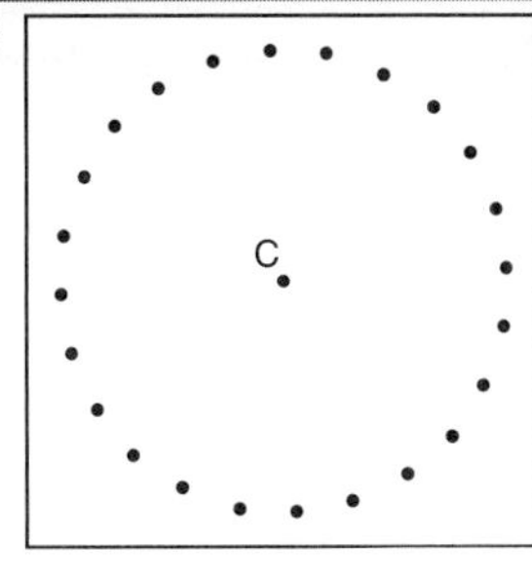

Materials

- circular geoboard with rubber bands or circular geoboard template with straightedge, one per pair
- protractor or goniometer (angle measuring tool), one per pair

1. A **central angle** is an angle whose vertex is the center of a circle. Draw a central angle. Name it and measure it.
2. A **minor arc** of a circle is the smaller of the two arcs formed by a central angle. A minor arc is usually named by its endpoints and another point on the arc, such as $\overset{\frown}{AXB}$. Name the minor arc associated with the central angle from item 1.
3. The measure of a minor arc is the same as the measure of the central angle that intercepts it. What is the measure of the minor arc in item 2?
4. A **chord** is a line segment whose endpoints are on the circle. Draw a chord whose endpoints share the endpoints of the minor arc in item 2.
   - **a.** Name the chord.
   - **b.** What do you notice about the triangle that is created by the endpoints of this chord and the center of the circle?
   - **c.** State all of the relationships that you know exist in the type of triangle created in item 4b.
5. Use some of the information you found in problem 4c to determine the center of the circle at right. Show your work and explain why your process always works.

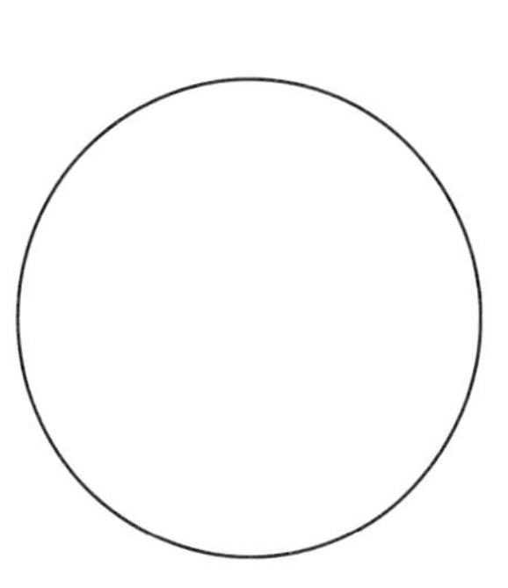

EXPLORE

In the **Launch**, you determined several uses of circles in your daily life. In order to use the relationships arising from circles to solve problems, it is helpful to know the relationships among special angles, triangles, and tangent lines.

3. Use the *Jigsaw* cooperative learning technique so that each member of your home team joins a different expert group to complete one of the student pages:

   Special Triangles on a Geoboard

   Finding Tangent Circles

   Circles and Tangent Lines

   Complete the student page in your expert group. Be prepared to share what you learn about circles with your home team. Once back in your home team, you will use what each of you learn to solve a problem.

TEACHING AND LEARNING HIGH SCHOOL MATHEMATICS © 2010 John Wiley & Sons, Inc.

Special Triangles on a Geoboard

Use a circular geoboard or a circular geoboard template to complete this activity.

Materials

- circular geoboard with rubber bands or circular geoboard template with straightedge, one per pair
- protractor or goniometer (angle measuring tool), one per pair

1. a. Create a triangle on a circular geoboard with one vertex at the center of the circle and the other two vertices on the circle.

b. What is true about all such triangles?

2. a. Create a triangle on a circular geoboard, one side of which is a diameter of the circle.

b. What seems to be true about the triangle?

c. Repeat items 2a and 2b for several more triangles. Write a conjecture.

d. Prove your conjecture from problem 2c. (Hint: Dissect the triangle into two triangles with central angles and use your observations from problem 1b.)

3. a. Create a central angle $\angle ACB$ with points A and B on the circle. Create $\angle AVB$ whose vertex, V, is also on the circle so that the angle intercepts the same arc as $\angle ACB$. Measure both angles.

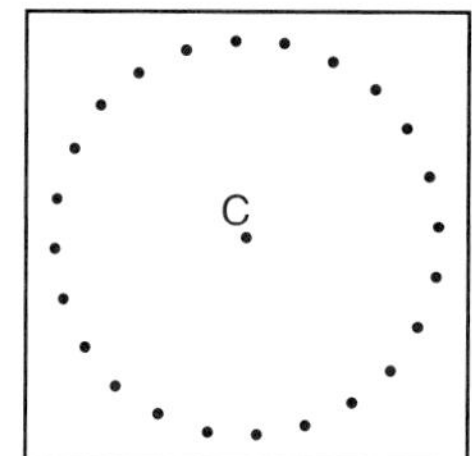

b. What relationship seems to exist between the measures of these angles?

c. Repeat the construction in problem 3a using the same central angle but other choices for the vertex on the circle. How do the angle measures relate to each other?

d. Repeat the construction in problem 3a for several other pairs of related angles. How do the angle measures relate to each other?

e. From your work in items 3a–3d, write a conjecture about the relationship between the measure of the angle whose vertex is on the circle and the measure of the central angle when these angles intercept the same arc. Prove your conjecture; you will need to consider three cases. For the first case, let one side of the triangle be a diameter.

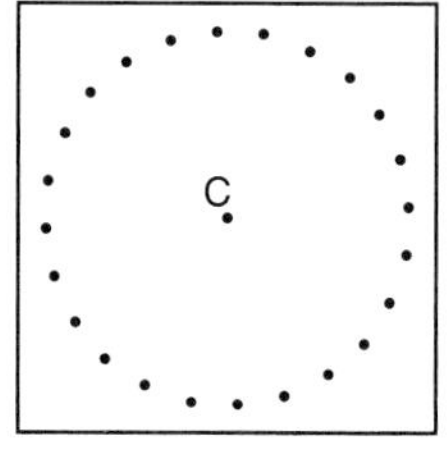

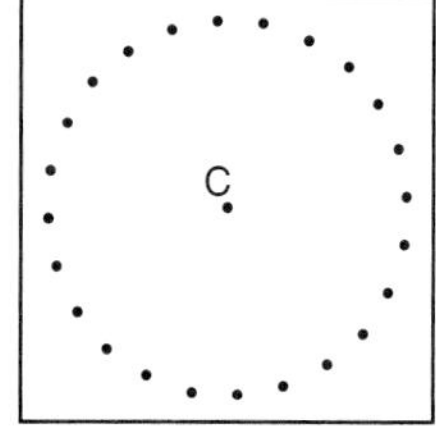

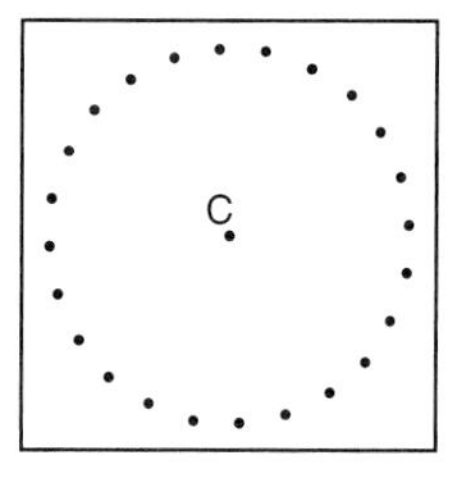

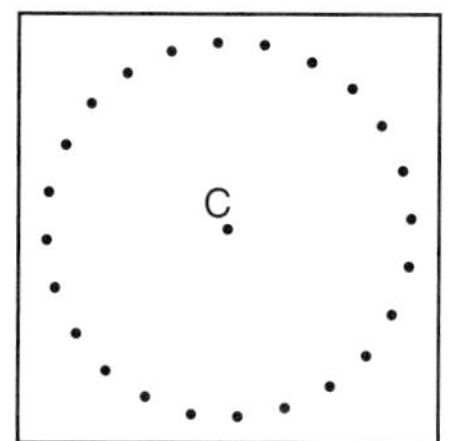

f. Recall that the measure of the minor arc is the same as the measure of the central angle that intercepts it. Write your conjecture from problem 3e in terms of the intercepted arc the two angles have in common.

4. Revisit problem 2. Do your results in problem 2 agree with your results from problem 3? Why or why not? Explain any similarities or differences.

TEACHING AND LEARNING HIGH SCHOOL MATHEMATICS © 2010 John Wiley & Sons, Inc.

Finding Tangent Circles

Use the materials provided and the figure below to investigate angles whose rays are both tangent to the same circle.

Materials

- compass, straightedge, and MIRA, one per pair
- flat circular objects from a half-inch to three inches in diameter (e.g., coins, bingo or poker chips, buttons, washers), 5 objects of different diameters per group

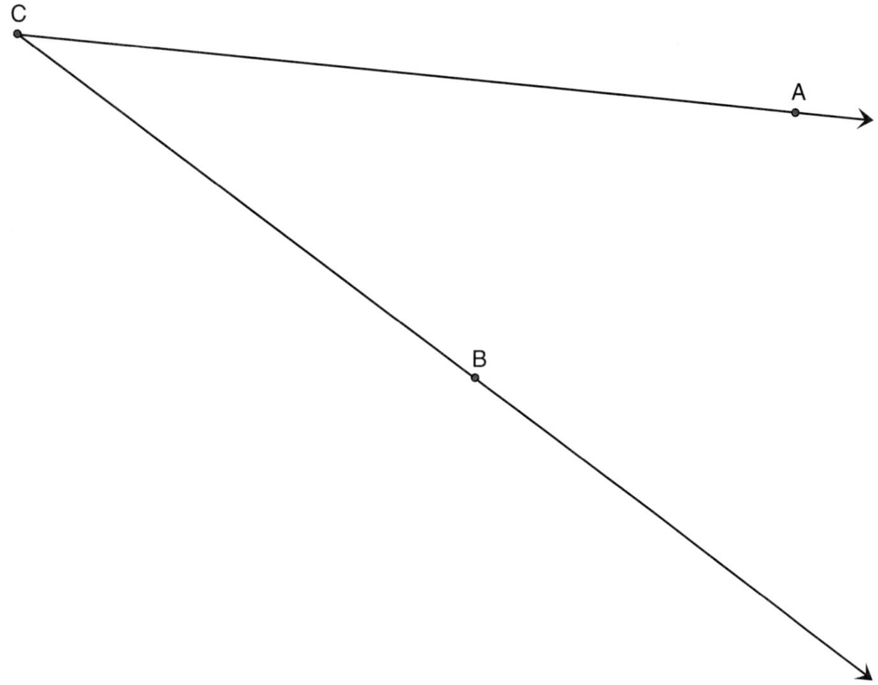

C

A

B

1. Can you construct a single circle that is tangent to the rays of $\angle ACB$ at both points A and B? If so, show such a circle. If not, explain why not.
2. One at a time, position and trace five circular objects with different diameters so that each one is tangent to both rays of the angle.
   - **a.** Construct the center of each circle.
   - **b.** What conjectures can you make about the relationship between the center of a circle and an angle whose rays are both tangent to the circle?
3. For each circle, mark the points of tangency and connect them to the center of the circle.
   - **a.** For a single circle, what conjectures can you make about the relationships between the points of tangency and the vertex of $\angle ACB$?
   - **b.** What conjectures can you make about the measure of the angle created between the radius and the tangent line at the point of tangency?
4. **a.** How can you construct the center of a circle inscribed in a triangle?
   - **b.** How can you determine the length of a radius of a circle inscribed in a triangle? (What is the shortest distance between a line and a point not on the line?)
   - **c.** Construct a triangle and its inscribed circle.
5. Repeat problem 4 for a square.
6. For what quadrilaterals can a circle be inscribed (all sides of the quadrilateral must be tangent to the circle)? Explain your choice(s).

TEACHING AND LEARNING HIGH SCHOOL MATHEMATICS © 2010 John Wiley & Sons, Inc.

Circles and Tangent Lines

Materials

- compass and straightedge, one per person
- protractor or goniometer (angle measuring tool), one per person
- interactive geometry software (optional)

1. If you have access to interactive geometry software, try the following:

a. Construct a circle with center C and points T and W on the circle.

b. Construct $\overline{TW}$.

c. Measure $\angle CTW$.

d. Move point W along the circle until it appears to coincide with point T. What do you notice about the measurement of $\angle CTW$?

2. In the figure at right, a diameter and tangent to the circle intersect at point T. Point C is the center of the circle.

a. What is the measure of the arc intercepted by $\angle BTC$? What is the measure of the arc intercepted by $\angle ATC$? How do you know?

b. At what angle does the tangent line appear to intersect the diameter?

c. What is the relationship between the angle measure in item 2b and the intercepted arc measure in item 2a?

3. Consider the following:

a. Assume that $\angle CTB$ is not a right angle. Locate a point X on $\overleftrightarrow{AB}$ so that $\overline{CX} \perp \overleftrightarrow{AB}$. Why is this possible?

b. Find a point Y so that X is the midpoint of $\overline{TY}$. What do you know about the relationships between ΔCXY and ΔCXT?

c. What do the relationships in item 3b tell you about the assumption in item 3a?

d. At what angle must a tangent line intersect a diameter (or radius) of a circle?

4. a. How can the relationship you determined in problem 3d help you construct a tangent line to a circle from a point on the circle?

b. What additional information must you have?

c. Construct a tangent line to the circle at right through point T on the circle.

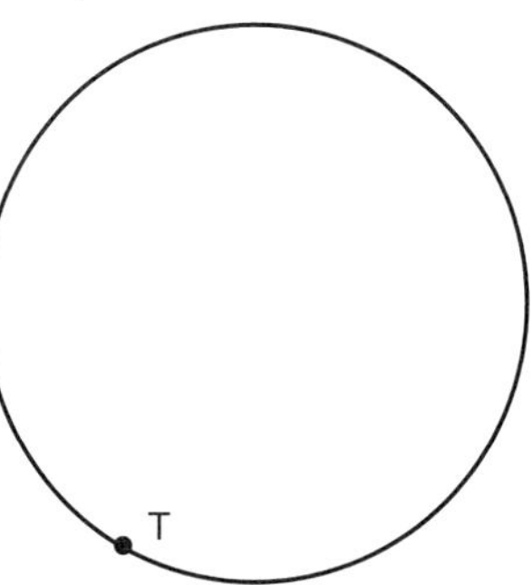

5. To find a tangent line to a circle from a point outside the circle, what additional information do you need?

4. While still in your expert group, take a few minutes to summarize what you learned about circle relationships.

SHARE

5. Return to your home team. Share what you learned while completing the student page in your expert group. What different information about circles did each team member learn?

TEACHING AND LEARNING HIGH SCHOOL MATHEMATICS © 2010 John Wiley & Sons, Inc.

EXPLORE MORE

6. Consider the student page, **Sprinkling around Obstacles**. The intent of the student page is to determine what students learned about circles from working on the student pages in the **Explore** by having them solve a problem in a real context. To solve the problem, students must employ several circle relationships. They have various tools available: MIRAs, patty paper, compass and straightedge, circular geoboards, and interactive geometry software. As you work on items 6a, 6b, and 6d, record your solution processes in a **QRS Guide**.
   - **a.** Use what you learned in your expert groups to solve the problems on the student page.
   - **b.** Solve the problems again using different physical or technological tools than you used for your first solution. (If you used a MIRA or patty paper, choose compass and straightedge or interactive geometry software for this problem.)
   - **c.** Compare differences in solutions based on the chosen tools.
   - **d.** Try to predict how high school students would solve the problems, assuming they had worked through the student pages in the **Explore**. Think about different approaches students might use with different tools.

Sprinkling around Obstacles

When designing an underground sprinkling system, the designer must accommodate obstacles like trees, large bushes, garden walls, etc. to ensure the sprinkling system waters the desired areas. The designer must consider where to position sprinkler heads to ensure the entire lawn is adequately covered while also minimizing the number of underground pipes and sprinkler heads.

Consider the diagram at right depicting part of a yard. A garden surrounds a large tree. A driveway restricts the location of sprinkler heads. Suppose a sprinkler head is positioned at point *S*, and the tree is positioned as shown.

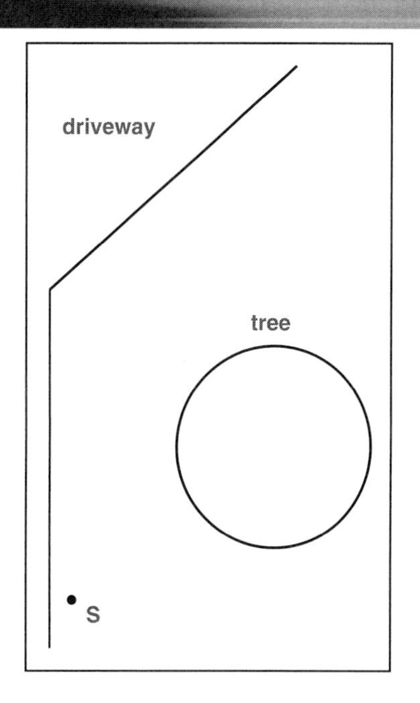

1. The circle represents the tree. Construct the center of the circle. Label the center *T*. Show and explain your process and describe how you know it works.
2. Construct tangent lines to the circle from the sprinkler head. Label the points of tangency *A* and *B*, respectively. Show and explain your process for finding the points of tangency and describe how you know your process works.
3. Relative to the context of the problem, what do the tangent lines represent?

TEACHING AND LEARNING HIGH SCHOOL MATHEMATICS © 2010 John Wiley & Sons, Inc.

7. Consider the **Sprinkling around Obstacles** student page, the solution processes you have used, and those you expect high school students to use. What questions should the teacher be prepared to ask to help students address their misconceptions or errors and make progress on the problem? (Recall the Stages of Questioning from Lesson 1.1.) Record teacher support in the **QRS Guide**.

SHARE AND SUMMARIZE

8. Share your responses to items 6 and 7 with the class.
9. When presented with problem 1 on the student page, **Finding Tangent Circles**, a student drew a "circle" that intersected both points. What questions would you like to ask to get the student to think more deeply about the problem?
10. **a.** How was the use of the *Jigsaw* cooperative learning strategy helpful in learning or reviewing many concepts about circle relationships?

    b. What might be some issues to consider when using *Jigsaw* in a high school mathematics classroom?
11. In this lesson, students began working with circles using nontechnological tools. They were allowed to use the tools if they wished, but were also given other choices. What are benefits and concerns of starting with physical tools rather than technological tools?

DEEPENING MATHEMATICAL UNDERSTANDING

1. *Major and Minor Arcs*.

    a. How can you find the measure of a major arc from the measure of its related minor arc?

    b. Using the pegs on a circular geoboard or points on the circular geoboard template, what are the measures of all the possible central angles you can make?
2. *Congruent Chords and Central Angles*. A *chord* is a line segment whose endpoints are on the circle.

    a. On a circular geoboard, create a chord. Create a second chord congruent to the one you just drew. Connect the endpoints of each chord to the center of the circle, forming two triangles. Measure the central angles for each triangle. What do you notice about these angle measurements?

    b. What does your result in problem 2a tell you about the measure of the arcs formed by congruent chords in the same circle?
3. *Cyclic Quadrilaterals*.

    a. Construct a circle and inscribe a quadrilateral in the circle (the vertices of the quadrilateral are on the circle). Such quadrilaterals are called *cyclic quadrilaterals*.

    b. What relationship exists between the measures of the opposite angles of a quadrilateral inscribed in a circle? Prove your conjectures. (Consider major and minor arcs created by opposite angles of the quadrilateral.)

    c. If the diagonals of a cyclic quadrilateral contain the center of the circle, then what relationship exists among the angles of the quadrilateral? Prove your conjecture. What relationships exist among the angles of the quadrilateral if only one of the diagonals contains the center of the circle?

    d. Which of the quadrilaterals in Lesson 2.3 are cyclic? Prove your conjectures.

    e. Under what conditions will kites be cyclic?
4. *Revisiting Sprinkling around Obstacles*. Consider the quadrilateral formed by connecting the points of tangency, A and B, with the center of the tree, T, and the sprinkler head, S.

WWW

TEACHING AND LEARNING HIGH SCHOOL MATHEMATICS © 2010 John Wiley & Sons, Inc.

a. Construct quadrilateral $SATB$ and diagonal $\overline{ST}$.

b. Determine any relationships among angles, sides, symmetries, etc. in quadrilateral $SATB$. Explain your findings.

c. What type of quadrilateral is $SATB$? How do you know?

5. *Proving Conjectures*. Revisit the student page **Finding Tangent Circles**. Prove the conjectures you made in problem 2b on the student page. You may use what you learned on other student pages to do so.

6. *Fermat Point*. In acute ΔFLY, the Fermat Point P, named after the famous mathematician Pierre Fermat, is the point for which the sum of the distances from each of the vertices to the point ($FP + LP + YP$) is a minimum.

   a. One way to construct the Fermat Point is to construct equilateral triangles on each of the sides of the original ΔFLY. Then for each of the equilateral triangles, construct the segment joining the opposite vertex of the original triangle and the opposite vertex of the respective equilateral triangle. For example, on side $\overline{FL}$, construct equilateral triangle ΔFLT and then construct segment $\overline{TY}$. Do the same on the other sides of the original triangle. The point of intersection of the three constructed segments is the Fermat Point. Use *The Geometer's Sketchpad*™ or another interactive geometry drawing tool to construct the Fermat Point for an acute scalene triangle. Show that your point satisfies the given condition.

   b. The Fermat Point is also the intersection point of the circumscribed circles to the three equilateral triangles constructed in problem 6a. Find the Fermat Point using this method. What other relationships exist in your construction? Justify your answers. (Reference: http://www.pballew.net/fermatpt.html, retrieved May 7, 2009).

DEVELOPING MATHEMATICAL PEDAGOGY

Preparing for Instruction

7. a. In what sports are circles found on the playing field?

   b. What circle relationships could be studied through the sports you identified?

8. Complete the student page **Finding Centers of Circles**. Comment on the amount of help provided to students in this student page as compared with the **Angles, Arcs, and Chords in Circles** student page in the **Launch** of this lesson.

TEACHING AND LEARNING HIGH SCHOOL MATHEMATICS © 2010 John Wiley & Sons, Inc.

Finding Centers of Circles

Circular objects have been used by humans for centuries. Pieces of these objects turn up at archeological digs. How might an archeologist determine the original size of a circular object?

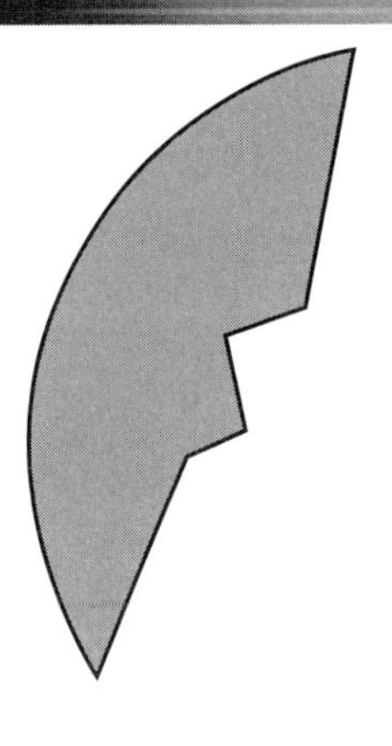

1. The figure at right represents a pottery shard found at an archeological dig. Draw the rest of the plate. Discuss how you determined the size of the circle that completes the plate.
2. A potter is interested in reconstructing the plate and needs its exact measurements. The thickness can be measured from the shard. How might the potter determine the diameter of the object from this piece?
3. Consider the circle at right. Its center is not provided.

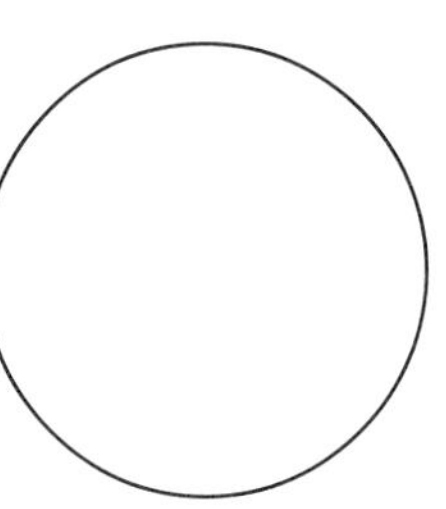

   a. Draw a chord.

   b. Construct the perpendicular bisector of the chord. Why is the perpendicular bisector important?

   c. Repeat the construction in 3a–b for another chord. What do you know about the point of intersection of the two perpendicular bisectors of the chords? Explain.
4. Revisit problem 2. Construct the circle using your work from problem 3.

Analyzing Students' Thinking

9. Study the **Sample Student Responses for Sprinkling around Obstacles** (students Abe, Bev, Colin, Darma, and Everest) found in the Unit Two Appendix. Answer the following questions for each student's response.

   a. Is the student's solution correct?

   b. What errors has the student made? Are the errors clerical or conceptual in nature? Clerical errors include careless errors that do not change the fundamental understanding of the problem, the difficulty of the problem, or the mathematics required to solve the problem (e.g., missing a label or sign, miswriting a number). Conceptual errors relate to misunderstanding the mathematical aspects of the problem.

   c. Find the **QRS Guide for Sprinkling around Obstacles** (see the Unit Two Appendix) which contains the task, some possible student responses, and some related teacher support. Incorporate the student responses you just analyzed into the **QRS Guide** along with any responses determined by your group during the **Share**. Write appropriate Teacher Support questions for the various student responses using the Stages of Questioning from Lesson 1.1 as a guide.

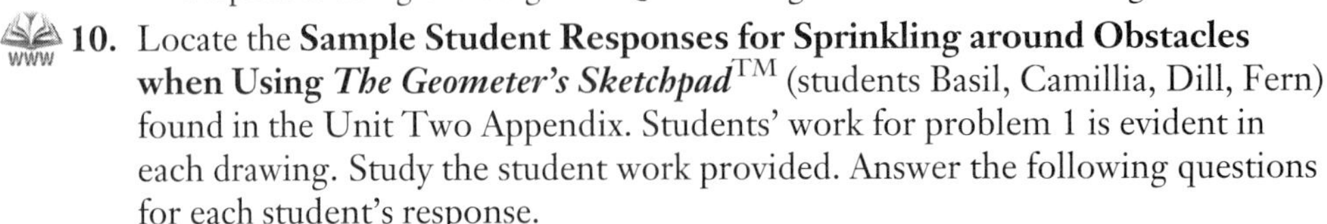

10. Locate the **Sample Student Responses for Sprinkling around Obstacles when Using *The Geometer's Sketchpad*™** (students Basil, Camillia, Dill, Fern) found in the Unit Two Appendix. Students' work for problem 1 is evident in each drawing. Study the student work provided. Answer the following questions for each student's response.

    a. Use the diagrams to explain how the student constructed the center of the circle.

    b. Is the student's solution for problems 1 and 2 correct?

TEACHING AND LEARNING HIGH SCHOOL MATHEMATICS © 2010 John Wiley & Sons, Inc.

c. What errors has the student made? Are the errors clerical or conceptual in nature?

d. What questions would you like to ask the student:

  i. To learn more about what he or she was thinking while solving the problem?

  ii. To help the student reconsider his or her work and get beyond conceptual difficulties?

  iii. To help the student monitor his or her work to correct any clerical errors?

11. Continue to consider the student page and related student work for **Sprinkling around Obstacles** (refer to the two sets of student responses in the Unit Two Appendix).

a. For problem 2 on the student page, what techniques did students use when technological tools were available that were not feasible when using other tools?

b. When students are allowed to use technology, their solution strategies can be unconventional.

  i. Which solution strategies resulted in a solution that could give correct results but was not a construction as requested in problem 2 of the student page? Explain.

  ii. Which solution strategies resulted in correct solutions and constructions? Explain.

c. Compare and contrast the nontechnology student work with the student work when using *The Geometer's Sketchpad*™.

d. What additional student responses and teacher support would you consider adding to the **QRS Guide for Sprinkling around Obstacles**? (See the Unit Two Appendix.)

Preparing for Instruction

12. Consider the student page **Sandlot Baseball** given to students during their study of the geometry of circles. The teacher's intent was to have students determine how to find inscribed and circumscribed circles of quadrilaterals. She also wanted students to determine which quadrilaterals could be inscribed in a circle and which quadrilaterals could have a circle inscribed in them.

a. Complete the student page **Sandlot Baseball**. Solve the problems on the student page in as many ways as you can.

b. Which of the solution strategies would you expect high school students to use?

c. What misconceptions do you expect high school students to have?

d. What errors might high school students make?

e. For each of the misconceptions or errors you determined in parts 12c and 12d, what questions should the teacher be prepared to ask to assist students to help them move beyond their difficulties? (Recall the Stages of Questioning from Lesson 1.1.)

f. Use your responses in items 12b–e to complete a **QRS Guide** for this student page.

TEACHING AND LEARNING HIGH SCHOOL MATHEMATICS © 2010 John Wiley & Sons, Inc.

Sandlot Baseball

Vacant lots sometimes become impromptu baseball fields. These lots rarely are the most convenient shapes and often contain obstacles that require less than convenient arrangements for infields. The infield of a regulation baseball field is a square with the pitcher's mound positioned roughly at the center of both an inscribed and a circumscribed circle (field dimensions vary by league: professional, little league, or pony league). The baselines for several playfields are shown, with *H* indicating home plate.

1. For the following nonsquare "sandlots":

   a. Determine the location for the pitcher's mound. Show your constructions (do not just draw the location of the pitcher's mound).

   b. State your reasons for the constructions you used to locate the pitcher's mound. What advantages are derived from using your process to locate a pitcher's mound?

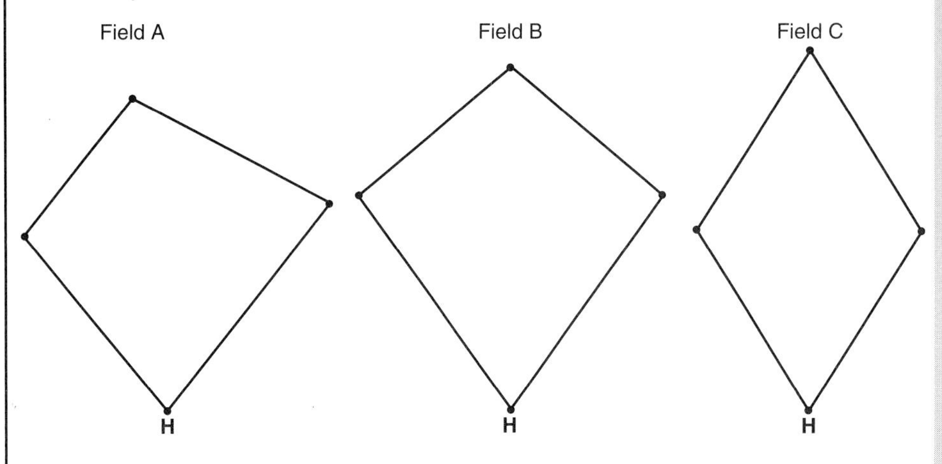

2. *Quadrilaterals Circumscribed by a Circle.*

   a. In terms of a baseball field, what is an advantage of locating the pitcher's mound at the center of a circumscribed circle?

   b. Of trapezoids, isosceles trapezoids, parallelograms, rectangles, kites, rhombi, and squares, which types of quadrilaterals can always be circumscribed by a circle? How do you know?

   c. Assume a quadrilateral can be inscribed in a circle. How do you construct the center and radius of the circle?

3. *Quadrilaterals with an Inscribed Circle.*

   a. In terms of a baseball field, what is an advantage of locating the pitcher's mound at the center of an inscribed circle?

   b. For which of trapezoids, isosceles trapezoids, parallelograms, rectangles, kites, rhombi, and squares, can a circle always be inscribed? How do you know?

   c. Assume a circle can be inscribed in a quadrilateral. How do you construct the center and radius of the circle?

13. Students' work for problem 1 on the **Sandlot Baseball** student page is provided on page 131.

   a. Determine the level of student understanding in each response. Record any new student responses in the **QRS Guide** from **DMP** problem 12 as misconceived, partial, or satisfactory.

TEACHING AND LEARNING HIGH SCHOOL MATHEMATICS © 2010 John Wiley & Sons, Inc.

b. Use the Stages of Questioning to help you determine which questions to ask each student to help them move to a deeper level of understanding. Record your teacher support in the **QRS Guide** for the student page.

__Zoe:__ For Field C, I constructed the diagonals and placed the pitcher's mound, P, at the intersection of the diagonals. This way, the pitcher is halfway between 1st and 3rd and also halfway between home and 2nd. He can throw the ball to each base easily.

P

H

__Ysabela:__ I drew the diagonals and put the pitcher at the intersection of these. That seemed to work OK for Field C, the rhombus, but made the pitcher too far from home for Field B, the kite, and in a really strange location for Field A, the trapezoid. I think it would be better for the pitcher to be the same distance from each base, but that isn't possible for these shapes.

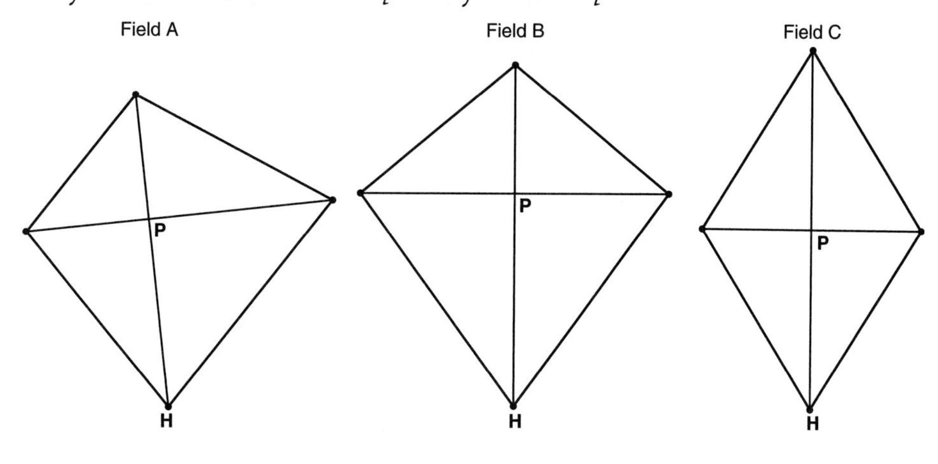

__Xavier:__ I think it would be best if the pitcher could be the same distance from each base. I can just look at the kite and rhombus and see that it's not possible to do that. No point will be the same distance from each of the vertices. But the isosceles trapezoid will work. If I take the perpendicular bisector of any two of the sides, their intersection will be the center of the circle. I tried out my idea with Sketchpad and realized I had to pick two sides that are not parallel. An isosceles trapezoid has a line of symmetry that is the perpendicular bisector of the parallel sides. Now, why does my process work? If I take any point on the perpendicular bisector of a segment and connect it to the endpoints of the segment, I get an isosceles triangle. So the center of the circle must be on the perpendicular bisector of each side of an inscribed figure. If I use the point of intersection formed by two perpendicular bisectors of two nonparallel chords, I must have the center of the circle. This is the only point that can connect endpoints of both chords to the center with all four lengths being equal.

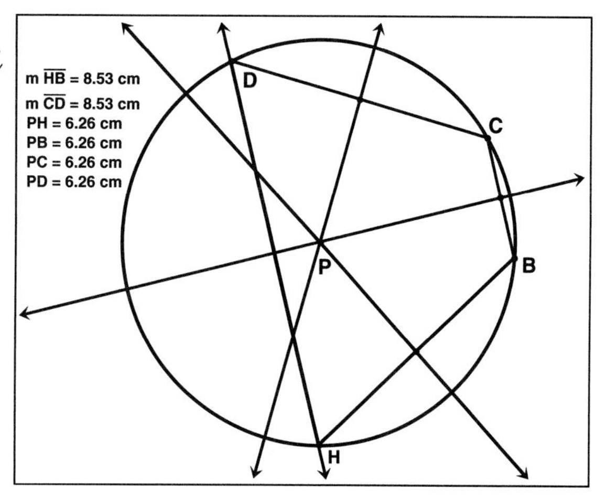

TEACHING AND LEARNING HIGH SCHOOL MATHEMATICS © 2010 John Wiley & Sons, Inc.

Communicating Mathematically

14. Understanding vocabulary and symbols is essential to success in mathematics. One strategy for helping students internalize vocabulary is a *personal dictionary* or *glossary*. Many students benefit from creating their own mathematics dictionary in which they write a definition in their own words, provide an example and indicate why it is an example, and provide a nonexample and indicate why it is a nonexample.

   a. Choose two terms from this lesson (e.g., tangent, circumscribed, inscribed, chord, central angle, intercepted arc), and write a personal dictionary entry for each.

   b. How might such a communication strategy be encouraged in your mathematics class?

Teaching for Equity

15. a. As mathematics classrooms incorporate more discourse and communication in the classroom than in the past, both oral and written, what are some issues that teachers need to consider when students in the class are not proficient in English?

   b. The cooperative learning strategy *Jigsaw* requires a considerable level of discourse among students as they share in their expert groups and then again in their home groups. How might teachers use this strategy in a class with several students with limited English proficiency?

   c. Moschkovich (1999) offers four suggestions for designing instruction for Latino students:

   - Recognize the diversity that exists within the Latino culture. Students from different groups within the broader Latino population may have different experiences to bring to the classroom (e.g., Caribbean culture may be different from Mexican culture or from South American culture).
   - Learn about your students and their culture.
   - Do not assume that difficulties are due to students' culture or home resources.
   - Build opportunities within the classroom for students to engage in mathematical discourse.

   How do you anticipate her suggestions influencing your instruction? (For more insights, read "Understanding the Needs of Latino Students in Reform-Oriented Mathematics Classrooms" by Judit N. Moschkovich. In *Changing the Faces of Mathematics: Perspectives on Latinos*, edited by Luis Ortiz-Franco, Norma G. Hernandez, and Yolanda De La Cruz (pp. 5–12). Reston, VA: National Council of Teachers of Mathematics, 1999.)

Reflecting on Professional Reading

16. a. Zbiek (2005) makes a distinction between a drawing and a construction in a technology environment. What do you think is the difference between drawing and constructing when using technological tools?

   b. Find at least three different geometry textbooks. Locate a topic related to circles other than the intersection of chords. Compare the presentations of the topic in the three textbooks, specifically considering the sequence of activities related to the topic and the statement of theorems related to the topic.

   c. Create one activity using some type of physical tool to help students understand the concept in problem 16b.

   d. If possible, create one activity using technology to help students understand the concept in problem 16b.

TEACHING AND LEARNING HIGH SCHOOL MATHEMATICS © 2010 John Wiley & Sons, Inc.

e. Read the article "Using Technology to Show Prospective Teachers the 'Power of Many Points'" by Rose Mary Zbiek. In *Technology-Supported Mathematics Learning Environments*, edited by William J. Masalski and Portia C. Elliott (pp. 291–301). Reston, VA: National Council of Teachers of Mathematics, 2005. Compare the characteristics of your technology activity to the technology activities created in the article. (See the course website for the article.)

2.5 Tasks with High Cognitive Demand: Measurement in the Plane and in Space

Many high school measurement topics have been explored in elementary school. However, students' elementary school knowledge may be procedural and may lack a conceptual base. Although students need to develop fluency with measurement formulas, they also need to understand the concepts underlying those formulas. When such conceptual understanding is lacking, students often apply formulas incorrectly while solving particular problems. To develop mathematically proficient students, teachers need to help students make connections among measurement formulas and the relevant concepts.

To help students grow mathematically, they need to engage in tasks with high levels of thinking and reasoning, also called tasks with high cognitive demand (Stein and Lane, 1996). So, teachers must select or create challenging tasks. Selecting good instructional tasks is a first step in designing effective instruction.

In this lesson, you will:

- Investigate area formulas for circles, and surface area and volume formulas for spheres, cylinders, prisms, and pyramids.
- Analyze a variety of two- and three-dimensional measurement tasks.
- Revise tasks to increase their level of challenge.

LAUNCH

Two high school students, Alex and David, were talking about the connections between prisms and cylinders and their volumes. The teacher overheard the following conversation:

David: *Prisms and cylinders have a lot in common. To find the volume, you can imagine having the shape of the base and then stacking it up as high as you need to reach the top of the prism or cylinder. So, to find the volume, you just find the area of the base and you multiply that area by the height, which is how many times you had to stack the base. You really don't have to memorize a volume formula, because you just find the area of the base and multiply by the height.*

Alex: *I thought that volume was always length times width times height.*

David: *That formula only works for a box, what the book calls a rectangular prism. If you think about the base, it's a rectangle. So, the area of the base is length times width. Now multiply that area by the height and you get the formula length times width times height. So, what I said before still works.*

1. Comment on David's explanation to Alex. What are any strengths or weaknesses of his explanation?

One difficulty many students face is learning a variety of formulas for different types of figures; these formulas often become confused with each other when

TEACHING AND LEARNING HIGH SCHOOL MATHEMATICS © 2010 John Wiley & Sons, Inc.

memorized rather than learned in a conceptual manner. David's explanation provides a unifying way to consider the volume formulas for prisms and cylinders that is consistent regardless of the shape of the base. Hence, David does not have to memorize additional volume formulas. Rather, his understanding of area formulas permits the volume formula to be quickly developed from his conceptual understanding that uses the shapes of the bases and the areas of those shapes.

One instructional goal for teachers is to help all students develop the type of understanding evident in David's explanation.

EXPLORE

2. Following are six mathematical tasks all centered on the same mathematics topic, the volume of a cone. Assume that students have not previously had formal work with volumes of cones but have worked with volumes of prisms and cylinders. Complete each task and then decide with your team whether you believe the task is at a high or low level of cognitive demand (intellectual challenge). Provide a rationale for your decision.

   Task A You have an empty cone, an empty cylinder of the same radius and height, sand, a scoop, and a bucket. Use the tools provided and what you already know to develop a volume formula for a cone.

   Task B A manufacturer of sno cones is interested in comparing the amount of sweetened crushed ice that can fill different sizes of cones. He charges c cents for a basic sno cone with height h and base with radius r. Provide a report to the manufacturer to help him determine how to adjust the price based on a change in the radius of the base.

   Task C What is the formula for the volume of a cone?

   Task D Find the volume an ice cream cone will hold (inside the cone, not rounded on top) if the cone has a height of 4 inches and a base with radius of 1 inch. Use the formula

$$V = \frac{\pi r^2 h}{3}.$$

   Task E Two different cones have the same volume, 80 cubic inches. What might be the dimensions of the cones?

   Task F Use the Internet to research the volume of a cone.

3. Look back in this book and identify tasks that you would evaluate as cognitively demanding. What makes a task intellectually challenging?

SHARE

4. Share with other groups your decisions about the cognitive demand of the six tasks and your rationales.
5. Based on your evaluation of the six tasks and of tasks you found challenging in this text, identify some criteria that make a task cognitively demanding.

EXPLORE MORE

6. Using the kind of thinking your teams have been doing and the thinking of many researchers and educators, Smith and Stein (1998) developed a framework of qualities of tasks at four *Levels of Demand*. Low-level tasks involve memorization or procedures without connections. High-level tasks involve procedures with connections or doing mathematics in challenging ways. Study the framework at the end of this lesson showing Smith and Stein's categorizations of tasks.

   a. What, if anything, on the list surprised you?

TEACHING AND LEARNING HIGH SCHOOL MATHEMATICS © 2010 John Wiley & Sons, Inc.

b. Some people believe an activity with real-world connections automatically is at a high level of cognitive demand. Are all real-world activities categorized this way in the framework?

c. Some people believe an activity involving technology or materials must also be at a high level of cognitive demand. Discuss these ideas in light of the framework.

d. Re-evaluate your analysis of the six tasks earlier in this lesson. What changes to your categorization, if any, would you make? Explain.

7. Textbook materials often are written at a lower level of cognitive demand than many educators would desire. So, one aspect of teaching is to rescale tasks to a higher level of cognitive demand. Below are two tasks in the realm of 2D and 3D measurement. In each case, identify the mathematics the task addresses. Then use the Smith and Stein framework to rescale the task to a higher level of cognitive demand. Prepare to share your task and the thinking you used to rescale it.

Task G A rectangular prism has a length of 12 inches, a width of 6 inches, and a height of 5 inches. Find the surface area of the prism.

Task H A candy container is shaped like the one shown at right and holds a solid piece of chocolate. It is 6 inches long. The base is an equilateral triangle with a side length of 1 inch. Use the formula $V = Bh$ to determine how much candy the container will hold if it is totally filled.

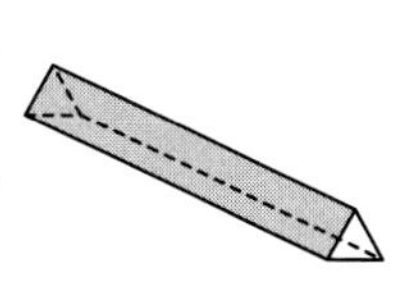

SHARE AND SUMMARIZE

8. Share with other groups your responses to problem 6.

9. Compare the different revisions of tasks G and H from various teams. What similarities and differences exist among the revisions?

10. Based on your discussions of the Levels of Demand and the revisions to tasks G and H, what difficulties do you envision in designing tasks at intellectually challenging levels?

Levels of Demand

Lower-level demands (memorization)

- Involve either reproducing previously learned facts, rules, formulas, or definitions or committing facts, rules, formulas, or definitions to memory.
- Cannot be solved using procedures because a procedure does not exist or because the time frame in which the task is being completed is too short to use a procedure.
- Are not ambiguous. Such tasks involve the exact reproduction of previously seen materials, and what is to be produced is clearly and directly stated.
- Have no connection to the concepts or meaning that underlie the facts, rules, formulas, or definitions being learned or reproduced.

Lower-level demands (procedures without connections)

- Are algorithmic. Use of the procedure either is specifically called for or is evident from prior instruction, experience, or placement of the task.
- Require limited cognitive demand for successful completion. Little ambiguity exists about what needs to be done and how to do it.

(*continued*)

TEACHING AND LEARNING HIGH SCHOOL MATHEMATICS © 2010 John Wiley & Sons, Inc.

(*continued*)

- Have no connection to the concepts or meaning that underlie the procedure being used.
- Are focused on producing correct answers instead of on developing mathematical understanding.
- Require no explanations or explanations that focus solely on describing the procedure that was used.

Higher-level demands (procedures with connections)

- Focus students' attention on the use of procedures for the purpose of developing deeper levels of understanding of mathematical concepts and ideas.
- Suggest explicitly or implicitly pathways to follow that are broad general procedures that have close connections to underlying conceptual ideas as opposed to narrow algorithms that are opaque with respect to underlying concepts.
- Usually are represented in multiple ways, such as visual diagrams, manipulatives, symbols, and problem situations. Making connections among multiple representations helps develop meaning.
- Require some degree of cognitive effort. Although general procedures may be followed, they cannot be followed mindlessly. Students need to engage with conceptual ideas that underlie the procedures to complete the task successfully and develop understanding.

Higher-level demands (doing mathematics)

- Require complex and nonalgorithmic thinking—a predictable, well-rehearsed approach or pathway is not explicitly suggested by the task, task instructions, or a worked-out example.
- Require students to explore and understand the nature of mathematical concepts, processes, or relationships.
- Demand self-monitoring or self-regulation of one's own cognitive processes.
- Require students to access relevant knowledge and experiences and make appropriate use of them in working through the task.
- Require students to analyze the tasks and actively examine task constraints that may limit possible solution strategies and solutions.
- Require considerable cognitive effort and may involve some level of anxiety for the student because of the unpredictable nature of the solution process required.

These characteristics are derived from the work of Doyle on academic tasks (1988) and Resnick on high-level thinking skills (1987), the *Professional Standards for Teaching Mathematics* (NCTM 1991), and the examination and categorization of hundreds of tasks used in QUASAR classrooms (Stein, Grover, and Henningsen 1996; Stein, Lane, and Silver 1996). From Smith, Margaret Schwan and Mary Kay Stein. "Selecting and Creating Mathematical Tasks: From Research to Practice." *Mathematics Teaching in the Middle School*, 3 (February 1998): 344–350. Reprinted with permission from *Mathematics Teaching in the Middle School*, ©1998 by the National Council of Teachers of Mathematics.

DEEPENING MATHEMATICAL UNDERSTANDING

1. *Soup Cans*. A cylindrical can has radius r and height h.
   a. As the radius stays the same and the height changes by a factor of k, what happens to the volume?

TEACHING AND LEARNING HIGH SCHOOL MATHEMATICS © 2010 John Wiley & Sons, Inc.

b. As the height stays the same and the radius changes by a factor of s, what happens to the volume?

c. Suppose you want to change the dimensions of the can but leave the volume unchanged. For the volume to remain constant, what relationship must exist between the changes to the radius and the changes to the height?

2. *Isoperimetric Properties*. Several real-world situations make use of various isoperimetric properties in two- and three-dimensional space. *Isoperimetric* comes from roots meaning "same perimeter."

   a. Consider a square, a regular hexagon, a regular octagon, and a circle, each with perimeters of 600 feet. Determine which of these figures has the greatest area. This relationship is called the isoperimetric property.

   b. Consider a tepee used by Native American Plains tribes in the 1800s. How does the isoperimetric property from problem 2a connect to the shape of tepees?

   c. The Mandan people lived in permanent dwellings with cylindrical bases and domed roofs. How does the isoperimetric property from problem 2a connect to the shape of the Mandan's dwellings?

   d. In problem 2a, you were given a fixed perimeter and made a statement about the greatest area. Suppose instead that you start with a given fixed area. Restate the isoperimetric property for a perimeter given fixed area.

3. *Extending Isoperimetric Ideas to Three Dimensions*.

   a. Find dimensions of a cube, a cylinder, a cone, and a sphere so that each has a surface area of 100π square meters. Then find the volume of each figure. Which figure has the greatest volume for the given surface area? Will your result always be the same? Justify your answer.

   b. Inuits in Canada and Eskimos in Alaska often used igloos in the winter. Investigate the shape of igloos. How does the shape relate to the isoperimetric concept you explored in problem 3a?

   c. Suppose you now start with a fixed volume. State a conjecture of a 3-dimensional analogy of the isoperimetric property related to surface area.

4. *Cavalieri's Principle and the Volume of a Sphere*. Cavalieri's Principle, named after Bonaventura Cavalieri (1598–1647), can be applied to any two solids included between parallel planes P_1 and P_2: *If all planes parallel to P_1 and P_2 intersect the two solids in plane sections with the same area, then the solids have the same volume*. This principle can be used to justify the volume formula of a sphere without using calculus.

   a. Label the drawings below so that the height of the cylinder is twice its radius, r, and the sphere has a diameter equal to the height of the cylinder. Explain why Cavalieri's Principle can be applied.

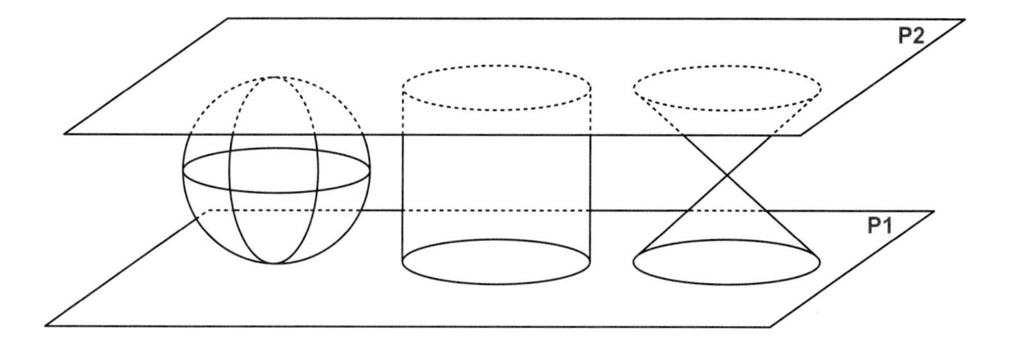

   b. *A great circle* of a sphere is a circle on the sphere that lies in a plane containing the center of the sphere. Find the area of a great circle of the sphere.

TEACHING AND LEARNING HIGH SCHOOL MATHEMATICS © 2010 John Wiley & Sons, Inc.

c. Sketch a circular cross-section of the sphere parallel to the great circle in problem 4b. Suppose the circular cross-section is at a distance of *h* from the center of the sphere. Find the area of this circular cross-section.

d. Return to the cylinder. Find the area of a circular cross-section of the cylinder at its center.

e. Imagine the two cones on p. 137 nested into the cylinder, so that the base of each cone is one of the bases of the cylinder and the vertices of the cones intersect at the center of the cylinder. Find the area of a circular section of the cone that is *h* units from the center of the cylinder where the vertices of the two cones meet.

f. Now consider the area of a cross-section that lies between the cylinder and the cone at a height of *h* units from the center of the cylinder. Find the area of this ring.

g. Use your results from problems 4c and 4f to explain why the volume of the sphere is the difference between the volume of the cylinder and the volume of the two cones nested in the cylinder.

h. Use problem 4g to find a formula for the volume of a sphere.

5. *Surface Area of a Sphere.* To develop a formula for the surface area of a sphere from the formulas of the volumes of spheres and pyramids, consider the following process.

   a. Start with the volume of a sphere.

   b. Imagine a sphere sectioned into "approximate pyramids" with vertices at the center of the sphere and congruent bases on the surface of the sphere. Find the sum of the volumes of all of these approximate pyramids.

   c. Set the expressions in problems 5a and 5b equal to each other. Simplify and use the sphere's volume to find a formula for the surface area of a sphere.

6. *More with Area and Volume.*

   a. Model cars often are made using one of the scales: 1 to 64, 1 to 24, or 1 to 16. For each scale model, compare the surface area and volume of the model to the surface area and volume of the real car.

   b. Suppose two figures are similar and the linear measurements in the larger figure are *k* times the corresponding measurement of the smaller figure. How do the areas and volumes of the two figures compare? Justify your response.

   c. The relationship in problem 6b can help explain other area formulas in which the two figures are not similar. For instance, one way to think about an ellipse is to consider it as a circle that has been stretched horizontally by a factor of *a* and vertically by a factor of *b*. That is, for a unit circle centered at the origin, the related ellipse results from stretching the unit circle horizontally and vertically so that the points $(\pm 1, 0)$ map onto the points $(\pm a, 0)$ and the points $(0, \pm 1)$ map onto the points $(0, \pm b)$. Use your work in problem 6b to justify a formula for the area of this ellipse.

7. *Lateral Areas of Pyramids and Cones.* There are many similarities in finding the lateral areas of pyramids and cones.

   a. Consider a square pyramid with base of side length *s*, height *h*, and slant height *l*. (Recall that the slant height is the length of the segment on a face of the pyramid drawn perpendicular from the vertex to the base.) Derive a formula for the lateral area of the pyramid.

   b. How would the formula for lateral area change if the pyramid's base were hexagonal?

   c. As the number of sides in the polygonal base increases, the pyramid approaches a cone. Use your work in problems 7a and 7b to find a formula for the lateral area of a cone.

TEACHING AND LEARNING HIGH SCHOOL MATHEMATICS © 2010 John Wiley & Sons, Inc.

DEVELOPING MATHEMATICAL PEDAGOGY

Reflecting on Our Own Learning

8. Think about your own learning of area and volume formulas for two- and three-dimensional figures.

a. In what ways did you learn formulas? Did you learn formulas as connected to each other through mathematical relationships or individually?

b. What are the pros and cons of learning formulas with connections rather than just learning formulas without connections?

Analyzing Instructional Activities

9. As indicated in the lesson, many high school students have used area formulas for various two-dimensional figures but may not be able to justify why the formula works. The student page, **Area Formulas for Familiar Two-Dimensional Figures,** reviews the formulas from a conceptual perspective and makes connections among the figures. (See pp. 140–142.)

a. Work through the **Area Formulas for Familiar Two-Dimensional Figures** student page.

b. Categorize each problem by its Level of Demand.

c. Where do you anticipate students needing help?

d. In the area formula for a parallelogram, many students confuse the height of the parallelogram with the length of the sides not designated as the base. How can you help students distinguish between these two values?

e. The student page guides students through two methods to justify the formula for the area of a trapezoid. Discuss the pros and cons of providing students with multiple justifications of a result.

f. Find another way to derive the area formula for a trapezoid.

g. In justifying the formula for the area of a circle, some teachers will have students draw a circle, highlight the circumference, and then cut the circle into sectors and physically form the "bumpy parallelogram." Why might such an approach be helpful when justifying this formula?

10. Choose one or more of the problems from **Area Formulas for Familiar Two-Dimensional Figures** to try with high school students.

a. Review your responses to the problems on the student page (see **DMP** problem 9). For each problem, anticipate students' work at three levels: misconceived, partial, and satisfactory. Record these responses in a **QRS Guide**.

b. What teacher support in the form of questions will you be ready to provide for each recorded solution? Record appropriate teacher support questions in the **QRS Guide**.

c. Try the problem(s) with high school students. How did they respond? Add any new responses to your **QRS Guide**.

d. Where did students have difficulties? Include teacher support to help students overcome these difficulties. What reactions or difficulties surprised you?

TEACHING AND LEARNING HIGH SCHOOL MATHEMATICS © 2010 John Wiley & Sons, Inc.

WWW

Area Formulas for Familiar Two-Dimensional Figures

1. ***Parallelograms.*** On the geoboard template below, draw a nonrectangular parallelogram with a base of 4 units and a height of 5 units. Recall that the base is the length of one of the sides and the height is the length of the perpendicular segment drawn from the opposite side to the base. Label your parallelogram *QUAD*.

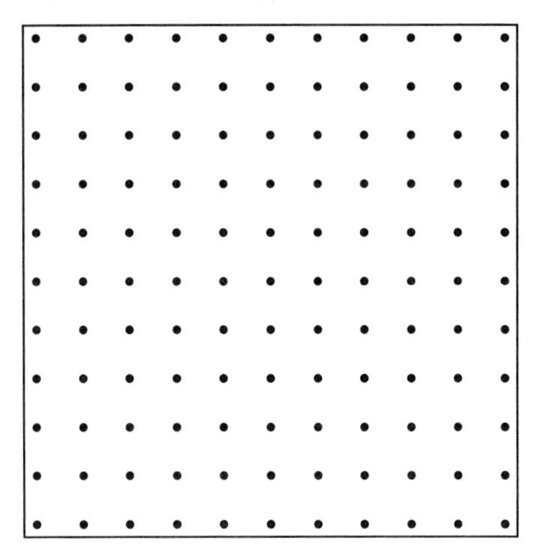

   a. Find the area of *QUAD* in square units.

   b. Draw a segment from one vertex perpendicular to the base; choose a vertex so that the segment you draw is in the interior of the parallelogram. Show that the triangular piece formed when this segment is drawn can be translated in such a way that the parallelogram becomes a rectangle.

   c. Find the area of the rectangle.

   d. Explain why the process in items 1b and 1c works for any parallelogram and how this process justifies the formula for the area of a parallelogram.

   e. Provide a rationale for using b and h rather than l and w in the area formulas for parallelograms and rectangles.

2. ***Triangles.*** Draw parallelogram *QUAD* again on the following template.

   a. Draw one of the diagonals of *QUAD*.

   b. Use the resulting figure to justify a formula for the area of a triangle, both in this specific case and in general.

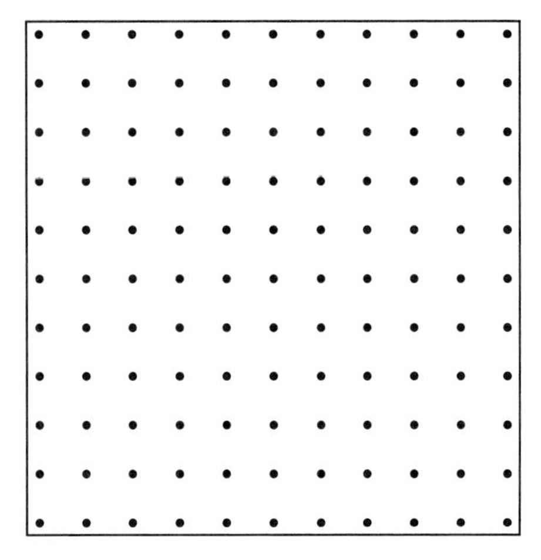

(*continued*)

TEACHING AND LEARNING HIGH SCHOOL MATHEMATICS © 2010 John Wiley & Sons, Inc.

3. ***Trapezoids.*** On the following template, draw trapezoid *RIPE* with parallel bases of length 2 units and 3 units and a height of 5 units.

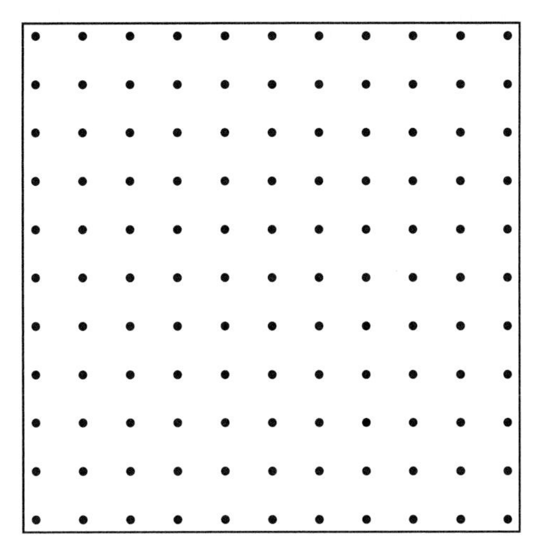

a. Estimate the area of trapezoid *RIPE*.

b. Draw one of the diagonals. You have now split the trapezoid into two triangles. Use the resulting drawing to justify the area formula for a trapezoid, both in this specific case and in general.

c. Draw trapezoid *RIPE* again on the following template. Imagine rotating another copy of *RIPE* 180° so that the two copies share a common side and form a parallelogram. Show your resulting parallelogram formed from two copies of trapezoid *RIPE*.

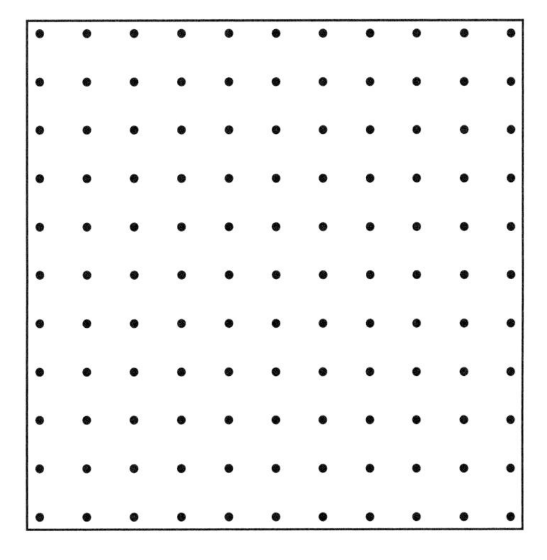

d. Identify the base and height of the parallelogram you formed in item 3c. Find the area of this parallelogram. Use this area to justify a formula for the area of a trapezoid, in this specific case and in general.

4. ***Circles.*** Consider the circle on p. 142 into an even number of sectors that are rearranged to form a shape that is close to a parallelogram (some teachers call this a "bumpy parallelogram").

(*continued*)

TEACHING AND LEARNING HIGH SCHOOL MATHEMATICS © 2010 John Wiley & Sons, Inc.

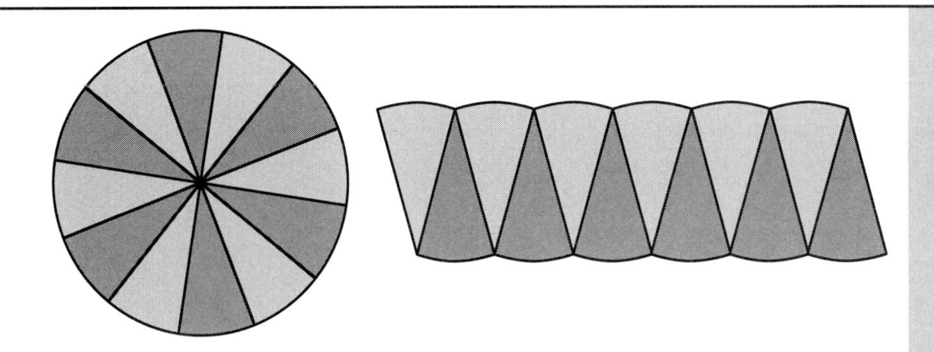

a. Identify the base and height of this "bumpy parallelogram" in terms of the original circle.

b. How can this diagram be used to justify the area formula for a circle?

c. Does it matter if the circle is cut into an even or an odd number of sectors? Explain.

11. Study the student page **Lateral Area of a Cone.**

a. Complete enough of the student page to solve problems 2e, 3, and 4 on the student page.

b. What Level of Demand is this task for high school geometry students? Why do you think so?

c. Anticipate a student having difficulty with problem 2 and suggest incremental help using the Stages of Questioning.

d. How does the student page exemplify Bruner's ideas of enactive, iconic, and symbolic representations (refer to the reading by Rubenstein in Lesson 2.3)?

e. In **DMU** 7, you derived the lateral area of a cone by considering pyramids with bases having more and more sides and then summing the areas of the triangular faces of the pyramids. Use algebra to show that the lateral surface area formula generated in this way is equal to the formula generated on the student page.

f. The total surface area of a cone is the sum of the lateral area and the area of the base. Write a formula for the surface area of a cone.

Lateral Area of a Cone

Materials

- several circular coffee filters per student

1. Use a circular coffee filter. Fold to find its center. Cut the circle to make a sector (as shown) and roll it to make a cone.

a. What part of the cone is the radius of the sector you rolled?

b. What part of the sector is the circumference of the base of the cone?

2. Verify that the labels in the figure at right agree with your findings in item 1. Complete the following to determine relationships between the cone and the circle if the cone is created from the sector of the circle shown.

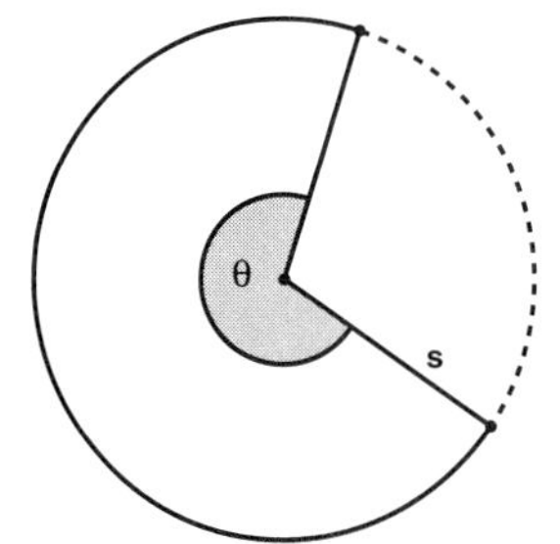

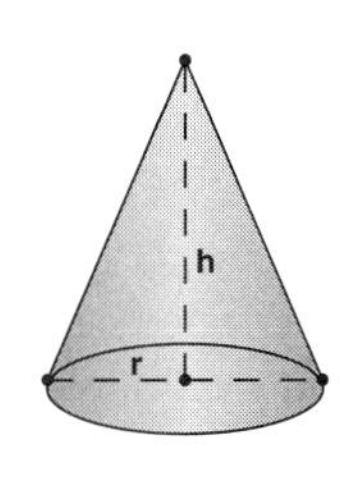

(*continued*)

TEACHING AND LEARNING HIGH SCHOOL MATHEMATICS © 2010 John Wiley & Sons, Inc.

a. Complete the table below:

Angle Measure	**Area of Sector**	**Circumference of Sector**
2π		
π		
$\frac{\pi}{2}$		
$\frac{\pi}{4}$		
$m\angle\theta$		

b. In terms of the measure of angle θ and radius s, what is the arc length of the circle used to create the cone?

c. In terms of the cone and its height h and radius r, what is the slant height of the cone?

d. What is the length of the arc of the circle that makes up the base of the cone in terms of the radius of the cone?

e. Use the relationships you found in items 2b–2d to write an equation showing the measure of angle θ in terms of the radius and height of the cone.

3. a. What is the area of the sector of the circle shown in terms of its radius s and measure of angle θ?

b. Rewrite your equation from item 3a in terms of the radius and height of the cone.

c. Use the measure of angle θ from item 2e to rewrite your equation in terms of the radius and height of the cone.

4. Use your work from problems 2 and 3 to find an equation that shows the lateral area of a cone.

Preparing for Instruction

12. Refer again to the Levels of Demand table. In planning instructional activities, teachers need to plan tasks at higher cognitive levels. For one of the following three areas, design a task that is *procedures with connections* and one that is *doing mathematics*:

    a. transformations of geometric figures

    b. coordinate proofs

    c. surface area of a sphere

13. Consider the following tasks. Revise each one so that the revision is *procedures with connections* or *doing mathematics*, according to the Levels of Demand framework.

    a. A cylinder has a height of 10 inches and a radius of 6 inches. Find the volume and surface area of the cylinder.

    b. The surface area of a sphere is given by the formula $SA = 4\pi r^2$. Find the surface area of a basketball with a diameter of 12 inches.

TEACHING AND LEARNING HIGH SCHOOL MATHEMATICS © 2010 John Wiley & Sons, Inc.

14. Kathryn Lasky's book *The Librarian Who Measured the Earth* (Boston: Little, Brown and Company, 1994) describes the work of Greek mathematician Eratosthenes (276 B.C.–194 B.C.), the first person credited with measuring the circumference of the Earth. Obtain a copy of the book. How might you design a lesson for high school students using this book as a springboard? What related task would you pose for students?

15. Steve Jenkins's book *Actual Size* (Boston: Houghton Mifflin, 2004) contains pages showing the actual size of various animal features. For instance, one page contains the actual size of a gorilla's handprint. Another page contains the actual size of a giant squid's eyeball. Suppose you are interested in designing a lesson related to proportional reasoning as well as area and volume of similar figures. Locate and read a copy of the book.

    a. On a transparency of a piece of grid paper, trace your handprint. Estimate the area of your hand. Use the transparency to estimate the area of a gorilla's handprint. Compare the two areas. How would the length and width of your hand and the gorilla's hand compare? Explain.

    b. Design a student activity that will help students understand this important area relationship between similar figures.

    c. Suppose you wanted to compare the volume of your eyeball to the volume of a giant squid's eyeball. Use the information in the book to design an activity that would help students understand the volume relationship between similar figures.

    d. Complete a **QRS Guide** for one or more of your activities. Try the activities with high school students. Revise the activities and the **QRS Guide** based on student responses. How did they react to the book? How did they react to your activities?

16. Teachers sometimes believe that hands-on materials, such as models or manipulatives, cannot be used with high school students.

    a. What are some benefits of using physical materials when exploring area and volume relationships with three-dimensional figures?

    b. What are some concerns?

17. Constructing tasks for students often involves careful attention to the particular numbers involved. Analyze each of the tasks in 17a–c. If there is something problematic about the numbers, identify the problem and suggest better numbers.

    a. Find the perimeter of a rectangle with dimensions 3 cm and 6 cm.

    b. Find the volume of a pyramid where the height is 5 cm and the base is square with edge length 3 cm.

    c. Find the volume and surface area of a sphere with a radius of 3 inches.

    d. Rethink problem 17a. In general, what relation between length and width of a rectangle creates potentially problematic tasks?

Communicating Mathematically

18. A *concept map* often is an effective communication strategy, particularly near the end of a unit. With this strategy, students build a web of relationships.

    a. Following is the beginning of a concept map for surface area. Add appropriate connections to each of the circles to show formulas and how they arise. In particular, consider where you might place cones and cylinders on the map to show their connections to other solids.

TEACHING AND LEARNING HIGH SCHOOL MATHEMATICS © 2010 John Wiley & Sons, Inc.

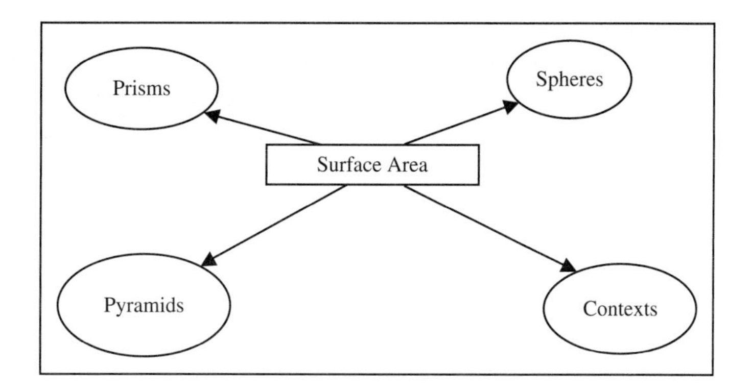

b. What are some ways you might use concept maps in the classroom?

c. What differences in the concept maps would you expect depending on when in a unit you used this strategy?

d. Compare concept maps to Link Sheets described in Lesson 2.2. What are some similarities between the two strategies? What are some differences?

e. Compare concept maps to the Frayer Model described in Lesson 2.2. What are some similarities between the two strategies? What are some differences?

19. a. Complete the following table to record surface area and volume formulas for each figure using B as the area of the base, L as the area of the lateral surface, and h as the height.

	Prism	**Pyramid**	**Cylinder**	**Cone**
Volume				
Surface Area				

b. Compare the graphic organizer in problem 19a with the concept map from problem 18. What relationships are clearer in the table?

Teaching for Equity

20. How does using worthwhile mathematical tasks (those of higher level cognitive demand) build equity in mathematics classrooms?

Reflecting on Professional Readings

21. Read "Measurement Standard for Grades 9–12" in *Principles and Standards for School Mathematics* (pp. 320–323).

a. Use unit analysis to convert 70 miles per hour to kilometers per hour. How does understanding fraction operations influence success with unit analysis?

b. Suppose the cone on the bottom of page 323 in the reading is cut into slices 1 cm thick. Find the upper and lower bounds for the volume.

c. What do you anticipate your greatest challenges to be in teaching measurement at the high school level?

22. Read "The Volume of a Pyramid: Low-Tech and High-Tech Approaches" by Masha Albrecht (*Mathematics Teacher*, 94 (January 2001): 58–61, plus lab pages).

a. Work through the student activities.

b. Another way students learn the volume formula for a cone is with relational solids. These are hollow cones and cylinders with the same base and height.

TEACHING AND LEARNING HIGH SCHOOL MATHEMATICS © 2010 John Wiley & Sons, Inc.

Students fill the solids with water or sand to discover a constant ratio between the volumes. Compare this approach to the one in Albrecht's article.

c. This article integrates the use of spreadsheets to explore geometric formulas. What are some challenges teachers might face when integrating such technology into the mathematics curriculum? What are some benefits of the use of spreadsheets in this way?

23. Read "Multiple Solutions: More Paths to an End or More Opportunities to Learn Mathematics" by Rose Mary Zbiek and Jeanne Shimizu (*Mathematics Teacher*, 99 (November 2005): 279–287).

a. Reflect on the different approaches used in Example 1 to find the amount of skin that covers the human body. Which solution approach did you prefer? Explain.

b. Consider the melon ball problem of Example 2. Which solution approach did you prefer in this situation? Explain.

c. What kinds of classroom expectations between teachers and students are needed to use open problems such as those in Examples 1 and 2?

2.6 Doing Mathematics: Axiomatic Systems

In Lesson 2.5, you analyzed tasks according to their level of cognitive demand and revised them as needed to increase their level of challenge. This lesson investigates more fully how some problems that Smith and Stein classify as *higher-level demands* (doing mathematics) can help students invent mathematics and investigate the logical conclusions of the assumptions they make. This work is very much like that of mathematicians in the sense of exploring concepts, processes, and relationships. Such problems engage students in complex tasks in which there is no clear approach and for which constraints and symbolizations need to be considered. These mathematical problems provide contexts for students to consider why undefined terms are necessary, definitions are reasonable, axioms are required, and how theorems arise from definitions, axioms, and other theorems.

You have already worked with definitions, axioms, and theorems in your engagement with reasoning in Unit One. Consider, for example, the second puzzle from the student page in Lesson 1.3, **DMU** problem 6, **Pandora's Boxes**. When solving problems that Smith and Stein classify at *higher levels of cognitive demand* (doing mathematics), one needs to decide what the problem is asking. This first step sometimes requires students to define unfamiliar terms or accept some terms as undefined. In **Pandora's Boxes** puzzle 2, no attempt was made to define the words, *evil*, *happiness*, *contains*, *box*, and *label*—these serve as the undefined terms of the system. In Lesson 2.3, you learned the properties of good definitions and observed that how one defines a term can be a matter of choice. For example, one can choose to define a rhombus as an equilateral quadrilateral or equivalently as a quadrilateral whose diagonals are perpendicular bisectors of each other. The choice of definition, in this case, gives rise to the other "definition" becoming a result (theorem) of this choice.

Axioms arise in puzzles as clues. They arise in games as rules. The clues or rules are the initial assumptions (i.e., axioms) that are taken as true without proof. Axioms describe how undefined and defined terms behave. In geometry, the line postulate states: *There is exactly one line through any two distinct points*. Point and line are undefined terms. The axiom describes a relationship between these terms.

In **Pandora's Boxes** puzzle 2, the two axioms for the system are:

- Both boxes are labeled with statements whose truth value is the same.
- A box contains either evil or happiness but not both, and neither box is empty.

TEACHING AND LEARNING HIGH SCHOOL MATHEMATICS © 2010 John Wiley & Sons, Inc.

The second axiom describes the relationship between evil and happiness; they cannot both be in the same box. This axiom also states that no box is empty. Students often add the assumption (axiom) that both boxes must contain something different, that is, one must contain happiness and the other must contain evil. With this additional axiom, the system is internally contradictory (called *inconsistent*) and no solution exists.

Finally, students look for and prove logical consequences of the axioms. These are called theorems. **Pandora's Boxes** puzzle 2 is a very simple axiomatic system; it only allows one consequence or theorem: *Happiness is in both boxes*. Euclidean geometry, in contrast, is a complex axiomatic system. There are multiple consequences that arise from the choices of undefined terms, definitions, and axioms of the system.

The following are elements of axiomatic systems:

- An axiomatic system contains a set of technical terms deliberately chosen as undefined terms and subject to the interpretation of the reader.
- All other technical terms of the system are ultimately defined by means of the undefined terms. These terms are the definitions of the system.
- The axiomatic system contains a set of statements, dealing with undefined terms and definitions, which are chosen to remain unproved. These are the axioms of the system.
- All other statements of the system must be logical consequences of the axioms and definitions. These derived statements are called the theorems of the axiomatic system.

Axiomatic systems often also have symbol systems that help mathematicians efficiently refer to and describe the elements of the system. In Lesson 2.1, the symbols of coordinate geometry were used to aid in proving relationships among polygons. In **Pandora's Boxes** puzzle 2, a dictionary of symbols made the theorem/solution easier to justify. This lesson provides an opportunity to create and use a symbol system to discover and study intersections between circles and squares.

In this lesson, you will:

- Build on work with the Euclidean geometry axiomatic system based on analysis of a problem involving intersections of circles and squares.
- Investigate the implications of your choices of undefined terms, definitions, and assumptions (axioms), make conjectures, and prove some of the conjectures (theorems).
- Use compass and straightedge and an interactive geometry drawing tool to make conjectures, and determine how to prove some of the conjectures.
- Analyze students' understanding of axiomatic systems through their work.

Materials

- interactive geometry drawing tool on a computer or graphing calculator, one per group
- compass and straightedge, one per person
- graph paper, two sheets per person

LAUNCH

During a course on Euclidean geometry in which students were building intuition for the Euclidean geometry axiomatic system, students investigated definitions for triangles, quadrilaterals, and circles; built intuition for the Euclidean postulates; and made conjectures and proved some theorems related to triangles, quadrilaterals, and circles. The following problem arose from this work:

TEACHING AND LEARNING HIGH SCHOOL MATHEMATICS © 2010 John Wiley & Sons, Inc.

Assuming that both figures lie in the same plane, how many points of intersection can exist between a circle and a square? (Canada & Blair, 2007) (Reprinted with permission from *Mathematics Teacher* © 2006/2007 by the National Council of Teachers of Mathematics.)

1. Work on the problem. Keep track of your use of aspects of axiomatic systems as you work.
   - **a.** What undefined terms are you assuming?
   - **b.** What defined terms are you using?
   - **c.** What terms must you define?
   - **d.** What assumptions or axioms are you accepting as true without proof?
   - **e.** What theorems have you learned in the past that might apply to this problem?
2. Disregarding number of intersections, in what different ways can a circle intersect a square? Explain.
3. How many points of intersection are possible between a circle and a square? Share your findings with the class.

EXPLORE

When attempting to solve the problem in groups, high school students found examples of circles and squares intersecting in zero through eight points, though they had some difficulties with finding examples for 5, 6, or 7 intersections. The teacher asked members of each group to share their examples on the board. Some examples for 0 through 2 intersections are provided in the following figure.

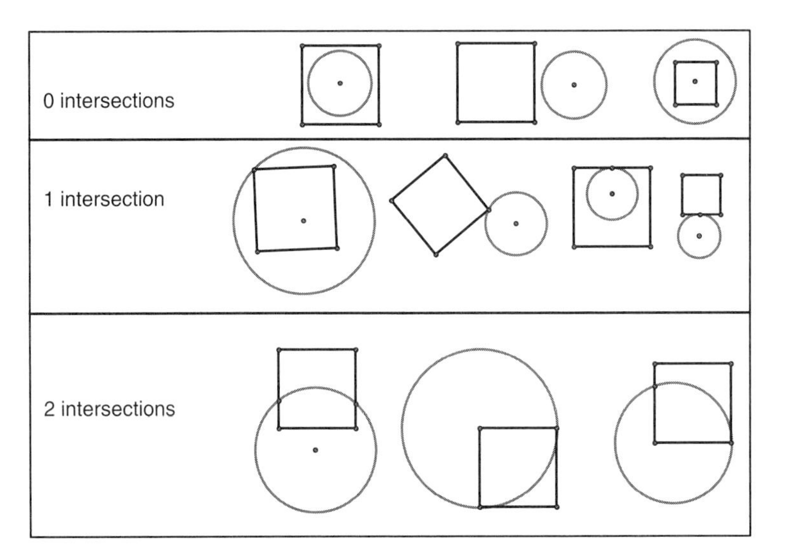

4. **a.** Have students found all the possibilities for each of the numbers of intersections shown? If not, what other intersections are possible?
   b. What additional questions would you like to ask class members now that this work has been shared publicly?
5. Referring to the students' examples for one intersection, the teacher began a discussion:

   You have shown two examples where the circle and square can intersect at a vertex, but in these examples, the circles are outside the square. Is it possible for a circle to intersect a square at a vertex where the circle is inside the square?

   - **a.** Answer the teacher's question.
   - **b.** Study the three student responses to the teacher's question. Determine which of the responses are correct. Why do you think so?

TEACHING AND LEARNING HIGH SCHOOL MATHEMATICS © 2010 John Wiley & Sons, Inc.

c. What theorems are the students attempting to prove, if any?

Adriana: *If the circle was small enough, it could fit into the corner and intersect at only the vertex. My circle is a little too big, but I think it will work.*

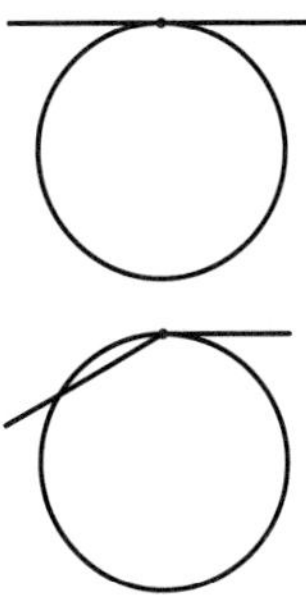

Anika: *When the square and the circle intersect at a tangent to one of the sides of the square, the angle formed at the intersection point is 180°. If we bend the line, even a little bit, at the tangent point, one of the segments has to intersect the circle at another point. The second segment is not tangent to the circle but intersects the circle in two points, its endpoint, and another point on the segment.*

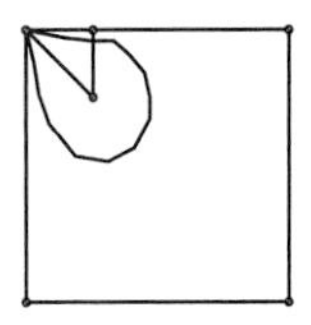

Ariel: *Suppose there is a circle that can fit inside a square, intersecting the square at only a vertex. Then we can connect the center of the circle to the vertex of the square. This length is a radius of the circle. If we construct a perpendicular to one of the sides of the square from the center of the circle, then the length of this perpendicular segment must be longer than the radius because the circle lies entirely inside the square. But the radius is the hypotenuse of a right triangle and the other longer side is one of the legs. This can't happen. In a right triangle, the hypotenuse is the longest side.*

6. For each correct student response in item 5, answer the following questions.

a. What defined or undefined terms is the student using?

b. What axioms is the student assuming true?

c. What theorems is the student using?

d. If the student's response is not a proof, what more does she need to show for her work to be a proof?

7. Referring to the work the students generated for 0 intersections and 1 intersection, the teacher commented:

You have shown several ways that a circle and a square can intersect either in no points or in a single point. For the examples of 0 intersections, two of the cases show one figure surrounding the other and the other case shows the figures side by side. Two of your 1-intersection examples show a circle and a square intersecting at a tangent point, one where the circle is inside the square and one where the circle is outside the square. Are these different?

a. Answer the teacher's question.

b. One group of students chose to categorize the three 0-intersection cases as different. Create symbolism that the group could use to distinguish each case. Use the symbolism to characterize the 1-intersection cases.

c. A second group of students chose to ignore the distinction between a circle being mostly inside a square or a square being mostly inside a circle in the process of finding 2-intersection cases. Does this decision reflect the students' choice of an undefined term, a definition, an axiom, a theorem, or none of these? Why do you think so?

d. Compare and contrast consequences of the students' decisions in items 7b and 7c for the 2-intersection cases. What conjectures might students make regarding numbers of intersections for these cases?

TEACHING AND LEARNING HIGH SCHOOL MATHEMATICS © 2010 John Wiley & Sons, Inc.

SHARE AND SUMMARIZE

8. The problem in this lesson was categorized as *higher level demands* (doing mathematics). Do you agree with this categorization? Why or why not?
9. Share your responses to items 4–7. How does the work completed in this lesson help students become more familiar with axiomatic systems?

   Note: Special thanks to Stephen Blair for his assistance sharing student work for this lesson.

DEEPENING MATHEMATICAL UNDERSTANDING

1. *Undefined terms.*
   a. Look up the term *line* in a dictionary. Continue to look up words used in subsequent definitions. What do you notice?
   b. Find another word to research in the dictionary so that, after looking up the terms used to define your word, the definition eventually uses the word with which you began your search.
   c. What does your work in problems 1a and 1b tell you about the importance of undefined terms?
2. *Symbolizing Intersections of Circles and Squares.*
   a. Define *intersection*.
   b. What are all the ways that a circle and a square can intersect?
   c. Decide on a way to symbolize each type of intersection that can occur between a circle and a square.
3. *Categorizing Intersections of Circles and Squares.*
   a. How many ways can a circle and a square intersect in two points?
   b. Use your symbolizations from problem 2c to distinguish among the different types of intersections.
   c. Refine intersection descriptions and symbolizations based on work in problems 3a and 3b.
4. *Two-Point Intersections of a Circle and Square.*
   a. Meridel was trying to see if she could find any other 2-point intersections for a circle and a square. She remembered work with finding circles tangent to an angle and completed the following construction:

      I constructed a square. Then I constructed a diagonal and a point on the diagonal. From this point, I dropped a perpendicular to one of the sides of the square and then created a circle using the point on the diagonal as the center and the intersection of the perpendicular with the square's side as the radius.

      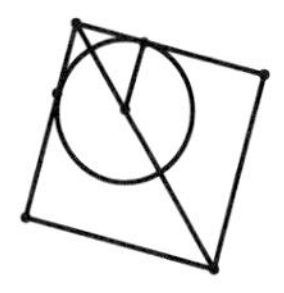

      Meridel found that, if she moved the center along the diagonal, several different intersections between a circle and a square were possible. What are they?
   b. Refine Meridel's directions in problem 4a so the new directions only construct one class of 2-point intersections between a circle and a square.
   c. Prove that the construction in problem 4b results in the 2-point intersection described and no other class of intersections.
   d. Outline a proof of your conjecture for problem 4a.
5. *Constructions of Circle and Square Intersections.*
   a. Use a compass and a straightedge to construct one of your classified drawings from **DMU** problem 3 (different from the possibilities described in **DMU** problem 4).

TEACHING AND LEARNING HIGH SCHOOL MATHEMATICS © 2010 John Wiley & Sons, Inc.

b. Give explicit, step-by-step directions to complete the construction in problem 5a. Give the directions to a friend. Ask your friend to construct your drawing from just these directions. Is more than one type of intersection between a circle and a square possible with your directions? If so, how can you make your directions more explicit? If not, prove that your directions provide only one type of intersection.

6. *More Constructions of Circle and Square Intersections.*

a. Determine how a circle and a square can intersect in more than 5 points. Classify the drawings.

b. Using a compass and a straightedge, construct one of your classified drawings from problem 6a.

c. Repeat problem 5b for your construction in problem 6b.

7. *Theorems Arising from Intersections of Circles and Squares.* Students in one geometry class decided that it would be "opening a can of worms" to try to find all the intersections between a circle and a square if they also had to decide if the circle was mostly on the outside of the square or mostly on the inside. They decided to forego this distinction as they continued finding four or more intersections between a circle and a square. They revisited the 1-point intersections and decided there were only two ways a circle and a square could intersect in one point. A theorem they proved, by cases, was: *Assuming that a circle and a square lie in the same plane, there are only two ways they can intersect in a single point. One way is through a point tangent to a circle along a side of a square. A second way is through a vertex point of a square.*

a. A student asked if a simpler theorem might be stated. She wondered if the following could be considered a theorem: *A circle can intersect a square in a single point if the intersection is a vertex of a square.* Comment on this student's statement. As stated, does it qualify as a theorem? Justify your response.

b. What other theorems might students attempt to prove that arise from the circle and square intersections problem?

c. Students suggested *point*, *line*, and *plane* as undefined terms and *circle*, *square*, and *intersection* as defined terms. What additional undefined or defined terms, axioms, or theorems will students need to prove their results?

8. *Extending an Axiomatic System.* As you solved the circle and square intersections problem:

a. What symbolizations or conventions helped you categorize your findings?

b. What consequences arose from the defined and undefined terms, axioms, and theorems you used?

c. List three consequences (theorems) that arise from ignoring size and position of square and circle, relative to each other.

d. List three consequences (theorems) related to but different from those in problem 8c that arise from distinguishing which shape is mostly inside the other shape.

e. Comment on the extent to which decisions about the rules of the game—axioms—affect the end result.

WWW **9.** *Revisiting Knightly Knews.* Locate the student page **Knightly Knews** in Lesson 1.4. List any undefined terms, defined terms, axioms, and theorems that arise from the puzzle. Determine whether the puzzle can be viewed as an axiomatic system. If not, what changes must be made for the puzzle to be considered an axiomatic system?

WWW **10.** *Revisiting Ring of the Lords.* Locate the student page **Ring of the Lords** in Lesson 1.4. List any undefined terms, defined terms, axioms, and theorems that arise from the puzzle. Determine whether the puzzle can be viewed as an axiomatic system. If not, what changes must be made for the puzzle to be considered an axiomatic system?

TEACHING AND LEARNING HIGH SCHOOL MATHEMATICS © 2010 John Wiley & Sons, Inc.

11. *Revisiting Exploring Quadrilaterals and their Definitions and Quadrilaterals on Spheres.* Locate the student page **Exploring Quadrilaterals and Their Definitions**, and its extension **Quadrilaterals on Spheres**, in Lesson 2.3.

- **a.** When based on their diagonals, compare the quadrilaterals constructed in a plane with those constructed on a sphere. What similarities exist? What differences exist?
- **b.** What undefined terms are important when working with quadrilaterals in a plane?
- **c.** What undefined terms are important when working with quadrilaterals on a sphere?
- **d.** What defined terms are important when working with quadrilaterals in a plane?
- **e.** What defined terms are important when working with quadrilaterals on a sphere?

12. *More Work with Quadrilaterals in the Plane and on a Sphere.* Consider your work on the student page **Exploring Quadrilaterals and Their Definitions** and its extension, **Quadrilaterals on Spheres,** in Lesson 2.3.

- **a.** Which axioms apply to quadrilaterals, regardless of the surface on which they are constructed? Explain.
- **b.** Which additional axioms are necessary for each surface that do not apply to the other?
- **c.** Consider definitions of quadrilaterals based on relationships between their sides and angles. Write these definitions in their most general sense. For example, rather than define a rectangle as a quadrilateral with four right angles, define the rectangle as a quadrilateral with four congruent angles.
- **d.** Write a definition for a parallelogram that does not depend on opposite sides being parallel.
- **e.** Determine which of the quadrilateral definitions are equivalent to their diagonal versions when the quadrilateral is constructed on a sphere.

DEVELOPING MATHEMATICAL PEDAGOGY

Reflecting on Our Own Learning

13. **a.** What prior experience have you had working with an interactive geometry drawing tool?

- **b.** What do you perceive to be the benefits of having such a tool to explore geometry?
- **c.** What are potential downsides?

14. **a.** What prior experiences have you had in working with axiomatic systems?

- **b.** How might games or puzzles have assisted you in understanding axiomatic systems?
- **c.** How might they help your students?

Communicating Mathematically

15. *Anticipation Guides* (Barton and Heidema, 2002) are a reading strategy that can be used at a lesson or unit level. Teachers write several statements related to the concepts that students would read about in their textbook; some of the statements are true, some are false. For each statement, there is a column labeled "Me" and a column labeled "Text." Students read each statement and indicate whether they believe the statement is true or false, recording their response in the column labeled "Me." After reading the lesson or unit, students review each

TEACHING AND LEARNING HIGH SCHOOL MATHEMATICS © 2010 John Wiley & Sons, Inc.

statement again, recording in the column labeled "Text" whether the statement is true or false based on the reading. For instance, in a lesson on circles, a sample statement might be: *Any segment drawn from the center of a circle to the intersection of a tangent line with the circle makes a 90° angle with the tangent line.*

a. Find a high school geometry text. Choose a unit (chapter or several lessons) dealing with circles, quadrilaterals, triangle congruence, or surface area and volume. Write at least five statements that could be used as part of an anticipation guide.

b. How might an anticipation guide be used as an advance organizer to orient students to concepts that will be studied in an upcoming unit?

c. How might an anticipation guide be used as a culminating activity near the end of a unit?

d. By the time students arrive in high school, they have been exposed to many geometric concepts. How might an anticipation guide be used by a teacher in designing instruction?

e. Write at least five statements that would be appropriate for an anticipation guide prior to the start of a lesson on axiomatic systems.

Analyzing Students' Thinking

16. While trying to answer the teacher's question, "Is it possible for a circle to intersect a square at a vertex where the circle is inside the square?" (see **Explore** problem 5), students kept trying to use the Pythagorean Theorem. Analyze the student work below.

a. How effective was the use of the Pythagorean Theorem? Suggest ways that the solution could have been more succinct.

b. In stage four of the Stages of Questioning, teachers encourage students to think about their thinking. What questions would you like to ask Azita?

Azita: *I wonder if we can use the Pythagorean Theorem to show that this kind of circle can't happen. In a right triangle,* $a^2 + b^2 = c^2$ *where a and b are the lengths of the legs and c is the length of the hypotenuse. I can't see how the circle could fit entirely inside the square and still only intersect the square at its vertex. The hypotenuse has to be the radius of the circle whose length is c, and the other two sides with lengths a and b have to be shorter than the radius. I constructed my circle and square with the computer and can see that the circle just can't be squeezed inside the square and still intersect the vertex of the square.*

Working with Students

17. Read "How Well Do Students Write Geometry Proofs?" by Sharon L. Senk (*Mathematics Teacher*, 78 (September 1985): 448–456).

a. Although a number of years old, this classic article addresses persistent concerns that student achievement at writing proofs is rather low. Share the items with a current geometry teacher. What is the teacher's perception about potential achievement on the proof items in the article?

b. If possible, have some students who have already studied geometry complete the proofs. How successful were the students? What misconceptions, if any, arose in the students' work?

c. The article makes several recommendations for teachers to consider in helping their students be successful at proof writing. Briefly summarize these recommendations. Which ones seem easiest to implement in your own classroom?

TEACHING AND LEARNING HIGH SCHOOL MATHEMATICS © 2010 John Wiley & Sons, Inc.

Reflecting on Professional Reading

Problems 18–20 refer to "Intersections of a Circle and a Square: An Investigation" by Dan Canada and Stephen Blair (*Mathematics Teacher*, 100 (December 2006/January 2007): 324–325).

18. Refer to the article by Canada and Blair.

a. On page 325, the authors report that a number of preservice teachers who worked on the problem admitted they did not think certain numbers of intersections were possible. What questions would you like to ask these preservice teachers to help them move beyond their difficulties (recall the Stages of Questioning from Lesson 1.1)?

b. Try out the applets created by the authors to investigate the circle and square intersections problem (see www.ewu.edu/prime, downloaded May 8, 2009). What are advantages of using these applets over using paper and pencil? What are advantages of using these applets over using compass and straightedge? What are disadvantages of using the applets before students have explored the problem by hand?

c. Compare the ways the intersections were classified in the article to the ways you classified them. What similarities did you find between classification schemes? What differences exist between classification schemes? What results are different based on more specific or less specific classification schemes?

19. Refer to the article by Canada and Blair.

a. You were not asked to construct intersections of circles with squares during the lesson. What more would students learn by completing constructions of some of the types of intersections between circles and squares?

b. Two different constructions are provided in the article. One began with a circle and the other began with a square. Attempt at least one of these constructions beginning with the other figure, that is, for the construction that began with the circle, see if you can construct the same type of 5-point intersection beginning with a square.

20. When the problem in the article by Canada and Blair was used with one group of students, they classified the types of intersections as *vertex tangent*, *vertex crossing*, *side crossing*, and *side tangent*.

a. Draw an illustration of each of these classifications.

b. How well do they agree with your classifications from **DMU** problems 2 and 3?

c. Some teachers argue against the use of the term *vertex tangent* because of concerns that students might develop misconceptions about the meaning of tangent that would create difficulties in more advanced courses. Where might such concerns arise? In what ways do these concerns support the importance of context when using definitions?

d. What language might be used in place of *vertex tangent* so that such potential misconceptions are avoided?

Preparing to Discuss Observations

21. In preparation for Lesson 2.7, review your notes from your observations in high school mathematics classes. Also, review the articles you read in preparation for the observations and in preparation for reflections. Be prepared to discuss your findings with your peers.

a. How did the reading from *How Students Learn: Mathematics in the Classroom* relate to the Learning Principle from *Principles and Standards for School Mathematics*?

b. Consider the examples used during Observation 2.1 to introduce students to the concepts studied, whether used by the teacher or referenced in the textbook. In what ways did the examples refer to students' prior experiences? How

TEACHING AND LEARNING HIGH SCHOOL MATHEMATICS © 2010 John Wiley & Sons, Inc.

motivating were the examples? In what ways did the examples focus on conceptual understanding, procedural fluency, or connections between the two?

c. In what ways did the teacher make students aware of the goals and/or objectives of the lesson? What approaches did the teacher use to monitor students' understanding during the lesson? In what ways were students encouraged to think about their own thinking?

Preparing to Discuss Listening to Students

22. Review students' responses to the problems on the student page **Connecting Midpoints of Sides of Polygons.**

a. Make certain your solutions and all the expected and observed student responses to the problems are recorded in a **QRS Guide** for the student page.

b. Categorize each response as misconceived, partial, or satisfactory.

c. Categorize each response according to the van Hiele Levels of Geometric Thought.

d. For each student response, record any appropriate teacher support.

e. What, if anything, surprised you about students' responses? What difficulties, if any, did they have that you had not anticipated?

f. Reconsider your reading of "Linking Theory and Practice in Teaching Geometry" by Randall E. Groth (*Mathematics Teacher*, 99 (August 2005): 27–30). Based on the students' responses to **Connecting Midpoints of Sides of Polygons** and your understanding of the van Hiele Levels of Geometric Thought, what issues or challenges would you expect to face in a high school geometry course?

2.7 Summarizing Classroom Observations and Listening to Students

At the beginning of this unit, you were asked to observe at least three class periods. Two of your observations were to focus on students' prior knowledge or experiences and how their everyday experiences could provide a bridge to abstract concepts. One observation was to focus on how teachers monitor students' progress during a lesson and how teachers help students learn to monitor their own progress. Successful students make connections among concepts and learn to think about their own learning; teacher support for students in these endeavors is a critical part of their success.

In addition, you were asked to work with a pair of high school students on a series of geometry tasks, **Connecting Midpoints of Sides of Polygons**, and to consider challenges their knowledge or lack of knowledge might pose for a teacher's instruction. In particular, the van Hiele Model of Geometric Thought provides insights into the types of opportunities that students need in order to develop the ability to engage in deductive thought.

In this lesson, you will:

- Compare the results of your observations with those of your peers.
- Reflect on what your observations suggest about students' learning of mathematics.

LAUNCH

1. Share your overall impressions about the classes you observed. What impressed you? What concerned you? What surprised you? How did these observations help you think about your own future classroom?

TEACHING AND LEARNING HIGH SCHOOL MATHEMATICS © 2010 John Wiley & Sons, Inc.

EXPLORE RESPONSES TO OBSERVING MATHEMATICS CLASSROOMS

2. Consider the examples teachers used during Observation 2.1.
   a. How did the examples motivate students or connect to their everyday experiences?
   b. In what ways did the examples focus on conceptual understanding, procedural fluency, or connections between the two?
3. Consider your comments about Observation 2.2.
   a. How did teachers make students aware of the goals and/or learning outcomes of the lesson?
   b. What techniques did teachers use to monitor students' understanding during the lesson?
   c. What approaches did teachers use to provide opportunities for students to think about their own thinking and monitor their understanding of the lesson's goals?
   d. What aspects of the lesson supported equity, that is, all learners being engaged in learning? What might have enhanced all students' learning?

EXPLORE RESPONSES TO LISTENING TO STUDENTS

4. Share students' responses to the student page **Connecting Midpoints of Sides of Polygons**.
   a. What, if anything, surprised you about students' responses?
   b. How were students able to justify their conjectures?
   c. With your peers, compare the student responses and teacher support you prepared prior to and after the interview.
   d. Include additional student responses and teacher support in your **QRS Guide**.
5. Critique the **QRS Guide** you prepared for the interview tasks with the guides created by your peers. Identify strengths and weaknesses in the types of teacher support you generated. What did you learn from your use of the guide that will help you prepare a **QRS Guide** for another student task?

SHARE AND SUMMARIZE

6. As a class, compare the responses from the different groups to the items in the **Explore**. Discuss the ways in which the Learning Principle was evident in the classes you observed.
7. Discuss the readings "The Learning Principle" from *Principles and Standards for School Mathematics* (pp. 20–21) and "Introduction" by M. Suzanne Donovan and John D. Bransford from *How Students Learn: Mathematics in the Classroom* (pp. 1–21).
   a. Based on the readings, what is a concept? What is a procedure?
   b. In the classes you observed, in what ways were teachers helping students understand concepts? Give examples and share how teachers helped students build conceptual understanding.
   c. In the classes you observed, in what ways were teachers helping students develop conceptual understanding for procedures or become proficient with procedures? Give examples.
   d. In what ways do the two readings reinforce each other?
   e. In what ways do the readings agree with the summary of learning theory in the reading by Rubenstein in Lesson 2.3?

TEACHING AND LEARNING HIGH SCHOOL MATHEMATICS © 2010 John Wiley & Sons, Inc.

8. Based on your observations and on the readings related to the observations, what appear to be some of the challenges you are likely to face in teaching high school mathematics? What is your current thinking about how to address one of those challenges?

DEVELOPING MATHEMATICAL PEDAGOGY

Reflecting on Our Own Learning

1. Think about your own experiences in mathematics classes, either at the high school level or college level.
   - **a.** What teaching strategies did effective mathematics teachers use to help you develop conceptual understanding? What made the teaching strategies effective for you?
   - **b.** What teaching strategies did effective mathematics teachers use to help you develop procedural fluency? What made the teaching strategies effective for you?
   - **c.** What teaching strategies did effective mathematics teachers use to help you become a proficient problem solver? How did these teachers help you learn to solve problems for which no known algorithm was available?
   - **d.** What additional teaching strategies do you think would help support a broader range of students?

Analyzing Instructional Materials

2. Locate at least three current geometry or integrated high school textbooks (copyrighted within the last ten years) that are for the same course level.
   - **a.** Choose one topic of study for the course you are observing that is found in all three textbooks. Investigate how the topic is addressed in each textbook. Compare and contrast the treatment of the topic across all three books. What are the similarities? What are the differences? How is the treatment balanced between developing procedural understanding and conceptual understanding?
   - **b.** What aspects of the topic, if any, seem to be missing?
   - **c.** In what ways do the materials support the learning of students from the concrete level to the abstract level?
   - **d.** In what ways do the materials provide worthwhile tasks (high cognitive demand)?
   - **e.** Find a task, if possible, that can be accessed by students in a variety of ways. What different ways of thinking do you anticipate students would use?
3. Find a set of textbooks from grades 7–12; if possible, obtain your set of textbooks from the same publisher.
   - **a.** Choose a topic that would be taught at some level in most of these grades. If possible, choose a topic with some connections to geometry. Investigate how this topic develops over grades 7–12 and write a synthesis of your investigation. Here are some questions to guide your synthesis:
     - How is the topic introduced in grade 7?
     - How do introductions in subsequent grades build on the grade 7 introduction?
     - How does the depth of study with the topic change and/or grow over time?
     - What is the focus of the topic of study over time: skills, applications, concepts, representations?
     - How does the development of the topic over the grades align with recommendations for that concept in *Principles and Standards for School Mathematics*?

TEACHING AND LEARNING HIGH SCHOOL MATHEMATICS © 2010 John Wiley & Sons, Inc.

b. Reflect on your investigation. What aspects of your investigation surprised you?

c. If you had access to teachers' editions of the materials, what additional resources seem to be available to teachers to scale the concepts up or down to meet the needs of students with varying ability levels?

Debating Issues

4. Suppose a colleague asks you to develop a policy brief for the school board to provide financial support for a computer lab for use with geometry classes. Identify some of the main points you would use to lobby for such a lab.

Focusing on Assessment

5. Throughout this unit, use of an interactive geometry drawing tool has been encouraged to complete several activities. In addition, there have been several readings about the use of technology in teaching geometry. Many graphing calculators have interactive geometry drawing tools as one of their applications.

   a. What technology access related to geometry have you observed in the schools you visited?

   b. If students are asked to construct and modify geometric objects to develop their own conjectures, a timed environment may or may not be appropriate. That is, if teachers want students to explore figures and consider various relationships, students need time to explore and reflect on their findings. Hence, some assessments might need to occur as out-of-class projects. What are some issues that would need to be considered in using assessments in this type of format?

Communicating Mathematically

6. Throughout this unit, a number of **Developing Mathematical Pedagogy (DMP)** questions focused on mathematical communication. Neither teachers nor students can usually sustain focusing on a new strategy each week. Rather, it is often more productive for a teacher to find a few powerful strategies (e.g., link sheet, anticipation guide, concept map) and to use them regularly. Students can then become comfortable with a particular strategy and learn what makes a quality response. Of the strategies presented in this unit, which are you likely to consider using in your own classroom? Which are you less likely to use? Provide a rationale for your choice.

Teaching for Equity

7. How does a teacher's understanding of how students learn mathematics help him or her create an equitable classroom?

8. Walker and McCoy (1997) write about the perceptions of African American students that influence their participation and achievement in mathematics. Here are some of the issues that arose during their interviews with students:

   - Students were concerned that others understood the mathematics better than they did and so did not want to speak up.
   - If students were the only minority student in the class, they often did not want to answer incorrectly because they felt they represented their entire group.
   - If students thought teachers cared about them and related to them, they were more likely to perform well in the teacher's class.
   - Students found sports to be a diversion, particularly if they were planning to attend college on a sports scholarship.
   - Students did not realize that mathematics was important to their future plans.

TEACHING AND LEARNING HIGH SCHOOL MATHEMATICS © 2010 John Wiley & Sons, Inc.

Based on the insights from students, what actions might you take in your own classroom to ensure that all students engage with the class and none become invisible? (For more insights, read "Students' Voices: African Americans and Mathematics" by Erica N. Walker and Leah P. McCoy. In Janet Trentacosta and Margaret J. Kenney (Eds.), *Multicultural and Gender Equity in the Mathematics Classroom: The Gift of Diversity* (pp. 71–80). Reston, VA: National Council of Teachers of Mathematics, 1997.)

Reflecting on Professional Readings

9. Read "Learning Mathematics Vocabulary: Potential Pitfalls and Instructional Strategies" by Denisse R. Thompson and Rheta N. Rubenstein (*Mathematics Teacher*, 93 (October 2000): 568–574).

a. The authors highlight several categories of difficulties with mathematics vocabulary. Choose two of these categories and identify at least five additional mathematics terms that fit the category.

b. Compare the issues raised by these authors with the strategies you have experienced in the **Communicating Mathematically** questions throughout this unit.

10. Read "Helping Teachers Connect Vocabulary and Conceptual Understanding" by A. Susan Gay (*Mathematics Teacher*, 102 (October 2008): 218–223). Gay describes several classroom activities that help individuals use vocabulary to address conceptual understanding.

a. How do the ideas suggested by Gay relate to conceptual understanding that has been the focus of this unit?

b. Which of the ideas suggested by Gay are you interested in using in your classes?

TEACHING AND LEARNING HIGH SCHOOL MATHEMATICS © 2010 John Wiley & Sons, Inc.

SYNTHESIZING UNIT TWO

Mathematics Strands: Geometry and Measurement

This unit focused on the learning of mathematics, including the importance of building conceptual understanding, the use of multiple perspectives when introducing a concept, the use of physical and technological tools, the use of worthwhile mathematical tasks with appropriate levels of cognitive demand, and the revision of lower level tasks to make them more intellectually challenging. Throughout the unit, you have been encouraged to think about potential student responses and appropriate teacher support to help students make sense of geometry and measurement.

The principles of learning discussed in this unit are important—regardless of the mathematical content studied. In addition, in geometry, mathematics educators have a particular theoretical framework on which to build students' geometrical understanding. This framework, the van Hiele Levels of Geometric Thought, consists of five levels, from visualization to rigor, through which students progress sequentially. For students to be successful in most high school geometry courses, they need to reach the level of deduction.

Geometry can be accessed through visual and hands-on physical tools, including geoboards, patty paper, MIRAs, compasses, and straightedges. Geometry can also be accessed through interactive geometric drawing tools. The variety of tools enables students to engage in investigating geometric figures, make their own conjectures, and then attempt to prove them.

Thinking about how students learn mathematics is a critical part of a teacher's daily work. Planning for instruction requires anticipating students' correct responses as well as their potential misconceptions and errors. By thinking about how students are likely to engage with mathematical tasks, teachers can prepare questions that give "just enough" help so that students move forward if they struggle with a concept.

Having arrived at the end of this unit, it is time to reflect on your understanding of the lessons learned throughout the unit. The following items provide an overview of the unit. For each item, you should be able to pose appropriate questions and respond meaningfully, providing support for your answers with specifics drawn from the various lessons, discussions, readings, homework problems, or observations in high school classes or interviews of high school students.

Website Resources

- van Hiele Levels of Geometric Thought
- **Where Do I Stand?** survey

MATHEMATICAL LEARNINGS

1. **a.** Given a geometric figure described with coordinates on the plane, use the coordinates to make conjectures about relationships within the figure (e.g., slopes of segments, lengths of segments).

   b. Construct a proof of your conjectures using coordinates, in a specific case as well as in a general case.

2. **a.** Given a set of side lengths and/or angle measures for two triangles, determine whether the triangles are congruent, similar, or neither.

   b. Determine what information about two triangles must be known to determine if the pair of triangles is congruent.

   c. State and explain the SAS triangle congruence axiom.

   d. State and explain the triangle congruence theorems.

TEACHING AND LEARNING HIGH SCHOOL MATHEMATICS © 2010 John Wiley & Sons, Inc.

3. a. Define each of the special quadrilaterals in at least two different ways and prove the definitions equivalent.
   b. Construct each of the quadrilaterals in the plane.
   c. State similarities and differences among quadrilaterals in Euclidean geometry. Categorize quadrilaterals according to inherited properties.
4. Identify relationships in circles among central angles, intercepted arcs, and tangents.
5. Construct basic geometric figures using straightedge and compass, MIRAs, an interactive geometry technological tool, or patty paper. Identify differences in the needed information for constructions with the different tools.
6. Derive formulas for two-dimensional and three-dimensional measures.
   a. Specifically, derive formulas for the areas of parallelograms, trapezoids, and circles.
   b. Derive formulas for the lateral area of a cone and of a pyramid as well as the surface area of a sphere.
   c. Derive formulas for the volume of a sphere and of a cylinder.
7. Identify the elements of an axiomatic system. Provide a rationale for why some terms need to remain undefined.

PEDAGOGICAL LEARNINGS

8. a. Identify the five van Hiele Levels of Geometric Thought and briefly describe the understandings students have at each of the levels.
   b. For a given geometry topic, describe an activity at each of the first four levels that would be appropriate for high school students.
9. Describe several strategies that can be used to help students become successful at writing proofs. Discuss the pros and cons of using the different strategies in a classroom.
10. a. Describe the four levels of cognitive demand for mathematics tasks and categorize tasks according to these levels.
    b. For a given topic in geometry or measurement, create a task at the two highest levels of cognitive demand.
    c. For a given low level task, rescale the task to a higher level of cognitive demand.
11. Describe a variety of activities that can help all students develop conceptual understanding of triangle congruence theorems.
12. Describe activities that can be used with high school students to help them develop conceptual understanding of area formulas for a parallelogram, a trapezoid, and a circle.
13. Describe activities that can be used with high school students to help them develop conceptual understanding of formulas for the surface area and volume of a sphere and the volume of a cone.
14. Given two similar three-dimensional figures with a scale factor of k between them.
    a. Determine how their perimeters, surface areas, and volumes are related.
    b. Describe one or more activities that could be used with high school students to help them understand these relationships among similar figures.
    c. Explain how your activities provide challenge for students having different strengths.
15. Given a problem, construct variations of the problem that provide support for struggling students as well as variations that provide more challenge for students who find the initial problem too easy.

TEACHING AND LEARNING HIGH SCHOOL MATHEMATICS © 2010 John Wiley & Sons, Inc.

16. Given a mathematics problem and sample student responses, provide suggestions for questions that teachers can pose to help students move beyond their response to a complete solution of the problem.
17. Discuss some of the pros and cons of having high school students explore different definitions of quadrilaterals and determine their equivalence.
18. In this unit, you completed constructions using low-tech tools (straightedge and compass, MIRA, or patty paper) and technology (an interactive geometry drawing tool).
    a. Compare and contrast these approaches to constructions. What are students likely to learn from one approach that they are not likely to learn from others? What are the strengths and limitations of each approach? How do particular tools support students with special needs?
    b. Many graphing calculators now contain interactive geometry drawing tools. Hence, teachers are no longer limited by lack of access to computer labs. What are some challenges teachers must consider when students have access to such technology on a continual basis?
19. Revisit your responses to the survey **Where Do I Stand?**, for items 4, 11, 12, 14, and 17 (see the course introduction). In what ways have your responses changed, deepened, or grown?
20. Revisit your responses to item 24 on the survey **Where Do I Stand?** Are there any additional challenges you expect to face as a high school mathematics teacher?

TEACHING AND LEARNING HIGH SCHOOL MATHEMATICS © 2010 John Wiley & Sons, Inc.

UNIT TWO INVESTIGATION: TRANSFORMATIONS

In the investigation for Unit Two, you will explore transformations through real-world examples, paper folding, use of a MIRA, an interactive geometry tool on a computer or graphing calculator, or matrix operations. You will also have an opportunity to create or analyze an artistic design that uses at least three different types of transformations.

Following are questions to guide your investigation. Choose some that interest you and take them as far as you can. Explain your solutions in coherent sentences. Include diagrams to make your arguments clear. Provide complete citations for all resources used to complete this investigation (including names of persons you consulted). Attach copies of examples of transformations from printed resources, including newspaper ads, wallpaper, gift wrap, photographs, and/or artwork.

1. *Defining Transformations and Related Terms.* In a high school geometry textbook (do not just rely on the Internet), look up each of the following words related to transformations. Define each. Determine the notation that shows the relationship between a preimage and an image under each transformation. List the full reference in your paper.

a. reflection	b. translation	c. rotation
d. glide reflection	e. dilation	f. orientation
g. isometry	h. preimage	i. image
j. center of rotation	k. magnitude of rotation	l. composition of transformations

2. *Transformations in Real Life.* Choose a high school of interest to you.
   a. Identify the high school as urban, rural, or suburban. Find out as much as you can about the cultural/ethnic background of the students in the school. Briefly summarize what you have learned about the students.
   b. Find three examples of each type of transformation (listed in items 1a–1e) that are likely to be of interest to students in the chosen high school. Examples must be real life, coming from media sources such as magazines, the newspaper, or the Internet, from photographs or artwork, or from wallpaper or gift wrap. Other sources are also possible, but examples should not be hand drawn unless you are a talented artist. Cut and paste examples into your paper, citing all resources fully under each image.
      i. Show by outline and notation the relationship between the preimage and image for each example.
      ii. Label at least 3 points on the preimage and show the transformed points on the image.
      iii. Below each picture, briefly state why each example is of interest to students in your school.
3. *Transformations by Hand.* Use paper folding or a MIRA to investigate what happens to a concave asymmetric two-dimensional object under the following compositions of reflections. Label at least three non-collinear vertices and their reflected images. For example, the reflection of point P across line 1 is denoted by P'; the reflection of P' across line 2 is denoted by P''. In each case, discuss the relationship between the final image and the preimage in terms of size, orientation, position, direction and magnitude of translation, center and magnitude of rotation, etc. What is similar? What is different? Explain.
   a. Reflection of a preimage across a pair of parallel lines
   b. Reflection of a preimage across a pair of nonperpendicular intersecting lines

TEACHING AND LEARNING HIGH SCHOOL MATHEMATICS © 2010 John Wiley & Sons, Inc.

4. *Transformations with Technology*. Use an interactive geometry drawing tool, such as *The Geometer's Sketchpad*™ or *Cabri Geometry*, to investigate what happens to a concave asymmetric two-dimensional object under the following transformations. Label each vertex point and its transformed image. Predict how the preimage is related to the final image. For items 4a–4d, justify each conjecture by finding a single transformation that will take the preimage to its final image, showing that this transformation works, and explaining why it works. Print the results for each case.

   a. Repeat the investigation in item 3.

      i. Determine the center and magnitude of rotation compared to the relationship between the lines of reflection from which the rotation arises. Explain why your findings are sensible.

      ii. Determine the direction and magnitude of translation compared to the relationship between the lines of reflection from which the translation arises. Explain why your findings are sensible.

   b. Transform a preimage by reflecting it across a line and then translating the result parallel to the line of reflection.

   c. Transform a preimage by reflecting it across a line and then rotating the result through a point on the line of reflection. Use an acute angle of rotation.

   d. Transform a preimage by reflecting it across three parallel lines, one at a time.

   e. Choose a point as a center of dilation and transform a preimage by dilating it to create a figure whose side lengths are 1.5 times the corresponding lengths of the preimage. Determine the relationship between the areas of the preimage and the image. Explain your findings.

5. *Transforming Points*. Determine the mapping of each point (x, y) to accomplish the transformation requested. Show the mapping using the notation $(x, y) \rightarrow (a, b)$, where a and b are replaced by the appropriate transformed coordinates.

   a. Translation of the point (x, y) by h units horizontally and k units vertically.

   b. Reflection of the point (x, y) across:
   i. the x-axis **ii.** the y-axis **iii.** the line $y = x$ **iv.** the line $y = -x$

   c. Rotation of the point (x, y) counterclockwise about the origin $(0, 0)$
   i. 90 degrees **ii.** 180 degrees **iii.** 270 degrees

   d. Dilation by a factor of n, resulting in similar figures with the origin $(0, 0)$ as the center of dilation.

6. *Matrices and Transformations*. The dimensions of a matrix are written as *number of rows* × *number of columns*. To multiply matrices together, matrices must be appropriate dimensions. Although it is possible to multiply a matrix of dimensions $m \times n$ by a matrix of dimensions $n \times p$ resulting in a matrix with dimensions $m \times p$, it is not possible to multiply a matrix of dimensions $n \times p$ by a matrix of dimensions $m \times n$. The multiplication of two matrices is shown below. Study the example to determine how entries are combined to produce the product.

$$\begin{bmatrix} q & r \\ s & t \end{bmatrix} \times \begin{bmatrix} a & b & c \\ d & e & f \end{bmatrix} = \begin{bmatrix} qa + rd & qb + re & qc + rf \\ sa + td & sb + te & sc + tf \end{bmatrix}$$

   a. Show that the 2×2 identity matrix for multiplication is

$$\begin{bmatrix} 1 & 0 \\ 0 & 1 \end{bmatrix}.$$

   b. Is matrix multiplication commutative? Explain.

   c. The matrix

$$\begin{bmatrix} a & b & c \\ d & e & f \end{bmatrix}$$

TEACHING AND LEARNING HIGH SCHOOL MATHEMATICS © 2010 John Wiley & Sons, Inc.

can be used to denote the triangle whose vertices are (a, d), (b, e), and (c, f). Draw a nonconvex, asymmetric pentagon in the viewing rectangle [0, 5] by [0, 3]. Create a matrix denoting the vertices of your pentagon.

d. Determine how to enter the matrix into your graphing calculator. Multiply the matrix

$$\begin{bmatrix} 1 & 0 \\ 0 & 1 \end{bmatrix}$$

by your pentagon matrix. Verify that you get the same pentagon matrix in return.

e. What is the result when you multiply the matrix

$$\begin{bmatrix} 0 & 1 \\ 1 & 0 \end{bmatrix}$$

by your pentagon matrix? Which of the transformations in item 5 gives this result? Explain why this result is sensible.

f. Without looking up the remaining transformation matrices in a textbook or any other resource, try to discover the matrices needed to transform your pentagon for each of the transformations in items 5b, 5c, and 5d. Show the matrix multiplication and sketch the preimage and image on the same coordinate axes. Explain why each matrix works as you say it does.

g. Determine the matrix that will rotate the point (x, y) counterclockwise about the origin for the angles shown. Show the matrix multiplication and sketch the preimage and image on the same coordinate axes. Explain why each matrix works as you say it does.

i. 45 degrees **ii.** θ degrees

7. *Connections with Group Theory*. In "Group Symmetries Connect Art and History with Mathematics" by Anthula Natsoulas (*Mathematics Teacher*, 93 (May 2000): 364–370), the notation r represents the least number of degrees in the rotational symmetry of a figure. So, for a square, r is 90° because a square has 90° rotational symmetry; r^2 represents two 90° turns, or a combined turn of 180°. Use the following notation for the lines of symmetry: v for vertical line; h for horizontal line; d_1 for the diagonal that starts in the upper left corner and ends in the lower right corner; and d_2 for the remaining diagonal.

a. Color the following squares to show what happens to the square when each of the symmetries is applied to it. The first square is the preimage (also the identity). Note that all rotations are in the counterclockwise direction.

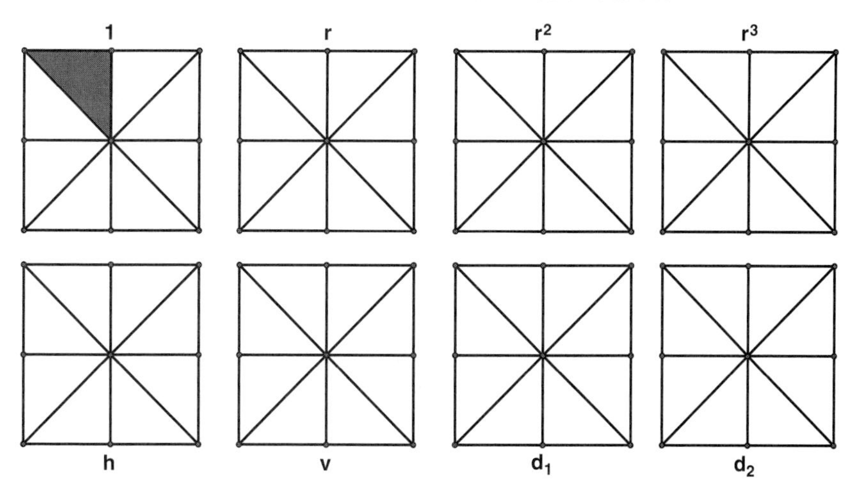

b. The notation $r \diamond h$ means reflection over the horizontal line midway through the square followed by a rotation of the square labeled 1 by 90° counterclockwise; the result is equivalent to d_2, a reflection over diagonal 2 (see the table). Complete the

TEACHING AND LEARNING HIGH SCHOOL MATHEMATICS © 2010 John Wiley & Sons, Inc.

table; write the result as a single symmetry, not as a composition of symmetries as was done in the article.

◇	1	r	r^2	r^3	h	v	d_1	d_2
1								
r					d_2			
r^2								
r^3								
h								
v								
d_1								
d_2								

c. Does this set of symmetries have an identity element?

d. Is this set of symmetries of a square closed under the operation of composition of symmetries?

e. Determine which elements have an inverse. Indicate each element with its inverse and how you know you are correct.

f. Is the operation of composition of symmetries commutative? How do you know?

g. Does the associative property seem to hold for this set of symmetries? Check associativity for several examples. How many combinations would you have to check to determine if associativity holds for this set of symmetries?

h. Recall the mathematical meaning of the term *group* (see the article). Does the set of symmetries of a square under composition comprise a group? Explain.

i. How are compositions of symmetries related to compositions of functions?

j. Find and read the article mentioned at the beginning of the problem. What relationships does the author make evident between geometry, art, and culture?

8. Designs from many cultures incorporate concepts about transformations and symmetry.

a. Find a piece of artwork from a cultural heritage of interest to your students that could have been created through the use of transformations. Show through labeling, outlining, and verbal description the transformations that could have been used to create the piece. The artwork should use more than one transformation.

b. Create your own artistic design, either with an interactive drawing tool or using transformation matrices, that uses at least three of the transformations you have studied in this investigation (translation, reflection, rotation, glide reflection, dilation). Show the preimage and each transformation of the preimage or subsequent image used to complete your final design. Your design must be creative and decorative; it must be something you would be willing to display on the bulletin board in your classroom with your name proudly attached.

c. Locate and read "Hmong Needlework and Mathematics" by Joan Cohen Jones. In *Changing the Faces of Mathematics: Perspectives on Asian Americans and Pacific Islanders*, edited by Carol A. Edwards (pp. 13–20). Reston, VA: National Council of Teachers of Mathematics, 1999. How do the Hmong needlework designs in the article connect to concepts about transformations and symmetry that are the focus of this unit investigation?

9. *Your Personal Transformation*. Individually, reflect in writing on your knowledge and understanding of transformations and the variety of ways they can be represented, the variety of tools that can be used to explore them, and the variety of places they arise.

TEACHING AND LEARNING HIGH SCHOOL MATHEMATICS © 2010 John Wiley & Sons, Inc.

UNIT three

PLANNING FOR INSTRUCTION

Mathematics Strands: Algebra and Functions

Effective classroom instruction takes significant preparation. Master teachers guide students toward mathematical understanding and are deliberate in their planning for instruction. They have a solid understanding of the curriculum for the course they are teaching, they know how topics build on and connect to each other, and they consider carefully the nature of the tasks in which they engage students. Effective teachers also reflect on their teaching regularly to determine what aspects of lessons went well and what changes need to be made to meet their students' needs. They also keep track of students' misconceptions and errors to determine ways to address these in future lessons.

Donovan and Bransford (2005), in a reading in Unit Two, detailed four lenses to determine the effectiveness of classroom environments. The community-centered lens was the focus of Unit One. The learner-centered lens was the focus of Unit Two. The assessment-centered lens is the focus of Unit Five. The focus of Units Three and Four is on the *knowledge-centered lens* as teachers think carefully about the selection and enactment of instructional activities.

As teachers plan for instruction, many issues must be considered. What should be the balance between conceptual understanding and procedural fluency? How should representations be used to help students make sense of mathematics? What applications of mathematics are important for students to see? What are the most important ideas for students to study in the limited time available? What is an appropriate level of complexity for the tasks and problems posed? What differentiation might be needed to modify lessons to accommodate students at varying levels and to ensure that all students have access to challenging mathematics? What technology could enhance learning and how might it best be used?

The focus of this unit is on elements for planning effective instruction. How do teachers determine appropriate goals and learning outcomes for instruction? High school students have been exposed to many of the concepts to be studied. So, how do teachers determine what students already know well so that review is minimized or omitted? How do teachers determine what students know marginally and plan so that upcoming lessons strengthen those underdeveloped areas? How do they recognize, select, or design tasks with high cognitive demand? How do teachers plan in order to both increase accessibility of tasks for struggling students as well as increase challenge for those who may be ready for more? How can teachers anticipate students' responses so they can guide students during the course of the lesson? How do teachers ensure that the mathematics classroom is an active environment rather than one where students are passive recipients of knowledge? All of these elements are important considerations as teachers plan effective lessons. In this unit, we focus on these fundamentals, giving teachers the opportunity to try out elements as they plan lesson segments, and then use this preparation in Unit Four to write complete lesson plans.

The mathematical strands for this lesson are algebra and functions. According to the Algebra Standard of *Principles and Standards for School Mathematics*, students need to explore relations and functions—both symbolically and graphically—and model real-world situations algebraically. Specifically, students in grades 9–12 should be able to generalize patterns; understand functions and be able to convert from one representation to another; compare properties of classes of functions; understand and write equivalent forms of equations, expressions, and inequalities; use symbolic algebra; and draw conclusions about a situation modeled by a function. Much of the content addressed in the Algebra Standard typically is addressed in Algebra I or Algebra II courses.

Algebra and functions, the mathematical foci of this unit, have roots in some of the earliest recorded history of mathematics. For example, in the Rhind Mathematical Papyrus, an Egyptian manuscript dating from around 1650 B.C., problems are stated and solved that are essentially linear equations. One such problem requires a number to be found such that the sum of the number, two-thirds of the number, one-half of the number, and one-seventh of the number is 33. Similarly, in the Chinese manuscript known as the *Nine Chapters of the Mathematical Art*, a problem requires five deer to be shared among five individuals proportionally as 5:4:3:2:1 (Katz, 1998, pp. 14–15). These two problems give rise, respectively, to the following linear equations:

$$x + \frac{2}{3}x + \frac{1}{2}x + \frac{1}{7}x = 33 \text{ and } 5x + 4x + 3x + 2x + x = 5.$$

Although neither the Egyptians nor the Chinese had the notation to solve these problems algebraically as we do today, they did provide verbal algorithms that could be followed to find solutions.

The Babylonians and the Chinese also developed techniques for solving systems of linear equations. Much of our knowledge of Babylonian mathematics comes from cuneiform tablets dating from around 1700 B.C. A sample problem from the Babylonian tablets, but written with modern units, is the following:

The area of two fields together is 1,800 acres and the first field yields 500 more bushels than the second. The first field yields $\frac{2}{3}$ bushel per acre and the second yields $\frac{1}{2}$ bushel per acre. What is the size of each field?

$$(x + y = 1800, \frac{2}{3}x - \frac{1}{2}y = 500)$$

(Katz, 1998, pp. 16–17)

Even the solution of quadratic equations was apparently known to the ancients. Among the Babylonian texts are problems requiring the solution to a system of the form $x + y = b$ and $xy = c$. Such problems likely arose in the context of problems relating the area and perimeter of rectangles, probably dealing with land issues (Katz, 1998). Rather than a general formula, Babylonian tablets contain numerous problems solved with specific values assigned to x, y, b and c.

For hundreds of years, algebraic problems were solved using a prose style. As printing became popular in the mid-1500s, mathematicians began to use symbols in their work. François Viète, a French mathematician, contributed significantly to symbol development. For example, he used consonants for known amounts and

TEACHING AND LEARNING HIGH SCHOOL MATHEMATICS © 2010 John Wiley & Sons, Inc.

vowels for unknowns. The notation he introduced helped spur development of algebraic ideas and the notation commonly used today.

Another important development in the history of algebra was the blending of algebra and geometry introduced by René Descartes (1596–1650), who is credited with the creation of the Cartesian coordinate system. According to Boyer (1991), Descartes had two purposes: "(1) through algebraic procedure to free geometry from the use of diagrams, and (2) to give meaning to the operations of algebra through geometric interpretation" (p. 339). The graphical representation of equations and functions is a natural outgrowth of the methods introduced by Descartes. Graphs often make abstract concepts and properties more concrete by illustrating relationships.

Although functions had been studied in some form by mathematicians for many years, much of our current understanding of functions is based on a work by Leonhard Euler published in 1748, *Introductio in Analysin Infinitorum* (*Introduction to Analysis of the Infinite*). As a work to focus on precalculus, Euler began with a definition of the term *function* and classified functions as algebraic or transcendental. Algebraic functions are those formed by basic operations on variables and constants (e.g., adding, subtracting, multiplying, dividing, taking roots, raising to numeric powers) (Weisstein, n.d.). Transcendental functions are those that are not algebraic (Weisstein). In this latter category, Euler focused on exponential, logarithmic, and trigonometric functions. He defined exponential functions as we do today and was the first to define logarithms in relation to exponential functions, essentially as their inverses. His work with trigonometric functions was the first to consider them as functions with numerical values and not solely as elements in a circle with a given radius (Katz, 1998).

The study of functions has been part of the high school mathematics curriculum for many years. But the introduction of graphing calculators into the school curriculum beginning in the late 1980s enhanced the ability of teachers and students to make connections between algebra and geometry, many of which were simply too difficult before the advances in graphing calculator technology. For instance, graphing calculators have simplified the process of graphing and table generation, making it possible to graph complicated functions easily, and providing a tool for exploring multiple representations of functions. Graphing calculators with computer algebra systems enable the user to perform complicated symbolic manipulations with just a few key strokes. The task for learners becomes interpreting these results to build conceptual understanding and to solve complex problems.

The availability of powerful technology in the high school curriculum has the potential to open mathematical doors for many students. At the same time, it poses challenges to teachers as they consider how to use that technology to enhance—rather than supplant—learning. As teachers begin to plan their mathematics instruction, the appropriate use of technology is just one of many elements that needs to be considered to design instruction to help all students be successful in mathematics. The consideration of many of those elements is the goal of this unit.

During the observation for this unit, you will have an opportunity to consider how teachers help students develop understanding of concepts as well as how they incorporate technology into their lessons. These insights will be helpful as you begin planning tasks for inclusion in lessons you develop.

Unit Three Team Builder: Find Your Function Family

In **Find Your Function Family**, each individual is given a card with an example from one of the following function families: linear, quadratic, cubic, exponential, rational, or square root. Each card displays a function example in one of these representations: equation, table, graph, children's story, or use in real context. In the equations and

TEACHING AND LEARNING HIGH SCHOOL MATHEMATICS © 2010 John Wiley & Sons, Inc.

tables, the independent variable is either x or t; the dependent variable is y. In all graphs, the horizontal axis is the x- or t-axis and the vertical axis is the y-axis.

Individuals are assigned new teammates by locating other classmates whose cards contain examples of functions in the same function family. Because there might be two teams with the same function family, each team's combined cards must include at most one of each of the following representations: graph, equation, table, and children's story or real life application. Complete the activity described on the **Find Your Function Family** student page to find your new team members.

Materials

- One **Function Card Sort** card per person

Website Resources

- **Find Your Function Family** student page
- **A Function RAFT** student page

Find Your Function Family

1. Your teacher will give you a function card showing a linear, quadratic, cubic, exponential, rational, or square root function represented as a graph, table, equation, children's story, or real life application. Look at your card; do not show it to anyone or talk about it for the moment. Identify the function family to which the example on your card belongs.
2. The teacher has also given other students in class a function card. Write three or four questions you will ask others to help you determine the function family on their cards. You may not ask directly to which function family the card belongs. For example, you may not ask questions such as, "Is your function an example of a quadratic function?" You may ask questions such as, "In equation form, would your function contain any exponents?"
3. Ask questions of your classmates to determine the function family of the examples on their cards.
4. Answer questions asked of you to help others determine the function family of the example on your card.
5. Keep track of those persons whose cards contain examples from the same function family as your card. Your teammates are other classmates whose examples are in the same function family as yours.

Materials

- one **Function Card Sort** card per person

After Locating Your Function Family

When teams have found each other, complete the student page **A Function RAFT**. RAFT is a strategy that sharpens students' writing by specifying a role (R) the writer plays, an audience (A) for the writing, a format (F) to describe the type of writing, and a topic (T) about which to write (Barton and Heidema, 2002).

TEACHING AND LEARNING HIGH SCHOOL MATHEMATICS © 2010 John Wiley & Sons, Inc.

A Function RAFT

1. Decide on a creative name for your team based on the assigned function family.
2. Write a creative and enlightening report about your function family using the following scheme and the table below:
   a. By rolling a six-sided die, randomly choose a **R**ole for the author of the report.
   b. By rolling a six-sided die, randomly choose an **A**udience for the report.
   c. Choose a **F**ormat for your report that is sensible for your Role and Audience.
   d. Your **T**opic is the function family your team has been assigned.

Roll of Die	Role	Audience	Format
1	Architect	Contractor	Advertisement
2	Engineer	TV audience	Letter
3	Politician	John Q Public	Diary entry
4	Environmentalist	School board	Newspaper article
5	Physician	Company manager	Petition
6	Designer	Customers	Advice column

Your report should provide the audience an example from your function family that is meaningful and related to the role of the author and of interest to the audience. The report needs to provide enough information that a novice can understand the example provided as well as why the example is from the function family assigned to your team.

The **RAFT** reflection strategy is adapted from *Teaching Reading in Mathematics* (2nd Ed.), by Mary Lee Barton and Clare Heidema, (Aurora, CO: Mid-continent Research for Education and Learning, 2002, pp. 139–140).

A master set of **Function Card Sort** cards can be found in the Unit Three Appendix. For further work with Families of Functions, see the Unit Investigation at the end of the unit.

Preparing to Observe Mathematics Classrooms: Focus on Curriculum and Technology

The curriculum and technology principles from *Principles and Standards for School Mathematics (PSSM)* are the focus of the observations in this unit. The mathematics curriculum is an essential factor in determining what students learn; the content students have an opportunity to learn is limited to what is offered. In addition, the nature of the curriculum influences the instructional approaches that teachers use. For instance, it often is easier for teachers to incorporate technology, writing, or

TEACHING AND LEARNING HIGH SCHOOL MATHEMATICS © 2010 John Wiley & Sons, Inc.

cooperative learning into their instruction if the curriculum—either the student text or the teacher materials—specifically includes those instructional approaches.

The Curriculum Principle from *PSSM* notes that curriculum needs to be coherent and needs to address important mathematics concepts sequenced within a course and from course to course. A set of activities or tasks alone do not make a curriculum. Are mathematics lessons focused on big ideas, or do they contain several unrelated concepts? How does the curriculum grow in complexity and depth across the grades? How do teachers or textbooks build on students' prior knowledge and experiences?

Technological tools are increasingly used in secondary mathematics classrooms. According to the Technology Principle from *PSSM*, appropriate technology should influence what and how students learn mathematics. Graphing calculators and computer software, especially dynamic tools for geometry and statistics, provide support for learning in algebra, geometry, data analysis, and other topics.

The observations for this unit are designed to help you consider issues of curriculum as they relate to planning for instruction and how technology might be appropriately incorporated into the high school mathematics curriculum.

1. *Prepare for the Observation.* In preparation for the observations, read the following selections:
   - **i.** "The Curriculum Principle" from *PSSM* (pp. 14–16).
   - **ii.** "The Technology Principle" from *PSSM* (pp. 24–27).

   a. What are the big issues in each principle? How are the principles related?

   b. Reflect on your observations from Units One and Two. Although your focus in those observations was not the curriculum, what overall sense do you have about the ways in which the curriculum in the mathematics classes you observed reflects the issues raised in "The Curriculum Principle"?

   c. Think about your own experiences in learning mathematics with technology. In what ways did you use technology to make sense of mathematics concepts? What was easier to understand by using technology? What was harder?

2. *Conduct the Observations.* Try to conduct all three observations in the same class; as much as possible, focus your observations at the prealgebra, algebra I, or algebra II levels (or comparable integrated classes). Observe at least three full class periods, preferably beginning your observations at the introduction of a new concept. If possible, try to observe for at least two consecutive days to investigate how lessons build on each other. Complete Observations 3.1 and 3.2.

3. *Observation 3.1: Curriculum Coherence.* Base your responses on visits to two class periods. During each visit, make note of the mathematical intent of the lesson, the concepts introduced or reviewed in the lesson, and the examples used to relate to the concepts. After the observations, consider the following questions.

   a. How many different concepts were introduced in the lessons? How were concepts related?

   b. If possible, look at the textbook used for the lessons. How well developed were the concepts? How did the textbook draw connections between concepts from lesson to lesson?

   c. Talk to the teacher. Find out his or her view about how the lessons you observed fit in the curriculum. On what concepts is he or she building? Toward what concepts is he or she building?

4. *Observation 3.2: Learning Related to Appropriate Use of Technology.* Begin the observation by noting the mathematical intent of the lesson.

   a. What opportunities did students have to use technology to explore mathematics?

   b. Did the use of technology appear to be appropriate or forced? Explain.

TEACHING AND LEARNING HIGH SCHOOL MATHEMATICS © 2010 John Wiley & Sons, Inc.

c. If technology was not used during the lesson, consider how technology could have been used to facilitate understanding of the concepts under study. If technology was used, what additional uses of technology might have been possible? Share a specific problem from the class and how a specific technological tool could have been helpful to solve the problem or illuminate the underlying concept.

5. *Prepare for Reflection.* Read "Standards for High School Mathematics: Why, What, How?" by Eric W. Hart and W. Gary Martin (*Mathematics Teacher*, 102 (December 2008/January 2009): 377–382).

   a. Comment on the authors' perspective relative to mathematics standards for high school.

   b. Locate your state or district mathematics standards. Review the standards appropriate to the classes you observed. What issues or challenges relative to implementing these standards do you anticipate based on your classroom observations?

   c. The article mentions recommendations by Achieve, Inc. (www.achieve.org). On the website for Achieve, go to the section called Benchmarks; under secondary mathematics benchmarks, find Achieve's recommendations for a model three-year high school sequence, either traditional or integrated depending on the type of curriculum used at the schools in which you observe. How do the recommendations in Achieve's curriculum sequence compare to the learning outcomes in the curriculum sequence at the school where you are observing?

   d. What similarities and differences are evident in the standards for your state or district and the standards in Achieve and/or *PSSM*?

6. *Reflect on the Curriculum.* Consider your observations together.

   a. Reflect on the ways in which your observations do or do not suggest a coherent curriculum.

   b. How did the lessons in close proximity connect to each other so that students explored concepts in greater depth over time? How did lessons on different days relate to each other?

   c. If what you observed of the curriculum needed greater coherence, what might you recommend to help students make connections across or within lessons?

7. *Summarize the Observations.* Start by providing a brief description of the level of the class you observed as well as the mathematical intent of each of the lessons.

   a. Summarize your comments about the readings and related questions from items 1 and 5.

   b. Refer to the questions associated with each of Observations 3.1 and 3.2. Use your observations to summarize your responses to the questions. Justify any conclusions with specific instances from the readings or the class observations.

   c. Consider your perspectives on curriculum and technology as described in the curriculum and technology principles, respectively, and in the classes you observed. What do you anticipate to be the major issues that you will need to consider in designing your mathematics instruction?

Be prepared to discuss your observations and reflections in the last lesson of this unit.

Listening to Students Reason about Functions

High school students should have had some experiences with patterning and functions while studying middle school mathematics. Though students might have worked with linear patterns and functions and possibly other families of functions, it

TEACHING AND LEARNING HIGH SCHOOL MATHEMATICS © 2010 John Wiley & Sons, Inc.

is not likely that they considered what happens when you combine functions through one of the arithmetic operations. The tasks for this interview provide an opportunity for you to engage students in thinking and reasoning about functions in a way that is likely new to them.

Website Resources

- **QRS Guide**
- **Climate Change** student page
- **Adding and Subtracting Functions** student page

1. In preparation for the interview, read "Understanding Connections between Equations and Graphs" by Eric J. Knuth (*Mathematics Teacher*, 93 (January 2000): 48–53).
   a. Identify some of the difficulties students had in making connections between equations and their graphs.
   b. What are some possible implications of the research for students' understanding of functions?
   c. What are potential insights teachers gain about students' thinking when they are asked to solve a task in more than one way?
2. Refer to the **Climate Change** student page.
   a. Work through problem 1, recording your work in the **QRS Guide** as you did during the Listening to Students tasks in Units One and Two. (Find a blank copy of the **QRS Guide** in the Unit One Appendix.)
   b. What issues do you anticipate arising as students complete the student page and write about their findings? Consider difficulties reported in Knuth (2000) and how those difficulties might influence their work with the tasks on the student page. Add these possible responses to your **QRS Guide**. Identify the responses as misconceived, partial, or satisfactory.
   c. Use the Stages of Questioning (see Lesson 1.1) to provide appropriate guidance to students, without giving too much help. Record your questions in the Teacher Support column of the **QRS Guide**.
   d. Repeat problems 2a–c for the second problem on the student page **Climate Change**.
3. Repeat problems 2a–c for the problems on the student page **Adding and Subtracting Functions**.
4. Identify a pair of high school students in an algebra class who have had some experiences with coordinate graphs. Have them work through the problems on the student pages **Climate Change** and **Adding and Subtracting Functions**.
   a. In the **QRS Guide**, record responses you observed from students that you did not originally anticipate.
   b. What, if anything, surprised you by students' responses?
   c. What difficulties, if any, did they have that you had not anticipated?
   d. What similarities and differences did you notice between your students' responses to the function tasks and students' responses to the tasks in the Knuth article?
   e. Update your **QRS Guide** to reflect additional students' responses, questions, and teacher support you would use with future students.

Be prepared to share your insights into how students reasoned about these function tasks during the last lesson of this unit.

TEACHING AND LEARNING HIGH SCHOOL MATHEMATICS © 2010 John Wiley & Sons, Inc.

Climate Change

Global warming is a major concern. As greenhouse gases accumulate in the atmosphere, they allow sunlight to enter the earth's atmosphere but prevent heat and reflected light from escaping, causing the temperatures on earth to rise. Reports of rising temperatures have concerned scientists for many years.

1. Consider the average monthly high temperatures for Barrow, Alaska, shown in the graph at right, with the months numbered 1 (January) to 12 (December). The temperatures are averaged over the 30 years from 1977 to 2006 (Resource: http://www.citydata.com/city/Barrow-Alaska.html downloaded February 10, 2007). Scientists report that these temperatures indicate a 4° F rise in average temperatures from the average temperatures for the 30 years from 1947 to 1976 (Resource: http://news.nationalgeographic.com/news/2004/12/1206_041206_global_warming.html downloaded February 10, 2007).

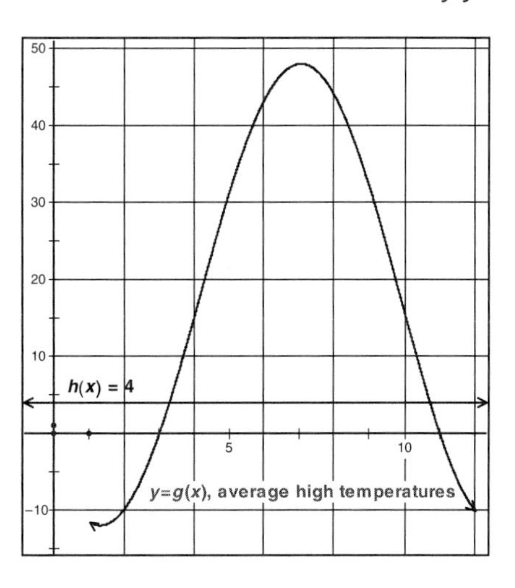

a. On the same axes, sketch the graph of the average monthly high temperatures for Barrow for the 30 years preceding 1977. Call this graph $y = f(x)$.

b. How is the graph of Barrow's current average high temperatures, $y = g(x)$, related to the graph of Barrow's high temperatures averaged over 1947 through 1976, $y = f(x)$?

c. Write an equation showing the relationship between the graphs of $y = f(x)$, $y = g(x)$, and $y = h(x) = 4$.

2. Suppose the average monthly high temperatures in Barrow rise over the next 10 years at the rate shown in the line, $y = h(x)$, in the following figure. The graph of $y = g(x)$ shows Barrow's expected average monthly high temperatures, if no climate change occurs.

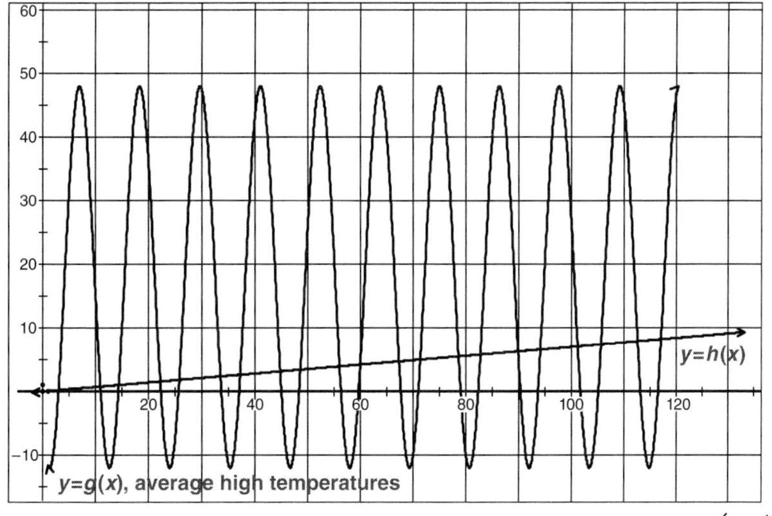

(*continued*)

TEACHING AND LEARNING HIGH SCHOOL MATHEMATICS © 2010 John Wiley & Sons, Inc.

a. Sketch the graph showing the average monthly high temperatures if the temperatures shown in the graph of $y = g(x)$ increase at the steady rate shown in $y = h(x)$.

b. Explain your new graph. Label it $y = k(x)$.

c. Write an equation showing the relationship between the graphs of $y = k(x)$, $y = g(x)$, and $y = h(x)$.

Adding and Subtracting Functions

1. Consider the pair of graphs shown below. Graphs of f and g are sketched for x in $[-5, 5]$ and for y in $[-7, 5]$. Grid lines are indicated every unit in both directions.

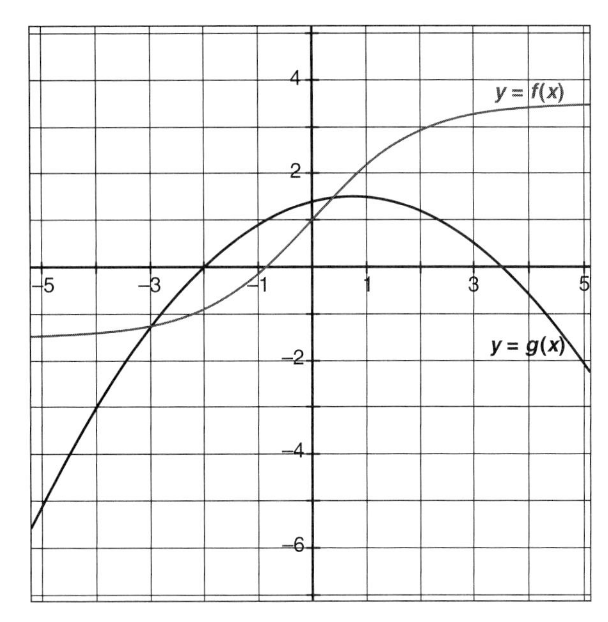

a. Create a table listing values of x, $f(x)$, $g(x)$, and $h(x) = f(x) + g(x)$.

b. Sketch the graph of $h(x) = f(x) + g(x)$ for the functions f and g provided.

2. a. Create a table listing values of x, $f(x)$, $g(x)$, and $h(x) = f(x) - g(x)$.

b. Using a different color, sketch the graph of $h(x) = f(x) - g(x)$ for the functions f and g.

c. Using a third color, sketch the graph of $y = g(x) - f(x)$.

d. How are the graphs of $y = f(x) - g(x)$ and $y = g(x) - f(x)$ related to each other? Explain.

3. a. Explain how to find the graph of $y = h(x)$ from the graphs of $y = f(x)$ and $y = g(x)$ when

i. $h(x) = f(x) + g(x)$ **ii.** $h(x) = f(x) - g(x)$ **iii.** $h(x) = g(x) - f(x)$

b. Without completing a table of values, how might you find the graph of $y = h(x)$ directly from the graphs of $y = f(x)$ and $y = g(x)$? Explain for the addition of two functions and then for the subtraction of two functions.

TEACHING AND LEARNING HIGH SCHOOL MATHEMATICS © 2010 John Wiley & Sons, Inc.

3.1 Building on Students' Knowledge and Experiences: Understanding Variables and Linear Functions

Algebra often serves as a gatekeeper; students who do not succeed in algebra have many educational and career opportunities curtailed. So success in algebra is a major equity issue in mathematics education. It opens doors to advanced mathematics. Yet algebra presents challenges for many students. One reason why algebra is difficult is because it deals with abstractions. In algebra, students need to be able to generalize beyond specific numbers, to see patterns, conjecture equations to fit the patterns, and describe the patterns abstractly. Moreover, this abstraction often involves the use of symbols.

In this lesson, activities stemming from everyday experiences provide a basis for making sense of variables that arise in linear patterns. The contexts provide a form of support so that students are eased into the use of symbols by building on what they know from their experiences in the world. As you engage with the activities, consider questions such as: How do students learn about variables? What difficulties arise as they learn about variables? How can teachers help students make sense of the meanings embedded within the contexts in which variables are used?

Variables and other symbols play an important role in learning mathematics. Hiebert (1988) describes five processes involved in developing competence with symbol systems:

1. *Connecting* individual symbols with contexts or referents that give them meaning;
2. *Developing* meanings for symbol manipulation procedures (i.e., associating actions with operations);
3. **a.** *Elaborating* procedures for symbols (i.e., working with more complex procedures);

   b. *Routinizing* the procedures for manipulating symbols; and
4. Using the symbols and rules as the context for *building* more abstract symbol systems.

Elaborating and routinizing processes are numbered together because they operate concurrently (Hiebert, 1988).

The first of the processes proposed by Hiebert involves attaching meaning to symbols by connecting them to a familiar context or referent. Once meaning is attached to the symbols, students learn to operate on the symbols by thinking about how they work in the context the symbols represent. For example, in this lesson you will learn how to help students build understanding for variables in contexts that are likely familiar. In the **DMU/DMP** questions, you will use algebra tiles—another tool to support algebra learning—to build understanding of processes needed to combine like terms and solve linear equations. The algebra tiles represent areas of rectangles whose side lengths are 1, or x, so that their areas are 1, x, and x^2. Length or area become the contexts for the symbols 1, x, and x^2. Students use these tiles to develop procedures for manipulating symbols. Eventually, they learn to manipulate symbols without the use of the materials but retain their meaning for future reference if needed. They become skilled at the manipulation of these symbols and eventually are able to use these symbols as the context for more complex symbol systems, such as analytic geometry and calculus.

In this lesson, you will:

- Learn strategies (e.g., contexts, algebra tiles) to make the concept of variable accessible to a broad range of students. The concept includes both variables as placeholders and as values that vary.

TEACHING AND LEARNING HIGH SCHOOL MATHEMATICS © 2010 John Wiley & Sons, Inc.

- Learn how to build students' understanding of linear functions by using variables to describe linear patterns.
- Use algebra tiles to simplify expressions and solve linear equations.
- Become familiar with some difficulties students have in using and understanding variables.

Materials

- calculator-based ranger (CBR) or other motion detector and graphing calculator, one per team

LAUNCH

When students first encounter algebraic ideas, the use of variables is most notably different from their previous mathematics experiences. Some students feel they are suddenly thrown into a world of arbitrary letters. How can instruction help students not only see patterns, but symbolize the patterns using variables? Consider the following:

1. How many eyes does one person have? Two people? Seven people? How many human eyes are in the classroom? How many in the school? How can we find the number of human eyes without knowing how many people there are? What assumptions must be made?
2. Consider at least three other body parts for which there is more than one body part per human. Create a table comparing numbers of humans to numbers of body parts for each of the body parts in your list. For each body part, write a complete English sentence to indicate how to find the number of body parts for an unspecified number of people.
3. Begin with the relationship indicated in your sentence in problem 2.
   - **a.** Write a phrase that gives most of the same information but uses fewer words. You may use familiar symbols to indicate arithmetic operations.
   - **b.** Write an equation that indicates words, arithmetic operations, and an equals sign to symbolize the relationship in item 3a.
   - **c.** Abbreviate the equation from item 3a, replacing words with meaningful letters.
   - **d.** Repeat items 3a–d for each of the body parts in your table from item 2.
4. How do your incremental "rules" in item 3 help naïve students move from concrete notions of patterns to using variables to symbolize simple patterns?

EXPLORE

Linear patterns arise in varied ways. The examples you generated with body parts in the **Launch** are all linear patterns.

5. List at least three contexts (other than body parts) from your everyday life in which linear patterns arise.
6. A motion detector—a tool for modeling distance, velocity, or acceleration versus time—operates by emitting a sound wave that bounces off the closest object, recording the time needed for the wave to return, and calculating the distance to the object. Set up a motion detector so that there are no other objects within a 15-degree conical alley of the motion detector's screen and so you can walk at least 3 meters unimpeded by other objects. Walk so that the CBR collects information about your movement and displays it graphically. What does the x-axis on the graph represent? What does the y-axis represent?
7. Use the motion detector and a graphing calculator to complete the following. In each case, describe how you walked and explain why the graph makes sense.

 © 2010 John Wiley & Sons, Inc.

a. Walk so that the graph you create is a straight line slanting upward.

b. Walk so that the graph is horizontal.

c. Walk so that the graph you create is a straight line slanting downward.

Children's stories, fairy tales, nursery rhymes, and songs provide numerous examples of linear patterns. Consider the following that are likely familiar to you or your students.

8. Recall the nursery rhyme, "Jack and Jill." Sketch graphs of each scenario on the same pair of axes. Label each axis with its meaning. Label the graph showing Jack's motion with *Jack*; label the graph showing Jill's motion with *Jill*.

   a. Jack and Jill ran up the hill. Both were running at a constant rate. Jill ran faster than Jack. Show graphs of their race.

   b. Jack and Jill rolled down the hill. Again, they each rolled at a constant rate. This time, Jack rolled faster than Jill. Show graphs of their movement down the hill.

   c. Describe the comparative steepness of the graphs in items 8a and 8b.

   d. Are the running graphs steeper or less steep than the rolling graphs? Why do you think so?

   e. Suppose Jack started rolling down the hill first. Assume Jill rolled down the hill shortly thereafter. Sketch graphs of this scenario. Explain your graphs.

   f. In each case above, what is the *domain*, the set of values possible for the independent variable?

   g. In each case above, what is the *range*, the set of values possible for the dependent variable?

9. The first line of each verse of the children's campfire song, "There's a Hole in the Bottom of the Sea," is listed:

   There's a hole in the bottom of the sea.

   There's a log in the hole in the bottom of the sea.

   There's a bump on the log in the hole in the bottom of the sea.

   There's a frog on the bump on the log in the hole in the bottom of the sea.

   There's a wart on the frog on the bump on the log in the hole in the bottom of the sea.

   There's a hair on the wart on the frog on the bump on the log in the hole in the bottom of the sea.

   There's a flea on the hair on the wart on the frog on the bump on the log in the hole in the bottom of the sea.

   In the first verse, there are no objects in the hole in the bottom of the sea. In the second verse, there is one object, a log, in the hole. Analyze the rest of the song as follows.

   a. Create a table that indicates the number of objects in each verse in terms of the verse number.

   b. Graph the number of objects versus the verse number. Is the graph connected or not? Should it be?

   c. What is the domain of the function that models this song? Explain.

   d. What is the range of the function that models this song? Explain.

10. Another familiar children's song can be analyzed two different ways. The first verse is:

    There were 10 in the bed and the little one said,

    "Roll over, roll over!"

    They all rolled over and one fell out!

    The second verse is the same with the exception that there are now 9 in the bed. The song continues in this way until, in the final verse, the last child falls out of bed.

TEACHING AND LEARNING HIGH SCHOOL MATHEMATICS © 2010 John Wiley & Sons, Inc.

a. Define the variables so that the independent variable is the verse number and the dependent variable is the number of children in the bed at the end of the verse. Create a table, sketch the graph, and symbolize the pattern that arises from these representations (including the song).

b. Redefine the independent variable to be the time it takes for a person to sing a verse. Let the dependent variable be defined as it was in item 10a. Do any of the representations from item 10a change with this definition of the independent variable? Why or why not?

c. What is the domain for the first scenario? for the second scenario?

d. What is the range for the first scenario? for the second scenario?

11. Organize your work from items 8–10. In each case, indicate the variables, the domain, the range, and whether the variables are discrete or continuous. Draw the graphs for the patterns in items 9 and 10. How do the domain and range affect the graphs of linear functions? Explain.

SHARE AND SUMMARIZE

12. Share your work from the **Explore** with your team. How can you recognize a linear pattern from each of the following representations?

a. a table
b. a graph
c. a context
d. an equation

13. How do the types of numbers in its domain affect the graph of a linear relationship? How do the types of numbers in its range affect the graph of a linear relationship?

14. Revisit the body parts scenarios. Are there any body parts for which the total number of body parts for some number of people is 16? Write an equation to show this relationship and solve the equation.

15. Some people would argue that the uses of the variable x in the equations $3x + 2 = 7$, $y = 4x + 3$, and $x + 4 = 4 + x$ are different. How are variables used differently in these equations?

16. How do the activities in the **Launch** and **Explore** of this lesson build on students' knowledge and experiences to help them understand variables, linear patterns, and linear equations? Explain.

DEEPENING MATHEMATICAL UNDERSTANDING

1. *Functions in Stories and in Context.*

a. Return to the **Function Card Sort** in the Unit Three Team Builder. (A set of cards can be found in the Unit Three Appendix.) Find the stories and real contexts that fit the linear function family. State how you know these cards can be represented by a linear function.

b. Write another story or determine another real context that fits this family of functions.

2. *Functions in Games.* Consider the game of *Monopoly*. Let x be the number of squares from GO, proceeding in order around the board (Boardwalk has the highest x value), and either number all of the squares consecutively or number only those squares that involve purchasable properties, excluding the railroads and utilities. Let y be the value of the property. Plot the pairs of values. To what family of functions does this set of data belong? Why do you think so?

TEACHING AND LEARNING HIGH SCHOOL MATHEMATICS © 2010 John Wiley & Sons, Inc.

3. *Variables as Pattern Generalizers*. One important use of variables in algebra is to generalize patterns or state properties.

   a. Consider the following instances of a pattern.

$$3 + 5 = 5 + 3$$
$$12 + 8 = 8 + 12$$
$$4\frac{1}{2} + 3\frac{4}{5} = 3\frac{4}{5} + 4\frac{1}{2}$$

   Use variables to generalize the pattern described by these three instances. What name is typically given to the pattern you generalized?

   b. A student in an Algebra I class attempted to generalize the pattern in problem 3a as follows:

$$a + b = b + c$$

   Comment on this student's work. How might you try to help this student?

4. *Linear Recursion*. Often, when students are building tables, they notice a pattern in how the dependent variable is changing and can find additional terms in the table by continuing the pattern, disregarding the values of the independent variable. Equations that model such relationships are called *recursive* equations.

   a. Find a pattern in the table below. Complete the table using the pattern. Determine the pattern based only on the change in the dependent variable. (It is also necessary that the independent variable is changing at a constant rate.)

x	0	1	2	3	4	5	6
y	8	5	2	−1			

   b. Write a rule for the pattern you found in problem 4a that states how to find the next value for the dependent variable in the table from the previous value.

   c. State the rule using notation relating the current entry, NOW, to the next entry, NEXT, and telling the START value.

   d. Use function notation to indicate the relationship between the current entry, $f(n)$, the next entry, $f(n + 1)$, and the initial value $f(1)$ or $f(0)$.

   e. Repeat problems 4a–d for the table below.

x	0	1	2	3	4	5	6
y	4	9	14	19			

   f. How does a linear recursive equation indicate the rate of change?

   g. How does the y-intercept arise in a recursive equation?

   h. For each example provided, how might you determine the x-intercept?

   i. How can you capitalize on students' natural tendencies to find recursive relationships to help them find explicit linear equations—equations that relate the independent and dependent variables directly?

5. *Piecewise Linear Graphs*. In some situations, a rule for a function varies over different intervals of the independent variable.

   a. The 2008 federal income tax rate for a single taxpayer is listed in the table on page 182. Plot the taxable income on the horizontal axis and the tax on the vertical axis. (Resource: http://www.house.gov/jct/x-32–08.pdf, downloaded June 2, 2009)

TEACHING AND LEARNING HIGH SCHOOL MATHEMATICS © 2010 John Wiley & Sons, Inc.

If taxable income is over:	But not over:	The tax is:
$0	$8,025	10% of the amount over $0
$8,025	$32,550	$802.50 plus 15% of the amount over $8,025
$32,550	$78,850	$4,481.25 plus 25% of the amount over $32,550
$78,850	$164,550	$16,056.25 plus 28% of the amount over $78,850
$164,550	$357,700	$40,052.25 plus 33% of the amount over $164,550
$357,700	no limit	$103,791.75 plus 35% of the amount over $357,700

b. Write a piecewise equation to fit the table and your graph from problem 5a.

c. Find a context of interest to you and write a scenario that would be represented by a piecewise linear graph.

6. *Solving Equations with Tables and Graphs.* Graphing calculators with table-generating features can be used as a tool for solving equations.

   a. Consider the equation $5x + 2 = 32$. Enter the equation $y = 5x + 2$ into the equation menu. Generate a table. How can you use the values in the table to find the solution to $5x + 2 = 32$?

   b. Use tables to find the solution to $7x + 10 = 4x - 17$. Describe your process.

   c. Solve each of the equations in problems 6a and 6b with graphs. How are graphs used to find the solution to each equation?

7. *Hoisting the Flag.* Consider each of the four graphs which describe raising or lowering a flag on a flagpole. In each case, the vertical axis represents the distance of the flag from the ground; the horizontal axis represents time. For each graph, write a story that describes how the flag was raised.

a.

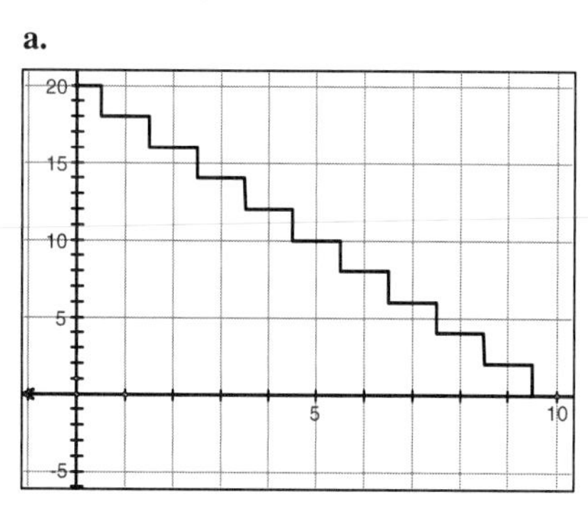

b.

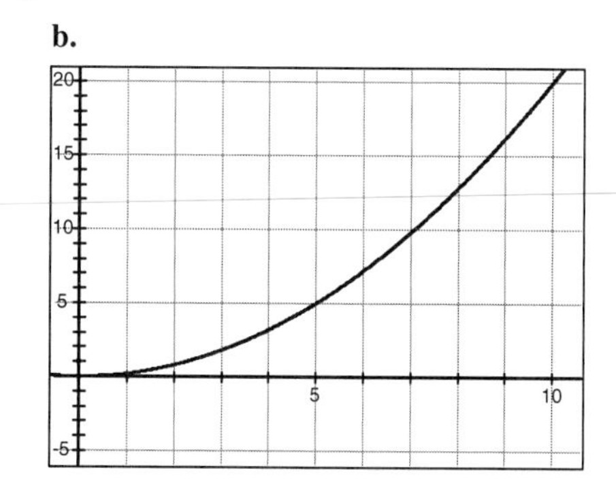

c.

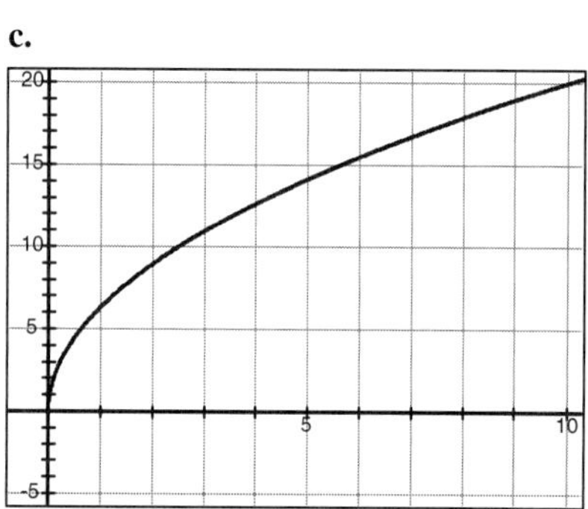

d.

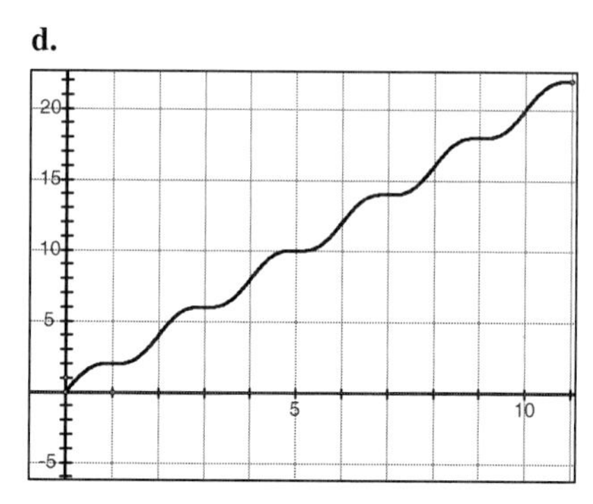

8. *Graphs with Discrete or Continuous Domains and Ranges.* For a given graph, the domain can be discrete or continuous depending on the context. Likewise, the range can be discrete or continuous.

   a. The table below contains a graph for which the domain and range are both discrete. Draw a graph for the other three cells. Briefly explain each graph.

		Domain	
		Discrete	**Continuous**
Range	**Discrete**	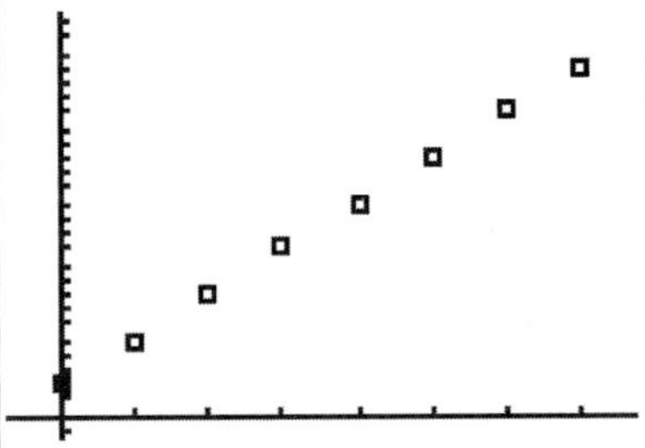	
	Continous		

   b. Write a context that describes a situation for each of the four cells. Which of these are functions?

   c. Refer to the children's stories or rhymes from the **Explore**. In which cell would each story or rhyme be placed? Note how the independent and dependent variables are defined in each case.

DEVELOPING MATHEMATICAL PEDAGOGY

Reflecting on Our Own Learning

9. Consider your own introduction to solving equations.

   a. Recall friends who struggled with the concept of variable. What approaches in this lesson may have helped them develop a more meaningful concept?

   b. What types of experiences did you have with concrete materials when learning about variables, patterns, and linear functions?

   c. In what ways have you used graphing calculators to study variables, patterns, and linear functions in algebra?

   d. Reflect on the extent to which you, or your students, might initially explore concepts via technology or other tools before learning symbolic approaches.

Making Historical Connections

10. Diophantus, a third-century mathematician who lived in Alexandria, Egypt, explored equations with integer solutions. A linear Diophantine equation is an equation of the form $ax + by = c$ for which a, b, and c are integers and only integer solutions are permitted for x and y.

   a. Show that the following problem gives rise to a linear equation whose form is that of a Diophantine equation: *Hot dogs come in packages of 8 or in packages of 12. What are the possible numbers of hot dogs that one can purchase?*

TEACHING AND LEARNING HIGH SCHOOL MATHEMATICS © 2010 John Wiley & Sons, Inc.

b. Find at least one integer solution to $15x - 21y = 114$.

c. Investigate the conditions under which a linear Diophantine equation has a solution.

Teaching for Equity

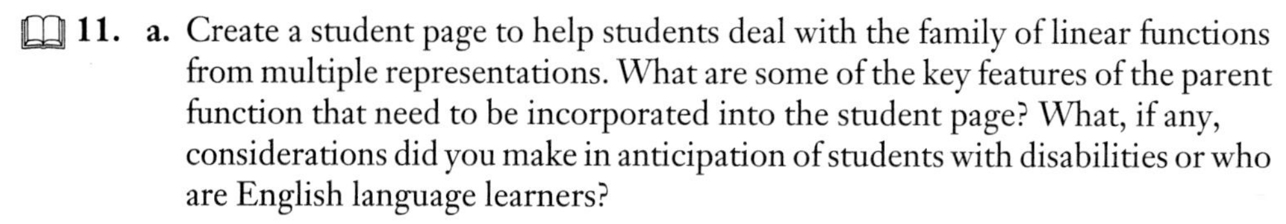

11. a. Create a student page to help students deal with the family of linear functions from multiple representations. What are some of the key features of the parent function that need to be incorporated into the student page? What, if any, considerations did you make in anticipation of students with disabilities or who are English language learners?

b. Complete a **QRS Guide** for the student page. Include possible student responses to the student page at three levels: misconceived, partial, and satisfactory. What teacher support or questions would you be prepared to use? Record these in the **QRS Guide**.

c. Try your student page with a high school student. What issues arose? What was easy for the student? What was difficult? What misconceptions, if any, did the student have?

d. Modify the student page as a result of trying the page with students. Make corresponding modifications to the **QRS Guide**. Explain how the modifications address the issues and misconceptions you mentioned in problem 11c.

12. Complete the student page **From Pictures to Variables**.

a. How might the visual models help those struggling with generating rules just from numbers or symbols?

b. What are potential strengths and weaknesses of such visual approaches for connecting variables to patterns?

c. What else might you ask related to the learning outcomes for this student page?

d. How might you modify or implement the page to accommodate English language learners or students with disabilities?

TEACHING AND LEARNING HIGH SCHOOL MATHEMATICS © 2010 John Wiley & Sons, Inc.

From Pictures to Variables

In each problem:

1. Find the number of pieces needed in case 3.
2. Draw the figures for cases 4 and 5, and find the number of pieces needed for each.
3. Describe the pattern in words.
4. Determine the number of pieces needed for the 10th case, 100th case, and the *n*th case.

	Case 1	Case 2	Case 3	Case 4	Case 5
1. boxes					
number of boxes	6	9			
2. hearts	♥ ♥ ♥ ♥ ♥	♥ ♥♥ ♥ ♥ ♥♥ ♥	♥ ♥♥♥ ♥ ♥♥♥ ♥ ♥♥♥ ♥		
number of hearts	5	8			
3. dots	● ●	●●● ●●●	●●●●● ●●●●●		
number of dots	2	6			

TEACHING AND LEARNING HIGH SCHOOL MATHEMATICS © 2010 John Wiley & Sons, Inc.

Analyzing Instructional Activities

13. Textbooks often use different notations to express recursive functions. These include the following (in each case an initial value must be identified):

- t_n and t_{n+1}
- t_n and t_{n-1}
- $f(n)$, with n an integer, $f(n+1)$ or $f(n-1)$
- NOW/PREVIOUS
- NOW/NEXT
- NEW/OLD
- NEXT/CURRENT

a. Indicate similarities and differences among these notations.

b. What are some issues, benefits, or concerns you have about using one or another of these notations?

14. **a.** Some sources note that in equations such as $3x + 2 = 8$, the use of x is not variable; x is a fixed, yet unknown value. Discuss this point of view.

b. The table feature on a graphing calculator can be used to help students recognize that $3x + 2$ has many values as x varies over a specified domain. Discuss the pros and cons of having students explore expressions such as $3x + 2$ before they solve specific equations such as $3x + 2 = 8$.

c. The use of x in $3x + 5 = 11$ differs from its use in $y = 3x + 5$. Yet, both occur in algebra, with its use in $3x + 5 = 11$ often occurring first. How might the notion of variable in $3x + 5 = 11$ influence students' understanding of variable in $y = 3x + 5$?

d. What difficulties might you expect high school students to have when working with variables? What steps might you take to alleviate the difficulties?

15. A set of algebra tiles consists of three types of pieces that are color-coded to be positive or negative (e.g., black is positive and red is negative). Side lengths are 1 and x. (A template of **Algebra Tiles** is in the Unit Three Appendix.)

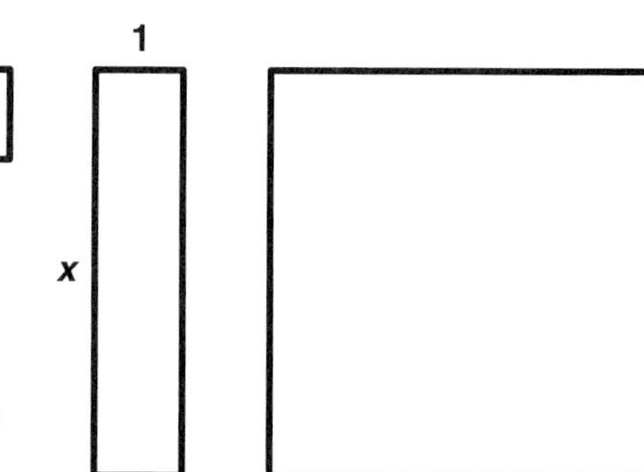

a. What is the area (value) of each piece?

b. Algebra tiles assume the *add opposites property*, that is, two pieces of the same area but with opposite signs balance each other to give a sum of 0. Use this property to write three different representations for 0 using the pieces.

c. How might the pieces be used to simplify each of the expressions: $-4 + 3$, $-4 + (-3)$, and $4 + (-3)$?

d. To simplify $-7 - (-3)$, a student can use the take-away interpretation for subtraction. That is, start with -7 units and take away -3 units to result in -4 units. For $-7 - 3$ a student would start with -7. There are no positive pieces to "take away." A student needs to add some form of zero to continue. Try this and record the process and results.

e. Use the results to summarize the rules for adding and subtracting integers.

16. Use a set of algebra tiles to complete the student page **Using Algebra Tiles to Simplify Expressions.** (A template of **Algebra Tiles** is in the Unit Three Appendix.) Problem 2 on the student page draws attention to a typical misconception that many beginning algebra students make. One instructional

TEACHING AND LEARNING HIGH SCHOOL MATHEMATICS © 2010 John Wiley & Sons, Inc.

strategy used by many teachers is to highlight such misconceptions for the entire class to address.

a. Comment on the extent to which you might use such strategies in your classroom.

b. The misconception highlighted in problem 2 on the student page is done in an open question format, that is, without indicating whether the student response is correct or not. The question could have been asked in a closed format as well: *Magdalena simplified $4x + 2x$ as $6x^2$. What did Magdalena do wrong?* What insights might a teacher get about students' thinking from the question in an open rather than closed format?

c. What are some benefits of using algebra tiles with students who are just beginning their study of simplifying expressions?

Using Algebra Tiles to Simplify Expressions

Materials

- one set of algebra tiles for each pair of students

Algebra tiles consist of three different-sized pieces; the smallest side length is ± 1 and the other side length is $\pm x$. The pieces are usually color-coded so that one color represents positive values and another color represents negative values. Identify the value of each of the pieces in the set of algebra tiles.

1. An expression consists of numbers or terms combined with the operations of addition, subtraction, multiplication, or division. Simplifying an expression means to write the expression as an equivalent expression with as few terms as possible. Represent each of the following expressions using algebra tiles and then simplify the expression.

   a. $3x + 1 + 2x + 4x + 3$

   b. $3x^2 + 2x + 5x + x^2 + 4$

   c. $2 + 4x^2 + (-2x^2) + 5x + 3 + (-7x)$

2. Magdalena simplified $4x + 2x$ as $6x^2$. Ruiz said Magdalena was wrong and that $4x + 2x$ simplifies to $6x$.

   a. Do you agree with Magdalena, Ruiz, or neither?

   b. Use algebra tiles to justify your answer to problem 2a and help either Magdalena or Ruiz or both find their mistake.

3. Terms that have the same variables to the same powers are called **like terms**. A student was absent from class for the lesson on adding and subtracting expressions with like and unlike terms. Use what you learned in problems 1 and 2 to explain to the absent student how to add and subtract expressions with like terms.

4. When you first studied multiplication, you learned that $4 \cdot 2$ means 4 groups with 2 in each group. Use this idea to represent each of the following with algebra tiles and then simplify the expression.

   a. $4(2x + 1)$

   b. $3(-4x + 2)$

   c. $2(x^2 + 4x + 3)$

   d. Show how the distributive property can be used to simplify each of problems 4a, 4b, and 4c.

5. Sometimes algebra tiles cannot be used to model simplifying an expression. Use what you have learned in problems 1–4 to simplify $3x^2 + 2x + 6y + 4y^2 + 3y + (-6x) + 9y$.

TEACHING AND LEARNING HIGH SCHOOL MATHEMATICS © 2010 John Wiley & Sons, Inc.

17. Work through the student page **Solving Equations with Algebra Tiles**.

a. How does the use of algebra tiles help students visualize the steps in solving equations?

b. Obviously, students will not always have algebra tiles. So, the purpose of the activity is to use the materials to help students build understanding for the process and to help them transition from the concrete to the abstract. What aspects of the student page were designed to help students bridge from using the materials to completing similar problems without materials?

c. Although concrete materials have many benefits, some limitations exist. What are some of the limitations you see with algebra tiles in terms of solving equations?

d. Initial operations and manipulations with variables in early algebra courses are often difficult for many students. How might the use of algebra tiles address some of those difficulties?

e. What issues do you believe might arise in using algebra tiles with early algebra students at high school? How might you try to address those issues?

f. Use algebra tiles to solve the following equations. Illustrate each step of your work with the tiles and show the accompanying symbolic algebra.

i. $2x + 7 = 5x - 2$ **ii.** $x - 3 = 3(2 - x) - 1$

Solving Equations with Algebra Tiles

Materials

- one set of algebra tiles for each pair of students

Equations can be useful to solve many problems, including the following: *A new package of oatmeal has 510 grams of oats. A typical serving takes 40 grams. If Elizabeth eats a typical serving of oatmeal every day, after how many days will she have 190 grams of oats left?*

Algebra tiles can help you learn how to solve such problems symbolically.

1. Use the algebra tiles to illustrate $4x = -12$. Think of division as equal sharing. Show how to use the algebra tiles to solve this equation.

2. An **equation** is a mathematical sentence in which the left side and the right side of the equals sign represent the same amount. The equals sign is like the fulcrum on a seesaw or a balance to show that the amounts on each side of the equals sign have the same value. So, any operations performed on one side of the equation must be performed on the other side in order for the equation to stay balanced. When working with algebra tiles, draw a vertical line on your work space to represent the equals sign.

a. Use the algebra tiles to illustrate $x + 3 = 7$.

b. Using two different approaches, find a value of x that makes this equation true:

i. Subtract the same amount from both sides of the equation.

ii. Add the same amount to both sides of the equation.

c. Which of the two methods in problem 2b did you prefer? Why?

3. Use algebra tiles to illustrate $2x + 3 = -7$.

a. Find a value of x that makes this equation true. What steps can you take to change this equation to one of the form $x =$ a number? How did your steps keep the equation balanced? Record your steps algebraically and with illustrations.

b. Verify that your solution is correct by substituting your value for x into $2x + 3$. You should get -7. If not, start over.

(*continued*)

TEACHING AND LEARNING HIGH SCHOOL MATHEMATICS © 2010 John Wiley & Sons, Inc.

c. Now solve each of the following using algebra tiles: $2x + 3 = 7$, $2x - 3 = 7$, $2x - 3 = -7$, $-2x + 3 = 7$, and $-2x + 3 = -7$. Record each of your steps with illustrations and algebraically with paper and pencil. Verify that your solutions are correct.

d. Compare the solution processes for each of the equations in problem 3c.

4. Consider an equation of the form $3x + 8 = -2x + 18$.

a. Illustrate the equation using algebra tiles.

b. How do you keep the equation balanced while changing it to the form $ax + b = c$? Record steps with a series of illustrations and the algebra that goes with each.

c. Continue solving until your equation is of the form $x =$ a number.

d. Verify that your solution works in the original equation, $3x + 8 = -2x + 18$, by showing that your value for x gives both sides of the equation the same value.

5. Use algebra tiles to illustrate the solutions to each of the following problems. Record your steps with paper and pencil.

a. $5x + 3 = -x - 9$

b. $2(x + 1) = 3(-x + 6) - 1$

6. Solve the word problem at the beginning of the page.

7. How do the algebra tiles help you with solving equations?

18. Revisit the student page **Solving Equations with Algebra Tiles** in problem 17.

a. Complete a **QRS Guide** for problems 5a and 5b. For each problem, anticipate three levels of student responses: misconceived, partial, and satisfactory.

b. Write teacher support questions to help students solve the problem, providing "just enough" help.

Preparing for Instruction

19. Algebra tiles may help students build conceptual understanding for a procedure, in this case, solving linear equations. The tool relies on students' familiarity with area and their ability to translate algebraic symbols into the tiles each symbol represents.

a. Predict different ways students will attempt to translate the expression $2(x + 6)$ into algebra tiles. How might you use algebra tiles to help students avoid the typical error of not applying the distributive law correctly when they simplify this expression?

b. Repeat problem 19a for the expression $-(x + 6)$.

c. A student's last two steps while solving a linear equation with algebra tiles were as follows: (Assume the student has used the algebra tiles correctly to this point.)

$-(x + 6) = 2$. I'll flip over the tiles on both sides of the equation and get $x + 6 = -2$. Then I add -6 to both sides of the equation to get $x = -8$.

This is the first use of the *flip rule* that has come up in class. Will the rule always work? Why or why not? Is this rule reasonable? How will you help students understand it? Will you allow students to use the rule, or is there another approach you would prefer they use? Explain.

20. In **DMU problem** 6 and in **DMP problem** 17, you solved equations with tables, graphs, or algebra tiles.

a. Compare solving linear equations with algebra tiles, tables, and graphs. Which do you prefer? Why?

TEACHING AND LEARNING HIGH SCHOOL MATHEMATICS © 2010 John Wiley & Sons, Inc.

b. What are some advantages and disadvantages to students of each of the three methods of solving equations?

21. In the Team Builder, each person was given a function card, asked to determine the function family to which it belongs, and requested to find the rest of their teammates by locating others whose cards are in the same function family. Following are rules for another approach to **Find Your Function Family**.

    i. A card is attached to your back. You are not allowed to look at the card or to look at a reflection of it.

    ii. Write three or four questions you will ask others to help you determine the function family to which the example on your back belongs. You may not ask directly to which function family the card belongs. For example, you may not ask questions such as, "Is my function a quadratic function?" You may ask questions such as, "In equation form, would my function contain any exponents?"

    iii. Ask questions of other classmates to determine the function family to which the example on your back belongs.

    iv. Answer questions of other classmates to help them determine the function family to which the examples on their backs belong.

    v. Once you have determined your function family, stand at the periphery of the room to wait until all classmates have determined their respective function families.

    vi. Your teammates are other classmates whose examples are in the same function family as yours.

    a. Discuss the extent to which the two approaches (the approach used in the Team Builder to this unit and the approach described above) promote communication about functions.

    b. Which approach do you think is easier for students? Why do you think so?

    c. The activity with function families requires students to have some knowledge of functions in order for students to engage successfully with it. When might you be able to use such an activity with Algebra I students? With Algebra II students? With precalculus students?

    d. For students who only know one function family, how could this activity be modified to help them gain a deep understanding of that function family?

Using Technological Tools

22. A computer algebra system (CAS) performs algebraic manipulations symbolically. That is, CAS accepts variables as inputs and returns variables as outputs.

    a. The two screen images from a TI-89 illustrate two different ways that a CAS could be used to solve $3x + 2 = 8x - 7$. In the first, the notation SOLVE($3x + 2 = 8x - 7$, x) indicates that the equation $3x + 2 = 8x - 7$ is to be solved with respect to x. In the second, the student entered his steps and the calculator simplified each. Compare and contrast these two approaches to the use of a CAS in an algebra I class.

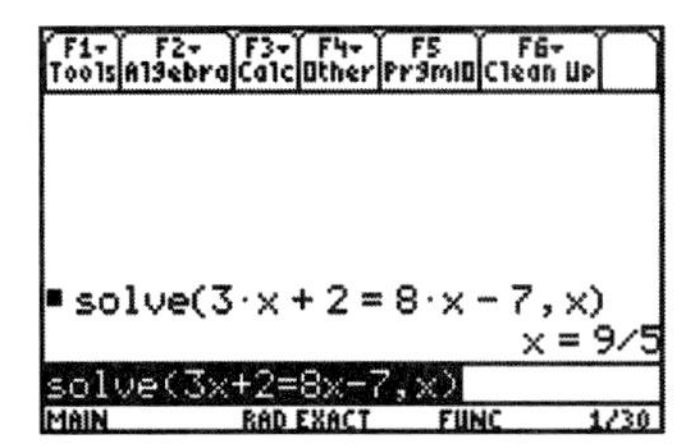

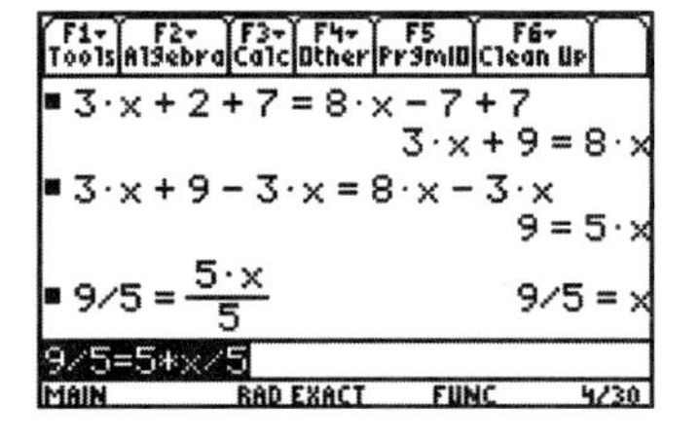

TEACHING AND LEARNING HIGH SCHOOL MATHEMATICS © 2010 John Wiley & Sons, Inc.

b. Create a problem for Algebra I students where a CAS would support student learning.

c. Create a problem for Algebra I students where a CAS would be inappropriate.

d. How might CAS technology relate to equity (high expectations and support for all students)?

Communicating Mathematically

23. Parentheses are used in many different ways in mathematics: They may indicate multiplication, as in $(5 + y)(x)$, or they may indicate the argument of a function, as in $g(x)$.

a. Identify another mathematical symbol that has more than one meaning.

b. Identify some strategies to help students avoid confusion when a symbol has more than one meaning.

24. Ms. Portier regularly has students work in cooperative groups on extended projects. When projects are completed, the group then prepares a joint presentation for the class and all students are to have an equal part in the oral presentation. However, during the presentation, Jackson, one of the group members, took over and prevented other group members from presenting their parts.

a. What actions might the teacher take?

b. Choose one possible teacher action and tell any benefits and concerns you have about it.

c. What are the pros and cons of waiting to address the problem after the group has finished its presentation?

d. If you are one of the students in the group other than Jackson, which approach would you recommend the teacher take?

Analyzing Students' Thinking

25. A student solved the equation $x - 3 = 3(2 - x) - 1$ using algebra tiles as shown at right. The student did not indicate the symbolic algebra that accompanied each step. What is the student doing? Is his work correct? If so, show the algebra that the student is accomplishing with each step. If not, indicate the error(s) the student is making and determine questions that you would like to ask the student to help him find his error.

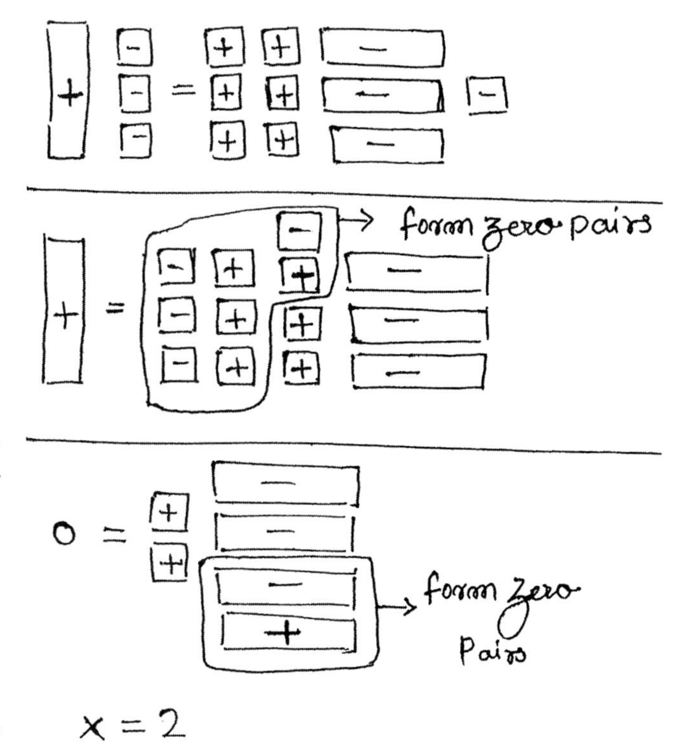

26. A student was asked to simplify $3x + 7x$. Below is the student's work.

3x + 7x = 10x, so x = 1

What do you think the student was thinking? How might you help this student address his or her misconception?

Reflecting on Professional Reading

27. Read the "Algebra Standard for Grades 9–12" from *Principles and Standards for School Mathematics*, pp. 296–306.

a. Reflect on the Team Builder that began this unit, **Find Your Function Family.** How does this activity relate to the recommendations in the algebra standard from *PSSM*?

TEACHING AND LEARNING HIGH SCHOOL MATHEMATICS © 2010 John Wiley & Sons, Inc.

b. What were some of the advantages and disadvantages of the different representations of functions you explored in the Team Builder?

c. The diagram accompanying Figure 7.11 in the article (p. 306) tells a story even though there is no scale on the vertical axis. Students need to consider the labels on the two axes to understand the diagram.

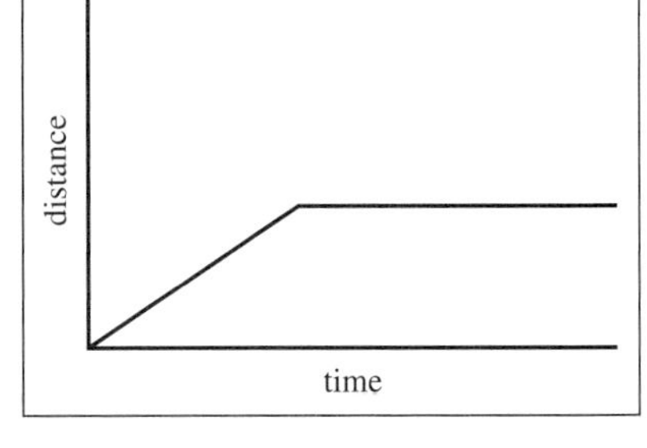

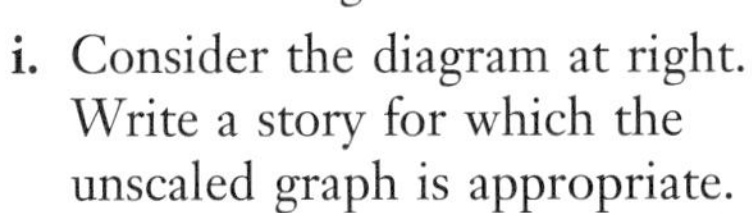

i. Consider the diagram at right. Write a story for which the unscaled graph is appropriate.

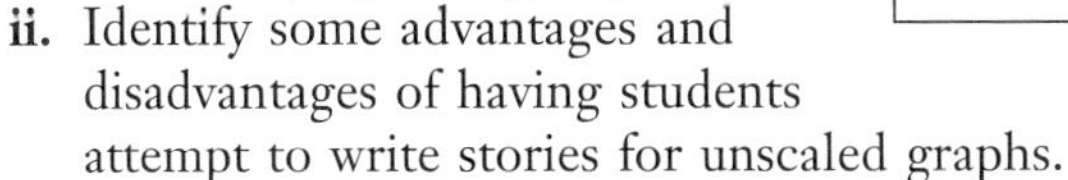

ii. Identify some advantages and disadvantages of having students attempt to write stories for unscaled graphs.

28. Read "Teaching and Learning Functions" by Mindy Kalchman and Kenneth R. Koedinger. (In M. Suzanne Donovan and John D. Bransford (Eds.), *How Students Learn: Mathematics in the Classroom* (pp. 351–393). Washington, D.C.: National Academies Press, 2005. The article is available online at http://books.nap.edu/openbook.php?record_id=11101&page=1).

a. How was the Technology Principle from *PSSM* incorporated in the sample lessons?

b. How does the work of this lesson relate to issues raised about the learning of functions in the reading from *How Students Learn*? If one or more of the principles was missing, what suggestions would you offer for changes to address the principles?

3.2 Thinking about Learning Outcomes: Exponential Functions

When planning lessons, teachers need to start with considering where they want their students to end. To decide how to teach a lesson, teachers first need to decide what they want students to understand and be able to do as a result of instruction. In this lesson, you will focus on determining these learning outcomes. (In some books or some state frameworks, learning outcomes are called expectations, goals, objectives, benchmarks, or standards. In this text, the term *learning outcomes* is used to be inclusive of those outcomes that are measurable as well as those that are not.)

Lesson planning begins with the teacher raising and addressing several questions about the lesson's learning outcomes. What mathematics will be developed or deepened? What is it that students will be able to do? What features of the new concepts must they be able to recognize, analyze, or use to solve problems? What will students be able to demonstrate, explain, or perform as a result of the lesson? All of these features are important in thinking about the learning outcomes of a lesson.

This course has emphasized instruction centered on tasks with high levels of cognitive demand. Such tasks entail significant mathematics, challenge and engage students, and require making connections among mathematical ideas. Teaching using such tasks is demanding for teachers. In designing instruction with such tasks, one of the first issues is identifying the core mathematical ideas the lesson should highlight. If students are to extract the important mathematics from an activity, it is essential for the teacher to have a clear sense of what that mathematics is and what its potential is for building students' understanding. A common pitfall for teachers who are first using programs with such tasks is to focus on the engagement and enjoyment of

 © 2010 John Wiley & Sons, Inc.

students without putting sufficient focus on the "big ideas" that the lesson includes. This is why it is important for teachers to be clear about the mathematical outcomes of the lesson as they plan instruction.

In this lesson, you will:

- Determine properties of exponential functions from different representations.
- Identify learning outcomes for a lesson on exponential functions.

Materials

- computer with software *Fathom*™, one per pair
- graphing calculator, one per person
- graph paper, several sheets per pair

Website Resources

- **Adding and Subtracting Functions** student page
- **Comparing Graphs of $y = B^X$** student page
- **Functions of the Form $y = AB^X + D$** student page
- *Fathom*™ file **Exponential Slider Graphs** in the software investigations folder
- **Experimenting with Growth and Decay** student page

LAUNCH

1. Refer to the student page **Adding and Subtracting Functions**, in the section **Listening to Students Reason about Functions**. What appear to be the learning outcomes, both mathematical and nonmathematical, for this student page?

EXPLORE

2. Three student pages related to exponential functions follow. With your group, choose one of the student pages. Ensure that each student page is chosen by at least one group.

   Comparing Graphs of $y = B^X$

   Functions of the Form $y = AB^X + D$

   Experimenting with Growth and Decay

   a. Complete the student page.

   b. Below is a list of learning outcomes pertaining to exponential functions. Determine which ones pertain to your student page.

   - Identify an exponential function from the general form of the equation.
   - Identify an exponential function from a graph.
   - Identify an exponential function from a table.
   - Relate exponential functions to real phenomena.
   - Understand and use the fact that the base of an exponential function determines if the function increases or decreases.
   - Determine how the base affects changes in the exponential function.
   - Determine the asymptotic behavior of an exponential function.
   - Determine the equation of an exponential function from a graph, table, or real life situation.
   - Interpret the meaning of the variable x in an exponential context.

TEACHING AND LEARNING HIGH SCHOOL MATHEMATICS © 2010 John Wiley & Sons, Inc.

c. Choose one of the learning outcomes you identified for your student page. As a teacher, what evidence would you accept that the student has achieved the learning outcome?

Comparing Graphs of $y = B^x$

Materials

- graphing calculator, one per person

1. Set the graphing calculator viewing rectangle for x in [–5, 5] and y in [0, 10]. Plot each of the following functions on the same screen.

   $y = 1.3^x$ $y = 1.5^x$ $y = 2^x$

   $y = 2.5^x$ $y = 3^x$ $y = 3.5^x$

   a. Compare the graphs. What do you notice?
   b. What points do the graphs have in common? How can you tell from the equations of each function?
   c. Describe the overall shapes of these graphs and their steepness relative to each other (indicate intervals over which you are comparing the graphs). What do you notice?
2. Set the graphing calculator viewing rectangle for x in [–2, 2] and y in [0, 2]. Repeat problem 1. What more can you tell about the graphs in this viewing rectangle?
3. Set the graphing calculator viewing rectangle for x in [–5, 5] and y in [0, 10]. Plot each of the following functions on the same screen.

   $$y = 0.3^x \quad y = 0.5^x \quad y = \left(\frac{2}{3}\right)^x \quad y = 0.8^x$$

   a. Compare the graphs. What do you notice?
   b. Plot each of the graphs again using a viewing rectangle set for x in [–2, 2] and y in [0, 2]. Compare the graphs. What do you notice?
   c. What points do the graphs have in common? Why?
   d. Compare the overall shapes of these graphs and their steepness relative to each other over several intervals. What do you notice?
4. Set the graphing calculator viewing rectangle for x in [–5, 5] and y in [0, 5]. Plot each pair of functions on the same screen.

   i. $y = 0.5^x$ $y = 2^x$

   ii. $y = \left(\frac{2}{3}\right)^x$ $y = 1.5^x$

   a. Compare each pair of graphs. What do you notice?
   b. How are the equations of each pair of graphs related?
5. Summarize what you learned about functions of the form $y = B^x$ from your work in problems 1–4.

TEACHING AND LEARNING HIGH SCHOOL MATHEMATICS © 2010 John Wiley & Sons, Inc.

Functions of the Form $y = AB^x + D$

Materials

- computer with software *Fathom*™, one computer per pair
- **Exponential Slider Graphs** *Fathom*™ file
- graph paper, one sheet per person

1. Locate the *Fathom*™ file **Exponential Slider Graphs**. Below the graph of $y = AB^x + D$, there are three sliders, tools that allow you to vary the parameters A, B, and D.
   - **a.** Set the starting values of the sliders so that $B = 2$, $A = 1$, and $D = 0$. You can enter the value directly into the slider by double clicking on the value listed, typing the desired value, and pressing Return. What is the equation of the function with these settings?
   - **b.** Experiment with the B slider. Click on the arrow to let B change values. What happens to the graph of $y = AB^x + D$?
   - **c.** For which values of B is $y = B^x$ an increasing function?
   - **d.** For which values of B is $y = B^x$ a decreasing function?
   - **e.** For which values of B is $y = B^x$ constant?
   - **f.** Why are your responses to problems 1c and 1d sensible?
2. Reset the B slider so that its value is 2.
   - **a.** Experiment with the slider for D.
   - **b.** Describe the shape and location of the graph of $y = AB^x + D$ as D is allowed to change.
   - **c.** What effect does D have on the graph of $y = AB^x + D$? How do you know?
3. Reset the slider for D to 0.
   - **a.** Experiment with the slider for A.
   - **b.** When $B > 1$, for what values of A is the graph of $y = AB^x + D$ increasing? decreasing? constant?
   - **c.** Why are your responses to item 3b sensible?
   - **d.** What effect does A have on the graph of $y = AB^x$? How do you know?
   - **e.** Set the B slider so that its value is 0.5. Repeat problems 3a–3d.
4. Without graphing the function electronically, sketch the graphs of each of the functions by hand. Explain how you know your work is correct.
   - **a.** $y = 3^x - 6$
   - **b.** $y = -0.5^x + 2$
   - **c.** $y = (-2)(3)^x$
   - **d.** $y = (-2)(3)^x - 6$
5. Check your results to item 4 by graphing the equations. If your results do not agree with your sketch, determine why there is a discrepancy.

TEACHING AND LEARNING HIGH SCHOOL MATHEMATICS © 2010 John Wiley & Sons, Inc.

Experimenting with Growth and Decay

Materials

- ball to bounce, one per team
- tape measure or meter stick taped to the wall, one per team

1. Messages sent on the Internet sometimes take on a life of their own. Consider a chain letter that requests you send the letter to 10 friends. Though in real life it is not good practice to forward such emails, assume that everyone who receives the message forwards it exactly as requested. Let r be the number of rounds and $p(r)$ the number of people who receive the message on round r.

a. Complete the table below. It has been started for you.

Number of rounds, r	Number of people $p(r)$ who receive the message during round r	$p(r) - p(r-1)$	$p(r) \div p(r-1)$
1	10	90	10
2	100	900	
3	1000		
4			
5			
6			

b. What patterns do you see in the table?

c. Which method, finding differences between consecutive terms or finding ratios between consecutive terms, is most enlightening and why?

d. Which of the methods helps you determine an equation for the function?

e. Determine an equation relating the number of rounds, r, and the number of people, p, who receive the message during round r.

2. In the movie *Pay It Forward* (Warner Brothers Pictures, 2000), a teacher challenges his students to propose an idea that will be significant enough to change the world. One student, Trevor, proposed that each person who is the recipient of a really big good deed, instead of paying it back, should pay it forward by doing something really important for three other people. Trevor explained that these three good deeds have to be really big and something that the recipients cannot do for themselves.

a. Suppose each recipient of a really good deed pays it forward the following day. Draw a tree diagram showing the number of people who are the recipients of good deeds each day for three days, with Day 1 being the day when the first three good deeds are done.

b. Organize the data into a table. Determine any patterns in the data.

c. Use one of the methods in problem 1 to explain how the graph changes.

d. Determine an equation relating day number to numbers of recipients.

3. Choose a ball to bounce. Drop the ball from a height of one meter (the bottom of the ball should begin 1 meter from the floor). Two observers should determine the rebound height of the ball (the greatest height the bottom of the ball achieves from the floor after one bounce). For each successive bounce, start the bottom of the ball at the height of the previous rebound.

(*continued*)

 © 2010 John Wiley & Sons, Inc.

a. Record the data in a table.

b. Determine any patterns in the data.

c. Use one of the methods in problem 1 to determine how the function changes as the bounce number increases.

d. Determine an equation relating bounce number with rebound height.

4. a. What mathematical similarities exist among the scenarios above? Explain.

b. How can you tell from a context when an exponential function will model the data?

SHARE AND SUMMARIZE

3. Share your choices of learning outcomes with other teams.
   a. What similarities or differences did you find?
   b. What additional learning outcomes are addressed by the student pages that are not on the list in **Explore** item 2b?
   c. What learning outcomes from the list do not seem to be addressed by the student pages?
4. What other outcomes related to exponential functions would be important for high school students to learn, perhaps in precalculus?
5. *PSSM* recommends that mathematics be taught so students can understand and use relationships among multiple representations of a concept. In what ways do the student pages **Comparing Graphs of $y = B^x$**, **Functions of the Form $y = AB^x + D$**, and **Experimenting with Growth and Decay**, help students learn about exponential functions from a variety of representations?
6. The algebra standard in *PSSM* includes a number of expectations (or learning outcomes) for grades 9–12 students under the broad objective "Understand patterns, relations, and functions." Review the expectations about functions in general.
   a. Determine how you would modify the expectations to be explicit for exponential functions.
   b. How does the list of learning outcomes at the beginning of the **Explore** relate to the list in *PSSM*?
   c. Which of the expectations from *PSSM* could be incorporated into the three student pages in this lesson? Provide a rationale and indicate how you would modify the student page to incorporate the additional expectations.

DEEPENING MATHEMATICAL UNDERSTANDING

1. *More Explorations with Exponential Functions.* During class, your group likely completed only one of the three student pages **Comparing Graphs of $y = B^x$**, **Functions of the Form $y = AB^x + D$**, and **Experimenting with Growth and Decay**.
   a. Complete the other two student pages.
   b. Create a concept map (see Lesson 2.5 **DMP** problem 18) indicating what knowledge these student pages contribute to students' understanding of exponential functions.
   c. List additional contexts you would expect to model with exponential functions.
2. *Playing with Exponential Functions.* Look up the Tower of Hanoi puzzle on the Internet. Determine a function that gives the least number of moves, m, needed

TEACHING AND LEARNING HIGH SCHOOL MATHEMATICS © 2010 John Wiley & Sons, Inc.

to move d disks from one peg to another. To what function family does this data set belong? Why is your answer reasonable?

3. *Recursive Exponential Functions*. Complete the student page **Recursion in Exponential Functions.**

   a. How do the recursive formulas help you find the base of the exponential function in the explicit formula?

   b. How do the recursive formulas help you find the horizontal asymptote for an exponential function?

   c. How does the horizontal asymptote arise in the explicit formula? Explain.

Recursion in Exponential Functions

1. Complete the following table.

n	$f(n) = 2^n$	$g(n) = 3 \cdot 2^n$	$h(n) = 3 \cdot 2^n + 5$
−3			
−2			
−1			
0			
1			
2			
3			
4			

a. Determine a way to find $f(n)$ from $f(n-1)$, where n is an integer. Write the relationship as an equation, $f(n) =$______ where $f(0) =$___.

b. Repeat problem 1a for $g(n)$ from $g(n-1)$, where n is an integer; $g(n) =$______, where $g(0) =$___.

c. Repeat problem 1a for $h(n)$ from $h(n-1)$, where n is an integer; $h(n) =$______, where $h(0) =$___.

d. Comment on any similarities between the recursive formulas you found in problems 1a–1c.

2. Repeat problem 1 for $f(n) = B^n$, $g(n) = A \cdot B^n$, and $h(n) = A \cdot B^n + D$, where $B > 0$ and $B \neq 1$ and n is an integer.

3. a. How do the recursive formulas relate to the base of the exponential function?

b. How do the recursive formulas relate to the horizontal asymptote of the exponential function?

c. What information does the starting value ($f(0)$, $g(0)$, or $h(0)$) provide about the function?

TEACHING AND LEARNING HIGH SCHOOL MATHEMATICS © 2010 John Wiley & Sons, Inc.

For problems 4 and 5, return to the **Function Card Sort** in the Unit Three Team Builder (these are also in the Unit Three Appendix).

4. *Exponential Functions in Stories and in Context.*
   a. Find the stories and real contexts that fit the function family you investigated in Lesson 3.2. State how you know these cards can be represented by the given function family.
   b. Write another story or determine another real context that fits this family of functions.
5. *Exponential Functions in Tables.* Find two tables for which the pattern of change in $y = f(x)$ for a constant change in x is the same as that for the function family explored in Lesson 3.2.
6. *The Population of China.* For many years, China had a policy that allowed urban married couples to have only one child or pay large fines to exceed that number. Rural families could have two children without fines. (Although this policy began in the late 1970s, there have recently been some changes in the policy.)
   a. Do you expect China's population to increase, decrease, or stay the same with this policy? Explain.
   b. Find population data for China in an almanac or on the Internet. Fit the data with an exponential function. Describe what you learn.
   c. Find China's population data for 1980–present. What function best fits this data? Explain your choice.

DEVELOPING MATHEMATICS PEDAGOGY

Analyzing Students' Thinking

7. For functions of the form $f(x) = b^x$, a student stated: $f(n) - f(n - 1) = f(n - 1)$ when n is an integer. Comment on the student's observation. Is it always true? If not, find a counterexample. Is it ever true? If so, find values of b for which the relationship is true.

Preparing for Instruction

8. Learning outcomes are categorized in many ways. The following table of educational objectives contains one list, part of which is based on the work of Benjamin Bloom and David Krathwahl in *Taxonomy of Educational Objectives* (New York: Addison-Wesley, 1984).
   a. For the three student pages **Comparing Graphs of $y = B^x$**, **Functions of the Form $y = AB^x + D$**, and **Experimenting with Growth and Decay**, identify how the learning outcomes you identified in the **Explore** align with those in the educational objectives table (page 200).
   b. Identify other learning outcomes you believe the student pages address after reading these categories.

TEACHING AND LEARNING HIGH SCHOOL MATHEMATICS © 2010 John Wiley & Sons, Inc.

Educational Objectives

knowledge—know, identify, recall, state

comprehension—understand, communicate, reason, use symbols and vocabulary correctly

skill—perform a procedure (e.g., solve a linear equation); perfect motor skills (e.g., use a compass, protractor, ruler); operate a calculator; use learnings in routine settings (exercises)

application—apply to contexts different from those presented in class but similar to the level of class examples

problem solving—apply previous learning in nonroutine settings (problems); transfer routine skills to novel situations; find patterns; generalize; prove; justify

evaluation—decide whether approaches used or explained by you or others are accurate, reasonable, general, useful

beliefs and affect—feel positively (about mathematics, learning mathematics, other learners); persist in learning; attempt problems; work creatively; reflect on learning

collaboration—work effectively in groups; value others' thinking; build on others' suggestions; communicate clearly; celebrate everyone's progress

In problems 9–11, refer to the three student pages **Comparing Graphs of $y = B^x$**, **Functions of the Form $y = AB^x + D$**, and **Experimenting with Growth and Decay**.

9. What ongoing learning outcomes would you have for students throughout their study of functions?

10. a. Modify one of the student pages as needed to make it appropriate for Algebra II students.

    b. What additional learning outcomes did you address when the student page was for Algebra II students rather than Algebra I students?

11. a. In the equation $y = A \cdot B^x + D$, identify the meaning of the base, exponent, and A in terms of the contexts for the chain letter, movie, and ball problems.

    b. Find another context of interest to students that can be represented by an exponential equation. How does the context help identify the equation by providing insight into reasonable values for A, B, and D?

12. Teachers write learning outcomes not in the abstract but with some knowledge about the level of their students. Many outcomes might be written about learning exponential functions, some that would be appropriate at Algebra I, some at Algebra II, and some at Precalculus. Find your state or local district descriptions for courses at these three levels. With regard to exponential functions, identify some objectives that students are not expected to address until Algebra II and/or Precalculus.

13. Choose one of the student pages from this lesson or from the **DMU/DMP** questions in this lesson.

    a. Complete a **QRS Guide** for the problems on the student page.

    b. Expand the **QRS Guide** to indicate the learning outcomes for the student page.

    c. Decide on a cooperative learning (CL) strategy you think would work well with the chosen student page. Indicate the CL strategy at the top of the **QRS Guide**.

TEACHING AND LEARNING HIGH SCHOOL MATHEMATICS © 2010 John Wiley & Sons, Inc.

14. Some schools expect teachers to write the learning outcomes for each lesson on the board each day.

    a. What are some potential benefits of this practice?

    b. What are some concerns you have about this practice?

Analyzing Instructional Materials

15. Complete the student page **Exploring Exponential Functions and Their Inverses**, using *The Geometer's Sketchpad*™ file **Exp Inverses** (see the Unit Three Software Investigations folder on the course website).

    a. Graphically, how does one find the inverse of an exponential function from the graph of $f(x) = B^x$?

    b. What learning outcomes are covered by this student page?

    c. Students were asked to explore inverses of exponential functions informally without naming the function. What are benefits of this approach?

    d. The student page was written for Algebra I students. What additional concepts would you like to pursue for Algebra II students? Suggest modifications to the student page to make it more appropriate for Algebra II students.

    e. Write the learning outcomes Algebra II students should achieve.

TEACHING AND LEARNING HIGH SCHOOL MATHEMATICS © 2010 John Wiley & Sons, Inc.

Exploring Exponential Functions and Their Inverses

Complete the following using *The Geometer's Sketchpad*™ file **Exp Inverses**.

1. a. Locate the slider at the bottom of the screen. Highlight only the point labeled *B*. Move the slider until the value of *B* is approximately 2.
   b. The point *P* appears on the graph of $f(x) = B^x$. Point *P′* is the reflected image of *P* across the line $y = x$. Make certain that only *P′* is selected, then turn on **Trace Point** in the Display menu.
   c. Drag *P* along the graph of $f(x) = B^x$. How does the traced graph compare with the graph of $f(x) = B^x$?
   d. Click on the button labeled **Show** $\boldsymbol{y = x}$. What relationship exists between $f(x) = B^x$ and the traced graph?
2. Use your work from problem 1 to complete the following:
   a. Is the traced graph a function? Why or why not?
   b. What is the domain of function *f* defined by $f(x) = B^x$? What is the range of *f*?
   c. What is the domain of the function represented by the traced graph? What is the range of the function represented by the traced graph?
   d. Describe the overall shape of the function $y = f(x)$ and its steepness over several intervals. Describe the overall shape of the traced graph and its steepness over several intervals. How are these two graphs related?
   e. Describe any symmetry or asymptotic behavior of $f(x) = B^x$. How are these properties evident in the traced graph? Explain.
   f. Determine any *x* or *y* intercepts of $y = f(x)$. Determine any *x* or *y* intercepts of the traced graph. What is the relationship between intercepts for each graph?
3. Move point *P* to coincide with the *y*-axis. Deselect all points, then select the point labeled *B* on the slider at the bottom of the page. Change the value of *B* so that it is between 0 and 1.
   a. For this value of *B*, describe the overall shape of the function $y = f(x)$ and its steepness over several intervals. Before tracing the graph, predict the graph of the image of $y = f(x)$.
   b. Drag point *P* along the graph of $f(x) = B^x$. Repeat problem 2 for this new function.
4. Move *P* to coincide with the *y*-axis. Deselect all points. Change the value of *B* and trace the graph for other values of *B*.
   a. For what values of *B* is $y = B^x$ increasing? decreasing? constant?
   b. For what values of *B* is the traced image increasing? decreasing?
   c. Describe the traced image when $y = B^x$ is a constant function.
   d. For each of problems 4a–c, explain why the graph behaves as it does.
5. From this exploration,
   a. Describe what you have investigated about the functions that arise from reflecting exponential functions across the line $y = x$.
   b. Create a concept map, Link Sheet, or Frayer Model to organize what you know about exponential functions.
   c. Create a concept map, Link Sheet, or Frayer Model to organize what you know about the functions that arise from reflecting exponential functions across the line $y = x$.

Materials

- computer with *The Geometer's Sketchpad*™ software, one computer per pair
- **Exp Inverses GSP**™ file

TEACHING AND LEARNING HIGH SCHOOL MATHEMATICS © 2010 John Wiley & Sons, Inc.

16. Find at least three different Algebra II textbooks with recent copyrights. Locate a section focusing on exponential functions. Analyze how multiple representations are used to help students make sense of the function family. Write a short reflection about your findings, supporting your conclusions with specific examples from the text.
17. Refer to *Student Math Notes*, "Are You Interested in Stretching Your Dollars?" (November 2001). (Go to the NCTM website, nctm.org, and look under the tab for Lessons and Resources to find *Student Math Notes*.)
    a. Work through the questions. Comment on the sequence of activities and how they build on each other.
    b. Write learning outcomes you believe are addressed in these *Student Math Notes* activities.
    c. Comment on the extent to which the tables are integral to the activities. What do you see as the purpose of these tables in terms of learning outcomes?

Using Internet Resources

18. a. Find data on the Internet that highlights the comparison of linear to exponential data. One example is the highest baseball players' salaries versus average salaries in major league baseball with time (years) as the independent variable.
    b. How might such data be used to develop student pages or other activities for students?

Teaching for Equity

19. Leslie Garrison and Jill Kerper Mora (1999) highlight four domains relating to language and concept development when dealing with English Language Learners (ELL). For each, they detail how learning of mathematics concepts and English are impacted.
    - Students attempt to learn an *unknown concept* in an *unknown language*. Learning is limited.
    - Students attempt to learn an *unknown concept* in a *known language*. Learning of the concept is possible.
    - Students attempt to learn a *known concept* in an *unknown language*. Learning of terminology related to the concept is possible in the unknown language.
    - Students attempt to learn a *known concept* in a *known language*. Learning advances toward the next conceptual or linguistic level.

    a. State your understanding of each of the domains in terms of its meaning for your mathematics instruction when ELL students are in your class.
    b. One communication strategy often used in mathematics classrooms is to have students construct their own personal dictionary of mathematics terms and symbols. How might such a dictionary be modified to help English language learners with the dual difficulties of learning mathematics in English and in their first language?
    c. Many textbook publishers now offer glossaries in languages other than English. Determine whether the textbooks used in high schools near you have such a resource available.
    d. Search the Internet for glossaries or free translation software. Try inputting a sentence in English, translating it to another language, and then translating back to English. How accurate was the translation?
    e. Consider the schools in which you have observed thus far or the school in which you teach. Investigate what resources exist for students who are not proficient in English. Also investigate the resources available to teachers who

TEACHING AND LEARNING HIGH SCHOOL MATHEMATICS © 2010 John Wiley & Sons, Inc.

have students in their mathematics classroom who are not proficient in English. (To read more, find "Adapting Mathematics Instruction for English-Language Learners: The Language-Concept Connection" by Leslie Garrison and Jill Kerper Mora (in *Changing the Faces of Mathematics: Perspectives on Latinos*, edited by Luis Ortiz-Franco, Norma G. Hernandez, and Yolanda De La Cruz (pp. 35–47). Reston, VA: National Council of Teachers of Mathematics, 1999.)

Using Technological Tools

20. The TI-nspire calculator has dynamic graphing capability, including sliders, similar to the dynamic capabilities on software such as *Fathom*™.

a. On a new page in the graph window, choose **Menu**, **Actions**, and then **Insert Slider**. Assign a variable name to the slider (say ***b***). Then, choose **Menu**, **Graph Type**, and **Function**. Input the function $f(x) = \boldsymbol{b}^x$, that is, insert the function $f(x) = \textbf{slider}^x$. Notice that the variable name for the slider shows in bold type. Graph the function. In the slider section of the window, use the hand to grab the slider. Move it and observe what happens to the graph.

b. In what ways is the dynamic capability on the calculator easier to use than on the computer software? In what ways is it more difficult to use?

c. What are some advantages and disadvantages of calculator versus computer technologies?

Reflecting on Professional Reading

21. Read "Seeing How Money Grows" by James Metz (*Mathematics Teacher*, 94 (April 2001): 278–280 plus lab sheets).

a. This article provides activities to compare and contrast linear growth models with exponential growth models. Comment on any advantages and disadvantages of looking at these two types of growth models simultaneously.

b. Compare the types of activities in this article with the student pages in the **Explore**. What similarities and differences do you notice?

c. What differences in learning outcomes would you expect to be written by teachers using the activities in this article compared to teachers using the student pages from the **Explore**?

d. Compare and contrast the activities in this article with the activities in the *Student Math Notes* in **DMP** problem 17. Identify any similarities and differences that may influence their appropriateness for use with high school students.

e. In what ways might context influence how well activities work with a particular group of students?

f. Identify any modifications or cautions you would want to keep in mind when doing this activity with English Language Learners (ELL) or students needing special accommodations.

3.3 Active Learning: Modeling Data through Experiments

Active learning environments help make mathematics accessible to all students. Active learning environments promote engagement, motivation, problem solving, and communication and are an important aspect of an equitable learning environment. One strategy that ensures students become actively engaged is to have them conduct experiments just as they do in their science classes. There are several

TEACHING AND LEARNING HIGH SCHOOL MATHEMATICS © 2010 John Wiley & Sons, Inc.

advantages of using experiments to engage students; in this lesson, two are highlighted:

- Students work together in pairs or small groups to conduct the experiment and collect the necessary data. In the real world, students must learn to work as members of a team.
- Students analyze the data and make sense of it, then communicate their results in writing. Writing across the curriculum is increasingly an important curricular requirement, and experiments help teachers meet such requirements.

If high school mathematics teachers lack classroom quantities of materials needed for experiments they can use learning centers, which are stations with different activities and materials. Only enough materials are needed for one or two groups of students at a time because groups can rotate through a series of different experiments. So, only a limited supply of equipment is needed.

Functions, one of the mathematical foci of this unit, provide a natural context for multiple experiments. Students collect data and attempt to model that data with an appropriate function. This inquiry-driven activity can help students realize that the scientific process applies to mathematics classes as well as to science classes.

In this lesson, you will:

- Conduct several experiments to collect data that can be modeled with linear, quadratic, cubic, exponential, square root, or rational functions.
- Model a set of data to determine an appropriate function.
- Engage in learning centers as a classroom instructional strategy.

Materials

Wave Activity (**Launch**)
- stopwatch

On a Roll

- 3 or 4 balls of different weights and diameters
- a piece of plastic rain gutter about 4 to 6 feet long
- CBR with RANGER program and a compatible graphing calculator
- graphing calculator screen-sized transparencies, several per group

Spot the Dot

- small mirror
- sticky notes
- centimeter tape measures or meter sticks, 2 per station

Building Pyramids

- set of small magnetic rods and steel ball bearings, at least 90 rods and 35 ball bearings (Alternatively, use toothpicks for the rods and gumdrops, mini marshmallows, or modeling clay for the ball bearings.)

Counting Cubes

- 60 cubes in each of 5 containers, with the container number corresponding to the number of faces on which a dot is found on each cube (For example, in container 3, all cubes have a dot on 3 faces of the cube.)

TEACHING AND LEARNING HIGH SCHOOL MATHEMATICS © 2010 John Wiley & Sons, Inc.

Tiling with Squares

- 2 sheets of graph paper per group

Website Resources

- *On a Roll* student page
- *Spot the Dot* student page
- *Building Pyramids* student page
- *Counting Cubes* student page
- *Tiling with Squares* student page

LAUNCH

Have you ever been to a football game when the crowd starts "the wave"? Have you ever wondered how long it takes for a wave to travel through the entire crowd?

1. What type of relationship, if any, would you predict exists between the number of people in a wave and the length of time it takes the wave to complete its travel?
2. Collect several pieces of data by having a given number of people participate in the wave. Use a stopwatch to record the time it takes the wave to travel. Generate the wave by raising and then lowering your arms. Increment the number of people in the wave by 5 each time. If necessary, have the wave go around the class two or three times in order to have a large enough number of people. Try to be consistent in how you raise and lower your arms. Record your data in the following table. (Adapted from "The Wave" in *Algebra Experiments I* by Mary Jean Winter and Ronald J. Carlson. Menlo Park, CA: Addison-Wesley, 1993.)

Number of People	Time for the Wave to Travel
5	
10	
15	
20	
25	
30	
35	
40	

3. **a.** Use a graphing calculator to make a scatterplot of the data.
   b. Which variable should be the independent variable? Which should be the dependent variable? Explain your choices.
   c. Which family of functions seems to best model the data? Why do you think so?
   d. Determine an equation to model the data.
4. Choose a number of people to generate a wave different from any of the numbers in the table. Use your model from item 3d to conjecture the time needed for the wave to travel. Test your conjecture by timing a wave generated by this number of people. Modify your model as needed.
5. What are some factors that might influence the accuracy of your model?

TEACHING AND LEARNING HIGH SCHOOL MATHEMATICS © 2010 John Wiley & Sons, Inc.

EXPLORE

The experiment in the **Launch** was likely conducted with the entire class in order to have a large enough number of students to form a wave. Many experiments, however, are easily completed by a small group working as a team. With your group, rotate through the five centers to gather data for each experiment. Use the accompanying student pages to guide your work at each center.

On a Roll	Counting Cubes
Spot the Dot	Tiling with Squares
Building Pyramids	

6. As you work through the centers, think about how high school students might respond to the questions on the accompanying student pages. In addition, consider the following questions:
   a. What function family arises from the experiment?
   b. What more do you know about the function family after having worked through the experiment?
7. a. What features of the graphs are important for students to know? How do teachers help students focus on these features?
   b. What aspects of technology should students know before conducting these experiments?
8. How important is it for students to engage actively in experiments such as the ones in this lesson?
9. After working through the five activities, discuss with the other members of your group pedagogical issues about the use of learning centers.
   a. Identify potential advantages and disadvantages of using this type of instructional approach in your classroom. Include in your considerations issues related to students with special needs (e.g., gifted, learning disabled, handicapped, English Language Learners).
   b. What are some potential obstacles to overcome?
   c. How might the time schedule for the class (e.g., 45 minutes, 60 minutes, or 90 minutes) influence how you might use learning centers in your classroom?
10. Identify some issues that teachers need to consider to manage the use of centers with high school students.

TEACHING AND LEARNING HIGH SCHOOL MATHEMATICS © 2010 John Wiley & Sons, Inc.

On a Roll

Problem

Investigate the graphical and symbolic representations of the motion of a ball rolling up and down a ramp through the following experiment.

1. Prop one end of a piece of rain gutter on the seat of a chair, creating a ramp. Tape the other end to the floor. Roll a ball up the ramp and let it roll back down.

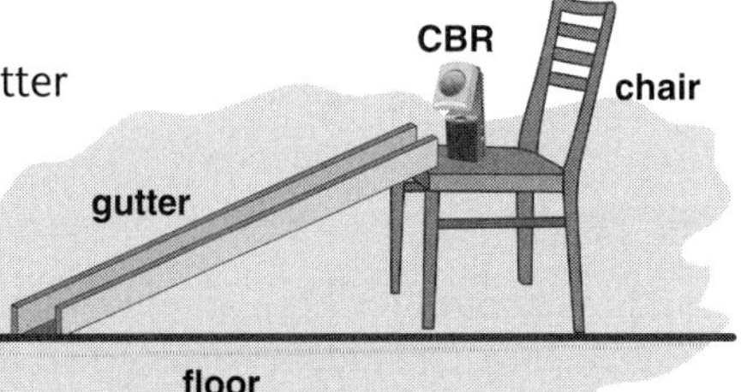

   a. Predict the shape of a graph that describes the motion of the ball. Sketch the graph of the height of the ball with respect to time.

   b. Write a short description about why you believe your graph is correct.

2. **a.** Place the CBR at the top of the ramp, facing down at the same angle as the ramp. Run the program RANGER on a graphing calculator connected to a CBR to collect data while the ball is rolling up and then down the ramp. Make certain the ball does not get closer than 0.5 meter (approximately 1.5 feet) from the CBR to avoid inaccurate data.

   b. Trace the graph of the motion of the ball onto a calculator screen-sized transparency. Does the graph show the motion as you expected it? Why or why not?

   c. Describe the graph's appearance when the ball was closest to the CBR.

   d. Run the program RANGER again, this time using the application BALL BOUNCE from the APPLICATIONS menu. Trace the new graph. What does BALL BOUNCE do to the data collected by the CBR?

   e. How can you tell from the graph when the ball was rolling up the ramp? rolling down the ramp?

   f. How can you tell from the graph when the ball was moving most slowly? most quickly? not moving? Explain.

3. Determine a function that models the data using the regression features on your graphing calculator as follows:

   a. Choose only the data pertaining to the motion of the ball as it was rolling on the ramp. From the PLOT MENU, choose SELECT DOMAIN and then follow the directions to use the cursor to select the appropriate data.

   b. QUIT the program and then use the regression capability of your calculator to fit the data.

4. If time permits, modify the experiment by starting the ball at the top of the ramp and letting it roll down the ramp. What, if anything, changes about the motion of the ball?

Materials

- 3 or 4 balls of different weights and diameters
- a piece of plastic rain gutter about 4 to 6 feet long
- CBR with RANGER program and a graphing calculator
- graphing calculator screen-sized transparencies, several per group

TEACHING AND LEARNING HIGH SCHOOL MATHEMATICS © 2010 John Wiley & Sons, Inc.

Spot the Dot

Problem

A dot on a wall is reflected into a mirror placed on the floor. In order to see the dot in the mirror, the distance you stand from the mirror depends on the height of the dot above the floor. Complete the experiment below to investigate this problem.

1. Place a mirror flat on the floor at some fixed distance from a wall. Mark a dot on a sticky note and place the sticky note on the wall.

a. Measure the distance of the mirror from the wall.

b. Measure the height of the individual locating the reflection of the dot in the mirror. Should you measure the full height of the person? Explain.

c. Measure the distance the same individual must stand from the mirror in order to see the dot on the wall reflected in the mirror. Repeat several times, each time moving the sticky note with the dot to a new spot on the wall and measuring the individual's distance from the mirror while still able to see the dot reflected in the mirror.

2. Make a conjecture about the relationship between the distance from the mirror and the height of the dot on the wall.

3. Complete the table at right, taking at least 7 pairs of measurements. The position of the mirror on the floor must remain constant. The same individual should locate the reflection of the dot in the mirror each time, attempting to observe the reflection in the same relative spot in the mirror (drawing cross-hairs on the mirror might be helpful). When finished collecting the data in the table, do NOT move the mirror. You will want to return to the lab setup to collect additional data points.

Height of Mark on Wall (Measured from Floor)	Distance of Observer from Mirror

4. a. Create a scatterplot of the data.

b. What function family seems to fit the data? Why do you think so?

c. Write a short paragraph about the data and your analysis of it.

5. a. Choose a new height to place the dot on the wall. Use your function to predict the distance the individual needs to stand from the mirror. Check your predicted value by returning to your lab setup.

b. Choose a new distance to stand away from the mirror. Use your function to predict the height at which the dot should be placed for the individual to see its reflection in the mirror. Check your predicted value by returning to the lab setup.

c. How accurate were your predictions?

d. If necessary, modify your function to find a better fitting model.

6. What is the underlying mathematics of the setup? Determine a theoretical function based on the underlying mathematics.

7. What real life uses might be made of your discoveries in this experiment?

Adapted from "Mirror, Mirror on the Floor" in *Algebra Experiments II: Exploring Nonlinear Functions* by Ronald J. Carlson and Mary Jean Winter, pp. 82–87. Menlo Park, CA: Addison-Wesley, 1993.

Materials

- small mirror
- sticky note
- centimeter tape measures or meter sticks, 2 per station

TEACHING AND LEARNING HIGH SCHOOL MATHEMATICS © 2010 John Wiley & Sons, Inc.

Building Pyramids

Problem

Triangular-based pyramids can be built using magnetic rods and ball bearings. Complete the experiment to determine a relationship between the side length of a pyramid and the number of rods or balls needed to create it.

1. Let the length of a magnetic rod (or a toothpick) be 1 unit. The steel ball bearings (or gumdrops) are used to connect the rods. With the magnetic rods and steel ball bearings, build regular triangular-based pyramids (each a tetrahedron) with sides of length 1 unit, 2 units, and 3 units. The picture below shows the first three pyramids built with rods and ball bearings.

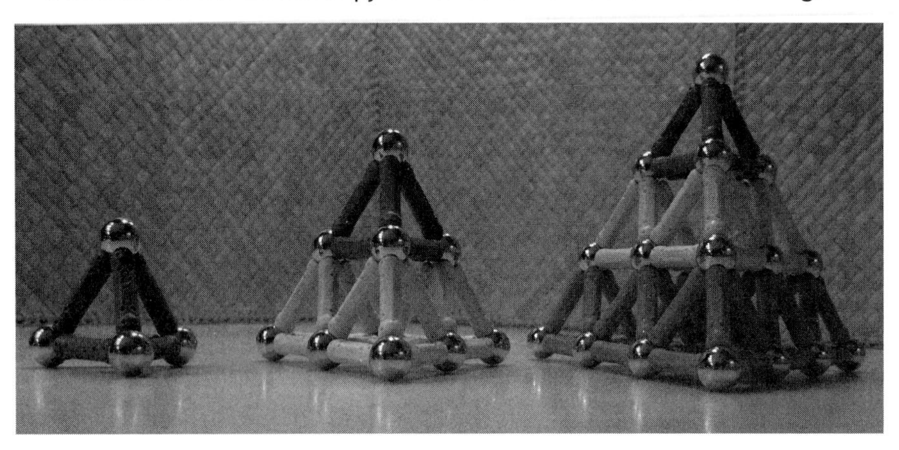

2. a. In the table below, record the number of rods and ball bearings needed to make the first four pyramids.

b. Make a scatterplot of pyramid side length versus number of rods.

c. Fit an appropriate function to the data.

d. Repeat items 2b and 2c for the pyramid side length versus the number of ball bearings.

e. How are the functions from 2c and 2d related?

Side Length of Pyramid	Number of Rods Needed	Number of Ball Bearings Needed
1		
2		
3		
4		
n		

3. One commercial manufacturer of magnetic rods and steel ball bearings packages a set with 36 rods and 34 balls. How many packages would be needed to make a pyramid with side length 6 units?

4. Suppose you had 1000 rods and 300 balls. What is the side length of the largest pyramid you could make?

Materials

- set of small magnetic rods all of the same length and steel ball bearings, at least 90 rods and 35 ball bearings (Alternatively, use toothpicks for the rods and gumdrops, mini marshmallows, or modeling clay for the ball bearings.)

TEACHING AND LEARNING HIGH SCHOOL MATHEMATICS © 2010 John Wiley & Sons, Inc.

Counting Cubes

Problem

Dots are painted on the faces of cubes. If you toss the cubes and a dot appears on the top face, remove the cube. To determine the relationship between the number of faces with dots on them and the number of cubes remaining after *n* rolls, complete the following experiment.

1. a. Take one of the containers. In the table corresponding to the number of dots on each cube, record the total number of cubes in the container as trial 0.

b. Shake the container and pour out the cubes on a sheet of paper or box lid so the cubes are one layer thick. If a dot is showing, set the cube to the side. Count the number of cubes remaining; this is the number associated with trial 1. Return the remaining cubes to the container.

c. Repeat item 1b until the container has only 1 or 2 cubes left.

1 Dot per Cube		2 Dots per Cube		3 Dots per Cube		4 Dots per Cube		5 Dots per Cube	
Trial Number	Number of Cubes Left	Trial Number	Number of Cubes Left	Trial Number	Number of Cubes Left	Trial Number	Number of Cubes Left	Trial Number	Number of Cubes Left
0		0		0		0		0	
1		1		1		1		1	
2		2		2		2		2	
3		3		3		3		3	
4		4		4		4		4	
5		5		5		5		5	
6		6		6		6		6	
7		7		7		7		7	
8		8		8		8		8	
9		9		9		9		9	
10		10		10		10		10	

2. Make a scatterplot of the data from problem 1 for the container. What family of functions models the data? Fit an appropriate function to the data.

3. Repeat the experiment for each of the five containers. How are the functions for each of the five containers related? Why is this reasonable?

4. Extension. Choose one of the containers from the set of materials. Pour all the cubes out. Put 5 cubes back in the container. Record this number in the table on page 212 for trial 0.

a. Shake the container and pour out the cubes into a single layer. For each cube that has a dot showing face up, add one more cube. Count the

(*continued*)

Materials

- 60 cubes in each of 5 containers with the container number corresponding to the number of faces on which a dot is found on each cube (For example, in container 3, all cubes have a dot on 3 faces of the cube.)

TEACHING AND LEARNING HIGH SCHOOL MATHEMATICS © 2010 John Wiley & Sons, Inc.

number of cubes and record this number for trial 1. Return the cubes to the container.

b. Repeat item 4a for at least six trials.

Trial Number	Number of Cubes Left
0	
1	
2	
3	
4	
5	

5. Make a scatterplot of your data. What family of functions describes the data? Try to fit an appropriate function to the data.
6. Repeat the experiment in items 4 and 5 for each of the five containers. How are the functions for each of the five containers related? Why is this reasonable?
7. What relationship exists between the functions when you removed cubes from the container and the functions when you added cubes to the container?
8. What real life uses might be made of your discoveries in this experiment?

Adapted from "Activities: Generating and Analyzing Data" by Jill Stevens (1993) and "Giving Exponential Functions a Fair Shake" by Jeffrey J. Wanko (2005).

Tiling with Squares

Materials

- 2 sheets of graph paper per group

Problem

Denny works with ceramic tile. Kelly gave him a large number of square tiles to use to make objects of different sizes. Denny is very clever with the tiles and is able to cut them in whatever way is needed to make the objects he wants. His current projects are all square coasters and trivets (an object put under a hot dish to protect a table). He makes them in various sizes to accommodate different sizes of cups and pots.

1. a. If Denny uses the numbers of tiles shown in the table at right to make square coasters or trivets, how many tiles will be along a single side of the coaster or trivet?
   b. Complete the table.
   c. How is the side length related to the total number of tiles Denny plans to use?

Number of Total Tiles Denny Will Use	Number of Tiles Along One Side of a Square Coaster or Trivet
1	
4	
9	
16	
25	
100	

For problem 2–4, assume tiles of dimensions 1 unit by 1 unit.

2. Denny would like different sizes of trivets than those for which he can easily arrange whole tiles. He plans to cut some of the tiles so he can make the coasters and trivets using different numbers of whole tiles. His cuts will be perpendicular to one of the sides of each tile he cuts.
   a. Use graph paper to show how Denny might cut the tiles to make square trivets with the number of whole tiles listed.

(continued)

TEACHING AND LEARNING HIGH SCHOOL MATHEMATICS © 2010 John Wiley & Sons, Inc.

b. Use your drawing to determine the approximate length of one side of the square trivet. Provide a range of side lengths if you cannot find an exact value from your drawing. Use the grid on the graph paper to measure as well as estimate lengths from your drawing.

Number of Total Tiles Denny Will Use	Number of Tiles Along One Side of a Square Coaster or Trivet
18	
20	
22	
27	
40	
200	

3. a. In each case in problems 1 and 2, what is the relationship between the number of tiles to be used and the side length of the coaster or trivet? Explain.

   b. How does your drawing help you understand the relationship between the area of a square and its side length?

   c. How can the side length be found exactly?

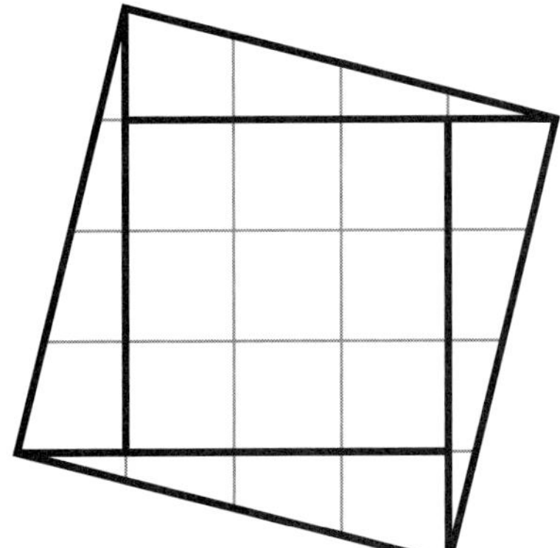

4. Suppose Denny does not restrict his cuts to be perpendicular to one of the sides. He might make cuts that result in coasters or trivets such as the one shown in the figure.

   a. What other arrangements can be made to make trivets or coasters using the numbers of tiles in the table below?

   b. Measure the side lengths using the grid on the graph paper. Use your conjecture from problem 3c to determine the side length exactly.

Number of Total Tiles Denny Will Use	Length of One Side of a Square Coaster or Trivet
17	
26	
37	
40	
200	

TEACHING AND LEARNING HIGH SCHOOL MATHEMATICS © 2010 John Wiley & Sons, Inc.

SHARE AND SUMMARIZE

11. a. Share the results from the five experiments with other teams. Explain any differences in your results.

    b. What types of situations give rise to quadratic functions? to cubic functions? to rational functions? to exponential functions? to square root functions?

12. How might these experiments help students understand modeling of data?

13. How might these experiments help students understand each function family?

    a. Which of the five experiments would be most useful when students are introduced to a function family? Explain.

    b. Which of the five experiments would be more helpful after students have learned characteristics of a function family? Explain.

14. Share your team's views about the importance of students actively engaging in experiments with the rest of the class. What similarities and differences exist among the perspectives from different teams?

15. a. As a class, share what each team considered advantages and disadvantages of using learning centers.

    b. Sometimes students need time limits when working with learning centers to keep them focused. What benefits and difficulties arise from setting time limits for students?

16. Share issues, identified by your teams, that teachers need to consider as they manage the use of centers with high school students.

DEEPENING MATHEMATICAL UNDERSTANDING

1. *Experiments with Thumb Tacks.* Revisit the experiment **Counting Cubes**.

    a. Repeat the basic experiment replacing cubes with thumb tacks. Set aside the tacks if they land point down. Find a function to model the data.

    b. How does the function change if you decide to set aside thumb tacks that land point up instead?

    c. Discuss similarities and differences between this experiment and the original experiment using cubes.

2. *Modeling Olympic Data.* The winning times for many Olympic events can be modeled by functions.

    a. Find the winning times for a swimming or track event (e.g., men's or women's times to swim the 100-meter freestyle, men's or women's times to run 10 000 meters). Plot the data and find a function to model the data.

    b. What might be some of the advantages and disadvantages to using such data in the classroom?

3. *Light Intensity.* The intensity of light tends to change as the distance from the light source changes. To investigate this relationship, find a partner and conduct the following experiment. You will need a flashlight with a strong light source and a piece of cardboard, cardstock, or construction paper, and grid paper mounted on cardstock or some other firm surface. Use binder clips to stand the cardboard upright.

    a. Make a pinhole in the piece of cardboard or construction paper. Stand the grid paper a few centimeters from the hole. Place the flashlight at a fixed distance behind the cardboard so the light shines through the hole onto the grid paper.

    b. Measure the diameter of the circle reflected onto the grid paper. Record both the diameter of the reflected circle and the distance of the grid paper from the cardboard. Move the grid paper a few centimeters farther away from the hole.

TEACHING AND LEARNING HIGH SCHOOL MATHEMATICS © 2010 John Wiley & Sons, Inc.

c. Repeat problem 3b until you have at least 8 data points.

d. Find a function that models the data.

e. Consider the basic physics of the situation. How do your common sense experiences with light intensity suggest that the function model you found makes sense?

4. *Still Rolling*. Students studying trigonometry can complete an extension for the experiment in the student page **On a Roll**.

a. Change the height of the inclined plane using uniform changes in the measure of the angle of incline. For each angle of incline, record the values of a in $y = ax^2 + bx + c$ obtained when you fit a quadratic regression equation to the graph of the rolling ball.

b. Compare the values of a with the sine of the angle θ measured between the floor and the inclined plane.

c. What relationship, if any, exists between a and $\sin\theta$?

d. Explain the reasonableness of your conjecture in problem 4c.

5. *Bouncing Balls*. At least two different experiments can be conducted using a ball that bounces and a motion detector attached to a graphing calculator. Try each experiment and determine which function family best models each. Choose a ball that is very bouncy!

a. Hold a motion detector directly over the place where you will bounce a ball. Use the program BALL BOUNCE in the APPLICATIONS menu. Determine the highest points on each bounce. Create a table with b as the bounce number, and r as the height of the bounce (at its highest point). What is the relationship between bounce number b and rebound height r? Explain.

b. Hold a motion detector directly over the place where you will bounce a ball. Use the program BALL BOUNCE in the APPLICATIONS menu. Record the starting height of the ball and its rebound height for a single bounce. Repeat using different starting heights to obtain at least 7 data pairs.

c. Using the regression capability of your calculator, determine an equation that best fits the data for each of the two experiments. To what function family does this model belong? Explain.

d. Which of these experiments could be conducted without the use of a motion detector? What materials would you need in this nontechnological environment?

6. *Regression Models on a Graphing Calculator*. Refer back to the student page **Building Pyramids**. Try to fit an equation to the data using just the data for the pyramids with side lengths 1, 2, and 3 units. What happens? Provide a mathematical explanation to explain why this occurs.

7. *Understanding Least Squares*. Complete the student page **Understanding Least Squares**.

a. Based on your work with the activity, why is a linear regression line often called a least squares line?

b. How did the use of a moveable regression line and models of squares around the line provide meaning for the concept of least squares?

c. Pearson's r is a correlation coefficient. What is the range of possible values for r and what meanings do these values have?

 © 2010 John Wiley & Sons, Inc.

Understanding Least Squares

In the table below, the mean monthly income is provided in terms of the number of years a person has been in school. The original data listed categories for the highest degree earned. The data below estimates the number of years for each degree category (after kindergarten).

Highest Degree Earned	Estimated Number of Years in School (after Kindergarten)	Rationale
Not a high school graduate	10	Most states require students to stay in school until age 16, usually grade 10
High school graduate only	12	
Some college, no degree	13	Estimated departure from college after 1 year
Vocational	13.5	Vocational programs tend to be 1–2 years in length beyond high school graduation
Associate's degree	14	Two years beyond high school
Bachelor's degree	16	Four years beyond high school
Master's degree	18	Two years beyond bachelor's degree
Doctorate	22	Four years beyond master's degree

Income Versus Education

	years_in_school	mean_monthly_salary	female_mean_monthly_salaries	male_mean_monthly_income
1	10.0	906	621	1211
2	12.0	1380	1008	1812
3	13.0	1579	1139	2045
4	13.5	1736	1373	2318
5	14.0	1985	1544	2561
6	16.0	2625	1809	3430
7	18.0	3411	2505	4298
8	22.0	4328	4020	4421

Source: *The American Almanac: statistical Abstract of the United States 1995–1996*. Austin, TX: The Reference Press, 1995 p. 158.

1. Open the *Fathom*™ document **Income vs. Education**. You will see the table above on the screen.
   a. Drag an empty graph from the tool bar to the screen below the table. Position the cursor over the attribute (column label) years_in_school. Holding the mouse button down, drag this attribute to the horizontal axis of the graph. Release the mouse button to drop the attribute onto the graph's horizontal axis. Drag and drop the attribute

(*continued*)

Materials

- *Fathom*™ software on a computer, one station per pair
- **Income vs. Education** *Fathom*™ file

TEACHING AND LEARNING HIGH SCHOOL MATHEMATICS © 2010 John Wiley & Sons, Inc.

female_mean_monthly_income to the vertical axis. You will see a scatterplot of the data. Resize the graph to make it as large as possible while the table is still visible.

b. With the graph highlighted (the frame will be showing), choose **Movable Line** from the Graph menu. Notice that you can rotate the line from both ends and move it up and down from the center. Manipulate the line until you are satisfied that it fits well the data on the scatterplot. Record the equation of this line: _________.

c. While the graph is highlighted, choose **Show Squares** from the Graph menu. Notice that squares are drawn. Look closely at these squares and try to determine how their dimensions are determined. Manipulate the movable line until you know how the squares relate to the data and to the line.

d. Notice that the value of the sum of squares is provided in the lower left corner of the graph. Manipulate the line to make the sums of squares as small as possible. Write down the equation that appears for the movable line: _________.

e. While the graph is highlighted, choose **Least Squares Line** from the Graph menu. Write down the equation that appears for this line: _________.

f. Compare the equations you found in problems 1b, 1d, and 1e. How close was your movable line to the line that minimized the sums of the squares?

2. Drag the attribute male_mean_monthly_income to the vertical axis, replacing female_mean_monthly_income. Repeat items 1b through 1f for this attribute.

3. Drag the attribute mean_monthly_income to the vertical axis, replacing male_mean_monthly_income. Repeat items 1b through 1f for this attribute.

4. Compare the least squares equations for each set of data.

a. What do you notice?

b. In each case, what does the slope represent?

c. In each case, what does the *y*-intercept represent?

d. Interpret the three equations. What would you like to tell your representative in Congress?

5. The years_in_school attribute was estimated. On the graph for mean monthly income, locate the data point indicating the mean income of persons who have a master's degree. Drag the point to a year earlier, a year later, three years later, etc. What effect does this change seem to have on the least squares fit line? (Undo **Drag Point** in the Edit menu to reset the data.)

6. What happens to the least squares fit line if the mean_monthly_income of persons achieving a master's degree is much higher or much lower? Drag the data point to see what happens. Notice what happens in the data table.

7. a. Summarize what you have learned about what the computer (or graphing calculator regression capability) is doing when it uses least squares fit lines to model linear data.

b. Write a question or two that you still have about the use of the least squares fit line.

TEACHING AND LEARNING HIGH SCHOOL MATHEMATICS © 2010 John Wiley & Sons, Inc.

8. *Linearizing Data*. When data is not initially linear, sometimes an equation to model the data is found by linearizing the data and then finding an equation for the least squares line as indicated below.

   a. Refer back to the data you collected as part of the **Counting Cubes** experiment. As part of the activity, you likely found that an equation that models the data is exponential. For one of the sets of (x, y) data, replace the data points with $(x, \log y)$ data points. Plot this new set of data points. Find an equation that models the new data. (Note: Your equation will be in the form $\log y = mx + d$.)

   b. Show that your linear equation from problem 8a is roughly equivalent to the exponential equation you found as part of the experiment.

   c. Use *Fathom*™ (or the TI-nspire calculator) to plot both (x, y) and $(x, \log y)$ on two separate graphs. On the (x, y) plot, create two sliders for the equation $y = ab^x$, one for a and one for b. While the graph is active, plot $y = ab^x$ using the global menu to insert a and b into the equation. Change the values of a and b until the graph appears to best fit the data. On the $(x, \log y)$ plot, use **Moveable Line** to plot the regression line that best fits the data. Compare the two plots. Transform $\log y = mx + d$ to obtain an equation of the form $y = ab^x$. Comment on similarities and differences between the log linear equation and the equation obtained when you found the curve of best fit for the exponential graph.

9. *Interpreting Graphs*. The student page **Vroom! Vroom!** helps students interpret graphs. The student page does not pertain to one particular function family but instead gives students an opportunity to think about what kinds of motions create what shapes of graphs.

   a. Use a motion detector attached to a graphing calculator to complete the student page.

   b. Characterize the graphs based on the function family to which each belongs.

   c. What type of propulsion results in each type of graph?

Vroom! Vroom!

Problem

For each different mode of propulsion, what shape graph best models the car's motion?

Materials

- Calculator Based Ranger (CBR) with RANGER program loaded into graphing calculator
- several types of toy cars, including ones propelled by springs, fly wheels, batteries, and pull strings
- two yardsticks or meter sticks
- transparencies cut to the size of a graphing calculator screen
- transparency or dry erase markers

Directions

Some cars tend to pull to one side or the other. If this happens, stack two yardsticks or meter sticks under each car or place a line of rain gutter along the side of the car so it will run in a straight line. Be sure the yardstick or gutter is positioned so that it will not be detected in the 15° beam of the CBR.

Stand the CBR on the floor or table and flip up the detection device to get it as far from the surface as possible. Without using the motion detector, determine how long it will take each car to travel 4 meters; record this number for each car. In the RANGER set-up menu, change Real Time to NO. Change the time setting to collect data for the number of seconds you determined for each car.

1. Play with each car until you are able to predict a graph that describes its motion. Do not run the CBR program until you have completed item 2.
2. Consider the following graphs. If you expect the motion graph to look like one of these graphs, write a brief description that explains why the motion of the car looks like that graph. If you expect the motion of the car to be

(continued)

TEACHING AND LEARNING HIGH SCHOOL MATHEMATICS © 2010 John Wiley & Sons, Inc.

different from any of these graphs, sketch a graph that you think models the motion and provide a brief description about why you think the graph will look that way.

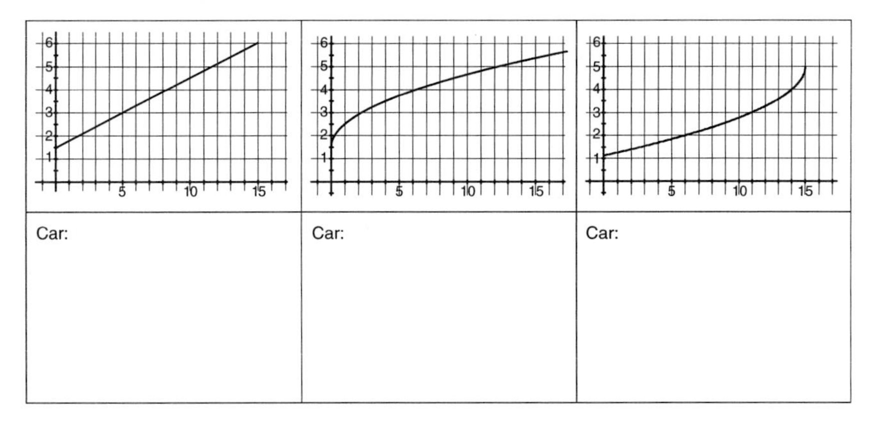

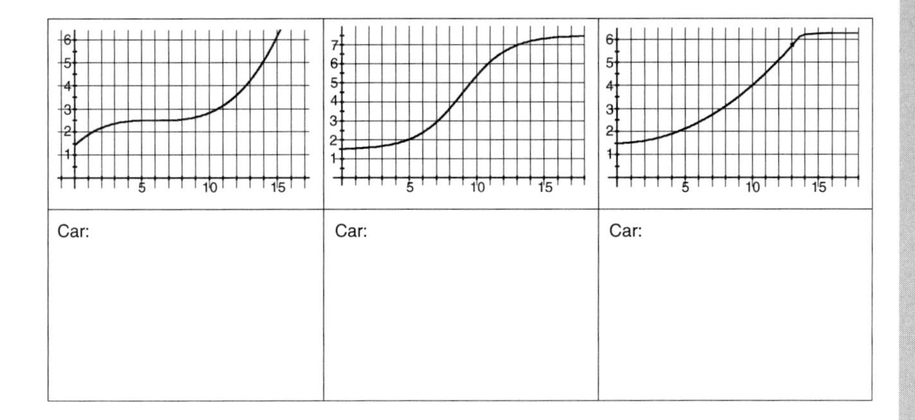

3. For each of the graphs in problem 2, is the car starting away from the CBR and moving toward it, or is the car near the CBR and moving away from it? How do you know?
4. For a car that is moving away from the CBR,
   a. How does the graph change when a car slows down?
   b. How does the graph change when a car speeds up?
5. For a car that is moving toward the CBR,
   a. If the graph is concave up (like a bowl facing upward), is the car moving faster or slowing down? Explain.
   b. If the graph is concave down (like a bowl upside down), is the car moving faster or slowing down? Explain.
6. Use the CBR for the following. Place the CBR on the floor or table so that it detects the motion of the car.
   a. Obtain a graph for each car using the CBR connected to a graphing calculator. Trace the graph onto a transparency that fits the calculator screen. Position the transparency over the graph for the car chosen in problem 2.
   b. For each car, explain the car's CBR graph and any discrepancies between it and the one you chose in problem 2. Explain any surprises or why your prediction makes sense.

TEACHING AND LEARNING HIGH SCHOOL MATHEMATICS © 2010 John Wiley & Sons, Inc.

10. *Getting to the Root of Square Root*. A common error that high school students make is to write $\sqrt{x^2 + y^2} = x + y$. Try the following experiment to investigate the relationship between the values on each side of this equation. You need three same thickness pipe cleaners and half a cup of rice to complete the experiment. (This experiment was created by Tim Steenwyk, a student who used a version of this text at Grand Valley State University in Fall 2005.)

a. Make a square with a pipe cleaner. Fill it with rice to the level of the pipe cleaner (use a stick or some flattening device to make certain the rice is uniformly deep with the pipe cleaner).

b. Divide the rice into two piles; the piles do not have to be the same size. Make two more squares with pipe cleaners so the rice fills each square to the same depth as it did the original square.

c. Arrange the three pipe cleaner squares so that one side of each square makes one side of a triangle. Describe the triangle.

d. Suppose x is the length of the side of the smallest square and y is the length of the side of the next largest square. Show the length $x + y$.

e. Using the lengths in item 10d, what is the length of the longest side of the triangle? How do you know?

f. Is $\sqrt{x^2 + y^2} = x + y$? Explain how the experiment helps you decide.

DEVELOPING MATHEMATICAL PEDAGOGY

Reflecting on Our Own Learning

11. Consider the various experiments you conducted in this lesson.

a. As a student, how often did you conduct experiments in your high school or college mathematics classes?

b. To what extent did the experiments help you distinguish situations that give rise to linear, exponential, rational, square root, cubic, or quadratic functions?

c. How did the use of technology enhance your understanding of each function family?

d. How did the use of technology enhance your thinking about the potential of technology and experiments to help high school students make sense of function families?

Preparing for Instruction

12. Consider the student page **Tiling Rectangles**.

a. Work through the student page.

b. Suppose you do not have enough tiles to conduct the experiment, either in a whole class or learning center environment. How might you use graph paper to conduct the experiment?

c. The student page relates two important geometric topics, area and similar figures, with the algebraic topic of functions.

i. Complete this sentence: As the side length of the tile increases ___, the number of tiles needed to cover the 12 by 12 rectangle ______.

ii. What are some advantages to students of relating algebra and geometry?

d. Modify the experiment to focus on a similar relationship related to the number of cubes needed to fill a box as a function of the side length of the cube.

TEACHING AND LEARNING HIGH SCHOOL MATHEMATICS © 2010 John Wiley & Sons, Inc.

Tiling Rectangles

Materials

- 150 tiles, 1 × 1 inch square
- 50 tiles, 2 × 2 inch square
- 25 tiles, 3 × 3 inch square
- 10 tiles, 4 × 4 inch square
- 6 tiles, 6 × 6 inch square

1. a. Draw a 12 × 12 inch rectangle. Cover the rectangle with 1 × 1 inch tiles without gaps and without overlaps. Count the number of tiles needed.

b. Repeat problem 1a with 2 × 2 inch tiles, 3 × 3 inch tiles, 4 × 4 inch tiles, and 6 × 6 inch tiles. Count the number of each size tile needed to cover the 12 × 12 inch rectangle.

c. Record your data in the table below.

Side Length of Square Tile	Number of Tiles Needed to Cover the Rectangle
1	
2	
3	
4	
6	

2. Plot the data in the table as a scatterplot. Find a function family that models the data.

3. How many 0.5 × 0.5 inch square tiles would be needed to cover the original 12 × 12 inch rectangle? How many 12 × 12 inch square tiles are needed? Add these two data points and replot the data.

4. Use the mathematics of the situation to explain why the function family you found in problem 2 is appropriate.

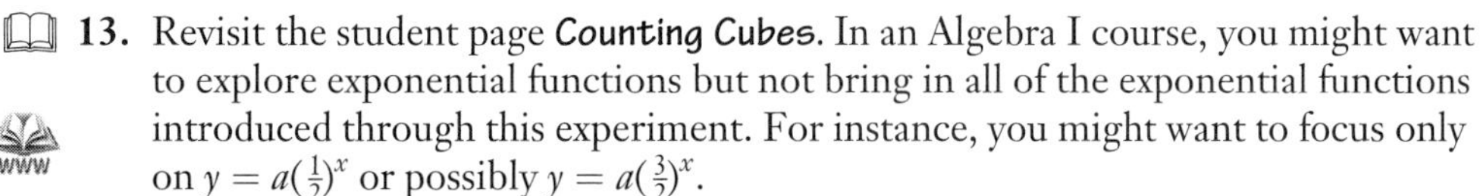

13. Revisit the student page **Counting Cubes**. In an Algebra I course, you might want to explore exponential functions but not bring in all of the exponential functions introduced through this experiment. For instance, you might want to focus only on $y = a(\frac{1}{2})^x$ or possibly $y = a(\frac{3}{2})^x$.

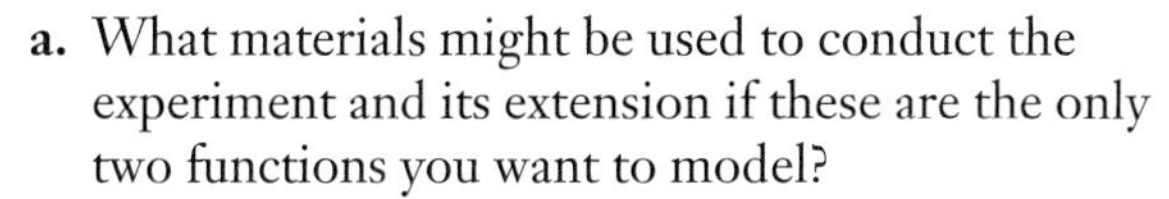

a. What materials might be used to conduct the experiment and its extension if these are the only two functions you want to model?

b. What are some reasons why Algebra I teachers might choose to limit the exponential functions with which their students engage?

14. Refer back to the student page **Building Pyramids**. The photo at right shows the second pyramid made with gumdrops and toothpicks. Use gumdrops and toothpicks and build the first three pyramids, with side lengths 1, 2, and 3 units respectively.

a. What are some advantages of using gumdrops and toothpicks in place of magnetic rods and steel ball bearings? What are some potential disadvantages?

b. How does the availability of low-cost materials address concerns about the feasibility of conducting experiments in mathematics classes?

TEACHING AND LEARNING HIGH SCHOOL MATHEMATICS © 2010 John Wiley & Sons, Inc.

15. Several children's books describe situations that can be modeled by functions. Locate at least three of the following children's books. Read the book and find a function that could be used to model the situation in the book. In each case, you will first need to think about what should be the independent and dependent variables.

 i. *The King's Chessboard* by David Birch. (New York: Puffin Pied Piper Books, 1988).
 ii. *One Hundred Hungry Ants* by Elinor J. Pinczes. (Boston: Houghton Mifflin, 1993).
 iii. *The 12 Circus Rings* by Seymour Chwast. (San Diego: Gulliver Books, 1993).
 iv. *Divide and Ride* by Stuart J. Murphy. (New York: HarperCollins Publishers, 1997).
 v. *Two of Everything* by Lily Toy Hong. (Morton Grove, IL: Albert Whitman & Company, 1993).
 vi. *Sea Squares* by Joy N. Hulme. (New York: Hyperion Books, 1991).
 vii. *Anno's Magic Seeds* by Mitsumasa Anno. (New York: Philomel Books, 1995).

16. Find a high school or basic college physics textbook. Review it to find some basic physics relationships that might be appropriate as experiments in the high school mathematics classroom. Design a student page for one of these experiments.

17. The article "Activities: Using Hooke's Law to Explore Linear Functions" by Chris McGlone and Gary M. Nieberle (*Mathematics Teacher*, 93 (May 2000): 391–398), describes a typical experiment using springs often conducted in science. Obtain a copy of the article and review the lab sheets that accompany the article.

 a. The function family from this article is linear functions. How might such experiments be useful in an Algebra I class?
 b. How might mathematics and science teachers work together to use this experiment or experiments similar to those in this lesson?

Analyzing Instructional Activities

18. Consider the two experiments **Vroom! Vroom!** and **On a Roll**. Does the order in which you do these experiments matter? Explain. Discuss the importance of sequencing activities depending on the instructional objectives of the lesson.

19. Find at least three Algebra II textbooks with recent copyrights. Investigate whether or not experiments are incorporated into the text. How is modeling data handled? Compare the approach used in the textbooks to the approach suggested in this lesson. Write a brief summary of your findings and comparisons. Cite specific examples as necessary.

20. Choose one of the student pages from this lesson, either from the text of the lesson or from the **DMU/DMP** questions.

 a. Anticipate students' interactions with the student page, including anticipated difficulties or misconceptions. Record these anticipated responses in a **QRS Guide** for the chosen student page.
 b. Determine appropriate teacher support for the student responses in problem 20a. Record these in the appropriate column in the **QRS Guide**.

 c. Try the student page with one or more high school students. Record any new student responses and appropriate teacher support in the **QRS Guide**.

TEACHING AND LEARNING HIGH SCHOOL MATHEMATICS © 2010 John Wiley & Sons, Inc.

Using Internet Resources

21. Vernier makes several devices that can be connected to graphing calculators or computers to gather data for experiments. Search the Internet for such lab equipment.
    a. Determine an experiment that can be conducted to help students become familiar with the periodic nature of sine or cosine functions.
    b. Find an experiment to help give students experience with logarithmic functions.

Reflecting on Professional Reading

22. Read "Using Data-Collection Devices to Enhance Student's Understanding" by Douglas A. Lapp and Vivian Flora Cyrus (*Mathematics Teacher*, 93 (September 2000): 504–510).
    a. The authors highlight several difficulties students often have when connecting graphs with the real world. Summarize these.
    b. What are some of the reasons the authors provide for using microcomputer-based laboratories?
    c. How does the information in the article connect with your experiences with experiments as in Lesson 3.3?
23. Read "Using Technology to Optimize and Generalize: The Least-Squares Line" by Maurice J. Burke and Ted R. Hodgson (*Mathematics Teacher*, 101 (September 2007): 102–107).
    a. Find the applet on the NCTM website referenced in the article. Work through the activities provided with the applet. Compare the activities with the applet to the activities completed with *Fathom*™ on the student page **Understanding Least Squares**. Identify the strengths and weaknesses of each.
    b. Apply the techniques described in this article to the least-squares line obtained as part of the *Wave Activity* from the **Launch** and from linearizing the exponential data in **DMU** problem 8.
24. Read "Kinematics and Graphs: Students' Difficulties and CBLs" by Patricia Hale (*Mathematics Teacher*, 93 (May 2000): 414–417). The author highlights some difficulties students face when using technology with experiments, particularly calculator-based laboratory devices. How do the difficulties highlighted by the author compare to issues you noted when conducting the experiments in this lesson?

3.4 Teaching with Technology: Geometry of Functions

Technology offers mathematics students opportunities to experiment with mathematical ideas in ways that were not possible formerly. What was once the goal of a problem, to graph a function, is now just its beginning. With technology, students can quickly graph several examples of a function, changing its equation in minor ways. They can start with a basic (*parent*) function, look at graphic and tabular results, and determine how changes in a single parameter in the equation $y = A \cdot f(Bx + C) + D$ affects these representations of the original function $y = f(x)$. (Parameters A, B, C, and D are considered to be constants in the equation though they are unknown values.) Students can also determine why the graph or table behaves as it does when a single parameter is changed. In time, students can predict the appearance of a graph of a function by recognizing how each of the parameters affects the graph of the related parent function; they can fit an equation to a set of data or a graph once they recognize the function family the graph or data represents.

TEACHING AND LEARNING HIGH SCHOOL MATHEMATICS © 2010 John Wiley & Sons, Inc.

As students explore functions with technology, one issue that arises is the paper trail that students might produce. How can students keep track of their investigations in order to make conjectures and generalizations based on the results? One strategy is to maintain a Learning Log, a running record in which they record their initial understandings about a concept and their updates as they learn more. (You had an opportunity to consider learning logs in **DMP** problem 17 in Lesson 2.1.) When students maintain a learning log, they can correct their own misconceptions and clarify developing conceptions. In the end, students reflect on and write what they learned about the concept (Barton and Heidema, 2002, pp. 132–133).

A second issue that arises during explorations with technology is how to make efficient use of the power of the technology to explore many aspects of a concept at one time. For instance, in studying families of functions, there are four parameters in the equation $y = A \cdot f(Bx + C) + D$; so explorations need to allow students to study the effects that small changes in each of the parameters A, B, C, and D have on a parent function $y = f(x)$, one at a time and later in conjunction with each other. Students also need to determine whether changes in the parameters have the same effect regardless of the parent function on which they are applied. Rather than have each student in a class investigate the effects of each parameter on each parent function, it is helpful to share the learning among class members. One way to do this is through the cooperative learning strategy *Jigsaw* (Davidson, 1990) which you used in Lesson 2.4.

In this lesson, you will:

- Determine how and why changes in the parameters A, B, C, and D in the equation $y = A \cdot f(Bx + C) + D$ affect the appearance and behavior of the function $y = f(x)$.
- Determine the equations of functions from a graph, table, or contextual setting.
- Use the strategy of Learning Logs to make sense of the relationships between parent functions and other family members.
- Experience the cooperative learning technique *Jigsaw* and reflect on its benefits.

Materials

- 7 pieces of string or yarn in different colors for **Launch** each about 5 meters long
- large sticky notes numbered $-5, -4, -3, \ldots, 3, 4, 5$ in one color to label the x-axis; -10 to 10 in another color to label the y-axis; and $-5, -2, 0, 2,$ and 5 in a third color for people to hold

Website Resources

- Exploring Functions of the Form $y = A \cdot f(x)$ student page
- Exploring Functions of the Form $y = f(x) + D$ student page
- Exploring Functions of the Form $y = f(x + C)$ student page
- Exploring Functions of the Form $y = f(Bx)$ student page
- Functions of the Form $g(x) = A \cdot f(Bx + C) + D$ student page

LAUNCH

What happens to a function when it is changed in small ways? Working with others who choose different values for x, you will form a graph of the function $y = |x|$. Then, when asked to do so, you will create a new graph as everyone changes their chosen x-value in specific ways.

TEACHING AND LEARNING HIGH SCHOOL MATHEMATICS © 2010 John Wiley & Sons, Inc.

1. Begin with the Cartesian coordinate plane laid out on the classroom floor. The axes should resemble those shown on the grid in the Unit Three Appendix. Make the axes large enough so there is enough room for participants to stand no closer than shoulder to shoulder along the x-axis. (As an alternative to constructing the graph with human participants, complete these steps using small objects to plot the points and string to connect them on the grid in the Unit Three Appendix.)
2. Minimally, five participants should represent the x-coordinates $-5, -2, 0, 2$, and 5. If desired, others can represent additional values of x. Locate yourself along the x-axis at the x-coordinate that represents your number.
3. Without changing your x-coordinate, move to a position with coordinates $(x, |x|)$ to create the graph of $y = |x|$ with your classmates. Use brightly colored string, yarn, or ribbon to mark the location of the graph on the floor. Tape down the string to indicate the graph. Label the graph with the equation $y = |x|$.
4. For each of items 4a–4d, move to the position on the coordinate plane that corresponds to the ordered pair for your chosen value of x. With your classmates, mark the graph on the floor using different colors of string, yarn, or ribbon for each graph. Label each graph with its corresponding equation.
   a. $(x, |x| + 1)$
   b. $(x, |x| - 2)$
   c. $(x, |x - 3|)$
   d. $(x, -|x|)$
5. Consider the equation $y = A|Bx + C| + D$. What are the values of A, B, C, and D for the graph created by class members in items 4a-d? Complete the table.

Item in Problem 4	*A*	*B*	*C*	*D*	Equation
4a					
4b					
4c					
4d					

6. Start with the parent function $y = |x|$.
   a. Explain how you would alter the graph of this equation to get each graph in items 4a–4d.
   b. Why is the new graph located where it is? Describe changes in x or y when moving from the graph of $y = |x|$ to the graph of $y = A|Bx + C| + D$.

EXPLORE

7. Use computer or graphing calculator technology to graph each of the following functions. Determine appropriate viewing windows that show the symmetry in each graph. Determine any symmetries of the function. Record a sketch of the graph and the symmetries for each in a learning log entry.
   a. $y = x^2$
   b. $y = \sin x$
   c. $y = \frac{1}{x}$
   d. $y = \sqrt{x}$
8. As an *expert* team, complete one of the following student pages. Use *Jigsaw* so that each member of the *home* team joins a different expert group to become the

TEACHING AND LEARNING HIGH SCHOOL MATHEMATICS © 2010 John Wiley & Sons, Inc.

expert on one of the function forms. Be ready to share what you have learned with your home team.

a. Exploring Functions of the Form $y = A \cdot f(x)$

b. Exploring Functions of the Form $y = f(x) + D$

c. Exploring Functions of the Form $y = f(x + C)$

d. Exploring Functions of the Form $y = f(Bx)$

9. While still in your *expert* group, take a few minutes to summarize what you learned about the form of the function your expert group explored. Be sure to address how the parameter A, B, C, or D affects the graph and why you think so. This is the first entry of your Learning Log. You will add to it in your *home* team. In particular, you should make note of the following:

   a. How is the graph affected when the parameter is greater than 1? between 0 and 1? less than 0?

   b. Why does the parameter cause that change? Describe changes in x or y when moving from the graph of $y = f(x)$ to the graph of $y = A \cdot f(Bx + C) + D$.

   c. In what contexts might the parameter change arise?

Exploring Functions of the Form $y = A \cdot f(x)$

Investigate the effect of changing the parameter A in the function $g(x) = A \cdot f(Bx + C) + D$ using the following functions. Unless otherwise noted, use a viewing window that gives x and y intervals in the same ratio (for most graphing calculators, this window is approximately [–10, 10] for x and [–6, 6] for y). Turn on the grid.

(1) $f(x) = |x|$ (2) $f(x) = \sin x$ (3) $f(x) = \frac{1}{x}$

1. a. Graph each of the following on the same screen: $f(x) = |x|$, $t(x) = 2|x|$, $h(x) = 0.4|x|$, and $k(x) = -|x|$.

b. Complete the table of values. (Use the table tool on your graphing calculator to assist you.)

x	$f(x) = \|x\|$	$t(x) = 2\|x\|$	$h(x) = 0.4\|x\|$	$k(x) = -\|x\|$
−3				
−2				
−1				
0				
1				
2				
3				

c. From the graphs and table, how do the functions t, h, and k relate to the function f? Why do the graphs appear as they do?

2. Repeat item 1a for parent functions (2) and (3), graphing $y = f(x)$, $t(x) = 2 \cdot f(x)$, $h(x) = 0.4 \cdot f(x)$, and $k(x) = -f(x)$ in each case. Clear the graphics screen before graphing each new set of related functions. Adjust the viewing window as needed to see each of the graphs

(continued)

TEACHING AND LEARNING HIGH SCHOOL MATHEMATICS © 2010 John Wiley & Sons, Inc.

and the relationships between them. For function (2), make sure the calculator mode is set to radian rather than degree and use the viewing window [–6, 6] by [–3, 3]. In each case, complete a table like the one in item 1b.

3. How does the change in A affect the graph of f? Describe the shapes and positions of the graphs of $y = t(x)$, $y = h(x)$, and $y = k(x)$ compared to that of $y = f(x)$.

4. **a.** How does the shape and position of the graph of $g(x) = A \cdot f(x)$ compare to that of the graph of $y = f(x)$ for each of the following:

i. $|A| > 1$ **ii.** $|A| < 1$ **iii.** $A = -1$

b. What effect does A have on x or y and on the graph of f? Consider each of the three cases in item 4a individually, then discuss their similarities.

c. Why does the graph of g appear as it does?

5. In an optimal environment, bacteria divide and form two new cells in 20 minutes. If one cell was present initially, a function giving the number of cells, c, based on the number of minutes elapsed, m, is $c = 2^{m/20}$. In a special laboratory environment, cell growth is retarded so that the number of cells present at time m is one-seventh of its normal growth.

a. Write an equation modeling the function that gives the bacterial cell growth in the special laboratory environment based on the amount of time elapsed.

b. How are these functions related to each other? How does your earlier work with studying the effect of parameter A help you understand the relationship between these functions?

c. Find a real life context that fits a different function family for which a change in the parameter A is reasonable.

TEACHING AND LEARNING HIGH SCHOOL MATHEMATICS © 2010 John Wiley & Sons, Inc.

Exploring Functions of the Form $y = f(x) + D$

Investigate the effect of changing the parameter D in the function $g(x) = A \cdot f(Bx + C) + D$ using the following functions. Unless otherwise noted, use a viewing window that gives x and y intervals in the same ratio (for most graphing calculators, this window is approximately [–10, 10] for x and [–6, 6] for y). Turn on the grid.

(1) $f(x) = |x|$ (2) $f(x) = \sin x$ (3) $f(x) = \frac{1}{x}$

1. a. Graph each of the following on the same screen: $f(x) = |x|$, $t(x) = |x| + 3$, and $h(x) = |x| - 4$.

b. Complete the table of values. Use the table tool on your graphing calculator to assist you.

x	$f(x) = \lvert x \rvert$	$t(x) = \lvert x \rvert + 3$	$h(x) = \lvert x \rvert - 4$
−3			
−2			
−1			
0			
1			
2			
3			

c. From the graphs and table, how do the functions t and h relate to the function f? Why do the graphs appear as they do?

2. Repeat item 1a for parent functions (2) and (3), graphing $y = f(x)$, $t(x) = f(x) + 3$, and $h(x) = f(x) - 4$ in each case. Clear the graphics screen before graphing each new set of related functions. Adjust the viewing window as needed to see each of the graphs and the relationships between them. For function (2), make sure that the calculator mode is set to radian rather than to degree.

3. How does the change in D affect the graph of f? Describe the shapes and positions of the graphs of $y = t(x)$ and $y = h(x)$ compared to that of $y = f(x)$.

4. a. How does the shape and position of the graph of $g(x) = f(x) + D$ compare to that of the graph of $y = f(x)$ for each of the following:

i. $D > 0$ **ii.** $D < 0$

b. What effect does D have on x or y and on the graph of f?

c. Why does the graph of g appear as it does?

5. The number of employees, y, needed for an afternoon shift at Delilah's Deli is given by $y = [0.6\sqrt{c}]$, where c is the number of customers expected per hour. (The function $y = \lfloor x \rfloor$ is called the floor function, or *greatest integer function*.)

a. In the evening, in addition to expected staff needs as given above, Delilah's Deli employs an extra person for security and a prep cook to prepare for the next day. What is the new equation?

b. How are these functions related to each other? How does your earlier work studying the effect of parameter D help you understand the relationship between these two functions?

c. Find a real-life context that fits a different function family for which a change in the parameter D is reasonable.

TEACHING AND LEARNING HIGH SCHOOL MATHEMATICS © 2010 John Wiley & Sons, Inc.

Exploring Functions of the Form $y = f(x + C)$

Investigate the effect of changing the parameter C in the function $g(x) = A \cdot f(Bx + C) + D$ using the following functions. Unless otherwise noted, use a viewing window that gives x and y intervals in the same ratio (for most graphing calculators, this window is approximately [–10, 10] for x and [–6, 6] for y). Turn on the grid.

(1) $f(x) = |x|$ (2) $f(x) = x^2$ (3) $f(x) = \sqrt{x}$ (4) $f(x) = \frac{1}{x}$

1. a. Graph each of the following on the same screen: $f(x) = |x|$, $t(x) = |x + 2|$, and $h(x) = |x - 4|$.

b. Complete the table of values. Use the table tool on your graphing calculator to assist you.

x	$f(x) = \|x\|$	$t(x) = \|x + 2\|$	$h(x) = \|x - 4\|$
−3			
−2			
−1			
0			
1			
2			
3			

c. From the graphs and table, how do the functions t and h relate to the function f? Why do the graphs appear as they do?

2. Repeat item 1a for parent functions (2), (3), and (4), graphing $y = f(x)$, $t(x) = f(x + 2)$, and $h(x) = f(x - 4)$ in each case. Clear the graphics screen before graphing each new set of related functions. Adjust the viewing window as needed to see each of the graphs and the relationships between them.

3. How does the change in C affect the graph of f? Describe the shapes and positions of the graphs of $y = t(x)$ and $y = h(x)$ compared to that of $y = f(x)$.

4. a. How does the shape and position of the graph of $g(x) = f(x + C)$ compare to that of the graph of $y = f(x)$ for each of the following:

i. $C > 0$ **ii.** $C < 0$

b. What effect does C have on x or y and on the graph of f?

c. Why does the graph of g appear as it does?

5. The population P of the United States for year t is approximated by the equation $P = U(t)$.

a. The following table gives the population P of the United States (in millions) at 10-year intervals in the nineteenth century.

(*continued*)

TEACHING AND LEARNING HIGH SCHOOL MATHEMATICS © 2010 John Wiley & Sons, Inc.

Year	t	U.S. Population (in millions)	Year Brazil's Population (in Millions) Is Hypothesized to Be Size Listed in Previous Column
1800	0	5.3	1850, or $t = 50$
1810	10	7.2	
1820	20	9.6	
1830	30	12.9	
1840	40	17.1	
1850	50	23.2	
1860	60	31.4	
1870	70	38.6	
1880	80	50.2	
1890	90	62.9	
1900	100	76.0	

Source: *The American Almanac: Statistical Abstract of the United States 1995–1996.* Austin, TX: The Reference Press, 1995. p. 8.

b. Plot the U.S. population data, with t as the independent variable and population size as the dependent variable. Describe the growth in population over time.

c. To which function family does the function seem to belong? Why do you think so?

d. Suppose Brazil experienced similar population size and growth as the United States, but 50 years later.

- **i.** Fill in the fourth column of the table in item 5a indicating the year that Brazil's population is expected to be the size of the population given for the United States based on the hypothesized relationship between the two countries. The first entry has been completed for you.
- **ii.** Sketch the data for Brazil on the same axes you used for item 5b.
- **iii.** Using the equation for the population of the United States, $P = U(t)$, what changes need to be made to this equation to give the hypothetical population for Brazil? Explain.

e. Find a real-life context that fits a different function family for which a change in the parameter C is reasonable.

TEACHING AND LEARNING HIGH SCHOOL MATHEMATICS © 2010 John Wiley & Sons, Inc.

Exploring Functions of the Form $y = f(Bx)$

Investigate the effect of changing the parameter B in the function $g(x) = A \cdot f(Bx + C) + D$ using the following functions. Unless otherwise noted, use a viewing window that gives x and y intervals in the same ratio (for most graphing calculators, this window is approximately [–10, 10] for x and [–6, 6] for y). Turn on the grid.

(1) $f(x) = |x|$ (2) $f(x) = x^2$ (3) $f(x) = \sin x$ (4) $f(x) = \lfloor x \rfloor$

(Note: $f(x) = \lfloor x \rfloor$ is the *greatest integer* or floor function; $f(x)$ is the greatest integer less than or equal to x.)

1. a. Graph each of the following on the same screen: $f(x) = |x|$, $t(x) = |2x|$, $h(x) = |0.4x|$, and $k(x) = |-x|$.

b. Complete the table of values. Use the table tool on your graphing calculator to assist you.

x	$f(x) = \|x\|$	$t(x) = \|2x\|$	$h(x) = \|0.4x\|$	$k(x) = \|-x\|$
−3				
−2				
−1				
0				
1				
2				
3				

c. From the graphs and table, how do the functions t, h, and k relate to the function f? Why do the graphs appear as they do?

2. Repeat item 1a for parent functions (2), (3), and (4), graphing $y = f(x)$, $t(x) = f(2x)$, $h(x) = f(0.4x)$, and $k(x) = f(-x)$ in each case. Clear the graphics screen before graphing each new set of related functions. Adjust the viewing window as needed to see each of the graphs and the relationships between them. To understand the relationships between the parent functions and other family members, graph the parent with one family member at a time. When graphing function (3), make sure that the calculator mode is set to radian rather than to degree mode; for this function and its family members, use the viewing window $[-10, 10]$ by $[-2, 2]$. When graphing function (4), set your calculator to dot mode rather than connected mode.

3. How does the change in B affect the graph of f? Describe the shapes and positions of the graphs of $y = t(x)$, $y = h(x)$, and $y = k(x)$ compared to that of $y = f(x)$.

4. a. How does the shape and position of the graph of $g(x) = f(Bx)$ compare to that of the graph of $y = f(x)$ for each of the following:

i. $|B| > 1$ **ii.** $|B| < 1$ **iii.** $|B| = -1$

(continued)

TEACHING AND LEARNING HIGH SCHOOL MATHEMATICS © 2010 John Wiley & Sons, Inc.

b. What effect does B have on x or y and on the graph of f? Consider each of the three cases in item 4a individually, then discuss their similarities.

c. Why does the graph of g appear as it does?

5. First-class flat mailer rates for the U.S. Postal Service in 2009 are given in the following table. As you know, postage rates increase periodically. Suppose the postal commission proposes to change the first-class rate designation for large envelopes by weight as shown in the table. (Source: http://www.usps.com/prices/first-class-mail-prices.htm downloaded June 1, 2009)

For Pieces Not Exceeding (oz.)	2009 Rate	Proposed Ounces	For Pieces Not Exceeding (oz.)	2009 Rate	Proposed Ounces
1	$0.88	0.5	7	$1.90	3.5
2	1.05	1.0	8	2.07	4.0
3	1.22	1.5	9	2.24	4.5
4	1.39	2.0	10	2.41	5.0
5	1.56	2.5	11	2.58	5.5
6	1.73	3.0	12	2.75	6.0
			13	2.92	6.5

a. Sketch a graph of both the current first-class mailer rates for large envelopes and the proposed first-class mailer rates on the same set of axes.

b. How are these functions related to each other? How does your earlier work studying parameter B help you understand the relationship between these functions?

c. Find another real-life context that fits a different function family for which a change in the parameter B is reasonable.

SHARE

10. Return to your *home* team. Using your tables and sample graphs, show what is happening to the graph of $y = f(x)$ to create the graph of $y = g(x)$ for the parameters A, B, C, or D that you studied in your *expert* group. Explain why your work is sensible based on changes in x or y.
11. Add statements to your Learning Log addressing each of the parameters, A, B, C, and D separately.
12. With your *home* team, complete the student page **Functions of the Form $g(x) = A \cdot f(Bx + C) + D$.** Apply what you learned about families of functions from each expert group's work.

© 2010 John Wiley & Sons, Inc.

Functions of the Form $g(x) = A \cdot f(Bx + C) + D$

1. The figure below shows the graph of a function $y = f(x)$ on the viewing rectangle [–4.5, 4.5] by [–2.5, 4.5].

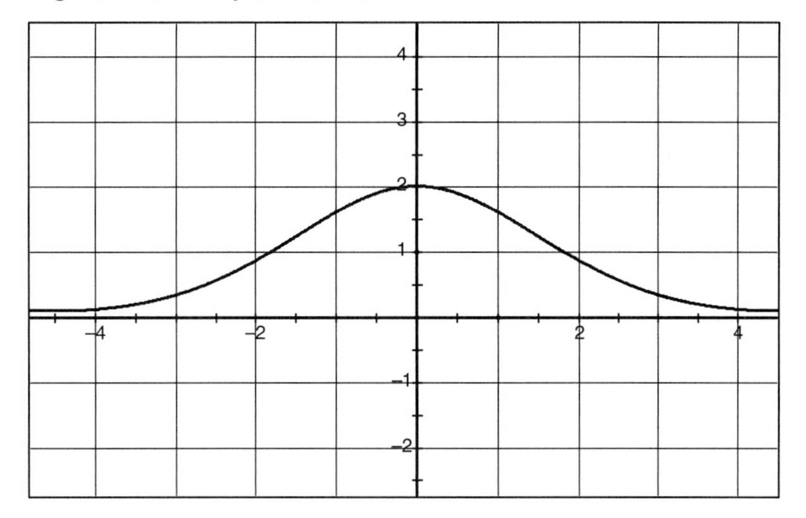

a. On the same axes, graph each of the following. Explain your work in each case.

i. $g_1(x) = f(x + 1)$

ii. $g_2(x) = 0.5 \cdot f(x + 1)$

iii. $g_3(x) = 0.5 \cdot f(x + 1) - 3$

b. Sketch the graph of $g_4(x) = -f(x)$. Explain your work.

c. Sketch the graph of $g_5(x) = -2 \cdot f(x)$. Explain your work.

d. Sketch the graph of $g_6(x) = f(-x)$. Explain your work.

e. How are the graphs you drew in 1b and 1d related? How do you know?

2. a. Use the function $f(x) = x^2$ and compare the following pairs of functions:

i. $g_1(x) = f(x) + 2$, $g_2(x) = f(x + 2)$ **ii.** $g_1(x) = f(x) - 1$, $g_2(x) = f(x - 1)$

b. Repeat problem 2a for the function $f(x) = \frac{1}{x}$.

c. Does $f(x) + k = f(x + k)$ for an arbitrary function and arbitrary choice of k? Why or why not?

3. a. Use the function $f(x) = x^2$ and compare the following pairs of functions:

i. $t_1(x) = 2 \cdot f(x)$, $t_2(x) = f(2x)$

ii. $t_1(x) = f(x)$, $t_2(x) = f(-x)$

b. Repeat problem 3a for the function $f(x) = \frac{1}{x}$.

c. Does $k \cdot f(x) = f(kx)$ for an arbitrary function and arbitrary choice of k? Why or why not?

4. The figure at right shows the graph of a function $y = f(x)$ on the viewing rectangle [–3, 3] by [–3, 3].

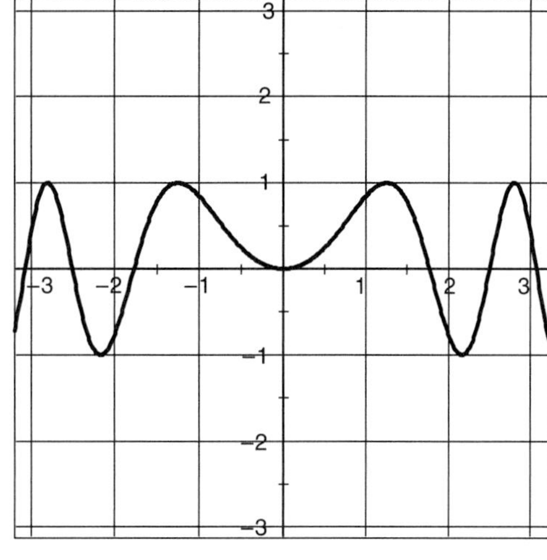

(continued)

TEACHING AND LEARNING HIGH SCHOOL MATHEMATICS © 2010 John Wiley & Sons, Inc.

a. Sketch the graph of $h(x) = f(2x) + 1$. Explain how you know your graph is correct.

b. Graph the function $y = f(-x)$. Compare the graphs of $y = f(x)$ with $y = f(-x)$. Explain your findings.

5. a. Are there functions for which $f(x) = f(-x)$ for all x? How do you know?

b. If there are functions $y = f(x)$ for which $f(x) = f(-x)$, list them. Explain your choice(s).

c. Are there any functions for which $f(-x) = -f(x)$ for all x? How do you know?

d. If there are functions $y = f(x)$, for which $f(-x) = -f(x)$, list them. Explain your choice(s).

SUMMARIZE

13. Now that you have had the chance to share in your *home* teams and to put together what you discovered in your *expert* groups, answer the following questions. Explain your responses in terms of the appearances of the related graphs and your knowledge of what happens to the values of x and $y = f(x)$ when one of these is changed by one of the parameters.

a. How are the graphs of $y = f(x)$ and $g(x) = A \cdot f(x)$ related?

b. How are the graphs of $y = f(x)$ and $g(x) = f(Bx)$ related?

c. How are the graphs of $y = f(x)$ and $g(x) = f(x + C)$ related?

d. How are the graphs of $y = f(x)$ and $g(x) = f(x) + D$ related?

e. High school students often state *what* happens to graphs when a parameter is changed, but they rarely explain *why* the graph appears as it does. List several questions you might ask students so that they are encouraged to explain their understanding more deeply.

14. The order in which the parameters are applied when dealing with functions is as important as the order of operations with numbers.

a. In what order would the parameters A, B, C, and D be applied? Why do you think so?

b. Identify a function for which the order in which the parameters are applied does not make a difference in the final image.

c. Identify a function for which the order in which the parameters are applied makes a difference in the final image.

15. The functions $y = \sin x$ and $y = \lfloor x \rfloor$ were used on the student pages and would likely be "black box" functions for many Algebra I and Algebra II students; that is, they are functions that students do not already know. The purpose here was solely to investigate the effects of the parameters on the functions. Discuss the pros and cons of having students explore "black box" functions.

16. How might the use of Learning Logs help students keep track of what they are learning as they explore mathematical concepts?

17. a. How did the *Jigsaw* cooperative learning strategy help you learn about the effects of the four parameters investigated in this lesson?

b. How does the *Jigsaw* strategy help build equity (support all learners to achieve high expectations)?

c. What might be some issues to consider when using *Jigsaw* in a high school mathematics classroom?

TEACHING AND LEARNING HIGH SCHOOL MATHEMATICS © 2010 John Wiley & Sons, Inc.

DEEPENING MATHEMATICAL UNDERSTANDING

1. *Converting Units*. An open top box is created by cutting out square corners from a 12×12 inch square piece of sheet metal and folding up the sides (see the figure at right).

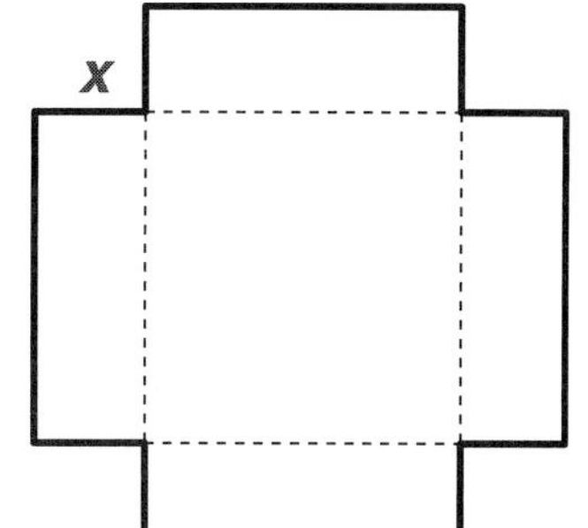

   a. Find an equation that gives the volume of the box in cubic inches.

   b. Determine a transformation of the equation in problem 1a that gives the volume of the box in cubic feet.

   c. What parameter is changed to convert from one measurement to the other in the equation $g(x) = A \cdot f(Bx + C) + D$, if the function in 1a is given by $y = f(x)$?

2. *More with the Geometry of Functions*.

   a. Complete the student page **The Geometry of Functions**.

   b. What is meant by a *one-to-one function*? How does this concept relate to issues on the student page related to obtaining a reflection of a function that is itself a function?

   c. In question 3 on the student page, the function is reflected over the y-axis and then over the x-axis. What single transformation is equivalent to the composition of these two reflections?

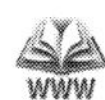

The Geometry of Functions

The graph of $y = f(x)$ is shown in each of the figures below. Follow the directions provided. Label a generic point $(x, f(x))$ (not on one of the axes). Find its image under the requested transformation. Answer the following for each of problems 1–4:

a. What point on the new graph corresponds to the point $(x, f(x))$ on the graph of $y = f(x)$?

b. Use the relationship between $(x, f(x))$ and its image to write an equation for the new relation in terms of the function $y = f(x)$.

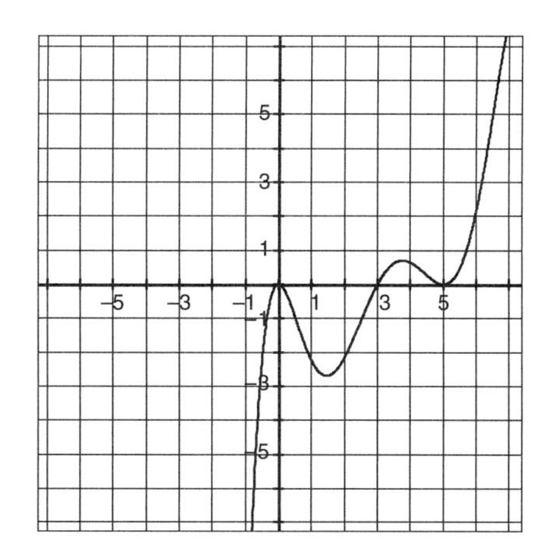

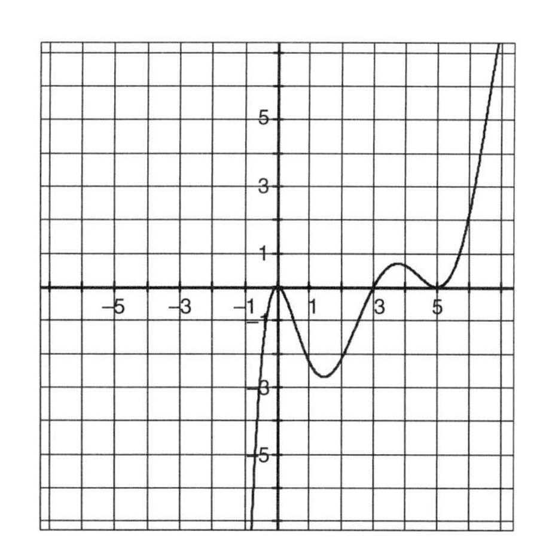

1. Reflect the graph of $y = f(x)$ across the x-axis.

2. Reflect the graph of $y = f(x)$ across the y-axis.

(continued)

TEACHING AND LEARNING HIGH SCHOOL MATHEMATICS © 2010 John Wiley & Sons, Inc.

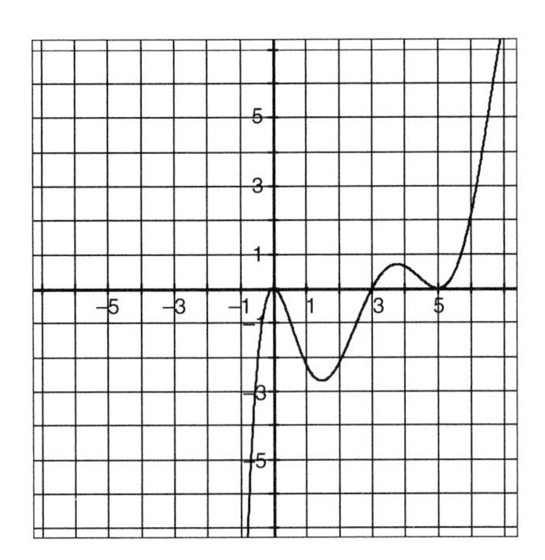

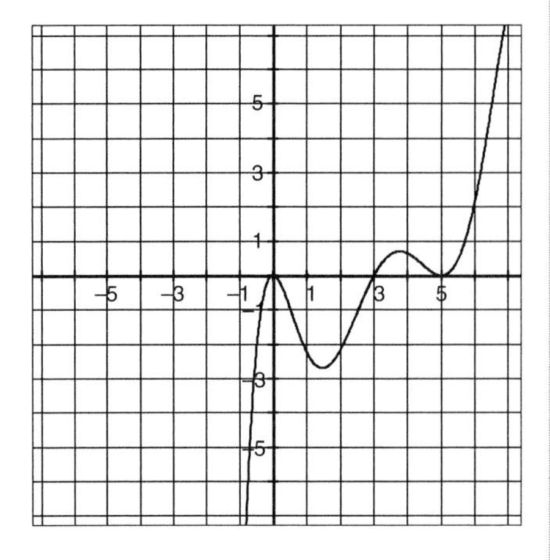

3. Reflect the graph of $y = f(x)$ across the y-axis and then across the x-axis.

4. Reflect the graph of $y = f(x)$ across the line $y = x$.

5. Revisit your work in problems 1–4.

a. State the definition of a function.

b. For which of these transformations on functions is the result always a function? Explain.

6. Revisit your work in problems 1–4. For which of these transformations on functions is the result not always a function? Explain.

a. For any transformations for which the new graph is not a function, what conditions must be placed on the graph of $y = f(x)$ in order for the image of the transformation to result in a function? Explain.

b. List as many functions as you can for which the image of this transformation is also a function.

7. Use a square grid (with x and y in the same scale) to complete the following for one of the columns of functions listed:

i. $f(x) = \frac{1}{2}x$	**ii.** $f(x) = -2x$	**iii.** $f(x) = \lvert\frac{1}{2}x\rvert$
iv. $f(x) = x^2$	**v.** $f(x) = -x^2$	**vi.** $f(x) = 10^x$
vii. $f(x) = x^3$	**viii.** $f(x) = \frac{1}{x}$	**ix.** $f(x) = \frac{1}{x^2}$

a. Sketch the graph of the line $y = x$.

b. Sketch the graph of the function.

c. Reflect the graph across the line $y = x$.

d. Determine whether or not the reflected image is a function.

e. If the reflected image is not a function, how must the preimage be restricted so that its reflected image is a function? Explain.

f. If possible, determine an equation of the reflection of the graph of $y = f(x)$ across the line $y = x$. Explain your work.

g. Describe the shape of the function $y = f(x)$. Describe the shape of the reflected image. How are these related?

h. Describe any symmetry or asymptotic behavior of $y = f(x)$. Describe any symmetry or asymptotic behavior of the reflected image. How are these related?

(*continued*)

TEACHING AND LEARNING HIGH SCHOOL MATHEMATICS © 2010 John Wiley & Sons, Inc.

i. Determine any x- or y- intercepts of $y = f(x)$. Determine any x or y intercepts of the reflected image of $y = f(x)$. How are these related?

8. In Michigan, when a person purchases a can or bottle containing soda, they are charged a 10-cent deposit. They receive the deposit back when they return the empty container to the store.

a. Complete the table on the left showing the relationship between numbers of bottles or cans of soda purchased and amount of deposit paid.

Number of Cans or Bottles Purchased	Amount of Deposit Paid
1	
2	
3	
4	
5	
6	
7	
8	
9	
10	

Amount of Money Received in Dollars	Number of Cans or Bottles Returned
1	
2	
3	
4	
5	
6	
7	
8	
9	
10	

b. School and church groups collect cans and bottles to raise money. Complete the table on the right showing the relationship between the amount of money a group will earn and the number of cans collected.

c. Find equations to model both scenarios.

d. What is the relationship between these two scenarios?

9. For two of the functions listed in problems 7i–7ix, find two related scenarios. What is significant about each pair?

3. *Freddie's Fast Food.* depending on how busy he expects the restaurant to be, the owner of Freddie's Fast Food schedules employees to work a shift. The restaurant capacity is 100 customers. The number of employees E that seems to work best for the number of expected customers C is shown in the table on page 238. Employees work partial shifts to meet the needs of the restaurant.

TEACHING AND LEARNING HIGH SCHOOL MATHEMATICS © 2010 John Wiley & Sons, Inc.

C	5	10	15	20	25	30	35	40	50	60	70	80	90	100
E	1	2	3	3	3	4	4	4	5	5	6	6	7	7

a. To what function family do the data seem to belong? Why do you think so?

b. Determine an equation that fits this function based on your work with parameters in this lesson.

c. Consider the relationship between $C = f(E)$ and $E = g(C)$. How can this relationship help you determine an equation to model the data?

4. *Even and Odd Functions.*

a. A function is *even* if $f(-x) = f(x)$ for all x. Which of the parent functions studied in Unit 3 are even? Prove your result.

b. What type of symmetry, if any, does an even function have?

c. A function is *odd* if $f(-x) = -f(x)$ for all x. Which of the parent functions studied in Unit 3 are odd? Prove your result.

d. What symmetry, if any, exists for odd functions?

5. *Composition of Functions.*

a. Complete the student page **Exploring Composite Functions**.

b. Using the technique in the student page, explain why $f(f^{-1}(x)) = x$.

c. Identify the inverse of any power function $y = f(x) = ax^b$.

d. Prove that your function in problem 5c is the inverse of the power function.

Exploring Composite Functions

Composite functions are functions of the form $y = f(g(x))$ where the output of one function, g, becomes the input of another function, f, provided the range of g is in the domain of f. They are helpful when trying to predict changes in one quantity based on another. For example, suppose that in a small lake the size of a population of small fish, S, is dependent on the amount of algae, A, present in a lake. Then S is a function of A, or $S = f(A)$. Suppose, also, that the size of a population of large fish, L, is dependent on the size of the population of small fish, so L is a function of S, or $L = g(S)$. Then the size of the population of large fish can be determined directly from the amount of algae present in the lake through the relationship $L = g(S) = g(f(A))$.

1. The graph of $y = g(x)$ is provided below. Assume that the function continues as shown, repeating every 10 units. Gridlines are shown in one unit increments for both x and y.

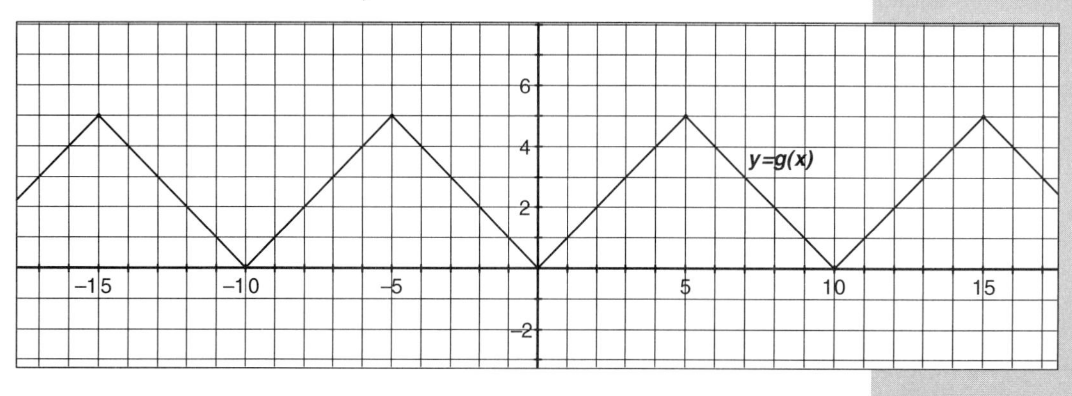

(continued)

TEACHING AND LEARNING HIGH SCHOOL MATHEMATICS © 2010 John Wiley & Sons, Inc.

a. Complete the table to determine several values of $y = f(g(x))$ where $f(x) = x^2$. Use the graph of $y = g(x)$ to determine the necessary values. Several values have been completed for you. How were these values determined?

b. Plot the points $(x, g(f(x)))$ on the grid provided.

c. What is the domain of $y = g(f(x))$? To which function, $y = f(x)$ or $y = g(x)$, is the domain of $y = g(f(x))$ most closely related? Why do you think so?

d. What is the range of $y = g(f(x))$? To which function, $y = f(x)$ *or* $y = g(x)$, is the range of $y = g(f(x))$ most closely related? Why do you think so?

x	$f(x) = x^2$	$y = g(f(x))$
−3	9	1
$-\sqrt{5} \approx -2.24$	5	5
−2	4	4
−1		
0		
1		
2		
	5	
3		
	10	
	15	
4		
	20	
	25	
	30	
	35	
6		

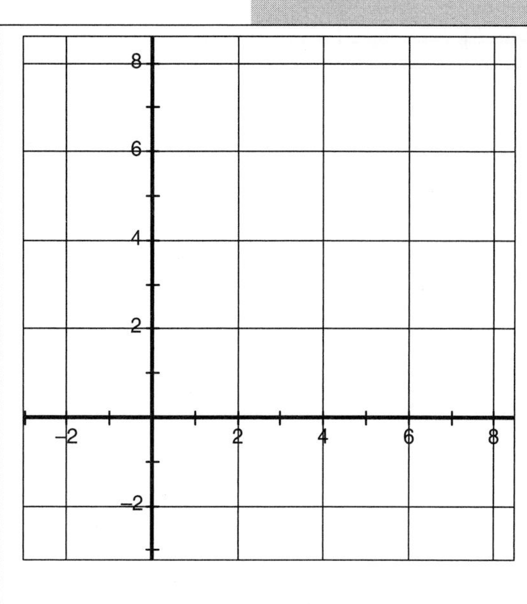

2. Repeat problem 1 for $f(x) = \sqrt{x}$.

a. What is the first x value greater than 0 for which $y = g(f(x)) = 5$?

b. Where is the next value of $g(f(x)) = 5$?

c. Locate all values of x between 0 and 5000 for which $g(f(x)) = 5$ and $g(f(x)) = 0$.

(*continued*)

TEACHING AND LEARNING HIGH SCHOOL MATHEMATICS © 2010 John Wiley & Sons, Inc.

x	$f(x)=\sqrt{x}$	$y=g(f(x))$
0		
100		
625		
1000		

3. Repeat problem 1 for $f(x)=\frac{1}{x}$.

x	$f(x)=\frac{1}{x}$	$y=g(f(x))$
-1		
$-\frac{1}{2}$		
$-\frac{1}{3}$		
$-\frac{1}{5}$		
$-\frac{1}{10}$		
0	undefined	undefined
$\frac{1}{10}$		
$\frac{1}{5}$		
$\frac{1}{3}$		
$\frac{1}{2}$		
1		

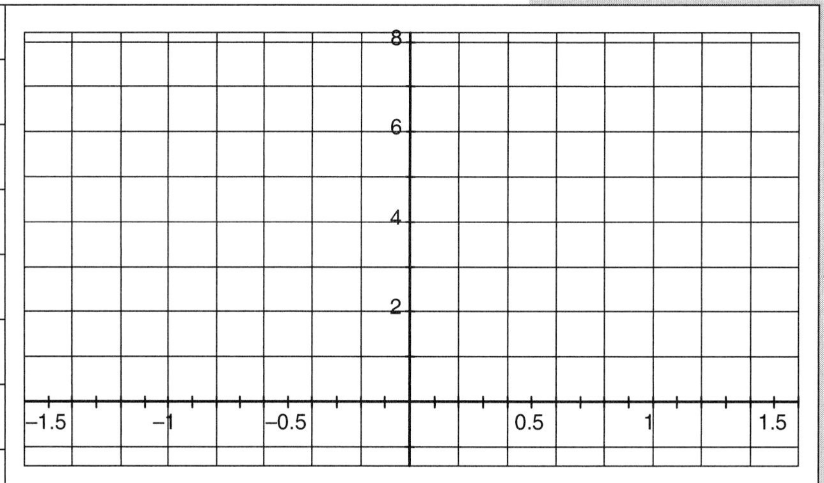

(continued)

© 2010 John Wiley & Sons, Inc.

4. Repeat problem 1 for $f(x) = 2^x$.

x	$f(x) = 2^x$	$y = g(f(x))$
−2		
−1		
0		
1		
2		
	5	
3		
	10	
	15	
4		
	20	
	25	
	30	
5		
	35	
6		
	40	
	45	
7		

6
4
2
−4 −2 2 4 6
−2

5. Compare each of the graphs you found for $y = g(f(x))$.

a. What is similar?

b. What is different?

c. Why do the graphs look as they do?

d. What effect does the function $y = f(x)$ have on the composite $y = g(f(x))$?

6. *More with Compositions of Functions.*

a. Complete the student page **Exploring Composites of Sine Functions.**

b. In what ways did use of the sine function help you understand what happened with composition of functions?

TEACHING AND LEARNING HIGH SCHOOL MATHEMATICS © 2010 John Wiley & Sons, Inc.

Exploring Composites of Sine Functions

Use a graphing utility on a computer or a graphing calculator to complete the following activity. Be sure your calculator is in radian mode.

1. a. Plot the functions $f(x) = \sin(x)$ and $g(x) = x^2$ on the viewing rectangle x in $[-2\pi, 2\pi]$ and y in $[-2, 5]$.

b. Plot the composite function, $y = f(g(x))$.

c. Plot the composite function, $y = g(f(x))$.

d. Compare the graphs created in items 1b–c. Why do they appear as they do? Explain in terms of the values of each of the parent functions, f and g.

e. Is composition of functions commutative? Why or why not?

2. Let $f(x) = \sin(x)$. Investigate the relationship among the graphs of $y = f(x)$, $y = g(x)$ and $y = f(g(x))$ for the following choices of $y = g(x)$. In each case, predict the appearance of the graph of the composite function $y = f(g(x))$ before plotting it. Choose appropriate viewing rectangles to plot all three functions on the same screen. Record your thoughts about why the composite function looks as it does based on the behaviors of the functions f and g. It might be helpful to turn off one or more of the functions f or g so that you can better determine the behavior of the composite function.

a. $g(x) = \frac{1}{x}$

b. $g(x) = \sqrt{x}$

c. $g(x) = |x|$

d. $g(x) = \lfloor x \rfloor$ (use dot mode)

3. Let $y = g(x)$ be defined as in problems 2a–2d and $f(x) = \sin(x)$. Investigate the relationship among the graphs of $y = f(x)$, $y = g(x)$, and $y = g(f(x))$. As in part 2, predict the appearance of the graph of the composite function $y = g(f(x))$ before plotting it. Choose appropriate viewing rectangles to plot all three functions on the same screen. Record your thoughts about why the composite function looks as it does based on the behaviors of the functions f and g. It might be helpful to turn off one or more of the functions f or g so that you can better determine the behavior of the composite function.

4. Classify each of the following functions as composite, sum, difference, product, or quotient of two functions. Explain your choice of classification.

a. $y = \sqrt{3x - 2}$

b. $y = \frac{1}{3x - 2}$

c. $y = |3x - 2|$

d. $y = \lfloor 3x - 2 \rfloor$

e. $y = \sqrt{x^2}$

f. $y = (\sqrt{x})^2$

7. *Explorations with Log Functions.* Consider the following generalized pattern:

$$\log(a \bullet b) = \log a + \log b.$$

a. Write three instances of this pattern.

b. Is this pattern true for $a, b > 0$? Use the laws of exponents to justify your response.

TEACHING AND LEARNING HIGH SCHOOL MATHEMATICS © 2010 John Wiley & Sons, Inc.

c. A problem similar to item 7a was given to Algebra II students in which they focused on whether or not $\log(x \cdot 5) = \log x \cdot \log 5$.

i. Determine if the pattern is true. If so, provide at least two different solutions students might provide with technology. If not, find a counter-example.

ii. Graphing calculators allow the user to enter $\log(x \cdot 5)$ and get a response. One student entered $\log(x \cdot 5)$ and got 1.17609. Another student entered $\log(x \cdot 5)$ and got 1.39794. Why might the two students have obtained different responses?

8. *More with Log Functions.* Refer to the **Function Card Sort** from the Unit Three Team Builder. (A full set of cards can be found in the Unit Three Appendix.) Share how you would revise the exponential cards to fit a logarithmic function family.

a. Find any stories or real contexts that can be altered to fit a logarithmic function family.

b. Find any tables or graphs that can be altered to fit the function family.

c. Create another table that fits this function family.

9. *Comparing Function Notation to Mapping Notation.* The student pages in this lesson all used function notation. Some books use mapping notation to represent a function. Consider the mapping notation $(x, y) \rightarrow (x + 2, y - 3)$, which means that each x-coordinate is replaced by $x + 2$ and each y-coordinate is replaced by $y - 3$.

a. Consider a function $y = f(x)$ which undergoes the mapping $(x, y) \rightarrow (x + 2,\ y - 3)$. Express an equation for the transformed function in the form $y = A \cdot f(Bx + C) + D$.

b. Consider a parent function $y = g(x)$ and its transformed function $h(x) = 3g(x - 5)$. Express the transformation using mapping notation.

10. *Cookies and Functions.* Consider the context of the children's story *The Doorbell Rang* by Pat Hutchins (New York: Mulberry Books, 1986). Suppose Mom made 20 cookies (and Grandma did not arrive with more). Initially, there are two children; every time the doorbell rings, more children arrive to share the cookies.

a. Determine an equation of the function that models the number of cookies each child will receive in terms of the number of children who will share the cookies.

b. Let the function in problem 10a be $y = f(x)$. Write cookie sharing scenarios and determine related equations for each of the following parameter changes:

i. $y = A \cdot f(x)$ ii. $y = f(x) + D$ iii. $y = f(x + C)$

iv. $y = f(Bx)$ v. $x = f(y)$

c. Choose your own context and repeat item 10b for your context.

DEVELOPING MATHEMATICAL PEDAGOGY

Reflecting on Our Own Learning

11. The work with parameters on the student pages in the lesson deals with transformations of functions.

a. What aspects of the work on these student pages were new for you?

b. What aspects were review but revisited from another perspective?

c. What aspects were extended to require deeper understanding to be exhibited?

TEACHING AND LEARNING HIGH SCHOOL MATHEMATICS © 2010 John Wiley & Sons, Inc.

Teaching for Equity

12. The **Launch** activity for this lesson was designed to help kinesthetic and visual learners become familiar with families of functions. In addition to working with the form $y = |x| + D$, students are also asked to consider what happens when A and C are changed in small ways. Each time, students need to explain not only how the graph should appear but why it moved, stretched, or shrank. They can check their predictions with a graphing calculator and discuss any differences they see between their predictions and the actual graphs.

   a. What difficulties do you expect students to have as they work through this activity? Record your thoughts on a **QRS Guide**.
   b. What questions should you be ready to ask to address the difficulties you identified in problem 12a? Record these questions as teacher support on your **QRS Guide** from question 12a.
   c. What are some possible benefits in using activities like those described here to launch Lesson 3.4? What disadvantages might there be?
   d. How do such activities address issues of equity in mathematics instruction?

13. Consider the student pages in this lesson.
   a. How might you modify these activities to increase their accessibility by students who have not yet studied function notation?
   b. How might you modify these activities to increase their challenge for students in Algebra II?
   c. Determine contextual problems for each student page that use logarithms or trigonometric functions.

Analyzing Instructional Activities

14. The student pages in this lesson focus on transformations of functions. Suppose students have studied transformations in a geometry course. How might that background help students understand transformations with functions?

15. The student pages in this lesson relied heavily on the use of graphing calculators or computer graphing software.
   a. Discuss the importance of having access to graphing calculator technology when completing activities such as those on the student pages in this lesson.
   b. How might you modify one of the activities if you lack access to graphing calculator technology?
   c. How does technology make mathematics more accessible to students with physical handicaps? With language learning issues?

16. In this text, families of functions have been addressed in the form $y = A \cdot f(Bx + C) + D$. In other texts, students might see the form $y = A \cdot f(B(x - C)) + D$.
   a. Sketch the graphs of $y = \sin(x)$ and $y = \sin(x + \frac{\pi}{4})$. How are they related?
   b. Sketch the graphs of $y = \sin(x)$ and $y = \sin(3x)$. How are they related?
   c. Sketch the graphs of $y = \sin(3x)$ and $y = \sin(3x + \frac{\pi}{4})$. How are they related?
   d. Is the relationship between $y = \sin(x)$ and $y = \sin(x + \frac{\pi}{4})$ exactly the same as the relationship between $y = \sin(3x)$ and $y = \sin(3x + \frac{\pi}{4})$? Explain.
   e. What are instructional advantages of using the form $y = A \cdot f(Bx + C) + D$?
   f. What are instructional advantages of using the form $y = A \cdot f(B(x - C)) + D$?

TEACHING AND LEARNING HIGH SCHOOL MATHEMATICS © 2010 John Wiley & Sons, Inc.

Preparing for Instruction

17. Refer to the student page **Exploring Square Root Functions.**
    a. Complete the student page.
    b. What difficulties, if any, do you anticipate students' having with the student page? What questions might you be prepared to ask to help them with these difficulties? Complete a **QRS Guide** for the student page, recording anticipated misconceived, partial, and satisfactory student responses and teacher support for each problem.

    c. Try the page with a group of two to four students at an appropriate level. What insights did you gain into the mathematics and into student thinking? Update the **QRS Guide** to record any additional responses you did not expect as well as appropriate teacher support.
    d. What experiences with functions would you suggest that students have prior to completing this student page? Explain.

    e. Look back at the **Function Card Sort** in the Team Builder for Unit Three. Based on your work in item 2 of the student page, find two tables and two graphs that fit the square root function family. Explain.

Exploring Square Root Functions

1. Donald is experimenting with making pancakes of different sizes. He has a cylindrical measuring cup that is 6 cm in diameter that he can adjust to fill to whatever level he chooses. He made a number of pancakes, noticing that regardless of their diameter, they were close to 0.4 cm thick, though there was some variability in thickness between the center and outer rim of each pancake. The data in the table below shows the relationship between the height, h, in cm of the pancake batter in the measuring cup and the diameter, D, in cm of the pancakes.

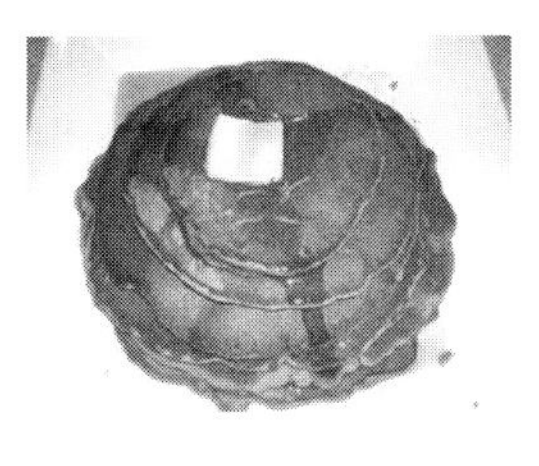

Height, h, of the Batter in the Measuring Cup (in cm)	Diameter, D, of the Pancake (in cm)
1	10
2	13
3	17
4	19
5	20
6	21

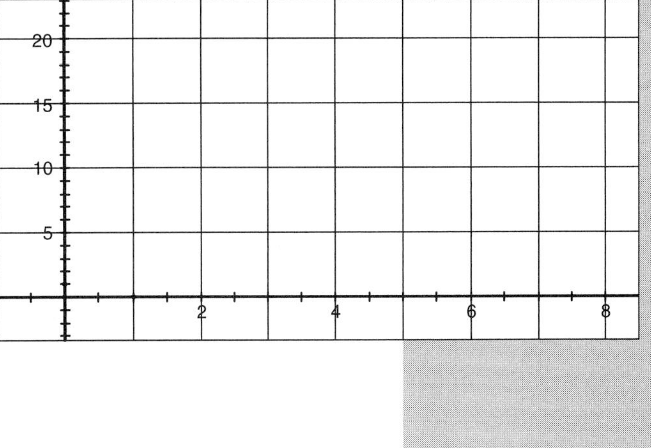

(continued)

TEACHING AND LEARNING HIGH SCHOOL MATHEMATICS © 2010 John Wiley & Sons, Inc.

a. Plot the data with h as the independent variable. Describe the shape of the graph.

b. Use a graphing calculator to determine a regression equation to fit the data. What relationship must you use to fit a regression formula to this data?

c. Plot the data and its regression equation on a graphing calculator. What are the dependent and independent variables in your plot?

d. Find a way to plot the data and the regression equation so that h is the independent variable.

e. Determine expressions for the volume of the batter in the measuring cup and the volume of the pancake. What do you know about these volumes?

f. Write the diameter of the pancake as a function of the height of the batter in the measuring cup.

g. Sketch the graph of the equation you found in item 1f on the same axes as the graphs in item 1d. Discuss how these graphs compare.

h. List potential causes for experimental error.

2. a. Complete the table of values.

x	$f(x) = \sqrt{x}$	$g(x) = f(x) - f(x-1)$
0		
1		
2		
3		
4		
5		
6		

b. Describe how $f(x) = \sqrt{x}$ is changing as x increases.

c. Without graphing the function, state what you know about the graph from your table and response to item 2b.

d. Using your response to item 2c, sketch a predicted shape of the graph. Do not plot any points or draw the graph with your graphing calculator.

e. Consider the table below:

x	-3	0	9	24	45	72	105	144
$y = h(x)$	0	3	6	9	12	15	18	21

Find first and second differences for x this time, rather than for y. What do you notice? Explain why your findings are sensible.

f. Considering your work in items 1 and 2e, how might you determine if a set of data fits the square root function family?

(continued)

TEACHING AND LEARNING HIGH SCHOOL MATHEMATICS © 2010 John Wiley & Sons, Inc.

3. a. Graph the parent function $f(x) = \sqrt{x}$.
   b. What are the coordinates of the y-intercept?
   c. What are the coordinates of the x-intercepts? How many x-intercepts are possible for this family of functions? How do you know and why is this reasonable?
   d. Describe the shape of the graph.
   e. Compare the values in the table in item 2a with the graph in item 3a. Why does the graph look as it does? Describe what happens to $f(x)$ as x changes.
   f. What new information is provided by the graph that the table does not show?
   g. What information is provided by the table that the graph does not show?

18. Choose one of the student pages from Lesson 3.4 or the **DMU/DMP** questions that accompany the lesson.
    a. Decide on the learning outcomes for the student page.

    b. Complete a **QRS Guide** for the student page, providing expected misconceived, partial, and satisfactory student responses for each numbered problem.
    c. Provide teacher support for each expected student response in the **QRS Guide**.
    d. Decide on a cooperative learning (CL) strategy that you think would best fit use of the student page with students in class. State why you chose that CL strategy. Write the chosen strategy at the top of the **QRS Guide**.

    e. Try the student page with one or more high school students. Add any new student responses to your **QRS Guide**.
    f. What new insights do you have about the students' conceptual understanding of the mathematical content of the student page from your work with students?

Communicating Mathematically

19. Consider again the Learning Log communication strategy used in this lesson.
    a. If you were to use the entries in the Learning Log as part of an assessment for this lesson, identify some concepts that you would expect to be addressed within the log.
    b. What particular accommodations might you make for English Language Learners in their entries to a Learning Log?

Teaching with Technological Tools

20. The TI-nspire calculator has dynamic graphing capability, including sliders.
    a. On a new page in the graph window, choose **Menu**, **Actions**, and then **Insert Slider**. Assign a variable name to the slider (say a). Repeat to insert a second slider, labeled b. Then, choose **Menu**, **Graph Type**, and **Function**. Choose one of the function families studied thus far in Unit Three. Graph the parent function $y = f(x)$ and the functions $y = a \cdot f(x)$ and $y = f(b \cdot x)$. Drag the sliders individually and observe what happens to each graph.
    b. Compare the structured work with the student pages to open this exploration using the graphing calculator and sliders. Under what conditions might you choose one approach over the other?

TEACHING AND LEARNING HIGH SCHOOL MATHEMATICS © 2010 John Wiley & Sons, Inc.

Reflecting on Professional Reading

21. Read "Connecting Procedural and Conceptual Knowledge of Functions" by Jon D. Davis (*Mathematics Teacher*, 99 (August 2005): 36–39).

  a. The author lists several issues that suggest students fail to make connections between their conceptual and procedural understandings of functions. Summarize the issues raised by the author.

  b. How do the ideas discussed by the author relate to concepts discussed throughout this unit as students have explored functions and their properties?

3.5 Increasing Challenge or Accessibility of Problems: Polynomial Functions

In today's classrooms one finds students with a wide range of experiences and ability levels. Teachers need to teach for equity, that is, teach so that all students are challenged and supported to engage in high level mathematics learning. So, teachers must design lessons thinking ahead to how problems may be increased in challenge for students ready for more. Likewise, teachers need to consider how to increase accessibility for students needing more support. Your work thus far with providing teacher support for anticipated student responses and recorded in **QRS Guides** is a start toward this type of lesson design.

Throughout this text, you have experienced how contexts, physical tools, cooperative learning, active environments, and technology can provide support and motivation so all students have opportunities to learn mathematics. Some curricula, but not all, are designed with these features. Teachers need to be ready to increase the challenge or accessibility of instructional tasks, whether or not their textbooks help with this work. This lesson provides opportunities for practicing this sort of *rescaling* of problems.

Before rescaling problems, teachers need to learn what students know about a mathematical topic. This may require some sort of pre-assessment. A simple pre-assessment can take the form of having students quickly brainstorm what they recall about a concept. A cooperative learning strategy that is helpful in this situation is *Roundtable* (Kagan, 1992), introduced in Lesson 2.2.

In this lesson, you will:

- Investigate mathematics concepts related to polynomial and power functions.
- Consider strategies that encourage students to share what they know about a mathematical concept before the concept is taught in class.
- Analyze algebraic tasks and determine ways to rescale them to increase challenge or accessibility based on students' needs.

Website Resources

- **Investigating Products of Linear Functions** student page
- **Variations on Investigating Products of Linear Functions**

LAUNCH

1. Use *Roundtable* to brainstorm what you know about quadratic functions.

The student page **Investigating Products of Linear Functions**, is intended as an introduction to quadratic functions for students who have studied linear functions previously.

2. In a small group, complete the student page. Be ready to share your observations with a larger group.

TEACHING AND LEARNING HIGH SCHOOL MATHEMATICS © 2010 John Wiley & Sons, Inc.

Investigating Products of Linear Functions

1. Graphs of lines are provided in the figures below. For the second pair of axes, $f(x) = g(x)$. (Note that grid lines are one unit apart for both x and y.) For each set of axes:

   a. Choose several integer and noninteger values of x. For each value of x, determine the values of $f(x)$, $g(x)$, and the product $h(x) = f(x) \cdot g(x)$.

   b. Plot the points $(x, h(x))$.

   c. Connect the points to show the graph of $y = h(x)$.

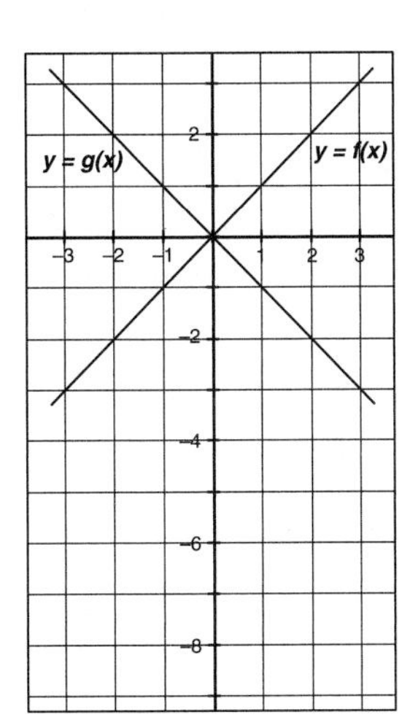

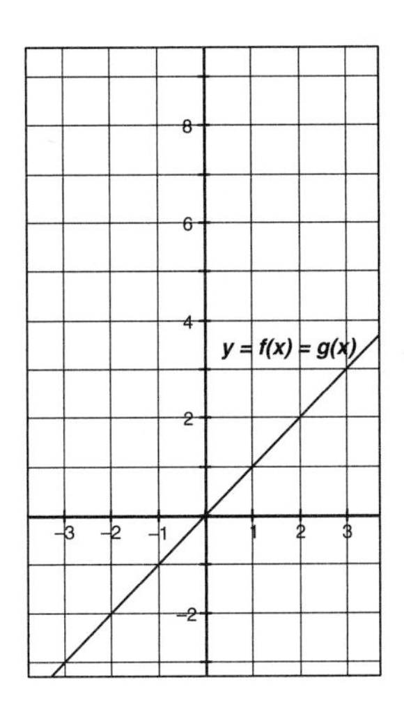

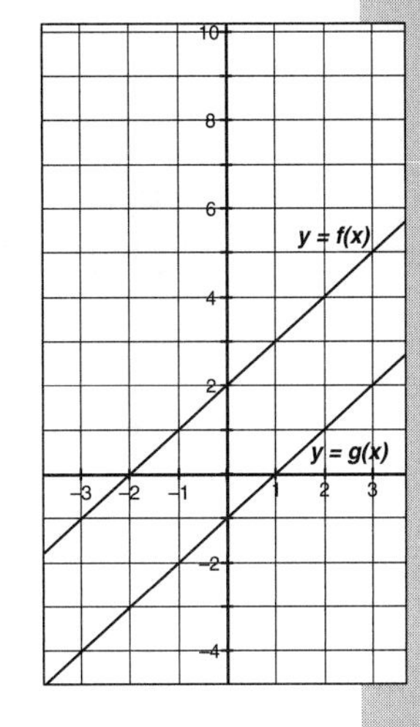

2. What observations can you make about the graph of $y = h(x)$ based on the graphs of $y = f(x)$ and $y = g(x)$? Explain why your observations make sense.
3. What kinds of functions arise from multiplying together two linear functions?

---

3. a. What are the learning outcomes of the student page?

   b. Consider the list you created during *Roundtable* while answering problem 1. Which of these items arose as you worked through the student page **Investigating Products of Linear Functions?**

   c. What prerequisites must a student have to be successful on the student page? Why do you think so?

4. Students are not asked to find the equations of the graphs of the linear functions. How would students' potential conceptual understanding change if they used the equations for $y = f(x)$ and $y = g(x)$ rather than just points on the graphs of these functions to find the graphs of the products?

EXPLORE

5. Consider the following variations on the problems in the student page **Investigating Products of Linear Functions.**

   a. Solve each variation.

   b. Which of the variations make the problem more accessible for students? What is it about this variation that makes the problem more accessible?

   c. Which of the variations make the problem more challenging? Why do you think so?

TEACHING AND LEARNING HIGH SCHOOL MATHEMATICS © 2010 John Wiley & Sons, Inc.

d. Which of the variations do not address the same mathematical learning outcomes as the original problem? What learning outcomes are addressed?

Variation 1.

a. For each pair of graphs, $y = f(x)$ and $y = g(x)$, graph the function $h(x) = f(x) \cdot g(x)$. (Note that grid lines are one unit apart for both x and y.)

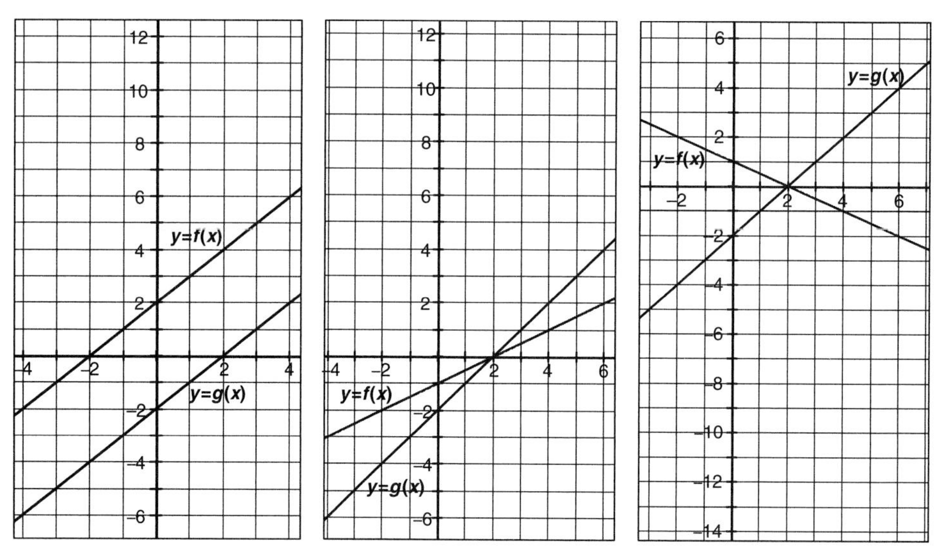

b. Describe the shape of the graph of $y = h(x)$.

c. From the graphs of $y = f(x)$ and $y = g(x)$, how can you tell over which intervals $y = h(x)$ will be positive? negative? Explain.

d. How can you tell where the zeroes of $y = h(x)$ will be from the graphs of $y = f(x)$ and $y = g(x)$? Explain.

e. Determine the point at which the graph of $y = h(x)$ changes direction. This point is called the vertex. How do the x-coordinates of the vertex and zeroes compare with each other for the function $y = h(x)$? Explain.

Variation 2.

a. Consider the graphs of functions shown in each figure below. Find the product of the functions shown in the graphs. (Note that grid lines are one unit apart for both x and y.)

b. What kinds of functions arise from multiplying together two or more linear functions?

c. For each pair of functions, what is the degree of the product function?

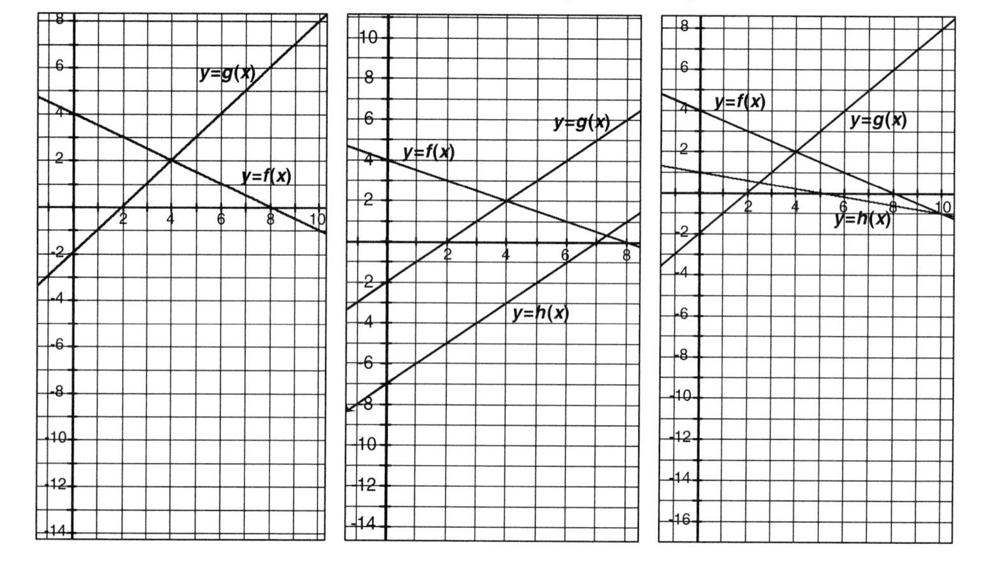

 © 2010 John Wiley & Sons, Inc.

d. How can you tell from the graphs of the linear functions when the product will lie above the x-axis? lie below the x-axis? intersect the x-axis? Explain.

e. How can you tell from the graphs of the linear functions whether or not the product function will have one or more of the following behaviors?

- As x tends to $-\infty$, y tends to $-\infty$.
- As x tends to $-\infty$, y tends to ∞.
- As x tends to ∞, y tends to $-\infty$.
- As x tends to ∞, y tends to ∞.

f. Determine an equation of a function that intersects the x-axis at $x = 2$, $x = -3$, and $x = 4$ and is positive for large values of x. Explain how you know your equation satisfies each requirement.

g. Determine an equation of a function that touches but does not cross the x-axis at $x = 2$ and is negative for all other values of x. Explain how you know your equation satisfies each requirement.

Variation 3.

a. For the graphs shown in the figure, complete the table of values. (Note that grid lines are one unit apart for both x and y.)

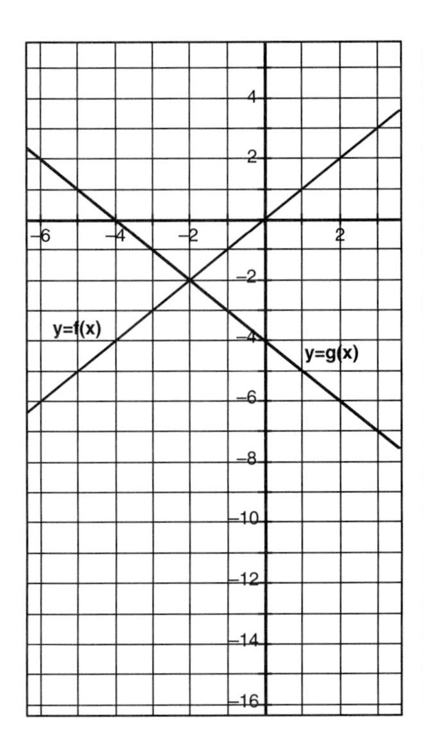

x	f(x)	g(x)	h(x) = f(x) • g(x)
–6			
–5			
–4			
–3			
–2			
–1			
0			
1			
2			
3			

b. Find the product of the two functions and graph the results.

c. What relationships exist between the original lines and the graph of the product?

Variation 4.

a. Some quadratic functions are not products of linear functions. Consider $y = x^2 + 1$. Why is there no pair of linear functions that multiply together to yield this function?

b. Multiply $y = x \cdot (x^2 + 1)$ together graphically. What do you notice about the graph of the product? Discuss its shape, direction, number of roots, steepness, etc. What type of function is the product?

6. Define polynomial functions. How does the work in this lesson help develop an understanding of polynomial functions?

TEACHING AND LEARNING HIGH SCHOOL MATHEMATICS © 2010 John Wiley & Sons, Inc.

SHARE AND SUMMARIZE

7. Share your responses to items 5 and 6.
   a. What are some ways you can make a problem more challenging for students who find a problem too easy?
   b. What are some ways to make a problem more accessible to students who find a problem too difficult?
   c. How can you be certain that a problem rescaled for accessibility still provides a challenge to students rather than giving too much away?

DEEPENING MATHEMATICAL UNDERSTANDING

1. *Exploring Quadratic Functions.* The student page **Exploring Quadratic Functions** is designed to be used early in students' study of quadratic functions.
   a. Complete the student page.
   b. Summarize what can be learned by working through the student page.

   c. Look back at the **Function Card Sort** from the Unit Three Team Builder (and found in the Unit Three Appendix). Based on your work in problem 2 on the student page, find two tables that exhibit the same pattern for the rate of change in y for a constant change in x as in the table in item 2a. Explain your choices.
   d. Look back at function cards showing graphs from the Unit Three Team Builder. Find two graphs that fit the same function family as $f(x) = x^2$. Explain your choices.

Exploring Quadratic Functions

1. Esther is a quilter. At right is one of her designs; the completed square is called a quilt block. She plans to make a square quilt using the same pattern and colors for each quilt block. She also plans to make different sizes of quilts for different purposes. To ensure that each quilt block matches, she purchases all of the fabric she needs at once.

Y	Y	W	W	R	B
Y	Y	W	W	B	R
W	W	Y	B	W	W
W	W	B	Y	W	W
R	B	W	W	Y	Y
B	R	W	W	Y	Y

   a. Determine the number of same- size pieces of each color Esther needs for the quilt block at right. This quilt block has dimensions 1×1.
   b. How many pieces of each color does she need for each of the different projects listed?

Quilt Dimensions in Blocks	Project Type	x	Number of Yellow Squares	Number of Blue Squares	Number of Red Squares	Number of White Squares
1×1	Pot holder	1	10	6	4	16
2×2	Small doll blanket	2		24		
3×3	Large doll blanket	3				
4×4	Baby blanket	4				
5×5	Lap robe	5				
6×6	Table cloth	6				
7×7	Twin bed cover	7				

(*continued*)

TEACHING AND LEARNING HIGH SCHOOL MATHEMATICS © 2010 John Wiley & Sons, Inc.

c. Graph the data in the table for all four colors on one set of coordinate axes. How are the graphs related?

d. Consider one color at a time. What patterns do you see in the relationship between the quilt dimensions in blocks and the number of squares needed to make the quilt?

e. For each color, determine an equation for each pattern you identified in item 1d.

2. a. Complete the table of values.

x	$f(x) = x^2$	$g(x) = f(x) - f(x-1)$	$g(x) - g(x-1)$
–3			
–2			
–1			
0			
1			
2			
3			
4			

b. Describe how $f(x) = x^2$ is changing as x increases.

c. Describe how quickly $f(x) = x^2$ is changing as x increases.

d. Describe any patterns or symmetry you see in the table.

e. Without graphing the function, state what you know about the graph from your table and responses to items 2b, 2c, and 2d.

f. Use your response to item 2e to sketch a predicted shape of the graph. Do not plot any points or draw the graph with a graphing calculator.

3. a. Complete a table like the one in problem 2 for each of the equations you determined in item 1e.

b. Complete a table like the one in problem 2 for $f(x) = ax^2 + bx + c$. What patterns do you notice in the first differences $[f(x) - f(x - 1)]$? What patterns do you notice in the second differences $[g(x) - g(x - 1)]$? What does this suggest about all quadratic functions?

4. a. Graph the parent function, $f(x) = x^2$.

b. What are the coordinates of the y-intercept?

c. What are the coordinates of the x-intercepts? How many x-intercepts are possible for this family of functions? How do you know?

d. Describe the shape of the graph.

(continued)

TEACHING AND LEARNING HIGH SCHOOL MATHEMATICS © 2010 John Wiley & Sons, Inc.

e. Compare the values in the table in item 2a with the graph in item 4a. Why does the graph look as it does? Describe what happens to $y = f(x)$ as x changes.

f. What new information is provided by the graph that the table does not show?

g. What information is provided by the table that the graph does not show?

5. a. What are the coordinates of the vertex of $f(x) = x^2$? Return to the quilt problem in item 1. What are the vertices for the graphs of each of the equations you determined in item 1e?

b. Two additional important points on $f(x) = x^2$ are (1, 1) and (−1, 1). Graph each of the equations from item 1e for x in [−5, 5]. For each of the graphs, what are the coordinates of the points that correspond with (1, 1) and (−1, 1) for these quadratic family members? Why do you think so?

c. For functions in the quadratic function family, why are the points that correspond to (0, 0), (1, 1), and (–1, 1) on $f(x) = x^2$ important? How can they help you determine the equation of the function represented by the table or graph? Explain.

2. *Multiplying Functions with Technology*. Using a TI-nspire calculator, it is possible to graph a function, then drag on the function to change its location or shape. To experiment more with multiplying linear functions, enter the equations as follows:

$$f1(x) = x + 1;\ f2(x) = x - 1;\ f3(x) = f1(x) \cdot f2(x).$$

   a. Graph the equations. List everything you notice about the relationships between the linear functions and the product of them.

   b. In the menu, select 1: **Actions**, then 1: **Pointer**. Select one of the lines by moving the pointer until the line is highlighted. Press the hand tool in the center of the arrow dial. Slowly drag on the line, allowing the line to be translated vertically or horizontally or rotated. Look carefully at the graph of the product. What do you notice?

   c. How are the x-intercepts of the product function related to the x-intercepts of the lines?

   d. How is the shape of the product function altered based on the steepness of the lines? How is the shape of the product function altered when the slopes of the lines have opposite signs?

   e. Where will you find the x-coordinate of the highest or lowest point of the product function? What information from the lines helps you find this point?

   f. Over what intervals is the product function positive? negative? How can you tell from the lines?

3. *Determining Polynomials from Points.*

   a. How many points determine a line?

   b. How many points determine a quadratic function?

   c. Is it possible for two points to determine a quadratic function? Why or why not?

   d. Can three collinear points lie on a parabola? Why or why not?

   e. A student said,

   I can determine a quadratic function from just two points if I know that one of the points is a vertex.

TEACHING AND LEARNING HIGH SCHOOL MATHEMATICS © 2010 John Wiley & Sons, Inc.

What do you think of the student's statement? Is it always true?

f. How many points are needed to determine a cubic function? Why do you think so?

4. *Recognizing Quadratic Functions.* In the article "The Many Uses of Algebraic Variables" by Randolph A. Philipp (*Mathematics Teacher*, 85 (October 1992): 557–561), an activity sheet is provided to encourage students to determine which examples are equations of quadratic functions of one variable. Some examples are listed.

$y = 7x^2 - 5x + 2$	$y - k = a(x - h)^2$	$E = mc^2$
$y = \frac{1}{2}\,\pi \bullet r \bullet d$	$S = v_0 \bullet t + \frac{1}{2}\,at^2$	$ax + by = c$

a. Determine which of the examples generally are considered equations of quadratic functions of one variable and state why you think so.

b. High school students become familiar with the quadratic form $y = ax^2 + bx + c$, where a, b, and c represent fixed amounts and $a \neq 0$. Some students do not sort this equation as a quadratic function.

i. What might a student be thinking if the equation is sorted as a linear function?

ii. What might a student be thinking if the equation is sorted as a cubic function?

c. A parameter is a value that is unknown but assumed to be a constant value. Which letters represent parameters in the six examples shown?

d. How do parameters make the task in 4a more challenging? How might the students in problem 4b be interpreting parameters?

e. Add at least five more examples to the list and indicate under what conditions the example could be quadratic.

5. *Functions in Stories and in Context.*

a. Return to the **Function Card Sort** from the Unit Three Team Builder (and found in the Unit Three Appendix). Find the polynomial stories and real contexts. State how you know these cards can be represented by the given function family.

b. Write another story or determine another real context that fits a polynomial function.

6. *Translation Difficulties.* The **Function Card Sort** card 6 that refers to the story *Bats on Parade* by Kathi Appelt (New York: HarperCollins, 1999) often is miscategorized by students. Distinguish between the following two interpretations and determine to which function family each interpretation belongs.

a. What is the total number of bats needed to make n formations?

b. What is the total number of bats needed to make the nth formation?

7. *Transforming Polynomial Functions.*

a. Is it possible to transform the function $f(x) = x^3$ into $g(x) = x(x - 1)(x + 1)$ by changing the parameters A, B, C, or D in the equation $g(x) = A \bullet f(Bx + C) + D$? Why or why not?

b. What shapes are possible with $g(x) = A \bullet (Bx + C)^3 + D$? Show all possible shapes and explain your choices.

c. Can all possible quadratic family members be written in the form $g(x) = A \bullet f(Bx + C) + D$ where $f(x) = x^2$? Are all of the parameters A, B, C, and D needed? If not, which one(s) are superfluous?

d. Are there other function families $y = f(x)$ for which some parameterization $g(x) = A \bullet f(Bx + C) + D$ does not give all possible family members? Explain your choice(s).

TEACHING AND LEARNING HIGH SCHOOL MATHEMATICS © 2010 John Wiley & Sons, Inc.

8. *Recursive Quadratic Equations.*

a. Complete the table below.

x	$y = x^2$	$y = 2x^2$	$y = 2x^2 + 5$
−3			
−2			
−1			
0			
1			
2			
3			
4			

b. Determine how each function changes from one term to the next.

c. For each function, write a recursive equation relating $f(n+1)$ to $f(n)$. Remember to indicate the starting value.

d. Repeat problems 8a–8c with the equations, $y = x^2$, $y = 3x^2$, and $y = 3x^2 - 4$.

e. What similarities do you see in the recursive equations?

f. Predict a recursive equation for the explicit equation $y = ax^2 + c$.

9. *Cubic Functions.*

a. Complete the table.

x	$f(x) = x^3$	$g(x) = f(x) - f(x-1)$	$h(x) = g(x) - g(x-1)$	$h(x) - h(x-1)$
−3				
−2				
−1				
0				
1				
2				
3				
4				

TEACHING AND LEARNING HIGH SCHOOL MATHEMATICS © 2010 John Wiley & Sons, Inc.

b. Describe how $f(x) = x^3$ is changing as x increases.

c. Describe how quickly $f(x) = x^3$ is changing as x increases.

d. Describe any patterns or symmetry you see in the table.

e. Without graphing the function, state what you know about the graph from your table and responses to items 9b, 9c, and 9d.

f. Using your responses to item 9e, predict a shape for the graph and sketch it. Do not plot any points or draw the graph with your graphing calculator.

g. Complete a table like the one in problem 9a for $f(x) = ax^3 + bx^2 + cx + d$. What patterns do you notice in the first differences? second differences? third differences? What does this tell you about all cubic functions?

h. Look back at the **Function Card Sort** from the Unit Three Team Builder. Based on your work in parts 9a–9g, find two tables that exhibit the same pattern for the rate of change in $f(x)$ for a constant change in x as in the tables in items 9a and 9g.

10. *Functions in Games*. In *Leapfrogs*, 10 pegs are aligned in 11 holes as shown to represent two types of frogs on a log. The object is to interchange the black pegs with the white pegs. There are two restrictions: A peg can move into an adjacent empty hole, or it can jump over a peg of the other color. (Adapted from Mason, Burton, and Stacey, 1985, p. 57).

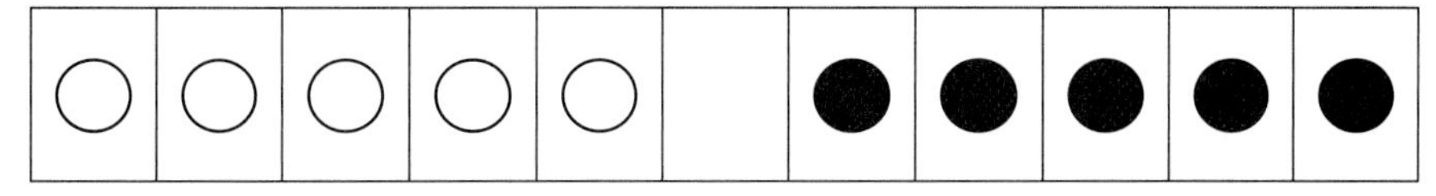

a. What is the least number of moves needed to exchange 5 "frogs" on each end of the log?

b. Record the least number of moves needed to exchange 1, 2, 3, 4, etc. frogs on each end of the log. To what function family do the data belong?

Number of frogs on each end of the log	1	2	3	4	5	10	n
Least number of moves needed to exchange frogs							

c. What is the least number of moves needed to exchange n "frogs" on each end of the log?

11. *Stacking Boxes*. A grocery store manager decided to feature cases of soft drinks in the front of the store. He wanted to make certain that shoppers would not miss the sale, so he decided to have the cases stacked in a square-based pyramid; that is, each level was made up of a square number of cases. He did not want to haul any more cases than needed into the store but instead would have stock brought in as needed to keep the pyramid several levels tall.

a. How many cases, C, of soft drinks are needed to create a square-based pyramid L levels high? Determine the number of cases needed for several values of L.

b. To what function family do the data you found in problem 11a belong? How do you know?

c. Determine an equation for C as a function of L.

12. *Multiplying Binomials and Factoring Trinomials with Algebra Tiles*.

a. Work through the student page **Multiplying and Factoring with Algebra Tiles**. (This student page assumes students have completed Lesson 3.1 **DMP** problem 15 and the student page **Using Algebra Tiles to Simplify Expressions**, in Lesson 3.1 **DMP** problem 16. A template for a set of algebra tiles can be found in the Unit Three Appendix.)

b. Use algebra tiles to find the product and simplify each of the following expressions. Show illustrations of the tiles and the accompanying algebraic

TEACHING AND LEARNING HIGH SCHOOL MATHEMATICS © 2010 John Wiley & Sons, Inc.

symbols the tiles represent at each step.

i. $2x(x-3)$ **ii.** $(x-1)(2x+3)$ **iii.** $(x+1)(2x+3)$

c. Use algebra tiles to factor each of the following expressions. Show illustrations of the tiles and the accompanying algebraic symbols the tiles represent at each step.

i. $2x^2+5x+3$ **ii.** $2x^2-7x+6$ **iii.** $2x^2+x-6$

d. Write your own problems for each category in problems 12b and 12c and solve the problems using algebra tiles. Ask another student to solve your problems. What difficulties did the student have when working on these problems?

e. What considerations did you have to make so that students can solve the problems using algebra tiles?

Multiplying and Factoring with Algebra Tiles

Materials

- one set of algebra tiles for each pair of students
- one algebra tile frame for each pair of students
- one sheet of graph or grid paper for each pair of students

Just as you can use algebra tiles to simplify expressions involving sums and differences, you can use algebra tiles to visualize the product of two expressions containing variables.

1. Use a piece of grid paper or graph paper to draw a rectangle with dimensions 6 units x 4 units. What is the area of the rectangle? How does the grid paper help you determine the area?

2. On the same piece of grid paper as in problem 1, draw a rectangle to show the area of a rectangle with one dimension of 7 units and the other dimension of $4+5$ units.

a. Show the area as the sum of two terms.

b. Show the area as a single result.

c. What property is illustrated by the example?

3. a. To find the product of x and 4, what algebra tile pieces are needed to represent x and 4? What is the area of a rectangle whose dimensions are x and 4? Show this area with algebra tiles and label the dimensions of the sides.

b. The algebra tile frame at right shows the dimensions of x and 3 on the outside of the frame. Notice that one factor is placed along the upper edge of the frame; the other factor is placed along the left edge of the frame. Fill in the rectangle so the dimensions of the pieces you use match with the side lengths of the pieces along the outside of the frame.

c. Find the sum of the areas of the pieces forming the interior of the rectangle to determine the product $x \cdot 3$.

d. Suppose you want to find the product of x and $x+2$. What algebra tile pieces are needed to represent x and $x+2$?

e. Use your work in item 3b to place the algebra tiles along the edge of the frame to find the product of x and $x+2$. Use algebra tiles to fill in a rectangle so the dimensions of the pieces you use match with the dimensions of the pieces along the outside of the frame.

f. Find the sum of the areas of the pieces forming the interior of the rectangle to determine the product of x and $x+2$.

g. Check that your product for x and $x+2$ makes sense as follows.

i. Substitute $x=4$ into x and $x+2$. Find the product of these two values.

(continued)

TEACHING AND LEARNING HIGH SCHOOL MATHEMATICS © 2010 John Wiley & Sons, Inc.

ii. Substitute $x = 4$ into the product you found in problem 3f.

iii. Do your results in problem 3g.i and 3g.ii agree? If not, revisit how you constructed your rectangle in problem 3e and check that your revised product works.

h. What property is illustrated by the example?

4. a. Outline a rectangle with dimensions $x + 4$ and $x + 3$. Fill in the rectangle with algebra tile pieces to find the product $(x + 4)(x + 3)$.

b. Check your result by choosing a value for $x \neq 0$ and substituting into $x + 4$, $x + 3$, the product of these two binomials, and the algebraic expression you obtained for $(x + 4)(x + 3)$.

c. Use algebra tiles to illustrate the product of $(2x + 1)(3x + 2)$.

5. Algebra tiles can also be used to work backwards from a product to find, if possible, two factors that generate that product.

a. Collect the pieces for $x^2 + 6x + 8$.

b. Arrange the pieces to form a rectangle. The dimensions of the rectangle are the factors of $x^2 + 6x + 8$. What are the factors?

c. Use the process you used in problem 5b to find the factors of $2x^2 + 5x + 2$.

d. How do the algebra tile pieces help you think about factoring?

13. *Factoring by Grouping*. Students often view factoring as solely a guess-and-check method. In **DMU** problem 12, you explored the use of algebra tiles to help students make sense of factoring a trinomial. Another strategy that can remove much of the guesswork is to factor by grouping. To understand this strategy, work through the following steps.

a. Expand $(ax + b)(cx + d)$.

b. Find the product of the two terms that comprise the coefficient of the x term. Find the product of the constant and the coefficient of the x^2 term. What do you observe?

c. Repeat items 13a and 13b, replacing the parameters a, b, c, and d with positive integers. Do not simplify products.

d. The results in problem 13b provide the basis for factoring by grouping. To factor $12x^2 + 11x + 2$, find two numbers whose product equals $12 \cdot 2$ (the product of the constant term and the leading coefficient) and whose sum is 11 (the coefficient of the x term). Use these numbers to rewrite $11x$ as a sum of two terms. Why is this helpful in determining how to decompose $11x$?

e. Then $12x^2 + 11x + 2 = 12x^2 + \_\_\_ x + \_\_\_ x + 2$. Group the first two terms and factor. Group the last two terms and factor.

f. Look at your result in problem 13e. There should be a common factor. Use the distributive property to factor again, obtaining the product of two binomials.

g. Apply the technique to factor $20x^2 + 23x + 6$.

h. Compare the factoring by grouping strategy to a systematic guess-and-check approach.

14. *Expanding $(a + b)^n$*. The coefficients in the expansion of $(a + b)^n$ relate to a famous pattern in mathematics. (Consider using algebraic symbol manipulation technology [CAS] on a computer or graphing calculator to expand each of the binomials.)

a. Expand each of the following binomial expressions: $(a + b)^2$, $(a + b)^3$, $(a + b)^4$, $(a + b)^5$, and $(a + b)^6$.

TEACHING AND LEARNING HIGH SCHOOL MATHEMATICS © 2010 John Wiley & Sons, Inc.

b. What patterns do you observe in the coefficients of the variables? What patterns do you observe in the exponents of the variables?

c. The coefficients in the polynomial expansion are related to what famous mathematical pattern?

15. *More Equation Solving with Tables and Graphs*. Consider the quadratic equation $3x^2 + 5x - 2 = 0$.

a. Create a table using a graphing calculator starting at some integer value and set the table increment to 0.25. One solution is an integer. Find it. The other solution must lie between which two values? How do you know?

b. How does the graph help you determine the number of solutions to expect?

c. Graphs and tables can sometimes be deceptive. Consider the quadratic equation $3x^2 + 5x + 2 = 0$.

i. The graph suggests there are how many solutions?

ii. Try to find these solutions with the table. Does the table suggest any other solutions? Explain.

iii. Use the quadratic formula to find the solution(s) to this equation. To what extent did the table or graph make it easy or hard to find the solution(s)?

16. *Solving Nonstandard Equations*. Tables and graphs can be helpful to solve equations for which symbolic methods are not readily available.

a. Use a graph to solve $4x^2 + 5x + 1 = 2^x$.

b. Check your solution to problem 16a by solving the equation with a table. What are some issues you must consider when attempting to use a table to solve the equation?

c. Teachers need to help students develop the flexibility to solve problems using the most appropriate method in a particular situation. How can the use of manipulatives, tables, or graphs help students who are struggling with symbolic approaches to equation solving?

17. *Multiplying Periodic Functions*. Though sine functions are not usually studied in Algebra I courses, they can be used as "black box" functions. In the student page **Multiplying Periodic Functions**, periodic functions are used to provide students more experience with multiplying two functions together so students see how the two functions interact with each other when multiplied.

a. Complete the student page.

b. What behaviors of functions become evident? Why do you think so?

c. What more do you understand about multiplying functions now that you have completed this student page?

TEACHING AND LEARNING HIGH SCHOOL MATHEMATICS © 2010 John Wiley & Sons, Inc.

Multiplying Periodic Functions

1. What happens when you multiply a function by a constant?

a. Graph $f(x) = \sin(x)$ on the viewing rectangle [–5, 5] by [–3, 3]. What do you notice about this function? Also, graph the functions $y = 1$ and $y = -1$ on the same window. How do these lines help you determine the behavior of $y = f(x)$?

b. Graph the functions $f(x) = \sin(x)$, $g(x) = A$, $y = -g(x)$ (so $y = -A$), and $h(x) = f(x) \cdot g(x)$ all on the same set of axes for each choice of A. (Use the viewing rectangle [–5, 5] by [–3, 3]. Use new coordinate axes for each choice of A.)
i. $A = 2$ **ii.** $A = 0.5$ **iii.** $A = -3$

c. Describe the relationship between the graphs of f, g, $-g$, and h for each choice of A. Explain why the graph of $h(x) = f(x) \cdot g(x)$ appears as it does based on the y-values of f and g for several values of x.

2. Suppose instead of multiplying the function $y = f(x)$ by a constant function $g(x) = A$, you multiply f by a function $y = g(x)$. (Assume $y = g(x)$ is not a constant function.)

a. How would you expect the new function, $y = f(x) \cdot g(x)$, to be related to each of the original functions?

b. Test your conjecture by graphing $h(x) = x \cdot \sin(x)$ on the viewing rectangle [–10, 10] for both x and y. Also graph $f(x) = \sin(x)$, $g(x) = x$, and $y = -g(x)$ on the same set of coordinate axes.

c. At what points does the graph of $y = h(x)$ coincide with the graph of $y = g(x)$ or $y = -g(x)$? Explain why these graphs coincide at these points.

d. How is the new function, $y = f(x) \cdot g(x)$, related to each of the functions $y = f(x)$, $y = g(x)$, and $y = -g(x)$?

3. Consider the effect on the function $f(x) = \sin(x)$ when f is multiplied by the function $g(x) = x^2$.

a. On a viewing rectangle [–10, 10] by [–100, 100], graph $g(x) = x^2$, $y = -g(x)$, and $h(x) = x^2 \cdot \sin(x)$.

b. How do the graphs of these functions compare with each other?

c. Overlay $y = 50 \cdot \sin(x)$. How do oscillations of $f(x) = \sin(x)$ compare with those of $h(x) = x^2 \cdot \sin(x)$?

4. Consider the effect on the function $f(x) = \sin(x)$ when f is multiplied by the function $g(x) = \sqrt{x}$.

a. On a viewing rectangle [0, 25] by [–5, 5], graph $g(x) = \sqrt{x}$, $y = -g(x)$, and $h(x) = \sqrt{x} \cdot \sin(x)$.

b. How do the graphs of these functions compare with each other?

c. Overlay $f(x) = \sin(x)$. How do the oscillations of $f(x) = \sin(x)$ compare with those of $h(x) = \sqrt{x} \cdot \sin(x)$?

5. From the functions in problems 1 through 4, how are $h(x) = A \cdot \sin(x)$ and $h(x) = g(x) \cdot \sin(x)$ similar? How are they different?

6. The graphs of $y = f(x)$ and $y = g(x)$ are shown on the coordinate axes for x in [–3.7, 3.7] and y in [–2.4, 2.4] with grid lines each half unit. Sketch the graph of $h(x) = f(x) \cdot g(x)$ on the same coordinate axes. Explain your result.

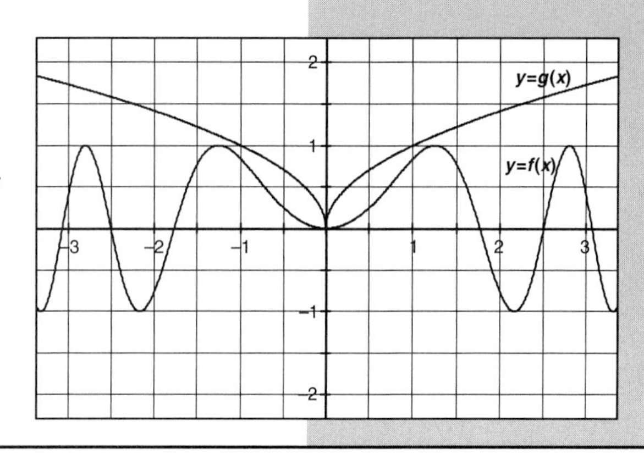

TEACHING AND LEARNING HIGH SCHOOL MATHEMATICS © 2010 John Wiley & Sons, Inc.

18. *Trigonometric Identities.*

a. Sketch the graph of $f(x) = \sin(x)$ and the product $g(x) = (\sin(x))^2 = \sin^2(x)$. What is the largest value of $g(x)$? Explain.

b. Sketch the graph of $f(x) = \cos(x)$ and the product $h(x) = (\cos(x))^2 = \cos^2(x)$. What is the largest value of $h(x)$? Explain.

c. Graphically, add $y = h(x)$ and $y = g(x)$. What is the result?

d. Look at a table of values for $g(x)$, $h(x)$, and $g(x) + h(x)$. Set the table to begin with $x = 0$ and to increment by values of $\frac{\pi}{12}$. What do you see?

e. Use the right triangle definitions of sine and cosine and the Pythagorean Theorem to explain why the results in item 18d are sensible.

f. Look up at least two trigonometric identities. Use a similar process to explore the relationships between other trigonometric identities.

19. *Power Functions.* A function is a *power function* if it is of the form $y = Ax^b$, where b is any real number and $A \neq 0$. In high school mathematics courses preceding calculus, the study of power functions is restricted to rational powers.

a. Graph each of the power functions $y = x$ and $y = x^2$. Determine the graph of $y = x^2 + x$ from these graphs. Is the new function a power function? Why or why not?

b. How are polynomial and power functions related? Sketch a Venn diagram to show this relationship.

c. How can a polynomial function with multiple terms be obtained from polynomial power functions?

d. For what values of b are power functions polynomials?

e. Determine the domain and range for polynomial power functions. What cases must you consider?

f. What effect does A have on the domain and range of the polynomial power function? How do you know?

DEVELOPING MATHEMATICAL PEDAGOGY

Reflecting on Our Own Learning

20. Reflect on your work with polynomial functions and representations in this lesson.

a. What, if anything, was new to you?

b. What, if anything, was review but revisited from a new perspective?

c. What were the benefits of looking at polynomial functions from the perspective of multiplying linear and quadratic functions together graphically?

d. What aspects of the development of understanding of polynomial functions do you expect to use in your own classroom?

Using Technological Tools

21. a. A student graphed $y = 3x^2$ with his graphing calculator and obtained the graph at right. When asked to describe the graph, the student responded

The graph is a line.

What appears to be the student's difficulty? What instructional approaches might you use to help the student?

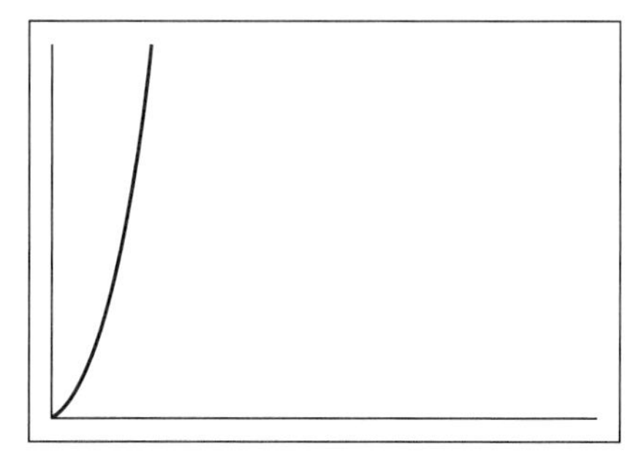

TEACHING AND LEARNING HIGH SCHOOL MATHEMATICS © 2010 John Wiley & Sons, Inc.

b. When students use a graphing calculator, one issue is what paper trail students should leave behind. That is, what should students write down in order to provide a record of what they have done so that the teacher knows what the student did and what the student thought? Consider the student response in problem 21a. Which aspects of the graph from the graphing calculator would you want to know and have students write down on a consistent basis?

22. Find several Algebra I and Algebra II textbooks with recent copyrights. Locate the sections on solving equations. Investigate how technology, specifically graphing calculators or computer algebra systems (CAS), is used in studying these concepts. Write a brief summary of how technology is used, citing examples from the texts as appropriate. If technology is not incorporated, provide some brief suggestions on ways in which technology could be used to help students understand equation solving in a deep, conceptual way.

23. Calculators with CAS provide another tool for exploring algebraic concepts.

a. Use a calculator with CAS to expand $(x + 1)^2$, $(x + 2)^2$, $(x + 3)^2$, $(x + 4)^2$, and $(x + 5)^2$.

b. Look at the patterns in the results. How might students use the patterns to conjecture a result for $(x + 7)^2$? What about for $(x + n)^2$?

c. Write a problem using CAS to help students develop strategies for expanding $(x - n)^2$, for any integer $n > 0$. What are some issues to consider? What are some possible extensions?

d. If possible, try your CAS activity with some high school students in an Algebra I class. Reflect on their interactions with the activity.

e. State any concerns you have about using CAS in early algebra instruction.

Analyzing Instructional Materials

24. Find at least three different Algebra II texts with recent copyrights. Locate a section focusing on polynomial functions.

a. Analyze how multiple representations are used to help students make sense of polynomials.

b. Did any of the texts present polynomials as products of linear and/or quadratic polynomials?

c. Were polynomial functions presented graphically?

d. Compare and contrast the presentations in the texts to those found in this lesson.

e. How might the presentations of polynomial functions in the Algebra II textbooks be enhanced?

f. What additional topics might be added to the presentation in the Algebra II textbooks to make it more complete?

g. Write a short summary of your findings, supporting your conclusions with specific examples from the Algebra II textbooks.

25. Complete the student page **Analyzing Polynomial Parent Functions**.

a. Determine prerequisite knowledge students must have to complete the page successfully.

b. Extend the student page so students consider all the possible shapes of polynomials of degrees 1 through 5.

c. Adapt the student page so students solve problems that require them to consider the end behavior of the polynomials.

WWW

d. What difficulties do you expect some students might have with the page? How might you make the task more accessible to those students while maintaining high cognitive challenges for them? Record anticipated student responses (misconceived, partial, and satisfactory) and teacher support in a **QRS Guide**.

TEACHING AND LEARNING HIGH SCHOOL MATHEMATICS © 2010 John Wiley & Sons, Inc.

Analyzing Polynomial Parent Functions

1. Sketch the graph of $y = x$. Graphically multiply $y = x$ by itself, obtaining the graph of $y = x^2$. Based on the relationship between the functions $y = x$ and $y = x^2$,
   a. Explain why the graph of $y = x^2$ appears as it does for x in the interval from 0 to 1.
   b. Explain why the graph of $y = x^2$ appears as it does for x in the interval from -1 to 0.
   c. Explain why the graph of $y = x^2$ appears as it does for $|x|$ greater than 1.
2. Sketch the graphs of $y = x^2$, $y = x^4$, and $y = x^6$ on the same set of axes. Complete the following using what you have learned about multiplying functions together graphically. Explain why the graphs appear as they do relative to each other.
   a. Compare the graphs for x in the interval from -1 to 1.
   b. Compare the graphs for $|x|$ greater than 1.
3. Repeat problem 2 for the graphs of $y = x$, $y = x^3$, $y = x^5$ and $y = x^7$.
4. Explain why the graphs in problem 2 touch but do not cross the x-axis.
5. Explain why the graphs in problem 3 cross the x-axis.
6. Determine equations of three polynomial functions that cross the x-axis at $x = 1$. How do you know?
7. Determine equations of three polynomial functions that touch but do not cross the x-axis at $x = 2$. How do you know?
8. Determine equations of three polynomial functions that touch the x-axis at $x = -1$ and cross the x-axis at $x = 4$ and $x = -2$. Explain how you know you are correct.

26. Consider the student page and variations in the **Explore** section of Lesson 3.5. Each was written using function notation. How might you modify the problems for use with students who have not yet studied function notation but are prepared to study concepts related to polynomial functions?

Teaching for Equity

27. Choose one of the student pages from another lesson in Unit Three and one or more of the problems on the student page. Rescale the task to make it more accessible. Rescale the task to make it more challenging. Provide the reasoning behind your changes.

28. Revisit the student page **Multiplying and Factoring with Algebra Tiles.**

a. All of the problems on the student page have positive coefficients of x^2 and x as well as positive values for the constant term. How might you use algebra tiles to show $(x - 3)(x + 2)$?

b. How might you use the algebra tiles to factor $x^2 - x - 12$?

c. Some people argue that when coefficients are all positive, algebra tiles can help students develop conceptual understanding for multiplication of two binomials or for factoring trinomials. However, when coefficients are negative, the algebra tiles are less helpful. Comment on this argument.

d. Based on your experiences with algebra tiles, how might you use them in your classroom?

e. How do algebra tiles support learning for a wide spectrum of students?

TEACHING AND LEARNING HIGH SCHOOL MATHEMATICS © 2010 John Wiley & Sons, Inc.

29. Although teachers are encouraged to have their students work in cooperative groups, there are times when groups do not function successfully. For instance, some members may come to class unprepared to participate in group work or fail to participate equitably in group assignments completed outside of class. Some teachers allow group members to "fire" a member who is not contributing; the "fired" member must then work alone.

   a. Comment on issues that might arise if students can be "fired" from their group.

   b. What other strategies might teachers use when group members are not contributing?

Preparing for Instruction

30. Choose one of the student pages from Lesson 3.5 or from the **DMU/DMP** questions accompanying the lesson.

   a. Determine the learning outcomes for the student page.

   b. Complete a **QRS Guide** for the student page, anticipating student responses at three levels for each major problem.

   c. Write teacher support questions for each student response and add these to the **QRS Guide**.

   d. Decide on a cooperative learning (CL) strategy that best fits use of the student page with a class of students. Why do you think this CL strategy is appropriate?

   e. Indicate the CL strategy and the learning outcomes at the top of the **QRS Guide**.

   f. Try the activity with one or two high school students. Record any additional student responses you did not anticipate and related teacher support in the **QRS Guide**. What more did you learn by using the student page and **QRS Guide** with students?

31. Consider the student page **Paper Folding for $(a + b)^2$ and $a^2 - b^2$**.

   a. Work through the student page. Compare multiplying binomials via paper folding and algebra tiles. What similarities and differences do you find?

   b. The activities on the student page just deal with $(a + b)^2$ or $a^2 - b^2$. What follow-up activities might you plan to solidify the concepts addressed on the student page?

TEACHING AND LEARNING HIGH SCHOOL MATHEMATICS © 2010 John Wiley & Sons, Inc.

Paper Folding for $(a + b)^2$ and $a^2 + b^2$

Paper folding can be used to illustrate the expansion of $(a + b)^2$ and the result when $a^2 - b^2$ is factored.

Materials

- 2 pieces of square paper per student

1. a. Begin with a piece of square paper. Fold one edge of the paper down several inches and make a crease all the way across the paper, parallel to one edge. Along an edge intersected by the crease, use the label b for the smaller distance and the label a for the larger distance. (Segments of length a and b do not overlap.)

b. Fold along the square's diagonal so you can mark the length b along the adjacent edge (do not crease the fold).

c. Make a crease all the way across the paper perpendicular to the fold in item 1a and b units from the edge of the paper. Your original piece of paper should now be sectioned into four rectangular parts.

d. Using the labels a and b from item 1a, identify the side lengths of each of the four rectangles into which your paper is sectioned. Find and label the area of each rectangle.

e. The original square has side length $a + b$. What is the overall area of the square?

f. Use the results in items 1d and 1e to write an algebraic sentence comparing two different expressions for the area of the square.

2. a. Start with another piece of square paper. This time, label the edge length of the square paper as a.

b. As in item 1a, fold one edge of the paper down several inches and make a crease all the way across the paper, parallel to one edge. Use the label b for the smaller distance from the edge of the square to the crease. What is the length of the larger distance from the crease to the other edge of the square?

c. Fold along the square's diagonal so you can mark the length b along the adjacent edge (do not crease the fold).

d. Make a crease all the way across the paper perpendicular to the fold in 2b and b units from the edge of the paper. Your original piece of paper should now be sectioned into four rectangular parts.

e. Using the labels a and b from items 2a and 2b, identify the side lengths of each of the four rectangles into which your paper is sectioned.

f. Using the side lengths of the square from item 2a, express the area of the total square. Tear off the portion of the square whose area is b^2. What is the area of the remaining portion of the paper?

g. Tear off one of the nonsquare rectangles remaining from item 2f, according to the creases in the paper. Rearrange these two pieces into a new rectangle. Identify the side lengths of this rectangle.

h. Use your results in items 2f and 2g to write a sentence comparing two different expressions for the area of the paper remaining after you tore off a piece with area b^2.

Focusing on Assessment

32. Consider the following item given to Algebra I students.

Is $(x + y)^2 = x^2 + y^2$?

How might you convince a friend that your response is correct?

a. Identify at least one way Algebra I students might respond to this item.

TEACHING AND LEARNING HIGH SCHOOL MATHEMATICS © 2010 John Wiley & Sons, Inc.

b. What might a teacher learn about students' thinking from such an item that a teacher would not learn from just asking students to expand a binomial such as $(x + 4)^2$?

Reflecting on Professional Reading

33. Read "Using Homemade Algebra Tiles to Develop Algebra and Prealgebra Concepts" by Annette Ricks Leitze and Nancy A. Kitt (*Mathematics Teacher*, 93 (September 2000): 462–466, 520).

a. Compare the activities introduced in the article with the activities on the student pages **Using Algebra Tiles to Simplify Expressions** and **Solving Equations with Algebra Tiles**, both from Lesson 3.1, and **Multiplying and Factoring with Algebra Tiles** from Lesson 3.5 What similarities and differences do you observe?

b. The article describes a box method to factor a trinomial. Use the box method to factor $x^2 - 8x + 15$.

c. In the article, there are examples in which algebra tiles are used with a negative coefficient for the variable term x or with a negative constant. What are some issues that need to be addressed when negatives are involved?

34. Read "Sequences and Polynomials Part I: Guidelines for Finding a Next Term and a General Term for Any Given Finite Sequence" by Dorothy T. Meserve and Bruce E. Meserve (*Mathematics Teacher*, 100 (February 2007): 426–429). The article provides a justification for the use of finite differences that were also explored in the student page **Exploring Quadratic Functions** and in **DMU** problem 9. In what ways did the article reinforce or support the work you did earlier in this lesson?

3.6 Accommodating Different Learning Styles: Rational Functions

The current focus on standards has led to higher expectations for all students. A look at many state standards or frameworks since the mid-1990s reveals that students are expected to accomplish much more than was expected of them prior to the reforms of the late 1980s and early 1990s. For teachers, the challenge is how to help all students reach this higher level of achievement. Students come with varying backgrounds. Their learning styles, interests, prior knowledge, and social skills can vary; they may speak a language other than English; and they may have different comfort levels with mathematics. The standards state what students need to know. However, it is up to teachers to incorporate a variety of instructional practices, often simultaneously, in order to help all students meet the high expectations of the standards.

In Unit One, the pedagogical focus was on creating a supportive learning environment, addressing students' socialization needs and comfort zones. In Unit Two, you found that students' learning is enhanced when new learning is based on what is already familiar from students' daily lives and previous work in school. In both units, you had an opportunity to consider issues related to equity and ways to ensure that instruction addresses specific needs of students from various ethnic groups or language backgrounds. This lesson focuses on another equity-supportive tool: attending to students' different learning styles, particularly reading/writing, visual, auditory, and kinesthetic.

In this lesson, you will:

- Explore relationships among the functions $y = f(x)$, $y = g(x)$, $h(x) = \frac{f(x)}{g(x)}$, and $k(x) = \frac{g(x)}{f(x)}$.
- Analyze elements of lessons that help students of diverse backgrounds develop mathematical understanding.

Website Resources

- **Investigating Rational Functions** student page

TEACHING AND LEARNING HIGH SCHOOL MATHEMATICS © 2010 John Wiley & Sons, Inc.

LAUNCH

Teachers might be tempted to teach from their dominant learning styles. However, students might not share the same learning styles as their teachers. An abbreviated learning style survey follows.

1. Take the survey to provide insight into your own learning styles. For each item, circle responses that best fit you. If a single answer does not match your perception, circle all responses that apply most to you. If none of the responses fit you, skip the item.
   - **a.** When I need directions to learn how to use a new tool, I prefer to
     - **V.** Be given pictorial directions with pertinent information labeled on the pictures.
     - **A.** Be told the directions out loud.
     - **R.** Read directions provided in words.
     - **K.** Be shown how to use the tool while I am concurrently practicing to use it.
   - **b.** When learning a new concept in mathematics class, I prefer to
     - **V.** See graphs and diagrams if possible.
     - **A.** Listen to the teacher explain what I should be learning.
     - **R.** Read about the new concept in a textbook.
     - **K.** Use materials like motion detectors connected to graphing calculators to experience the concept physically.
   - **c.** When I solve a mathematics problem, the thing I am likely to do first is
     - **V.** Draw pictures or diagrams.
     - **A.** Talk to myself out loud about the problem.
     - **R.** Solve the problem with algebra or other symbols.
     - **K.** Use physical materials or act it out to try to understand the problem.
   - **d.** When I have made a discovery on a mathematics problem I have been struggling with and am anxious to share results with my team members, I prefer to
     - **V.** Show graphs and diagrams to explain what I have found.
     - **A.** Call them on the phone to tell them orally about what I have discovered.
     - **R.** Write a note or email and explain my discovery in words.
     - **K.** Use physical materials to show what I discovered.
   - **e.** For recreation, I prefer
     - **V.** To watch movies, look at photographs, or sit in a quiet place and enjoy the scenery.
     - **A.** To listen to books on tape, music, or news reports, or talk to friends on the phone.
     - **R.** To read a book, write in my journal, or communicate with my friends through email, text messages, or instant messages.
     - **K.** To build something, dance, play sports, or in some other way be active.
2. Indicate the number of responses for each of the categories:

   V. ____________ A. ____________ R. ____________ K. ____________

   - **a.** What seems to be your dominant learning styles? (V = visual, A = auditory, R = reading/writing, and K = kinesthetic.)
   - **b.** Compare your learning styles to those of your peers. What seem to be the dominant learning styles of your peers?
   - **c.** How is this information useful in helping you think about the learning styles of your students?

TEACHING AND LEARNING HIGH SCHOOL MATHEMATICS © 2010 John Wiley & Sons, Inc.

EXPLORE

Human cognition tends to be pluralistic; that is, humans process information and learn through a variety of cognitive styles rather than a single style. The student page that follows encourages students to use more than one learning style (or modality) to begin to learn about rational functions.

3. **a.** Complete the student page **Investigating Rational Functions.**
   b. How does the student page accommodate different learning styles of students?
   c. How does it help students broaden their preferred learning styles?
   d. How would you suggest modifying the student page to be even more accommodating of one of the four learning styles?

Investigating Rational Functions

1. In the children's story *The Doorbell Rang* by Pat Hutchins (New York: Scholastic, 1986), two children were about to share a plate of cookies equally between them when the doorbell rang. They had determined that their shares would be 6 cookies each. Each time the doorbell rings, one or two friends are at the door; they are invited in to share the cookies.

a. In the story, a constant number of cookies is being divided by a changing number of children. This situation can be viewed as dividing a constant function $f(x) = 12$ by a linear function $g(x) = x$. In terms of the story,

i. What does the constant function represent?

ii. What does the linear function represent?

b. Consider a case where x is continuous. Sketch graphs of $f(x) = 12$ and $g(x) = x$ on the grid provided. Use values on each graph to complete the table, then sketch the graph of the quotient $h(x) = \frac{f(x)}{g(x)}$. (Grid lines are shown every 1 unit for x and every 2 units for y.)

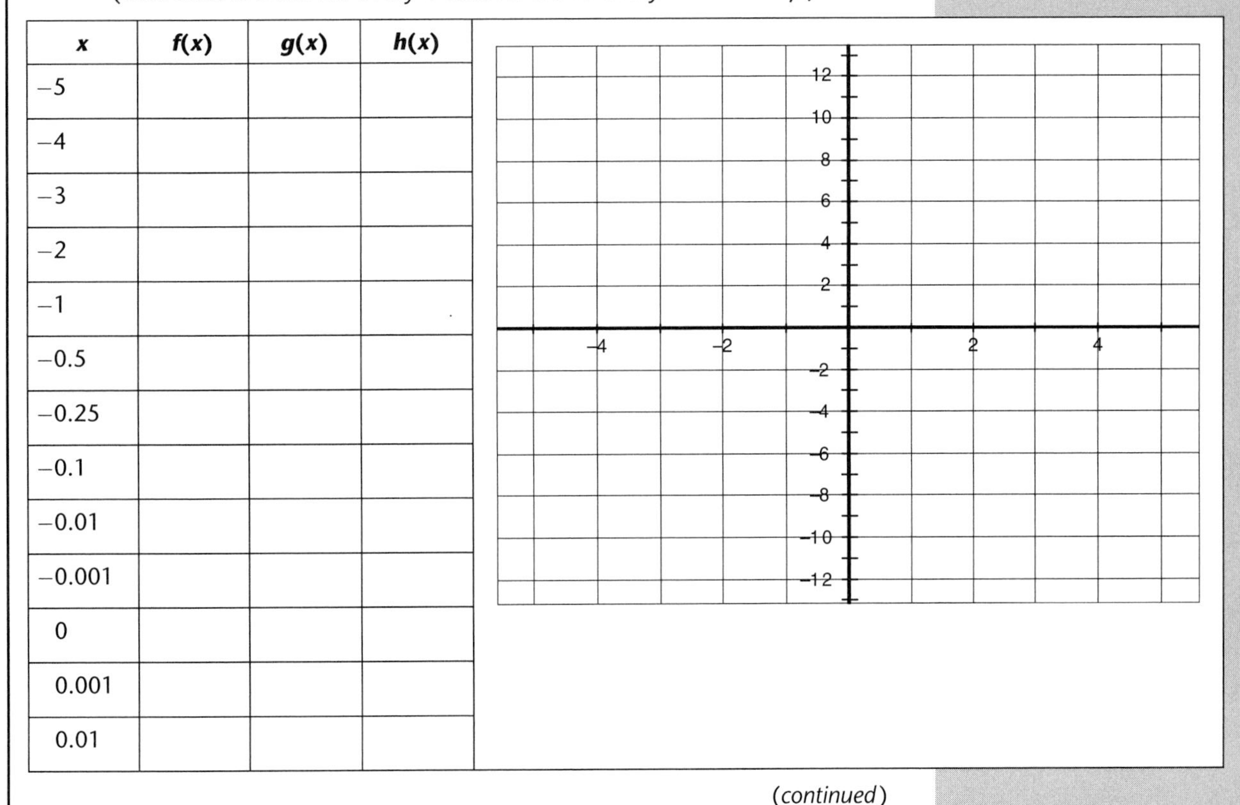

x	$f(x)$	$g(x)$	$h(x)$
−5			
−4			
−3			
−2			
−1			
−0.5			
−0.25			
−0.1			
−0.01			
−0.001			
0			
0.001			
0.01			

(continued)

TEACHING AND LEARNING HIGH SCHOOL MATHEMATICS © 2010 John Wiley & Sons, Inc.

0.1			
0.25			
0.5			
1			
2			
3			
4			
5			

c. Based on the values of $y = f(x)$ and $y = g(x)$, explain why the graph of $y = h(x)$ appears as it does for
i. x near zero **ii.** $x < 0$ **iii.** $x > 0$

d. Based on the values of $y = f(x)$ and $y = g(x)$, how do you expect the graph of $y = h(x)$ to appear as
i. x approaches ∞ **ii.** x approaches $-\infty$

2. Shown in the figure are the graphs of $f(x) = 2$ and $g(x) = x - 1$.

x	$f(x)$	$g(x)$	$h(x)$
−5			
−4			
−3			
−2			
−1			
0			
0.5			
0.75			
0.9			
0.99			
1			
1.01			
1.10			
1.25			
1.5			
2			
3			
4			
5			

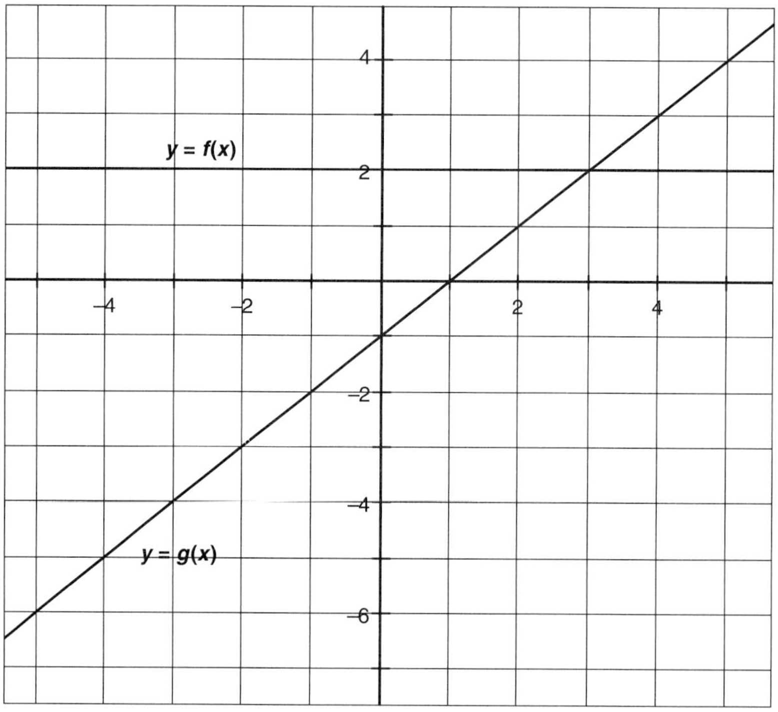

y = f(x)
y = g(x)
4
2
−2
−4
−6
−4
−2
2
4

(continued)

TEACHING AND LEARNING HIGH SCHOOL MATHEMATICS © 2010 John Wiley & Sons, Inc.

a. Complete the table of values for x, $f(x)$, $g(x)$, and $h(x) = \frac{f(x)}{g(x)}$ for several values of x.

b. Sketch the graph of $h(x) = \frac{f(x)}{g(x)}$ based on the table and graphs of f and g.

c. Based on the values of $y = f(x)$ and $y = g(x)$, explain why the graph of $y = h(x)$ appears as it does for

i. $x = 1$ **ii.** x near 1 **iii.** $x = 3$ **iv.** $x < 1$ **v.** $x > 1$

d. Based on the values of $y = f(x)$ and $y = g(x)$, how do you expect the graph of $y = h(x)$ to appear as

i. x approaches ∞ **ii.** x approaches $-\infty$

For problems 3–5, create tables as needed.

3. Sketch the graphs of $f(x) = 1$ and $g(x) = x^2$. Divide the functions using values from the graphs of f and g to obtain the graph of $h(x) = \frac{f(x)}{g(x)}$. Based on the graphs of f and g,

a. Explain why the graph of the quotient behaves as it does for x near 0.

b. Explain why the graph of the quotient behaves as it does for large $|x|$.

c. Based on your work in problems 1, 3a, and 3b, predict and explain the behavior of $y = \frac{1}{x^n}$ for x near 0 when

i. n is even **ii.** n is odd

d. Based on your work in problems 1, 3a, and 3b, predict and explain the behavior of $y = \frac{1}{x^n}$ for large $|x|$ when

i. n is even **ii.** n is odd

4. Sketch the graphs of $f(x) = x - 1$ and $g(x) = x$. Divide the functions using values from the graphs of f and g to obtain the graph of $h(x) = \frac{f(x)}{g(x)}$. Based on the graphs of f and g,

a. Explain why the graph of the quotient behaves as it does for x near 0.

b. Explain why the graph of the quotient behaves as it does for large $|x|$.

c. Divide the functions again using values from the graphs of f and g, this time obtaining the graph of $k(x) = \frac{g(x)}{f(x)}$. Based on the graphs of f and g, explain any vertical or horizontal asymptotic behavior.

5. Sketch the graphs of $f(x) = x - 1$ and $g(x) = x^2$. Divide the functions using values from the graphs of f and g to obtain the graph of $h(x) = \frac{f(x)}{g(x)}$. Based on the graphs of f and g,

a. Explain why the graph of the quotient behaves as it does for x near 0.

b. Explain why the graph of the quotient behaves as it does for large $|x|$.

c. Divide the functions again using values from the graphs of f and g, this time obtaining the graph of $k(x) = \frac{f(x)}{g(x)}$. Based on the graphs of f and g, explain any asymptotic behavior.

6. A **rational function** is the quotient of two polynomial functions and is defined for all values of x such that the denominator is not zero.

(continued)

TEACHING AND LEARNING HIGH SCHOOL MATHEMATICS © 2010 John Wiley & Sons, Inc.

Consider the degrees of the functions in the numerator and denominator for problems 1–5. What relationship exists between the degrees of the polynomial functions when the quotient function has an asymptote of

a. $y = 0$?

b. $y = k$ where k is a constant?

c. $y = mx + b$ where $m \neq 0$?

7. Determine an equation of a rational function with a vertical asymptote at $x = 4$ and a horizontal asymptote at $x = -1$.

8. Determine an equation of a rational function with a vertical asymptote at $x = 1$ and an oblique asymptote of $y = 2x$.

Source: Student page inspired by Beckmann and Sundstrom (1992) and Beckmann (1993).

SHARE AND SUMMARIZE

4. The student page is condensed in that it provides examples for several functions but perhaps not enough for all students. Keeping several learning styles in mind, suggest additional functions students might find helpful in order to complete problems 6–8 on the student page.

5. The student page begins by providing tables, grids, or graphs on which students can organize and record their work. It encourages students to continue in this vein depending on the student's needs.

- **a.** Comment on the level of help in the first two problems versus the level of help for the later problems.
- **b.** Would you recommend students be given more help, less help, or the same amount of help in organizing and recording their work in problems 3–5? Explain your choice.

6. It is not obvious how kinesthetic learners are aided by the questions in the student page.

- **a.** Suggest ways in which kinesthetic learners' needs are addressed by the student page.
- **b.** What other activities or questions might be included to address their needs more strongly? Explain.

DEEPENING MATHEMATICAL UNDERSTANDING

1. *Exploring* $y = \frac{k}{x}$.

- **a.** Complete the student page **Exploring Functions of the Form $y = \frac{k}{x}$**.
- **b.** What prerequisite knowledge is assumed?
- **c.** What are students expected to learn from completing the student page?

TEACHING AND LEARNING HIGH SCHOOL MATHEMATICS © 2010 John Wiley & Sons, Inc.

Exploring Functions of the Form $y = \frac{k}{x}$

1. United States currency is minted and printed in various denominations. Mary wants to give her great-granddaughter Tegan $100 for her birthday to be put in her college fund. She likes to give gifts of all the same denomination and in new condition.

a. Determine the number of coins or bills Mary will get from the bank for each denomination she could choose. Complete the table provided.

Denomination	Value of Denomination, *d* (One Coin or Bill)	Number of Coins or Bills, *n*, Needed to Make $100
Penny		
Nickel		
Dime		
Quarter		
Half-dollar		
Dollar		
Five-dollar bill		
Ten-dollar bill		
Twenty-dollar bill		
Fifty-dollar bill		
Hundred-dollar bill		

b. What patterns do you see in the table?

c. Graph the data in the table. Note any trends in the graph for different intervals.

d. Determine an equation relating the value of the coin/bill, *d*, and the number of coins/bills, *n*, needed to make $100.

2. a. Complete the table of values.

b. Describe how $y = \frac{k}{x}$ changes as *x* increases. Note that the change is different for different parts of the domain. Describe each separately.

c. Describe any patterns or symmetry you see in the table.

d. Without graphing the function, state what you know about the graph from your table and responses to items 2b and 2c.

x	$f(x) = \frac{k}{x}$	$h(x) = x \cdot f(x)$
−3		
−2		
−1		
0		
1		
2		
3		
4		

e. Using your responses to item 2d, sketch a predicted shape of the graph.

(*continued*)

TEACHING AND LEARNING HIGH SCHOOL MATHEMATICS © 2010 John Wiley & Sons, Inc.

3. **a.** Graph the parent function $f(x) = \frac{1}{x}$.

   b. What are the coordinates of the y-intercept? Explain.

   c. What are the coordinates of the x-intercepts? How many x-intercepts are possible for this family of functions? How do you know? Why is this reasonable?

   d. Describe the shape of the graph.

   e. Refer to the table in item 2a. Assume the value of k is 1. Compare the values in the table with the graph in item 3a. Why does the graph look as it does? Describe what happens to $f(x)$ as x changes for various x intervals.

   f. What new information is provided by the graph that the table does not show?

   g. What information is provided by the table that the graph does not show?

4. The rational function $f(x) = \frac{1}{x}$ exhibits asymptotic behavior, that is, as x grows large, $f(x)$ approaches a constant value. Also as x approaches 0, $f(x)$ grows without bound.

   a. Complete the tables. For each, describe what is happening to the values of $f(x)$ for successive values of x in the table.

x	$f(x) = \frac{1}{x}$
1	
0.1	
0.01	
0.001	
0.0001	
0.00001	
0.000001	

x	$f(x) = \frac{1}{x}$
1	
10	
100	
1000	
10 000	
100 000	
1 000 000	

x	$f(x) = \frac{1}{x}$
−1	
−0.1	
−0.01	
−0.001	
−0.0001	
−0.00001	
−0.000001	

x	$f(x) = \frac{1}{x}$
−1	
−10	
−100	
−1000	
−10 000	
−100 000	
−1 000 000	

   b. Describe why this behavior is sensible, based on the equation.

   c. Compare the table with the graph of $f(x) = \frac{1}{x}$. Describe the relationship between these two representations for this function.

(*continued*)

TEACHING AND LEARNING HIGH SCHOOL MATHEMATICS © 2010 John Wiley & Sons, Inc.

d. Consider the equation you determined in item 1d. Does similar asymptotic behavior occur in this example? Explain.

e. Consider the function $f(x) = \frac{1}{x-3} + 2$. Where do you expect asymptotes to appear? Why?

f. Graph the function in item 4e. Was your prediction in item 4e correct? Explain.

g. If needed, complete a table similar to that in item 4a to see the asymptotic behavior of the function, $f(x) = \frac{1}{x-3} + 2$.

2. *Production Costs.* The average cost of a product A is the total production cost C divided by the number of objects produced x:

$$A(x) = \frac{C(x)}{x}$$

The equation giving the average cost of a banner produced by Flagg Siblings, Inc. is given by

$$A(x) = \frac{C(x)}{x} = \frac{8x + 15000}{x}$$

a. Graph $y = A(x)$ on a viewing rectangle that shows the function for x in [0, 100].

b. Is the equation sensible in the case where no banners are produced? Explain.

c. Why does the graph of A appear as it does for x close to zero? Explain in terms of the meanings of $C(x)$ and x.

d. Why does the graph of A appear as it does for large values of x? Explain in terms of the meanings of $C(x)$ and x.

3. *Rational Functions.* A *rational function* is a function $r(x) = \frac{p(x)}{q(x)}$ where p and q are polynomials and $q(x) \neq 0$.

a. Sketch the graphs of $f(x) = x - 1$ and $g(x) = x + 1$.

b. From these graphs, sketch the graph of the quotient, $h(x) = \frac{f(x)}{g(x)}$ by hand.

c. From the graphs of f and g, determine the x-intercepts and any asymptotes for $y = h(x)$.

i. What characteristics of the graphs of f and g indicate the x-intercepts of h?

ii. What characteristics of f and g indicate the vertical asymptotes?

iii. What characteristics of f and g indicate the horizontal asymptotes?

d. How can you tell from the graphs of f and g where the graph of h will lie above the x-axis?

e. How can you tell from the graphs of f and g how the graph of h will behave as x tends to infinity?

4. *More Rational Functions.*

a. Repeat **DMU** problems 3a–3e for $f(x) = x - 1$ and $g(x) = 1 - x$. Note any differences that occur. Explain these differences.

b. Repeat **DMU** problems 3a–3e for $f(x) = x - 1$ and $g(x) = (x + 1)(x - 2)$. Note any differences that occur. Explain these differences.

c. Determine an equation for a rational function that has vertical asymptotes at $x = -3$ and $x = 2$, and is positive for large x. Explain how you know your rational function satisfies each requirement.

d. How many rational functions satisfy the conditions listed in problem 4c? Explain.

5. *Revisiting the Function Card Sort.* Return to the **Function Card Sort** from the Unit Three Team Builder (and found in the Unit Three Appendix). Find the cards

TEACHING AND LEARNING HIGH SCHOOL MATHEMATICS © 2010 John Wiley & Sons, Inc.

that fit the function families you investigated in Lesson 3.6. State how you know these cards can be represented by the given function family.

a. Find the stories and real contexts that fit the function family.

b. Write another story or determine another real context that fits this family of functions.

c. Find two tables that fit the function family.

d. Find two graphs that fit the function family.

6. *Domain, Range, and Symmetry of Related Functions*. For polynomial functions f and g, when $h(x) = \frac{f(x)}{g(x)}$ with $g(x) \neq 0$, how are the domain, range, and symmetries of function h related to the domains, ranges, and symmetries of functions f and g?

7. *Inverse Functions*. The term *inverse function* is sometimes applied to functions such as $y = \frac{1}{x}$ and $y = \frac{1}{x^2}$. This language causes confusion for students as they learn more about functions.

a. Complete a Frayer Model (see the **Launch** of Lesson 2.2 and the template in the Unit Two Appendix) for each of the following pairs of function families.

i. Inverse functions such as $y = x^2$ and $y = \sqrt{x}$.

ii. Rational functions such as $y = \frac{p(x)}{q(x)}$ and $y = \frac{q(x)}{p(x)}$ where $p(x)$ and $q(x)$ are both polynomial functions and $p(x) \neq 0$, $q(x) \neq 0$.

iii. Reciprocal functions such as $y = x^2$ and $y = \frac{1}{x^2}$.

b. Compare and contrast each function category using the information compiled on the three Frayer Models.

c. How might the language *inverse function* confuse students?

d. Two functions f and g are *inverses* of each other if and only if $f(g(x)) = x$ and $g(f(x)) = x$ for appropriate values in the domain of each function. For which of i, ii, or iii in problem 7a, if any, is it appropriate to use the language *inverse function* to describe the relationship between the two functions?

e. In mathematics, the use of -1 as a power can also cause difficulties. Categorize the use of the power -1 in the following examples as indicating reciprocal, inverse, or some other meaning.

i. $y = x^{-1}$	**ii.** $y = (f(x))^{-1}$	**ii.** $y = f^{-1}(x)$
iv. $y = sin^{-1}(x)$	**v.** $y = (\sin(x))^{-1}$	**vi.** 3^{-1}

f. Which examples in question 7e are most accurately named inverse functions?

g. To avoid confusion, what language would you recommend for the rest of the examples that do not arise in the category you listed in problem 7f?

8. *Direct and Inverse Variation*. *Direct variation* often is defined as a function of the form $y = kx^n$ for $k \neq 0$ and integer $n > 0$. *Inverse variation* often is defined as a function of the form $y = \frac{k}{x^n}$ for $k \neq 0$, $x \neq 0$, and integer $n > 0$.

a. Relate direct and inverse variation functions when $n = 1$ to the families of functions explored in Unit Three.

b. Identify three direct and three inverse variation real situations likely to be of interest to high school students.

c. Compare the graphs of direct and inverse variation functions.

d. Revisit the definition of power functions from Lesson 3.5 **DMU** problem 19. How are direct and inverse variation functions related to power functions?

e. Some Algebra I textbooks limit the discussion of direct and inverse variation functions to the case where $n = 1$. What might be some rationales for limiting the scope of functions in early courses?

f. Some Algebra II textbooks extend the discussion of direct and inverse variation functions to allow n to be any positive integer.

 © 2010 John Wiley & Sons, Inc.

i. Why is $n = 0$ not included in direct or inverse variation? What function family arises?

ii. Is $y = kx^n$ for $k \neq 0$ and $n = 0$ a power function? Why or why not?

9. *Combining Functions.* Consider the graphs of $y = f(x)$ and $y = g(x)$ shown in the viewing rectangle for x in [–2, 11] and y in [–3, 5]. For each of problems 9a–9d, sketch the graph (one on each grid provided and label the graph with its equation). Explain how you know your results are correct.

a. $h(x) = f(x) + g(x)$ b. $h(x) = f(x) \cdot g(x)$

c. $h(x) = \frac{f(x)}{g(x)}$ d. $h(x) = \frac{g(x)}{f(x)}$

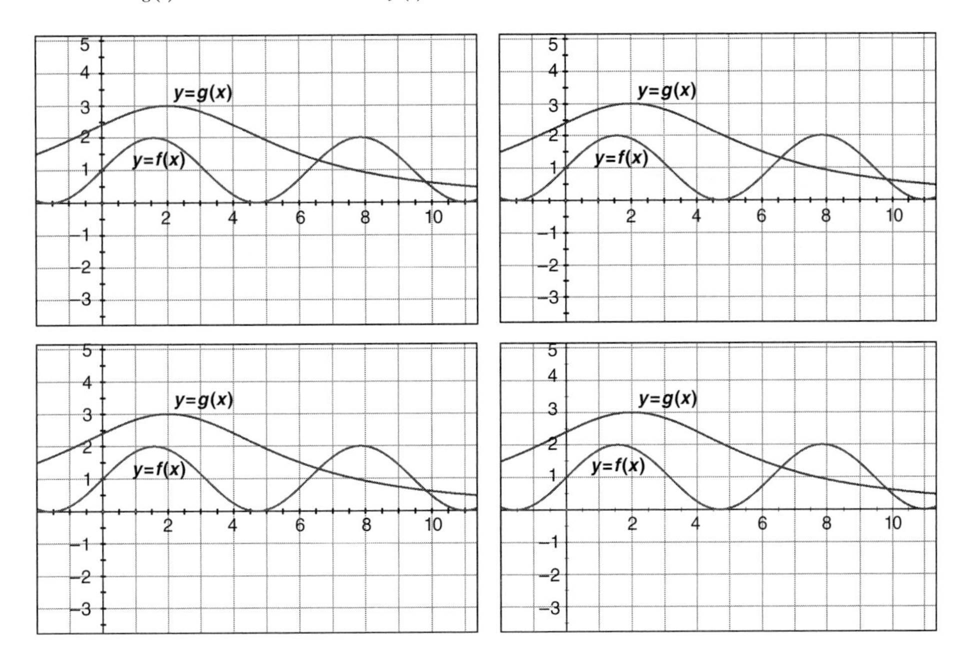

10. *Dividing Trigonometric Functions.* Complete the student page **Dividing Functions**.

Dividing Functions

Set your graphing calculator to dot mode and radian mode to complete the following questions.

1. Graph the function $f(x) = \frac{1}{x}$ on the viewing rectangle [–5, 5] by [–5, 5]. Describe and explain the behavior of $y = f(x)$ for x near 0 and as x grows large (use a larger viewing rectangle to explore this "end" behavior).
2. Graph the functions $f(x) = \sin(x)$ and $h(x) = \frac{1}{\sin(x)}$ on the viewing rectangle [–5, 5] by [–3, 3].
   - **a.** Recall that $\frac{1}{\sin(x)} = \csc(x)$. How are the graphs of $y = \sin(x)$ and $y = \csc(x)$ related?
   - **b.** Why do the asymptotes occur where they do? Explain.
   - **c.** At what points do these graphs intersect? Explain why they intersect at these points.
3. Graph the function $h(x) = \frac{\sin(x)}{x}$ on the viewing rectangle [–5, 5] by [–3, 3].
   - **a.** What is the value of $h(0)$? What does the value of $h(0)$ appear to be from the graph? Explain any differences.
   - **b.** Zoom in on the graph of $h(x) = \frac{\sin(x)}{x}$ a few times to determine the values of $y = h(x)$ for x near 0. Describe the appearance of h on this viewing rectangle. What could account for this behavior? Use the graphs of $f(x) = \sin(x)$ and $g(x) = x$ for x near 0 to explain why $h(x) = \frac{\sin(x)}{x}$ behaves as it does for x near 0.
4. **a.** Sketch the graphs of $f(x) = \sin(x)$ and $g(x) = \cos(x)$ on the same axes. From the graphs of f and g, predict the graph of the quotient $h(x) = \frac{\sin(x)}{\cos(x)}$.
   - **b.** For what values of x does $y = h(x)$ appear to have asymptotes? Explain why the graph of h behaves as it does for these values of x.
   - **c.** Note that $h(x) = \frac{\sin(x)}{\cos(x)} = \tan(x)$. From the graph of h, predict the graph of its reciprocal, $y = \cot(x)$. Explain how you know the graph of $y = \cot(x)$ looks as it does based on the graphs of $y = \sin(x)$, $y = \cos(x)$, and $y = \tan(x)$.
5. The graphs of functions f and g are provided in the figure at right for x in [–8, 10] and y in [–6, 10] with grid lines shown every unit.

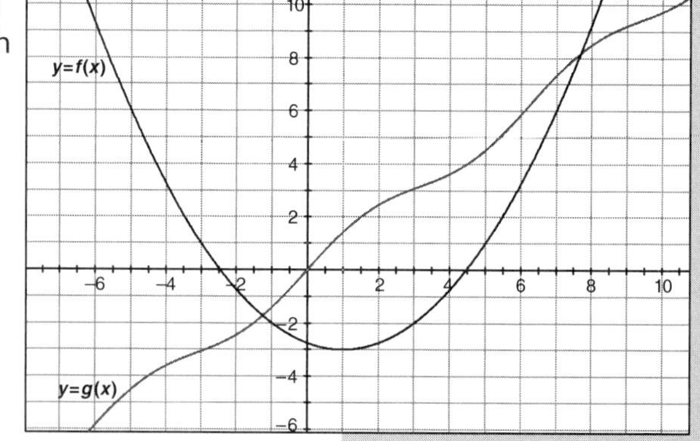

   - **a.** Sketch the graph of $h(x) = \frac{f(x)}{g(x)}$ on the same set of axes. Assume that f and g continue to grow as they appear as x gets larger in absolute value.
   - **b.** Describe the appearance of the graph of $y = h(x)$ for the values of x stated below. For each case, explain why h appears as it does in terms of f and g and the values of the quotients of these functions for x in the intervals given.
     - **i.** Near $x = 0$
     - **ii.** Near $x = -2.5$
     - **iii.** Near $x = -1.25$ (where $f(-1.25) = g(-1.25)$)
     - **iv.** Near $x = 4.5$
     - **v.** Near $x = 7.75$ (where $f(7.75) = g(7.75)$)
     - **vi.** As x approaches ∞
     - **vii.** As x approaches $-\infty$

Student page inspired by Beckmann and Sundstrom (1992) and Beckmann (1993).

TEACHING AND LEARNING HIGH SCHOOL MATHEMATICS © 2010 John Wiley & Sons, Inc.

11. *Operations on Functions*. Graph the following functions. By describing the behaviors of the parent functions $y = x$ and $y = \cos(x)$, explain why these graphs appear as they do on the intervals given.

a. $f(x) = \cos(x) + x$ for x in $[-10, 10]$ and y in $[-10, 10]$. Also graph $y = \cos(x)$ and $y = x$.

b. $f(x) = x \cdot \cos(x)$ for x in $[-6.3, 6.3]$ and y in $[-7, 7]$. Also graph $y = x$ and $y = -x$.

c. $f(x) = \sec(x)$ for x in $[-6.3, 6.3]$ and y in $[-10, 10]$. Also graph $y = \cos(x)$.

DEVELOPING MATHEMATICAL PEDAGOGY

Reflecting on Our Own Learning

12. Reflect on your work with the various functions and representations in this lesson.

a. What, if anything, was new to you?

b. What, if anything, was review but revisited from a new perspective?

c. What were the benefits of looking at operations on functions from multiple perspectives?

Using Internet Resources

13. Look up the Learning Styles Cognitive Preference Inventory on the website http://www.georgebrown.ca:80/saffairs/stusucc/learningstyles.aspx (downloaded July 19, 2009).

a. Take the inventory. From taking this inventory, what appear to be your dominant learning style(s)?

b. Read the information that follows the learning styles inventory. Comment on what parts of the information correctly describe you.

c. What useful information might an inventory of this type provide to help you plan for instruction?

d. Complete an Internet search to find another learning styles inventory. What additional learning styles do other inventories include?

Analyzing Instructional Materials

14. Revisit the student page **Investigating Rational Functions** from the **Explore** of this lesson.

a. What assumptions were made about students' prerequisite knowledge for the student page?

b. What *Roundtable* question might you ask of students to help them activate what they know of the prerequisites?

c. What appear to be the learning outcomes for this student page?

15. In this lesson, you developed conceptual understanding for combining functions using the operation of division. Describe how conceptual understanding was built.

16. Find at least three different Algebra II or precalculus textbooks with recent copyrights. Locate a section focusing on rational functions.

a. Analyze the development of rational functions. Consider questions like those bulleted here. Write a short summary about your findings, supporting your conclusions with specific examples from the texts; compare and contrast the presentations in the texts to those found in this lesson.

TEACHING AND LEARNING HIGH SCHOOL MATHEMATICS © 2010 John Wiley & Sons, Inc.

- How are multiple representations used to help students make sense of these types of functions?
- Did any of the textbooks present rational functions as quotients of polynomials? Did this presentation include looking at the graphs of the numerator and denominator and finding the quotient graphically?

b. In lessons related to rational functions, what learning styles are most accommodated by the textbooks you reviewed? Provide evidence.

c. How might the presentations of rational functions be enhanced? What are ways to accommodate the following learning styles?

i. reading/writing

ii. auditory

iii. visual

iv. kinesthetic

d. What topics related to rational functions are found in high school textbooks not addressed in Lesson 3.6? Which of the topics are necessary? How would you supplement Lesson 3.6 activities to include some of these topics?

17. Follow the directions for problem 16, replacing rational functions with trigonometric functions, to analyze the development of trigonometric functions in three high school textbooks.

WWW **18.** Revisit the student page **Dividing Functions**. The student page was written for students studying advanced functions in an Algebra II class or beyond. Students in Algebra I can also develop conceptual understanding for multiplying and dividing functions graphically and through the use of tables of values.

a. Discuss the reasonableness of using the function $y = \sin(x)$ with Algebra I students, assuming the function is a "black box" function for them. (That is, they just use the button on the calculator without understanding the nature of the function.)

b. Checking for understanding is accomplished by providing graphs without the accompanying equations.

i. Explain what understanding a successful student must have to solve a problem of this sort without using equations to first plot the results with a graphing calculator.

ii. Would you characterize a student's understanding as conceptual or procedural if they were able to complete problems 5a–b from the student page successfully? Explain.

iii. What evidence would you need to decide students' levels of understanding?

Teaching for Equity

19. Revisit one of the student pages from Lesson 3.6 or an earlier lesson in Unit Three.

a. Identify some ways to modify the student page to support additional learning styles (visual, auditory, reading/writing, or kinesthetic).

b. What are some general strategies for addressing each of the four learning styles in mathematics instruction?

Preparing for Instruction

20. Choose one of the student pages from Lesson 3.6 or from the **DMU/DMP** questions accompanying the lesson.

TEACHING AND LEARNING HIGH SCHOOL MATHEMATICS © 2010 John Wiley & Sons, Inc.

a. How are each of the learning styles (visual, auditory, reading/writing, or kinesthetic) supported in the student page?

b. For any learning styles not adequately supported, how might you modify the student page to support additional learning styles?

c. Determine the learning outcomes for the student page.

d. Complete a **QRS Guide** for the student page and recommended modifications. Anticipate student responses at three levels for each major problem.

e. Write teacher support questions for each student response and add these to the **QRS Guide**.

f. Decide on a cooperative learning (CL) strategy that best fits use of the student page with a class of students. Why do you think this CL strategy is appropriate? Indicate the CL strategy and the learning outcomes at the top of the **QRS Guide**.

g. Try the modified activity with one or more high school students. Record any additional student responses you did not anticipate and related teacher support in the **QRS Guide**. What more did you learn by using the modified student page and **QRS Guide** with students?

Reflecting on Professional Readings

21. In "Understanding the Significance of Context: A Framework to Examine Equity and Reform in Secondary Mathematics" by Celia K. Rousseau and Angiline Powell (*The High School Journal*, 88 (April/May 2005): 19–31), the authors raise the following issues related to opportunities to learn that support equitable mathematics instruction:

- Teachers with limited class time are reluctant to engage in activities that are more hands-on and conceptual. Instead, when time is limited, they focus on procedures.
- If teachers have limited planning time, they are less able to plan more active lessons. Limited planning time is more likely in school environments serving a population with a higher percentage of students from low-income families.
- The pressures of standardized tests may make teachers more reluctant to spend the time required for more active lessons, especially if standardized tests seem to focus more on computational skills.
- Although smaller class size can lead to better achievement for minority students, schools serving minority populations often are underfunded so that class sizes are larger than desired.
- Student mobility and high absenteeism often are prevalent in urban school environments.
- Financial conditions in urban schools may hinder the adoption of high-quality instructional materials.

a. Choose two of the indicated issues. Discuss how each of the issues might enable or hinder the use of instructional practices in the vein described in this text.

b. In what ways can these issues influence students' opportunities to learn important mathematics?

c. In what ways might teachers' expectations influence their instructional practices?

d. Locate the article. Read the vignettes of the two school districts. What additional instructional issues are raised from your reading?

22. Read "Queuing Theory: A Rational Approach to the Problem of Waiting in Line" by Thomas G. Edwards and Kenneth R. Chelst (*Mathematics Teacher*, 95 (May 2002): 372–376).

TEACHING AND LEARNING HIGH SCHOOL MATHEMATICS © 2010 John Wiley & Sons, Inc.

a. How might the context described in the article help students make sense of the domain for a rational function?

b. Find at least one other real-life context using rational functions likely to be of interest to high school students. What similarities and differences exist between your context and the context in the article?

Preparing to Discuss Observations

23. In preparation for Lesson 3.7, review your notes from your observations in high school mathematics classes. Also, review the articles you read in preparation for the observations. Be prepared to discuss your findings with your peers.

a. How many different concepts were introduced in the lessons? How were concepts related?

b. If you had an opportunity to review the textbooks used for the class, how were the concepts developed in the text? How did the textbook draw connections between concepts from lesson to lesson?

c. What opportunities did students have to use technology to explore mathematics? How might technology have been used to facilitate understanding of the concepts under study?

d. How did the curriculum in the classes you observed reflect the issues raised in the Curriculum and Technology Principles from *PSSM*?

e. Consider the reading associated with the observations: "Standards for High School Mathematics: Why, What, How?" by Eric W. Hart and W. Gary Martin (*Mathematics Teacher*, 102 (December 2008/January 2009): 377–382). What were the authors' perspectives about standards?

Preparing to Discuss Listening to Students

24. In preparation for Lesson 3.7, review your notes from your interviewing of students.

a. Consider your reading of the article "Understanding Connections between Equations and Graphs" by Eric J. Knuth (*Mathematics Teacher*, 93 (January 2000): 48–53). What were some of the difficulties he listed that students make when connecting equations and their graphs?

b. Summarize the responses of your high school students to the student pages **Climate Change** and **Adding and Subtracting Functions**. Update the **QRS Guide** you created for this interview to include student responses and teacher support you did not anticipate. Be ready to discuss:

i. What, if anything, surprised you by students' responses?

ii. What difficulties, if any, did students have that you did not anticipate?

iii. What similarities and differences did you notice between your students' responses to the function tasks and their responses to the tasks in the Knuth article?

3.7 Summarizing Classroom Observations and Listening to Students

At the beginning of this unit, you were asked to observe in at least three full-class periods, preferably in prealgebra, Algebra I, or Algebra II classes (or comparable integrated classes). During your observations, you were to focus on the curriculum—the number of different concepts introduced in one lesson, how the concepts were related within a lesson and from lesson to lesson, and how well the concepts were developed in the textbook. In addition, you were asked to consider what opportunities

TEACHING AND LEARNING HIGH SCHOOL MATHEMATICS © 2010 John Wiley & Sons, Inc.

students had to use technology to explore mathematics and the ways in which technology could be used to facilitate students' understanding.

In the introduction to the unit, we referred to the *knowledge-centered lens* (Donovan & Bransford, 2005) as one lens for determining the effectiveness of the classroom environment. A lens with this focus means teachers think carefully about their curriculum, how the topics build on each other and are connected, and what types of tasks are used to engage students in learning mathematics. A curriculum is not just a set of activities; it is a sequence of well-developed, coherent activities designed to engage students and support their development of significant concepts, processes, and applications.

In addition to curriculum content, teachers need to consider what tools are useful to help students learn the curriculum. With today's readily available technology, teachers need to consider how technology can be used to help students learn important algebraic concepts and how technology can help overcome computational and algebraic barriers many students have faced in the past. Taken together, these issues are important ingredients in implementing *PSSM's* Technology Principle.

You also had an opportunity to work with a pair of high school students as they solved problems on two student pages **Climate Change** and **Adding and Subtracting Functions.** What difficulties did students have as they reasoned about functions from a conceptual perspective?

In this lesson, you will:

- Compare the results of your observations or student interviews with those of your peers.
- Reflect on what your observations suggest about curriculum, its coherence in mathematics classrooms, and the extent to which it embodies the vision of *Principles and Standards for School Mathematics.*
- Reflect on the ways in which technology currently is used in high school mathematics classes to help students learn important algebraic concepts.

LAUNCH

1. What is your current thinking about what it means for a curriculum to be coherent?

EXPLORE RESPONSES TO LISTENING TO STUDENTS

2. Share students' responses to **Climate Change** and **Adding and Subtracting Functions.**
   a. Describe students' responses generally.
   b. What, if anything, surprised you by their responses?
   c. What difficulties, if any, did students have that you had not anticipated?
3. Critique the **QRS Guide** you prepared for the student pages with the guides created by your peers. Identify strengths and weaknesses in the types of teacher support you generated.
4. What similarities and differences did you notice between the responses of the students who completed the two student pages and students' responses to the tasks in the Knuth article?

EXPLORE RESPONSES TO OBSERVING CLASSROOMS

5. With the members of your group, share your overall summaries about the classes you observed. What surprised you? What concerned you? How did these observations help you think about your own future classrooms?

TEACHING AND LEARNING HIGH SCHOOL MATHEMATICS © 2010 John Wiley & Sons, Inc.

6. Consider the number of different concepts you and the members of your group observed in each lesson.
   a. On average, how many concepts were introduced each lesson during the classes you observed?
   b. Share your perceptions about how well the various concepts introduced in a lesson seemed to be related.
   c. How well developed were the concepts in the textbook? What connections did the textbook make between concepts from lesson to lesson?
7. Consider the sequence of lessons you observed. In responding to the following questions, you will need to consider the proximity of your observations to each other. Reflect on the ways in which your observations suggest a coherent curriculum.
   a. How did the lessons in close proximity connect to each other so that students explored concepts in greater depth over time? How did lessons on different days relate to each other?
   b. If what you observed of the curriculum needed greater coherence, what might you recommend to help students make connections across or within lessons?
8. a. What opportunities did students have to use technology to explore mathematics?
   b. How was technology used to help students understand the concepts studied?
   c. How did the lessons address the recommendations in the Technology Principle from *Principles and Standards for School Mathematics*?
   d. If opportunities for learning with technology were missed, identify a specific problem and suggest how a technological tool might have been useful when teaching the lesson.

SHARE AND SUMMARIZE

9. As a class, compare the responses from the different groups to the items in the two **Explore** sections of this lesson. Discuss the extent to which the Curriculum Principle and the Technology Principle were evident in the classes you observed.
10. Discuss the reading "Standards for High School Mathematics: Why, What, How?" by Eric W. Hart and W. Gary Martin (*Mathematics Teacher*, 102 (December 2008/January 2009): 377–382).
    a. Comment on the authors' perspective relative to mathematics standards for high school.
    b. What issues or challenges relative to implementing your district or state standards do you anticipate based on your classroom observations?
    c. What similarities and differences are evident in the standards for your state or district and the standards in *Principles and Standards for School Mathematics*?
11. Consider all of the classroom observations and interviews you have conducted thus far. What appear to be your greatest challenges in teaching high school mathematics?

DEVELOPING MATHEMATICAL PEDAGOGY

Reflecting on Our Own Learning

1. Reflect on your own mathematics classes, either at the high school or college level.

TEACHING AND LEARNING HIGH SCHOOL MATHEMATICS © 2010 John Wiley & Sons, Inc.

a. How was technology used to help you explore mathematics concepts?

b. What technology policies have been in place in your college mathematics classes? How have those policies helped or hindered your learning of mathematics?

2. Consider your own experiences learning about functions. In what ways were your experiences similar to or different from your experiences with functions in this unit?

Preparing for Instruction

3. Name one unifying concept in algebra that might be a basis for developing a more coherent curriculum in your classroom.

4. One issue teachers often face when using technology in the classroom is how to manage technology in ways that do not consume large amounts of classroom time. One strategy is to number calculators and assign students a number. Students pick up their numbered calculator from the storage area on entering the classroom. If anything is wrong with the calculator, it is the student's responsibility to notify the teacher immediately; otherwise, the student is held responsible. What are some other strategies that might be used to manage technology?

Focusing on Assessment

5. When students have continual access to technology such as graphing calculators, teachers need to reconsider the types of questions they ask students on exams and quizzes. It would not make sense, for instance, to ask students to graph $y = 3(x - 4)^2$ if they have a graphing calculator.

a. One strategy teachers use is to have an exam in two parts. On the first part, which contains basic graphing problems, students are not permitted to use a calculator. When that part is finished, students are given the second part of the test on which a calculator is permitted. Comment on this approach to dealing with assessment in a graphing calculator environment. What issues might arise using this approach? How might you address these?

b. A second strategy is to ask questions that focus on conceptual understanding. For instance, students might be given the graph at right but not the equation that generated it. Then the student might be asked, "Given the graph of $y = f(x)$ on the viewing window for x in [0, 12] and y in [0, 20], find x so that $12 = f(x)$." or "Given the graph of $y = f(x)$ on the viewing window for x in [0, 12] and y in [0, 20], determine the equation of the function. Explain your work." Another such example can be found in Lesson 3.6, **DMP** problem 10, item 5 on the student page **Dividing Functions**. What are the advantages of asking such questions? What does the teacher learn about the robustness of a student's understanding?

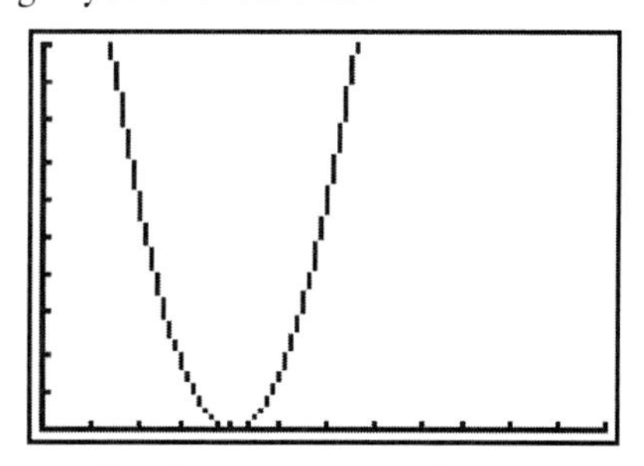

c. Create at least two other questions that focus on understanding of functions and for which having access to a graphing calculator does not make the item trivial.

Analyzing Instructional Materials

6. Find several textbooks at the Algebra I or Algebra II level with recent copyrights. If possible, try to obtain several books at the same level from different publishers.

TEACHING AND LEARNING HIGH SCHOOL MATHEMATICS © 2010 John Wiley & Sons, Inc.

a. Choose a unit in which you would expect that technology might be a useful learning tool. Compare and contrast the ways in which technology is integrated into the curriculum of that topic.

b. What additional instructional planning would be required for teachers to use technology as part of teaching from each textbook?

Debating Issues

7. Computer algebra systems (CAS) are readily available on many hand-held graphing calculators. CAS machines will perform symbolic algebra and return a symbolic result (e.g., expand($(x + 3)^2$) is returned as $x^2 + 6x + 9$). Some educators believe CAS should not be available to students prior to precalculus, at which time students should have well-developed algebraic manipulation skills. Others believe CAS should be available beginning in Algebra I to help students as they are first learning algebra symbolically.

   a. What are some of the issues you would expect to be raised on both sides of the debate?

   b. Choose a side of the debate and a specific lesson. Write a compelling argument supporting your point of view.

Teaching for Equity

8. Many school districts now expect students to have graphing calculators as a tool for use in the mathematics classroom on a regular basis, often beginning with Algebra I but definitely by Algebra II. One concern is how to ensure that students with limited finances are not disadvantaged by these expectations.

   a. Provide some suggestions for teachers and schools to make sure that financial ability does not become a barrier to technology access.

   b. Investigate what graphing resources are available free on the Internet.

   c. Check with one of your local schools. What access to the Internet is available to students beyond regularly scheduled class periods?

9. Review the high school textbooks you have examined in this course.

   a. Identify several features you found that support the creation of equitable classrooms.

   b. Identify a feature that does not promote equity. Suggest a modification that would better support equity.

TEACHING AND LEARNING HIGH SCHOOL MATHEMATICS © 2010 John Wiley & Sons, Inc.

Synthesizing Unit Three

Mathematics Strands: Algebra and Functions

This unit focused on elements involved in planning for equitable, coherent, well-developed, meaningful instruction. As indicated throughout the lessons of this unit, instructional activities can engage all high school students, through concrete manipulatives such as algebra tiles, through technology such as graphing calculators and computer algebra systems, through multiple representations in exploring functions, and through active approaches such as the use of experiments. These varied approaches allow multiple points of entry into mathematical concepts, an important feature for teachers to consider given the varying ability levels of students in most mathematics classrooms.

Too often, for many students algebra is a barrier to more advanced mathematics study. The abstract nature of algebra makes it challenging. Yet equity demands that all students have the opportunity to study algebra meaningfully. Hence, teachers need to expose students to rich tasks that will help them make sense of the mathematics. As the activities in this unit illustrate, active, student-centered modes of instruction in mathematics promote student interest, engagement, and understanding. Planning **QRS** segments with three levels of anticipated student responses and teacher support helps ensure these intentions are carried out in ways that support equitable classroom instruction.

The prevalence of powerful graphing calculator technology facilitates approaching functions from symbolic, tabular, and graphical forms as well as through real-world verbal contexts. Students need to develop the flexibility to approach a function from multiple representations and to choose the most appropriate approach in a particular situation.

Having arrived at the end of this unit, it is time to reflect on your understanding of the lessons learned throughout the unit. The following items provide an overview of the unit. You should be able to respond meaningfully to each item, providing support for your answers with specifics drawn from the lessons, discussions, readings, homework problems, observations in high school classes, or listening to high school students.

Website Resources

- Cooperative Learning Powerpoint
- **QRS Guide**
- **Where Do I Stand?** survey

MATHEMATICAL LEARNINGS

1. Given a situation represented by a real-world context, a table, or a graph, identify the appropriate function family. Determine rates of change for each function family. Identify the critical features of the graphs for each function family.
2. Discuss how different function families arise from combining other functions through the use of addition, subtraction, multiplication, or division. Create a Venn diagram showing relationships among various types of functions that arise from combining two or more functions. Create a concept map showing relationships among all the function families studied in this unit.
3. Describe how different parameters change a parent function. For instance, given the equation $y = A \cdot f\,(Bx + C) + D$, describe the influence of A, B, C, and D on the parent function $f(x) = x^2$.
4. Identify experiments that can be used to model a linear, quadratic, cubic, rational, square root, or exponential function. In conducting an experiment, find an appropriate equation that models the data.

TEACHING AND LEARNING HIGH SCHOOL MATHEMATICS © 2010 John Wiley & Sons, Inc.

PEDAGOGICAL LEARNINGS

5. Use contexts—including literature, science, popular culture, and other disciplines—to provide meaning for mathematics concepts.
6. Rescale problems to increase the level of cognitive demand or to increase accessibility.
7. **a.** Use algebra tiles to illustrate simplifying algebraic expressions and the distributive property.
   b. Use algebra tiles to illustrate multiplication of two binomials.
   c. Use algebra tiles to illustrate how to factor a trinomial.
   d. Discuss some of the advantages and limitations of using algebra tiles.
8. Discuss pros and cons of using various approaches, for example, concrete materials such as algebra tiles, in a high school mathematics classroom. How are such approaches related to equity?
9. **a.** Discuss some of the issues teachers must address when using technology, such as graphing calculators, in a high school mathematics classroom.
   b. How can graphing calculators facilitate the use of multiple representations when studying functions?
   c. What issues related to equity arise in relation to the use of technology?
10. Suppose a colleague questions you about the need to design instructional approaches that lead to active mathematics classrooms. Discuss some of the pros and cons of an active approach you would share with your colleague.
11. Describe a variety of teaching approaches to address students' different learning styles.
12. Identify several cooperative learning strategies that might be appropriate for use in a high school mathematics classroom. Discuss some of the pros and cons of using cooperative learning in mathematics. How is equity impacted by the use of cooperative learning?
13. Complete **QRS Guides** to support the teaching and learning of algebraic topics that include three levels of student responses and appropriate teacher support. Identify prerequisites and learning outcomes. Include cooperative learning strategies that support the **QRS** segment.
14. Revisit your responses to the survey **Where Do I Stand?** for items 2, 5, 6, 7, 9, 10, 12, 13, and 21. How have your responses changed, deepened, or grown?
15. Revisit your responses to item 24 on the survey **Where Do I Stand?** Are there any additional challenges you expect to face as a high school mathematics teacher?

TEACHING AND LEARNING HIGH SCHOOL MATHEMATICS © 2010 John Wiley & Sons, Inc.

Unit Three Investigation: Families of Functions

In the investigation for Unit Three, you will investigate a function family by finding examples from several different sources and determining why each example fits the function family.

Following are questions to guide your investigation. Choose some that interest you and take them as far as you can. Explain your solutions in coherent sentences. Include diagrams, as appropriate, to make your arguments clear. Provide complete citations for all resources used to complete this investigation. Attach copies of data sets and articles from their original sources.

You should obtain a variety of examples for your function family. Do not use the same example to answer more than one of the following questions. Find examples different from those presented throughout this unit.

Website Resources

- **Function Card Sort** student page
- Function Card Sort Answer Template
- **QRS Guide**

1. Sort the **Function Card Sort** from the Unit Three Team Builder into a large grid by function type and representation. (A set of cards and an answer template can be found in the Unit Three Appendix.)
2. Based on your sorting of the function cards,
   - **a.** Which representations are easiest to categorize? Why do you think so?
   - **b.** Which representations are most difficult to categorize? Why do you think so?

For the remainder of the items, provide responses based on your team's **function family**.

3. For only your function family, explain how you can tell that a function belongs to your function family when it is represented:
   - **a.** As a graph.
   - **b.** As a table.
   - **c.** As a children's story or real-life context.
4. Create a set of cards for your function family that is different from the cards on the **Function Card Sort** student page. State how you know each example fits your function family. Include one card from each of the following representations: equation, table, graph, children's story, real-life example.
5. Create a set of cards that do not fit your function family but are likely to be categorized by high school students as fitting your function family. Include one card from each of the following representations: equation, table, graph, children's story, real-life example. Provide a rationale that explains why you think each card might be categorized incorrectly by high school students. The cards do not have to contain functions but can instead contain relations that are not functions.
6. Choose a high school of interest to you.
   - **a.** Identify the high school as urban, rural, or suburban. Find out as much as you can about the school from the school's website. Investigate the interests of the student body by survey, through the school's website, or in some other way. Describe students' interests and how you learned about them.
   - **b.** Find three examples of functions that fit your assigned function family and that are likely to be of interest to students in the chosen high school. These examples must be set in a real-life context; arise from a newspaper, journal article, or data

TEACHING AND LEARNING HIGH SCHOOL MATHEMATICS © 2010 John Wiley & Sons, Inc.

set in an almanac or reputable Web resource; or arise from children's or other literature. Do not submit data found in mathematics textbooks. Attach copies of articles and data sets.

c. For each of the examples in item 6b, write two questions you would expect high school students to answer as they investigate each data set, article, etc. These questions must be designed to help students understand the function that represents the data set.

7. Find, adapt, or create an inquiry-based activity that uses technology or physical materials to help high school students develop conceptual understanding for your function family. Use the NCTM algebra, problem solving, connections, communication, representation, and reasoning and proof standards to help you determine objectives for your activity. State the objectives and include a descriptive paragraph of what you expect students to learn about your function family through this activity. Cite all resources. Include a copy of any handouts for students as well as a **QRS Guide** giving misconceived, partial, and satisfactory student responses for each problem in the activity. Also provide appropriate teacher support.
8. Prepare an organized plan to introduce each of your function family examples from this investigation (beginning with item 4) to your classmates. The plan should indicate in what order you would expect high school students to study each example to help them progress from a naïve understanding of the function family to a deeper understanding. Use the reading "Teaching and Learning Functions" from Lesson 3.1 to help you sequence the examples into a coherent order.
9. Reflect in writing on what you have learned about your function family through the work you completed in this investigation and throughout Unit Three.
10. Reflect in writing on what you learned about how students learn about functions. Consider the work you completed as part of this investigation and throughout the unit from the readings and from working with the activities in the lessons.

TEACHING AND LEARNING HIGH SCHOOL MATHEMATICS © 2010 John Wiley & Sons, Inc.

UNIT *Four*

LESSON PLANNING

Mathematics Strands: Data Analysis and Probability

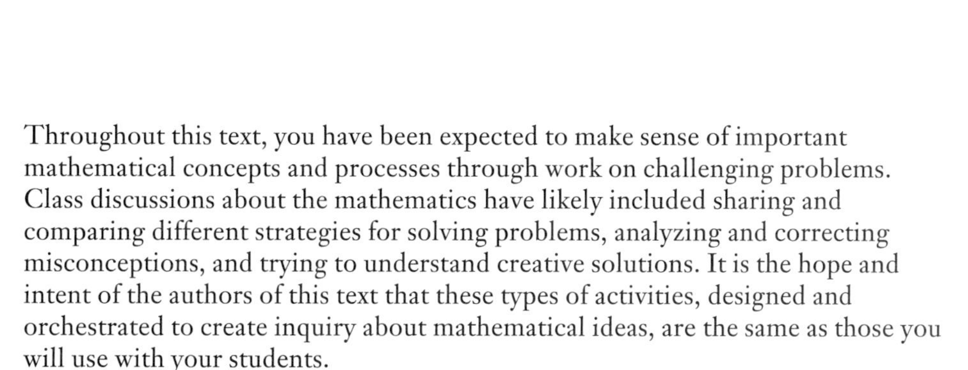

Throughout this text, you have been expected to make sense of important mathematical concepts and processes through work on challenging problems. Class discussions about the mathematics have likely included sharing and comparing different strategies for solving problems, analyzing and correcting misconceptions, and trying to understand creative solutions. It is the hope and intent of the authors of this text that these types of activities, designed and orchestrated to create inquiry about mathematical ideas, are the same as those you will use with your students.

In Unit Two, four lenses (Donovan and Bransford, 2005) were identified that can help determine the effectiveness of classroom environments. In previous units, you explored issues related to mathematics teaching and learning through some of these lenses, specifically developing strategies for creating an environment of shared learning (the community-centered lens in Unit One), understanding how students think about and learn mathematical ideas (the learning-centered lens in Unit Two), and studying elements prerequisite to planning a lesson (the knowledge-centered lens in Unit Three). In each of these units, you considered mathematical tasks, anticipated student solutions to those tasks (misconceived, partial, and satisfactory), and generated appropriate teacher support questions to provide "just enough" help to move students forward in their thinking on the task. All of this work was recorded in **QRS Guides**, an important element in planning effective classroom instruction. In Unit Four, you will continue to focus on the knowledge-centered lens as you integrate **QRS Guides** with planning elements from Units One, Two, and Three to design complete lesson plans for classroom instruction.

Planning instruction that is effective in leading to conceptual understanding of mathematics is challenging. Successful planning requires careful integration of strategies and techniques from the community-centered lens, the learning-centered lens, and the knowledge-centered lens as well as the assessment-centered lens to be studied in Unit Five. In particular, you will consider how teachers might launch a lesson, support student exploration, and orchestrate a summarizing discussion that helps students clarify important mathematics. Even when inquiry-based lessons are the norm, you will consider how teachers might blend direct instruction with inquiry when exploration is not feasible or practical. You will also consider how teachers might plan multi-day segments of instruction.

Teaching mathematics means supporting students as they learn to think more and more clearly about mathematical ideas, processes, applications, and relationships. The goal is to help students modify their thinking. But there is no direct route, so teaching operates indirectly. Students need to be engaged in solving problems that are well-sequenced and designed to introduce significant mathematics with increasingly mature ways of thinking. Then students need to share these ways of thinking publicly, both orally and in writing, through words, diagrams, tables, graphs, and symbols. Teachers need to use questions and subtasks to make evident the most important ideas and relationships. They also need to provide opportunities for students to record their understandings and to continue to use what they are learning. Teaching in this way is often referred to as *inquiry-based teaching*.

Inquiry-based teaching is far different from *lecture followed by guided practice*. Methods using lecture with guided practice work well only for some students and only for a limited number of topics. Being truly dedicated to helping all students learn significant mathematics requires more. Teachers need to know the mathematics at multiple levels. They also need to anticipate the many ways students make sense of the task at hand and to consider ways to provide needed support. You have done this part of planning when you completed **QRS Guides** before working with students on a task and then added to the **QRS Guides** after you observed students engage with the task. Teachers need to build environments where students learn to be open to challenges and committed to engaging with and supporting one another as they form a learning community.

The mathematical focus of this unit is data analysis and probability. According to the Data Analysis and Probability Standard of *Principles and Standards for School Mathematics*, students should have the opportunity to:

- pose questions that can be investigated by collecting, analyzing, and displaying data;
- analyze data through statistical methods appropriate to the data;
- use the data to make and evaluate predictions; and
- explore foundational concepts of probability.

Specifically, students in grades 9–12 should be able to understand different types of data displays and how to use them, compute basic statistics for a set of data, determine appropriate inferences that can be made from data, model a set of bivariate data with functions when possible, understand how sampling can be used to provide information about a population, and be able to compute probabilities of events including independent and dependent events. *The Guide to Assessment and Instruction in Statistics Education (GAISE) Report* recommends that students understand statistics as the study of variability (Franklin et al., 2007).

Data analysis and probability impact our lives in important ways. It is impossible to open a newspaper without being faced with statistics, whether on the stock market page, the sports page, or the daily news (e.g., rise and fall of oil or housing prices, unemployment rates, interest rates). Probability underlies decisions to play state lotteries, to understand the results of a medical test (false positive or false negative), or when making decisions with different levels of risk. Further, probability provides the models that support decision-making from data.

Basic statistics has a long history. For instance, the ancient Hebrews took a census to determine the numbers of fighting men in each of the 12 tribes, and the Roman Empire conducted censuses throughout their vast domain in the time of the Caesars. Since the founding of the United States, a population census has been conducted every 10 years and is mandated in the U.S. Constitution. The results of this census determine reapportionment of the U.S. House of Representatives every 10 years to ensure that state representation is based on each state's proportion of the country's population.

In the 19th century, statistics began to be used to study agriculture as well as the social sciences. In particular, Adrien Marie Legendre built on the work of

TEACHING AND LEARNING HIGH SCHOOL MATHEMATICS © 2010 John Wiley & Sons, Inc.

astronomers from the 18th century to develop his method of least squares to blend numerous data observations into a single estimate (Katz, 1998). Today's attempts to model a set of data with an appropriate function rely on minimizing the sum of the squares of the differences between the actual data values and the values provided by the function model.

In 1718, Abraham de Moivre first published *The Doctrine of Chances*, a probability text that contained rules and showed how those rules could be applied to various games of chance (Katz, 1998). In addition, through his work with sums of the binomial $(a + b)^n$, he later developed the normal distribution as an approximation to the binomial distribution. Mathematicians in the 19th century, specifically Adolphe Quetelet and Francis Galton, used the normal curve to study the many characteristics in people or plants that are normally distributed. Further work with the normal curve has shown how statistics and probability are connected in designing experiments and determining tests that shed light on whether results are due to chance or suggest a real difference in the effects of two different experimental procedures.

Today's high school students will need to have some basic understanding of data analysis and probability to be informed citizens of the 21st century. The statistical thinking they need to develop is different from algebraic or geometric reasoning; statistical thinking is inferential in contrast to algebraic and geometric reasoning, which are deductive. This unit provides an opportunity for you, as their teacher, to explore these concepts while simultaneously considering how to plan complete mathematics lessons that are effective for students. The observation for this unit, which focuses on teaching, is another avenue to investigate elements of tasks that support or hinder effective instruction.

Unit Four Team Builder: A Dream Team in Hockey

Preparing for the Team Builder

Everyone should receive a playing card from a modified standard card deck. The deck is comprised of one set of four of each kind for each team of four students, or a comparable deck fitting the number of students in the class. Shuffle the cards. Individuals select cards and find others with the same value (e.g., all the sixes form a team). Once teams have found each other, they should complete the activity described on the **Dream Team Mania** student page.

Materials

- deck of playing cards, one set of four of each kind for each team of four students
- graphing calculators or random number table
- spinners, dice, or coins

Website Resources

- **Dream Team Mania** student page

In the Unit Four Investigation at the end of this unit, teams build a dream team for a sport of their choice and continue to explore the computer game situation to obtain a full set of cards for a team of any size.

TEACHING AND LEARNING HIGH SCHOOL MATHEMATICS © 2010 John Wiley & Sons, Inc.

Dream Team Mania

Dream Team Mania is a new computer game. You get to design your own personal dream team for your favorite sport and pit yours against other players' dream teams. Like the baseball all-star teams, your dream team is built of professional players from any existing team (past or present!). The computer game is played as follows:

- Start by selecting a sport.
- Design your dream team, one player for each position that would be on the playing surface at one time.
- Enter your sport and players into the game file labeled Dream Team.
- Play against another person who has chosen the same sport.
- Take turns earning players. On each person's turn, the computer randomly generates a player from your personal Dream Team list. Each player on your defined list is equally likely to be generated on each of your turns, so the computer may generate a duplicate of a player you already have.

Continue taking turns earning players until you have your complete Dream Team. Whoever completes his or her Dream Team first is credited with winning the first game of the season. (Later Dream Teams play more season games in another aspect of the computer simulation.)

1. In hockey, six team members are on the playing surface at a time. Specify six of your favorite hockey players and tell why you chose them. Alternatively, create six fictional players using characteristics of your classmates to create and name the players. Name your hockey team in a way that has something to do with statistics or probability.
2. Be sure everyone in your group understands how Dream Team Mania works. Without calculating, estimate how many turns your group would need before you would have your entire Dream hockey team of six players. What reasons do you have for your estimate?
3. Design a way to simulate the number of turns to collect your entire Dream Team. Consider several possible tools: graphing calculators, random number table, spinners, dice, coins, playing cards, or computer simulation.
4. Use one of your plans from item 3 to simulate playing Dream Team Mania. How many turns did your group need to earn all the players for your dream hockey team?
5. **a.** Share with the rest of the class your group's estimate in item 2, along with its rationale.

   b. Share with the class your group's ideas for simulating the experiment of earning players.

   c. Compare your group's results in item 4 with those of other groups. How similar were your results?

Preparing to Observe Mathematics Classrooms: Focus on Teaching

An appropriate and coherent mathematics curriculum is an essential ingredient for effective mathematics instruction. But good curriculum is not enough. Teachers must also use that curriculum in ways recommended by the Teaching Principle from *Principles and Standards for School Mathematics* (NCTM, 2000). Effective teachers have knowledge about the mathematics students already know and what they still need to

TEACHING AND LEARNING HIGH SCHOOL MATHEMATICS © 2010 John Wiley & Sons, Inc.

learn and are able to design instruction that challenges and appropriately supports students.

In discussing this principle, the authors of *PSSM* note that teachers need several kinds of knowledge. Although knowing mathematics content is necessary, it is not sufficient. Teachers need a deep understanding of curriculum and the major ideas of the curriculum that are important for a given grade or level. Teachers also need *pedagogical content knowledge*, that is, knowledge about appropriate teaching strategies for particular content and how to sequence content for effective instruction as well as knowledge about understandings and misconceptions that students often bring to a learning situation. This pedagogical content knowledge is mathematics-specific and differs from general teaching knowledge (e.g., classroom management, general learning theories for adolescents) that all teachers need. Effective teachers integrate their content knowledge, curriculum knowledge, pedagogical content knowledge, and general pedagogical knowledge as they design instruction to support high levels of student learning.

In this observation you will focus on the following: How teachers use intellectually challenging tasks to engage students with mathematics, how these tasks enable students to approach mathematics from more than one perspective, how teachers manage student exploration with these tasks in ways that maintain the intellectual challenge, and how teachers help students understand and summarize the important mathematical ideas inherent in the tasks.

1. *Prepare for the Observation.* In preparation for the observations, read the "Teaching Principle" (pp. 16–19) in *Principles and Standards for School Mathematics.*
   a. Identify the major issues in this principle.
   b. Reflect on your observations from Units One, Two, and Three. What aspects of the Teaching Principle were evident in the instruction you observed? What aspects, if any, seemed to need more attention?
2. *Conduct the Observation: Focus on the Mathematical Tasks of the Lesson.* For the observation in this unit, try to observe at least three full-class periods in the same class. Begin by noting the mathematical intent of the lesson. In this observation, focus on and record the mathematical tasks in which students are expected to engage. After the observation, answer the following questions.
   a. Consider the tasks used during the lesson. Did they intrigue you? Did they appear to intrigue or challenge students? Provide evidence to support your comments.
   b. Describe how multiple perspectives were integrated into classroom instruction or how you could modify tasks so they could be approached from multiple perspectives. If the tasks used in the observations did not allow for multiple perspectives, design a new task that addresses the same mathematical concepts but can be approached in more than one way.
   c. Comment on how the teacher addressed students' emergent understandings of the mathematics in the tasks, including resolving misconceptions or responding to students' incorrect responses. How did the teacher help maintain the intellectual challenge for students if they struggled, or what happened to reduce the challenge? Provide evidence for your comments.
   d. How did the teacher help students understand and summarize the important mathematical ideas inherent in the tasks?
3. *Prepare for Reflection.* Middleton and Spanias (2002) synthesized research on motivation and identified five factors related to student motivation in mathematics:
   - Motivation is learned.
   - Motivation depends on how students see their own successes or failures.
   - Intrinsic motivation is better than extrinsic rewards.

TEACHING AND LEARNING HIGH SCHOOL MATHEMATICS © 2010 John Wiley & Sons, Inc.

- Inequities attributed to gender or ethnicity can often be explained by the different ways various groups are taught to view mathematics.
- Teachers are crucial in motivating students to succeed.

a. What aspects of teaching that you observed incorporated the motivational factors raised by Middleton and Spanias?

b. What aspects, if any, did you observe that were contrary to the factors raised by Middleton and Spanias?

4. *Summarize the Observations.* Start by providing a brief description of the level of the class that you observed as well as the mathematical intent of each of the lessons.

a. Summarize your comments about the readings and related questions from items 1 and 3.

b. Refer to the questions associated with the observation. Use your observations to summarize your responses to the questions. Justify any conclusions with specific instances from the readings or the class observations.

c. Consider your perspectives on teaching as described in the readings for this observation. What do you anticipate to be the major issues you will need to consider in designing your mathematics instruction?

d. What appear to be some difficulties in maintaining the mathematical challenge of a task as it is enacted in the classroom?

5. *Reflect on Teaching.* Consider all three of the lessons you observed for the observation. What aspects of the instruction you observed would you like to emulate in your own classroom? What aspects would you want to avoid? What did you observe that will help you maintain or raise, as appropriate, the level of mathematical challenge in your classroom? Justify your thinking.

Be prepared to discuss your observations and reflections in the last lesson of this unit.

Listening to Students Reason about Data Analysis and Probability

Data Analysis and Probability are relatively new additions to school curricula, gaining increased attention in the last two decades of the 20th century. Consequently, research on students' thinking in these areas is also relatively recent; research continues to be conducted and even more is needed. In this **Listening to Students Reason** activity, you will replicate some of this research informally using questions similar to those researchers have found useful in revealing significant student misconceptions about these content areas.

Website Resources

- Thinking about Probability student page
- **QRS Guide**

1. Refer to the student page Thinking about Probability (adapted from items in Shaughnessy, 2003).

a. Work through the problems. Record your solutions in a **QRS Guide**.

b. For at least three of the items anticipate student responses (misconceived, partial, or satisfactory). Record expected student responses in the **QRS Guide** for item 1a.

c. Add appropriate teacher support to the **QRS Guide**.

2. Ask several students to complete the Thinking about Probability student page. It is preferable to choose students with different learning histories regarding statistics

TEACHING AND LEARNING HIGH SCHOOL MATHEMATICS © 2010 John Wiley & Sons, Inc.

and probability; one or two of the students should have had little prior opportunity to learn these topics.

a. What, if anything, surprised you by their responses?

b. What difficulties, if any, did they have that you had not anticipated?

c. Speculate about the reasons why some misconceived responses are given.

d. What similarities and differences did you notice among your students' responses?

e. Update your **QRS Guide** to reflect these students' responses as well as appropriate teacher support you would use with future students who respond as they did.

3. Subsequent to the interview, read "Research on Students' Understandings of Probability" by J. Michael Shaughnessy (In *A Research Companion to Principles and Standards for School Mathematics*, edited by Jeremy Kilpatrick, W. Gary Martin, and Deborah Schifter (pp. 216–226). Reston, VA: National Council of Teachers of Mathematics, 2003). How did the misconceptions you experienced and found in your students' work compare with those found by researchers?

Be prepared to share your insights into how students reasoned about these probability tasks during the last lesson of this unit.

TEACHING AND LEARNING HIGH SCHOOL MATHEMATICS © 2010 John Wiley & Sons, Inc.

Thinking about Probability

1. A jar has 1,000 colored chips: 200 red, 300 blue, 400 yellow, and 100 green.
   a. Suppose you mix the chips and draw a handful of 100 chips. What would you expect to find?
   b. Suppose you withdraw 10 different samples of 100 chips each. Each time, you examine the sample, record the colors, return the 100 chips to the jar, and mix everything together before drawing the next sample. What would you expect the number of reds to be in each of these ten drawings?
   c. Explain your reasoning.
2. Suppose a coin is flipped eight times. Decide whether each sequence of heads and tails is feasible. Explain why or why not.
   a. HHHHTTTT
   b. HTHTTHHT
   c. HHHHTHHT
   d. HTHTHTHT
3. Here are two lists of 23 red (R) and black (B) cards from a deck with an equal number of red and black cards. In one case, the cards were actually shuffled, pulled and replaced; then the results were recorded. In the other case, a student made up what he thought was a list of random Rs and Bs. Which list do you think was created by the random shuffling? Explain your thinking.
   a. BBBRRBBRRBBBRRRRRRBBBRB
   b. RBRBRBRBBRBRBBRBRRBRRBR
4. One day, a data analysis firm kept track of all the people in a given city who traveled to work. Which do you think was more likely to occur?
   a. A person was in an accident with their vehicle.
   b. A person was in an accident and was under 25 years old.
5. Jim believes he has a 50% chance that both spinners will land on white if he spins them simultaneously. Do you agree or disagree with Jim? Explain.

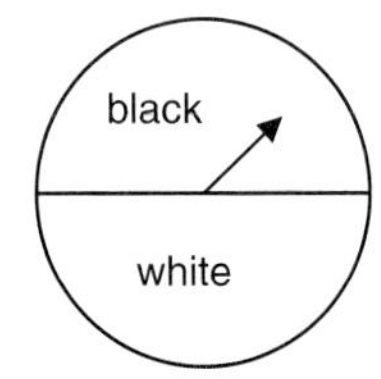

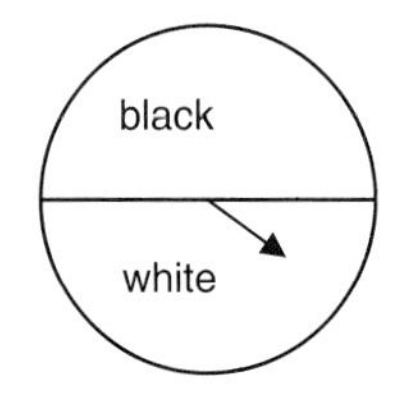

6. Two bags have green and yellow marbles as indicated.
   Bag A: 4 green and 1 yellow   Bag B: 12 green and 3 yellow
   Which bag gives the better chance of selecting a green marble when one marble is drawn from each bag? Explain your answer.
   a. Both give the same chance. **b.** Bag A **c.** Bag B **d.** I don't know.

The items on this student page are adapted from "Research on Students' Understandings of Probability" by J. Michael Shaughnessy in *A Research Companion to Principles and Standards for School Mathematics*, edited by Jeremy Kilpatrick, W. Gary Martin, and Deborah Schifter (pp. 216–226). Reston, VA: National Council of Teachers of Mathematics, 2003.

TEACHING AND LEARNING HIGH SCHOOL MATHEMATICS © 2010 John Wiley & Sons, Inc.

4.1 Planning a Lesson Launch and Explore: Data Analysis

Careful lesson planning is key to successful instruction. In previous portions of this text, you studied underpinnings of planning, such as identifying cognitively challenging tasks, considering how students might respond to problems or tasks, planning teacher support to help students move forward in their work, and rescaling tasks to increase their cognitive demand or to provide more support. In this unit, you will use these elements and synthesize them into complete lesson plans. Guidance for lesson planning builds on the *Thinking through a Lesson Protocol* (Smith, Bill, & Hughes, 2008; Smith, Hughes, Engle, & Stein, 2009). The protocol aids teachers as they think about the mathematical goals of a lesson and how to plan in ways that maintain the mathematical integrity of the task. There are three major parts to the protocol:

Part 1: Selecting and setting up a mathematical task

Part 2: Supporting students' exploration of the task

Part 3: Sharing and discussing the task

These three components likely remind you of the **Launch, Explore,** and **Share and Summarize** or **LES** format used as the instructional format in this text to foster inquiry learning. Throughout the first three units, you have had many opportunities to engage in lessons using this format and to reflect on the characteristics of each component.

Teaching an inquiry-based lesson is demanding because teachers need to think on their feet. Smith and her colleagues believe that by adopting certain planning and instructional practices, teachers can reduce the number of things they need to consider on the spot. Consequently, Smith and colleagues have identified five practices that contribute to orchestrating productive **Share and Summarize** discussions. These practices occur before and during a lesson, but all must be thought about in advance of the actual teaching.

- **Anticipating**—generating and recording an array of student strategies for the task. These anticipated responses should include common misconceptions. Sources for these responses are the teacher's own solutions, colleagues' solutions, research, or previously collected student work. In this text, expected student responses are recorded in the student responses (expected or observed) column in the **QRS Guide** introduced in Unit One.
- **Monitoring**—observing and listening carefully to students as they work and noting strategies, representations, and other ideas that are important to share in the whole group discussion. Observed but unanticipated responses of interest are added to the **QRS Guide**. In the margin, the teacher may note which students or teams are generating different solution strategies.
- **Selecting**—deciding which specific student strategies to share. These should be ones that help students overcome misconceptions and develop sound mathematical reasoning and processes. Some of these are selected for sharing with the entire class.
- **Sequencing**—deciding in what order to have solutions presented. In the Teacher Support column on the **QRS Guide**, teachers should indicate which strategy will be presented first, second, etc.
- **Connecting**—planning which solutions students need to compare or other questions that will be raised with an eye to forwarding the mathematical agenda of the lesson. In the **Lesson Planning Guide**, these questions should be included in the **Share and Summarize** portion.

TEACHING AND LEARNING HIGH SCHOOL MATHEMATICS © 2010 John Wiley & Sons, Inc.

Smith and her colleagues make the point that productive **Share and Summarize** discussions happen when teachers center teaching on challenging tasks with significant mathematics and make deliberate, carefully reasoned decisions about what student work to share and what questions to pose.

The focus of this lesson is on the first two parts of the protocol. The class will work together to think about setting up the task (**Launch**) and supporting students' exploration of the task (**Explore**). A necessary first step is to solve the problem yourself in order to anticipate student responses. Then, studying student work on the problem enables you to visualize more clearly what may arise in your classroom. You will use this student work and analysis to plan support that provides "just enough" help to move students forward without undermining their opportunity to think for themselves. In Lesson 4.2, you will think in more detail about Part 3 of the protocol, specifically how to orchestrate the **Share and Summarize** portion of a lesson. In the five phases described by Smith and her colleagues, anticipating comes during initial planning of the lesson; monitoring, selecting, and sequencing occur as students **Explore** the mathematical task. Connecting occurs during the **Share and Summarize** and depends on monitoring students' work and anticipating a sequence of solutions to be shared.

Lesson 4.1 introduces a **Lesson Planning Guide** that synthesizes the elements of teaching, many of which have been developed earlier in the text and some of which are introduced in this unit. The **Lesson Planning Guide** includes an expanded version of the **QRS Guide**.

The content strand of this lesson is data analysis, which is a broad topic of study. Data analysis includes looking for patterns in data, finding productive ways to display data, noting the shape of the data, looking for trends, identifying outliers, considering centers (e.g., mean, median, and mode) and spreads (e.g., range, variance, and standard deviation), and more. In statistical reasoning, students may first need to find ways to interpret a "messy" problem so they can analyze it. Then they may need to determine how to collect data, how to organize and present it, how to find significant information within it, and how to communicate their findings to others. The conclusions students draw from data are probable inferences, not provable deductions. The task in the lesson focuses on variability. Other data analysis concepts are addressed in **DMU** questions.

In this lesson, you will:

- Think about variation and its role in statistical reasoning through a specific task.
- Identify the learning outcomes related to a data analysis task.
- Consider ways to **Launch** a task.
- Consider multiple ways learners might approach a task.
- Integrate your own work and that of students to create a **QRS Guide** with expected student responses and teacher support that would occur during the **Explore** portion of the lesson.
- Begin to use a **Lesson Planning Guide** to help you think carefully about elements needed for effective lesson planning.

Materials (to be provided as needed)

- **Predicting Old Faithful Sample Student Work** (in the Unit Four Appendix)

Website Resources

- *Predicting Old Faithful* student page
- **QRS Guide**
- **Annotated Lesson Planning Guide**
- **Lesson Planning Guide**
- **Sample Lesson Plan for Predicting Old Faithful**

TEACHING AND LEARNING HIGH SCHOOL MATHEMATICS © 2010 John Wiley & Sons, Inc.

LAUNCH

Reflect on the various lessons in which you have engaged throughout this text.

1. What is the purpose of the **Launch**? What cautions do teachers need to consider during this phase?
2. What is the purpose of the **Explore**? What teacher moves seem to be critical during this phase? What cautions should teachers consider during this phase?

EXPLORE

3. Complete the student page **Predicting Old Faithful** and answer the following.
   a. Identify the learning outcomes for the task.
   b. Listen for the variety of ways of thinking and representations made about the problem that arise among your team members. Try to predict how high school students might approach the problem. Record expected student responses (misconceived, partial, and satisfactory) in a **QRS Guide**.

Predicting Old Faithful

Yellowstone National Park in Wyoming and Montana is part of the Yellowstone Volcano. It encompasses several natural thermal features, including geyser basins, thermal mud pots, hot springs, and acid lakes. One of its long-time major attractions is Old Faithful, a geyser that erupts multiple times each day. Imagine you have arrived to see Old Faithful but discover it has just finished erupting.

1. Discuss the following questions with your partners.
   a. How long would you expect to wait for the next eruption?
   b. What data would you want to have?
   c. How much data do you think you would need?
   d. How would you use the data?

Several sets of data are provided at the end of this page. Your instructor will assign each team a data set for a different day. The data for a given day show the numbers of minutes from the end of one eruption to the beginning of the next in the order that the eruptions occurred.

2. Looking at the data, estimate how long someone would expect to wait until the next eruption.
3. Use at least two different statistical tools, including at least one graph, to help you get a better sense of the wait time between eruptions.
4. How does your original estimate from question 1 compare with your analytical findings in question 3?
5. Outline a report in which you predict the next eruption based on the data you were provided.
6. Analyze at least two additional data sets and refine your report.
7. Share reports with the full class.
8. Compare and contrast different statistical tools or graphs that teams used. What is gained or lost by using different approaches? How are different graphs or tools related to one another?
9. What have you learned about making predictions? About variation? What more might you investigate related to this problem?

(continued)

TEACHING AND LEARNING HIGH SCHOOL MATHEMATICS © 2010 John Wiley & Sons, Inc.

Data Sets:

Day	Minutes Between Eruptions																	
1	86	71	57	80	75	77	60	86	77	56	81	50	89	54	90	73	60	83
2	65	82	84	54	85	58	79	57	88	68	76	78	74	85	75	65	76	58
3	91	50	87	48	93	54	86	53	78	52	83	60	87	49	80	60	92	43
4	89	60	84	69	74	71	108	50	77	57	80	61	82	48	81	73	62	79
5	54	80	73	81	62	81	71	79	81	74	59	81	66	87	53	80	50	87
6	51	82	58	81	49	92	50	88	62	93	56	89	51	79	58	82	52	88
7	52	78	69	75	77	53	80	55	87	53	85	61	93	54	76	80	81	59
8	86	78	71	77	76	94	75	50	83	82	72	77	75	65	79	72	78	77
9	79	75	78	64	80	49	88	54	85	51	96	50	80	78	81	72	75	78
10	87	69	55	83	49	82	57	84	57	84	73	78	57	79	57	90	62	87
11	78	52	98	48	78	79	65	84	50	83	60	80	50	88	50	84	74	76
12	65	89	49	88	51	78	85	65	75	77	69	92	68	87	61	81	55	93
13	53	84	70	73	93	50	87	77	74	72	82	74	80	49	91	53	86	49
14	79	89	87	76	59	80	89	45	93	72	71	54	79	74	65	78	57	87
15	72	84	47	84	57	87	68	86	75	73	53	82	93	77	54	96	48	89
16	63	84	76	62	83	50	85	78	78	81	78	76	74	81	66	84	48	93

This student page is adapted from J. Michael Shaughnessy and Maxine Pfannkuch. "How Faithful is Old Faithful? Statistical Thinking: A Story of Variation and Prediction." *Mathematics Teacher,* 95 (April 2002): 252–259. Table of data reprinted with permission from the *Mathematics Teacher* © 2002 by the National Council of Teachers of Mathematics.

SHARE

4. Share the strategies and representations that different teams used for **Predicting Old Faithful.**
5. What are the learning outcomes for this task?
6. Read the **Annotated Lesson Planning Guide** in the Unit Four Appendix. Note that items in italics are things to think about, not all of which apply to every lesson. Find the **Launch** portion on the guide. Discuss how you might **Launch** a lesson that includes the **Predicting Old Faithful** student page in a brief but engaging way.

EXPLORE MORE

7. Study the samples of student work for **Predicting Old Faithful** in the Unit Four Appendix.
   - **a.** What kinds of statistical thinking are different students using?
   - **b.** Which representations are most productive for addressing this task?

TEACHING AND LEARNING HIGH SCHOOL MATHEMATICS © 2010 John Wiley & Sons, Inc.

8. Study the **Explore** portion of the **Annotated Lesson Planning Guide.** Use your learning from completing the task, from identifying learning outcomes, and from studying the sample student work to complete the **Explore** portion of the **Lesson Planning Guide** found in the Unit Four Appendix. Use the **QRS Guide** you began in **Explore** item 3 as part of the **Explore** portion of the **Lesson Planning Guide**.

SHARE AND SUMMARIZE

9. Share with the class your team's responses to questions 7 and 8.
10. Study the **Annotated Lesson Planning Guide**. Although this guide is far more detailed than teachers use regularly, practicing with it for several lessons helps teachers think more carefully about all the lessons they teach.
    - **a.** How does the guide help you think carefully about what needs to be done to prepare for and implement a lesson?
    - **b.** The **Annotated Lesson Planning Guide** contains a section on continuity. What might have been some of the topics in lessons previous to this one? What lessons might follow this one?
    - **c.** Study the **Sample Lesson Plan for Predicting Old Faithful** in the Unit Four Appendix. How does the completed sample lesson plan help you understand the elements in the **Annotated Lesson Planning Guide**? How might you modify the sample lesson plan in light of your experiences with the task?
    - **d.** The lesson plan indicates notes the teacher made as students were exploring the mathematics. In particular, it indicates the order in which the teacher would plan to share responses. Comment on the sequence of solutions suggested by the teacher.

DEEPENING MATHEMATICAL UNDERSTANDING

1. *Selecting Graphs for Different Purposes*. **Predicting Old Faithful** likely brought forth a variety of types of graphs. Choose two types of graphs from the list below and compare their advantages and disadvantages for the **Predicting Old Faithful** data sets.

a. bar graph	**b.** box plot	**c.** circle graph
d. frequency plot	**e.** histogram	**f.** pictograph
g. scatterplot	**h.** stem-and-leaf plot	**i.** trend (or line) graph

2. *Mathematical vs. Statistical Reasoning*. Some students determine a measure of center (e.g., mean, median, or mode) and some measure of spread (e.g., range, interquartile range, or standard deviation) to analyze the data in **Predicting Old Faithful**. Others graph the eruption times sequentially with x being the eruption number and y the wait time. What are merits of each process? Which was most appropriate for **Predicting Old Faithful**? Why do you think so?
3. *Averages*. One way to increase the level of challenge of a problem is to reverse what is given and what is to be found, as recommended in Mokros and Russell (1995). When students do problems like the following, they reveal whether concepts are well understood. Solve and identify the thinking required.

   The average cost of seven meals at a restaurant was $15.99.
    - **a.** If only one meal cost exactly $15.99, what might be the prices of the other six meals?
    - **b.** Suppose five of the seven meals average $18.49 in cost. If the other two meals have the same cost as each other, what must that cost be? Solve using two different methods.

TEACHING AND LEARNING HIGH SCHOOL MATHEMATICS © 2010 John Wiley & Sons, Inc.

4. *Variance vs. Centers*. Consider the following quotation: "Although girls consistently earn higher high school grades, their SAT scores continue to lag behind boys', with the gap reaching 36 points in math." (Woo, Elaine and Doug Smith, "California and the West: SAT scores rise, but trouble spots remain." *Los Angeles Times*, August 27, 1997, p. A3.) (SAT scores for math can range from 200 to 800 points.)
   a. Does this statement mean that every boy does better on the SAT than every girl?
   b. Assume the statement is accurate. Could there be thousands of girls who score better than thousands of boys?
   c. Sketch possible graphs, one for boys' SAT math scores and one for girls' SAT math scores.
   d. True or false? Looking only at means and not at standard deviations can mask important information. Explain.
5. *Mean vs. Median*. Imagine a game where you roll a die. If you roll a five, you win \$1; otherwise you get nothing.
   a. Suppose you play 600 games. How much do you expect to win?
   b. In the 600 rolls, what would most likely be the median amount you would win?
   c. What would be a fair price to pay to play this game? Why?
   d. Statistics textbooks commonly state, "The median is more stable than the mean." Interpret this statement.
   e. Interpret the statement "Knowing the mean and number of data values allows you to reconstruct the grand total."
   f. Identify at least two different situations, one in which the mean has clear advantages and one in which the median has clear advantages. Explain why for each case.

DEVELOPING MATHEMATICAL PEDAGOGY

Preparing for Instruction

6. Refer to the **Annotated Lesson Planning Guide** in the Unit Four Appendix.
   a. What elements of the **Annotated Lesson Planning Guide** helped you recognize some of the "invisible" work of teaching? Explain.
   b. One element in the **Explore** section is *embedded assessment*. This refers to the teacher taking notes while students are engaged in an activity recording how they are responding to a task and what misconceptions or understandings they seem to have. Give an example where you can recall learning something about other students' understanding while an activity was in progress.
   c. *Tone-setting*, another element in the **Explore,** refers to setting the environment in the classroom so students are comfortable sharing partial or even misconceived work, know that struggling is part of their work, understand that ideas (not people) are to be critiqued, and more. Design a statement you might use with students to help them understand the value of sharing incomplete work or work with errors.

7. Find a student page from earlier in this text. Refer to the **Annotated Lesson Planning Guide** and design a **Launch** for that page. Explain why you designed it as you did.

For questions 8 and 9, refer to the **Sample Lesson Plan for Predicting Old Faithful** in the Unit Four Appendix.

8. Some future teachers were planning a **Launch** for the student page and decided to create a miniature volcano to introduce it. Comment on this idea.

TEACHING AND LEARNING HIGH SCHOOL MATHEMATICS © 2010 John Wiley & Sons, Inc.

9. What if you were teaching the lesson and no students thought about using a display that revealed the oscillating or alternating pattern in the wait times. What question(s) might you pose to move them forward without pre-empting their thinking?

10. Inquiry-based teaching is predicated on the use of challenging tasks that involve significant mathematics. Once a task is selected, the teacher must frame and maintain the task's cognitive demand. Students will often say, "Please, show us how to do a sample problem," or "Just give us the first step." Research from the Third International Mathematics and Science Study (TIMSS) video study revealed that U.S. teachers often posed challenging problems but nearly always showed students how to do the problems rather than letting students struggle and construct their own understanding. Suppose you are engaging students in the **Predicting Old Faithful** student page.

a. Where are places you would anticipate students would struggle and request support?

b. What can you do to resist giving too much support?

c. Why might teachers allow students to struggle?

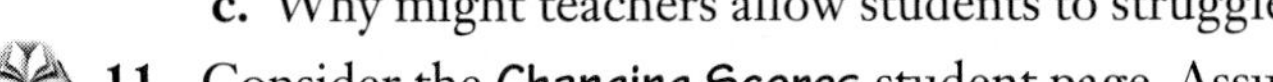

11. Consider the **Changing Scores** student page. Assume students have previously studied means and standard deviations. Record your responses to problems 11a–11f in a **Lesson Planning Guide** for a lesson in which this student page is used.

a. What are major learning outcomes for this student page?

b. How does the student page help students reach these outcomes?

c. How would you **Launch** a lesson for this page?

d. Suppose you have a pair of students who are puzzled by question 8 on the student page. What might be the source of their puzzlement? What responses might you anticipate for this question? What teacher support might you provide to help them address their confusion?

e. Plan the **Explore** portion of a lesson in which **Changing Scores** might be used.

f. Identify another context for data in which values might reasonably be translated or rescaled. Include this context as **Homework** or an **Extension** for the lesson.

TEACHING AND LEARNING HIGH SCHOOL MATHEMATICS © 2010 John Wiley & Sons, Inc.

Changing Scores

Materials

- graphing calculators, one per person

Mrs. Wilson was reviewing her students' scores on a test for their Data Unit:

58, 65, 66, 69, 71, 72, 75, 76, 76, 76, 78, 81, 81, 83, 83, 86, 87, 89, 89, 90, 90, 90, 91, 91, 91, 92

Enter the data into a list in your graphing calculator and use the calculator for the following analyses.

1. Make a histogram of the scores using 5 as the interval size (show the histogram on the x-interval [50, 100] with x-scale = 5).
2. Make a histogram of the scores using 10 as the interval size. Compare the two histograms. What is the effect of changing the interval size? What becomes more or less clear?
3. Find the mean of the scores. Where does this value appear in each histogram?
4. Find the standard deviation of the scores. What does the standard deviation tell you?
5. Make a box plot of the scores. (Use the box plot option that does not show outliers.) Is there information that the box plot reveals more clearly than the histograms? If so, explain what the information is and how the box plot better shows it. If not, how are the two displays comparable? What does each show that the other does not?

Mrs. Wilson thought her students had worked hard during the unit and that the test was challenging. She wanted to give the students a bonus on their work. She debated between adding 10 points to each score or increasing each score by 10% of the original score. In problems 6 and 7, you will compare these two possibilities.

6. **a.** Create a new list for the scores with 10 points added to each. What transformation is this?
   b. Create a histogram of the transformed data with an interval width of 5 (start at 50). Compare it to the histogram of the original data.
   c. Find the mean of the transformed data. How does it compare to the mean of the original data.
   d. Why does the result in problem 6c make sense?
7. **a.** Create a new list for the scores by adding 10% of each original score to the score. What transformation is this?
   b. Create a histogram of the transformed data. How does it compare to the histogram of the original data?
   c. Find the mean of the transformed data. How does it compare to the mean of the original data?
   d. Why does the result in problem 7c make sense?
8. The algebraic form for the mean of a set with n data values is $\bar{x} = \frac{x_1 + x_2 + \ldots + x_n}{n}$.
   a. Add a constant k to each x_i. What transformation is this? Solve for the new mean. State in words what your results show.
   b. Add p percent to each original value. What transformation is this? Solve for the new mean. State in words what your results show.
9. What are some good reasons why Mrs. Wilson might decide to add a constant to the scores? What are some good reasons why she might add a percentage of the original value to the scores?

TEACHING AND LEARNING HIGH SCHOOL MATHEMATICS © 2010 John Wiley & Sons, Inc.

12. Consider the student page **Homeruns for Major League Baseball Players**. Record your responses to problems 12a–12h in a **Lesson Planning Guide** for a lesson in which this student page is used.

a. Complete the analysis suggested.

b. What learning outcomes are expected from students who complete this student page?

c. How would you **Launch** this lesson? Complete the **Launch** portion of the **Lesson Planning Guide**.

d. Consider potential student responses to the task. Record these responses (misconceived, partial, or satisfactory) on a **QRS Guide**.

e. Design appropriate teacher support for the expected student responses in problem 12d. Note the teacher support in the **QRS Guide**.

f. Suppose some students finish the student page quickly. What might you say to or ask of these students?

g. What might you say to or ask of a student who is ready for more challenge?

h. Finish the **Explore** portion of the **Lesson Planning Guide** for a lesson in which this student page is used.

TEACHING AND LEARNING HIGH SCHOOL MATHEMATICS © 2010 John Wiley & Sons, Inc.

Homeruns for Major League Baseball Players

Homerun Records for Some Major League Baseball Players

	Roger Maris	Mickey Mantle	Hank Aaron		Ty Cobb		Babe Ruth
1951		13		1905	1	1914	0
1952		23		1906	1	1915	4
1953		21		1907	5	1916	3
1954		27	13	1908	4	1917	2
1955		37	27	1909	9	1918	11
1956		52	26	1910	8	1919	29
1957	14	34	44	1911	8	1920	54
1958	28	42	30	1912	7	1921	59
1959	16	31	39	1913	4	1922	35
1960	39	40	40	1914	2	1923	41
1961	61	54	34	1915	3	1924	46
1962	33	30	45	1916	5	1925	25
1963	23	15	44	1917	6	1926	47
1964	26	35	24	1918	3	1927	60
1965	8	19	32	1919	1	1928	54
1966	13	23	44	1920	2	1929	46
1967	9	22	39	1921	12	1930	49
1968	5	18	29	1922	4	1931	46
1969			44	1923	6	1932	41
1970			38	1924	4	1933	34
1971			47	1925	12	1934	22
1972			34	1926	4	1935	6
1973			40	1927	5		
1974			20	1928	1		
1975			12				
1976			10				

Source: http://www.baseball-reference.com/leaders/HR_season.shtml, downloaded June 8, 2009.

Sports provide a large amount of data to analyze. In baseball, homerun records are especially interesting to baseball fans. The table above lists the number of homeruns per season for five of baseball's great major league players.

1. Scan the data. What conjectures can you make?
2. Choose one or more ways to display the data so that you find some interesting comparisons of the players. If possible, look for evidence that addresses your earlier conjectures.
3. Think about the issue of variation over time or within a player's numbers of homeruns. What do the data reveal?
4. Of these players, who do you think has the best homerun record? Explain your choice.
5. Prepare to report your findings to other fans.

Materials

- graphing calculators, one per person

TEACHING AND LEARNING HIGH SCHOOL MATHEMATICS © 2010 John Wiley & Sons, Inc.

13. Consider **DMU** problem 1 where several types of graphs are listed. Assume you are trying to help students learn to differentiate when to use one type of graph versus another.

  a. Find or create a data set you might present to students as they work on this learning outcome. Create appropriate directions.

  b. What makes your data set challenging for students as they decide what type of graph to use?

  c. Pilot your data set and directions with a set of students. What did you learn about the task and students' understanding?

14. Consider the **Challenger Disaster** student page. Record your responses to problems 14a–14d in a **Lesson Planning Guide** for a lesson in which this student page is used.

  a. What is an important learning outcome for students who complete this student page?

  b. What might have been the focus of lessons previous to this one?

  c. Plan a **Launch** for this student page.

  d. Anticipate two different student responses to problem 3 on the student page. Write what you might say or do to support each student's thinking. Record expected student responses and teacher support in a **QRS Guide**.

  e. A presidential commission looked into the disaster and suggested a logistic regression model for the full data set: $y = 6(1 + e^{-(5.085-0.1156t)})^{-1}$. Enter this equation into your graphing calculator together with the full data set. Describe how well the curve fits the data.

  f. Locate and read the article "Analyzing Data Relating to the *Challenger* Disaster" by Linda Tappin (*Mathematics Teacher*, 87 (September 1994): 423–426). After reading the article, how might you modify the lesson for advanced students? Include your suggestions as **Extensions** on the **Lesson Planning Guide** for the lesson.

Challenger Disaster

Materials

- graphing calculators, one per person

A tragic event in American history was the mid-air destruction of the space shuttle *Challenger* on January 28, 1986. Within minutes of take-off, while Americans watched the launch of the first space shuttle that included a teacher astronaut, the shuttle exploded. Later analysis found the cause to be a failure of O-rings, rubber gaskets that seal the sections of the solid-fuel rocket boosters.

Imagine you are one of a team of scientists studying data related to the decision on whether to launch the shuttle. The night before the scheduled launch, there was concern that the anticipated temperature would be 31° F; the coldest previous launch was 53° F. Was the temperature too cold for a launch?

Data on many parameters had been collected from earlier launches. Of 24 previous launches, distress on primary O-rings was found in seven cases as shown in the table.

Flight	1	2	3	4	11	12	17
Launch temperature °F	53	57	58	63	70	70	75
Number of O-rings damaged	2	1	1	1	1	1	2

(*continued*)

TEACHING AND LEARNING HIGH SCHOOL MATHEMATICS © 2010 John Wiley & Sons, Inc.

1. Graph the data, preferably using a graphing calculator.
   a. Does there appear to be a relationship between the launch temperature and the number of O-rings damaged? If so, describe it.
   b. Use your analysis to estimate a value for the number of O-rings damaged for a launch temperature of 31° F.
   c. Does your analysis suggest that there should be concern about the temperature for launch?
2. Do you have any other thoughts about this data set as the basis for the decision to launch?

The data above were the flights when O-rings suffered damage. There were other cases, too, that included no O-ring damage. These cases (number of O-rings damaged equals zero) occurred for the following launch temperatures: 66, 67, 67, 67, 68, 69, 70, 70, 72, 73, 75, 76, 76, 78, 79, 81.

3. Append the above data to your graph.
   a. Re-evaluate your thinking about a relationship between the launch temperature and the number of O-rings damaged. How would you now describe the relationship?
   b. Does your re-analysis suggest there should be concern about the temperature for launch?
4. What does this investigation illustrate about data analysis?

This student page is inspired by "Analyzing Data Relating to the *Challenger* Disaster" by Linda Tappin. *Mathematics Teacher*, 87 (September 1994): 423–426.

15. Consider the following data set for a manufacturing company:

Job	Annual Salary	Number of People in Position
President	$600,000	1
Vice President—Marketing	$360,000	1
Vice President—Sales	$450,000	1
Vice President—Finance	$420,000	1
Office Employees	$64,000	10
Salespersons	$72,000	16
Warehouse Staff	$44,000	7
Computer Analysts	$90,000	4

Assume the data are the centerpiece of a lesson about measures of center—mean, median, and mode. Record your responses to problems 15a–15e in a **Lesson Planning Guide** for a lesson in which this data is used.

a. What learning outcomes might such a lesson have?
b. Design a student page for ninth-grade students around this data set for the learning outcomes you identified.
c. Plan a **Launch** for your lesson.

 © 2010 John Wiley & Sons, Inc.

d. Anticipate several student responses for this lesson and plan appropriate Teacher Support. Record the student responses and related teacher support in a **QRS Guide**.

e. Complete the **Launch** and **Explore** portions of the **Lesson Planning Guide**.

f. Try your student page with a small group of students. What did you learn about their thinking and the effectiveness of your student page to enhance student thinking? How might you modify your lesson plan based on the knowledge you gained from working with students?

Analyzing Students' Thinking

16. Following are two data sets with student graphs.

a. For each graph, determine if there are errors, and if so, tell the nature of the errors.

b. If the graph contains errors, tell how the error will create later mathematical difficulties.

c. What might be the benefits of having students analyze these graphs?

First data set and graph:

°C	°F
10	50
15	59
20	68

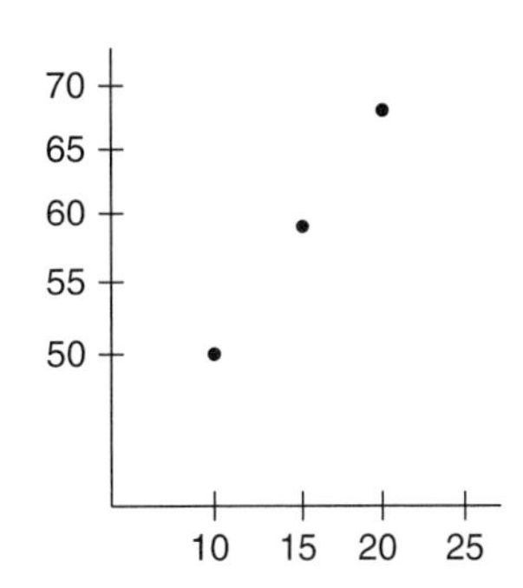

Second data set and graph:

Salaries at Limitless Productions	
Annual Salary	**Number of Employees**
17,000	30
25,000	20
45,000	12
90,000	2
150,000	1

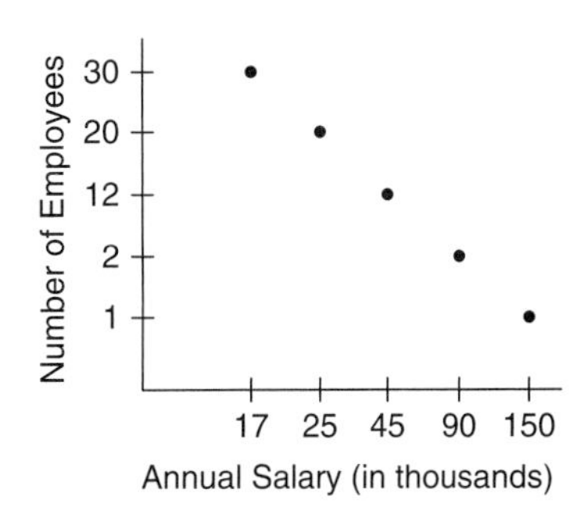

Communicating Mathematically

17. Many mathematical terms are also common English words with less technical meanings. For example, *average* is used in everyday language with broader meanings than in mathematics.

a. Identify a few more terms from data analysis and probability that are also common English words.

b. Choose one of your terms and give an instance where students might have misconceptions due to the differences in technical and everyday uses

TEACHING AND LEARNING HIGH SCHOOL MATHEMATICS © 2010 John Wiley & Sons, Inc.

of the same word. Tell how you would help students overcome their confusion.

18. One goal of this text is to help you develop teaching resources. In teaching data analysis, it is helpful to have examples of interesting data that are well-displayed and analyzed and examples of errors in data display or analysis. Begin a collection of at least half a dozen examples from newspapers, shopping, the Internet, magazines, or other sources that you believe might be of interest to high school students. For each, tell how you would frame the item as a problem, having students identify, among other things, any errors that exist, and how they can be corrected.

Making Historical Connections

19. As indicated in the unit introduction, the U.S. Constitution mandates a census of the U.S. population every 10 years.

a. Research different statistical methods used for reapportionment of the U.S. House of Representatives.

b. Which states have only one U.S. representative? For a state to have just one representative, what seems to be the maximum percentage of the U.S. population residing in that state?

c. Which states have been gaining seats in the U.S. House? Which states have been losing seats?

d. How might topics related to the U.S. census be used to develop some lessons that integrate mathematics and social studies?

Reflecting on Professional Reading

20. Read the "Data Analysis and Probability" Standard (pp. 48–51) from *Principles and Standards for School Mathematics.*

a. Which of the four major expectations did the **Predicting Old Faithful** student page address?

b. Which sub-bullets within the major expectations did the student page address?

c. Revisit the student pages in the **DMU** and **DMP**. Determine which of the expectations are addressed by each student page.

21. Read "How Faithful Is Old Faithful? Statistical Thinking: A Story of Variation and Prediction" by J. Michael Shaughnessy and Maxine Pfannkuch (*Mathematics Teacher*, 95 (April 2002): 252–259).

a. What are some advantages and disadvantages of the various graphs students created?

b. Shaughnessy and Pfannkuch have identified five fundamental elements of statistical thinking. Identify each and explain, if possible, what part of the **Predicting Old Faithful** activity did or could have addressed that element.

c. The authors suggest that good statistical teaching should focus on distributions, variation, and patterns. Contexts that support these topics need data that include variation. Suggest another data set that likely creates bimodal or multimodal data or data that have other unusual features.

d. The authors close with several questions to deepen and extend students' thinking during the lesson. Choose one question you think would make a particularly strong enhancement of the lesson and state why you would use it.

22. Read "Integrating Statistics into a Course on Functions" by James E. Schultz and Rheta N. Rubenstein (*Mathematics Teacher*, 83 (November 1990): 612–617). (See the course website for the article.)

TEACHING AND LEARNING HIGH SCHOOL MATHEMATICS © 2010 John Wiley & Sons, Inc.

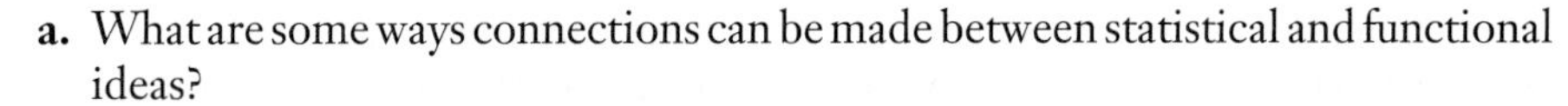

a. What are some ways connections can be made between statistical and functional ideas?

b. Revisit the **Changing Scores** student page in **DMP** problem 11. How are the transformations in this student page related to transformations of functions?

c. Calculators have randomizing functions, many of which can be customized. This article shows the transformations underlying the workings of one such custom function, for simulating rolling dice. Create a comparable function to simulate drawing randomly from the numbers 11–20.

d. What ideas did the article suggest that you wish you had had an opportunity to learn in your high school education?

4.2 Planning a Lesson Share and Summarize: Probability

This text has attempted to provide model lessons centered on worthwhile mathematical tasks with the following characteristics:

- Tasks involve significant mathematics.
- Tasks are at a high level of cognitive demand or challenge with entry points for all students.
- Tasks are connected to previous learning.
- Students need to find and use critical relationships between concepts while working on the task.
- Students do not have a known or evident procedure to find a solution when beginning the task.

Mathematical tasks are a major resource through which learning is fostered. A well-designed task moves students from what they already know to new ideas in a way that enables them to make sense of the new ideas for themselves. However, even with high-quality tasks, learning requires student engagement and teacher support. Appropriate teacher support is critical during the **Explore** portion of the lesson and then in the final **Share and Summarize** portion of the lesson as teachers help students clarify and consolidate their learning.

In the **Share and Summarize** portion, the teacher strives to help all students achieve the learning outcomes of the lesson. The teacher has decisions to make. He or she needs to decide which students to ask to share their work with the entire class. The work selected may be centered on common misconceptions. Or, the teacher may choose students to present solutions that are incomplete with the idea that the full class might continue to work toward a solution using lines of reasoning initiated in that incomplete start. As another alternative, the teacher may ask students to share solutions that involve insightful strategies that the teacher wants everyone to understand well. The teacher might also share important solutions that did not arise in class so that students consider a full spectrum of possibilities. During the **Share and Summarize**, the teacher continues to ask critical questions focused on the learning outcomes of the lesson, simultaneously watching and listening to determine what students understand. This careful observation during the lesson, as well as similar data collection during the **Explore**, is sometimes called *embedded* or *formative assessment*. In the midst of the lesson, the teacher continues to attend to student thinking and understanding.

In Lesson 4.1, you were introduced to a **Lesson Planning Guide** using the **Launch**, **Explore**, and **Share and Summarize** format. You thought about designing a **Launch** and an **Explore** to build a lesson for a student page. As you have done so

TEACHING AND LEARNING HIGH SCHOOL MATHEMATICS © 2010 John Wiley & Sons, Inc.

many times throughout this textbook, you wrote questions to ask students, anticipated their responses, and wrote Teacher Support. You also considered how teachers might monitor students' work in order to sequence solutions for later sharing in the **Share and Summarize** portion of the lesson. It is in this phase that teachers engage in the practice of *Connecting* (Smith et al., 2008; Smith et al., 2009); recall from Lesson 4.1 that this is the phase in which teachers plan questions that help forward the mathematical agenda of the lesson.

The mathematics of this lesson is probability, the theoretical underpinning on which statistical analyses are based. In this lesson, you will:

- Investigate a probability problem.
- Use elements from the **Annotated Lesson Planning Guide** to plan a **Share and Summarize** portion of a lesson.

Materials

- circular spinners with three equal areas, paper clips, one of each per person
- coins, one per person

Website Resources

- **Predicting Old Faithful** student page
- **Sample Lesson Plan for Predicting Old Faithful**
- **Monty's Dilemma** student page
- **QRS Guide**
- **Annotated Lesson Planning Guide**
- **Lesson Planning Guide**

LAUNCH

What does it take to design an effective lesson **Summary**? Consider, as one example, the previous lesson about **Predicting Old Faithful.**

1. What must the instructor do, think about, and plan to help students highlight major mathematical ideas in the closing discussion of the lesson?
2. Refer to the **QRS Guide** and the **Share and Summarize** from the **Sample Lesson Plan for Predicting Old Faithful**. In the **QRS Guide**, the teacher has made notes about the sequence in which student solutions will be shared with the whole class.
   a. Do you agree with the sequence for sharing solutions? Why or why not? What changes, if any, would you make in the sequence?
   b. Comment on the teacher's plans for the **Share and Summarize**. What changes or additions would you make?

EXPLORE

3. Work through the student page **Monty's Dilemma**, a classic problem. As you work and share your thinking with others, use this opportunity to draft a **QRS Guide** on which you record your solution processes and those of your team members. This sharing provides actual "student work" to support the lesson planning that will follow.

TEACHING AND LEARNING HIGH SCHOOL MATHEMATICS © 2010 John Wiley & Sons, Inc.

Monty's Dilemma

Materials

- circular spinner with three equal areas, paper clips, one of each per person
- coins, one per person

In a game show, contestants are shown three closed doors. Behind two of them are useless or inexpensive prizes; behind the third is a valuable prize. A contestant is asked to select a door. Then the host, Monty, dramatically opens one of the other doors to reveal an undesirable prize. The contestant now has the opportunity to stay with his original choice, to switch to the other unopened door, or to flip a coin to decide whether to stay (heads) or switch (tails). The contestant wins what is behind the final door he chooses.

1. a. If you were the contestant and wanted the valuable prize, what would you do? Think of your initial reaction and why you think that way.

b. Share your response to problem 1a with your team.

2. Investigate Monty's Dilemma by simulating each of the three strategies. Using a spinner with equal areas representing each of the three doors, spin to select a door. *Assume the prize is behind door A.* Monty shows you an undesirable prize behind one of the doors you did not select.

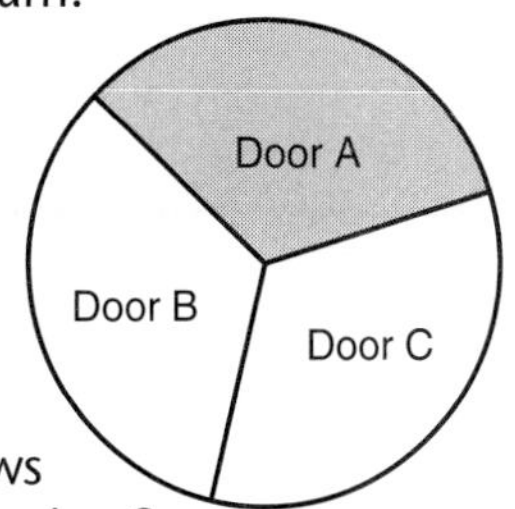

a. Stay strategy. Regardless of what door Monty shows you, you stay with your initial choice. Do you win or lose?

b. Switch strategy. You switch your initial choice. Do you win or lose?

c. Flip strategy. You flip a coin to determine if you stay with your initial choice (heads) or switch doors (tails). Do you win or lose?

3. For each strategy, run 25 trials. For this simulation, the prize is always behind Door A. Pool your findings with three other students to make a total of 100 trials for each strategy.

4. Pool your group's results with those of the class. What is the **relative frequency** (ratio of number of wins to number of trials) of winning when you use each strategy?

5. Analyze each strategy as follows. For each strategy, complete the table to determine whether you win or lose. Remember that the spinner is used to determine your initial selection of a door. Also, the valuable prize is behind Door A in all cases. The first entry is completed for you.

a. The Stay Strategy

Spinner lands on	What door does Monty show?	What door do you choose?	Win or lose?
Door A	B or C	A	W
Door B			
Door C			

b. The Switch Strategy

Spinner lands on	What door does Monty show?	What door do you choose?	Win or lose?
Door A			
Door B			
Door C			

(*continued*)

TEACHING AND LEARNING HIGH SCHOOL MATHEMATICS © 2010 John Wiley & Sons, Inc.

c. The Flip Strategy – A head means stay and a tail means switch.

Spinner lands on	What door does Monty show?	Result of flipping a coin	What door do you choose?	Win or lose?
Door A	B or C	H	A	W
		T	C or B	
Door B		H		
		T		
Door C		H		
		T		

6. For each strategy, what is the (theoretical) probability of choosing the door hiding the valuable prize? Why do you think so?
7. Compare your findings with your initial guesses. What do you now understand?

This student page is inspired by Shaughnessy, J. Michael and Thomas Dick. "Monty's Dilemma: Should You Stick or Switch?" *Mathematics Teacher,* 84 (April 1991): 252–256.

4. Imagine you are the teacher who has observed small groups working on Monty's Dilemma and you are now ready to plan the **Share and Summarize** portion of your lesson using the five practices described in Lesson 4.1: *anticipating*, *monitoring*, *selecting*, *sequencing*, and *connecting*.
   a. As a team, review your **QRS Guide** to determine the different ways of thinking about the problem you *anticipated* as well as any unexpected ways you noticed from *monitoring* during the lesson. Add any unexpected approaches to the problem to your **QRS Guide.**
   b. Identify the important mathematics students should develop or deepen with this lesson.
5. Plan a **Share and Summarize** lesson portion for Monty's Dilemma. Consult the **Annotated Lesson Planning Guide** in the Unit Four Appendix for questions to aid in your planning.
   a. *Select* the ways of thinking you believe are important for students to share. Identify, if relevant, misconceptions that need to be discussed in the full group.
   b. Decide how you will *sequence* what is shared, and record this in the Teacher Notes column of the **QRS Guide**. Clarify with your team the reasons for your selections and sequence.
   c. Plan how you will help students understand and *make connections* among the different ways of thinking about this problem. Compose two to four major questions you will pose in the **Share and Summarize** portion that will help students focus on the critical mathematical ideas of the lesson.

SHARE AND SUMMARIZE

6. Share and compare with other teams your responses to questions 4 and 5.
7. This lesson was based on an overall recommendation that students be introduced to solving probability problems through the following

TEACHING AND LEARNING HIGH SCHOOL MATHEMATICS © 2010 John Wiley & Sons, Inc.

sequence of activities. Identify the potential benefits for each of the following.

a. Students first make a guess or give their initial reaction to a problem.

b. Students simulate the problem.

c. Students do a large number of simulations or pool data with others.

d. Students informally analyze the solution to a problem.

e. Students analyze the solution formally once they have built conceptual understanding through the previous steps.

DEEPENING MATHEMATICAL UNDERSTANDING

Note: Two student pages focusing on basic concepts of probability appear in **DMP** problem 14. If needed, review those student pages as a refresher prior to completing the **DMU** questions.

1. *Monty's Dilemma Analytically*. **Monty's Dilemma** can also be solved analytically.

a. Use area models, tree diagrams, or conditional probability to produce an analytical solution.

b. For the flip strategy, modify the table from the student page in the lesson to analyze what happens if there are four doors instead of three. What is your probability of winning?

c. Repeat problem 1b for five doors.

d. Explain how you know your solutions in problems 1a, 1b, and 1c are correct beyond applying a known algorithm.

2. *Three Card Problem*. Consider the following problem: There are three cards. One is red on both sides. One is red on one side and white on the other. One is white on both sides. Suppose one of the three cards is selected at random and randomly placed down, showing red side face-up. What is the probability that the other side is red as well?

a. Tell your first hunch about the solution and why you think that way.

b. Find a way to simulate the problem many times. What does the simulation suggest?

c. Find an analytical method to solve the problem.

d. In what ways is this problem like **Monty's Dilemma**? In what ways is it different?

3. *One Son Policy*. Until some recent modifications, China has had a policy that limits families to one child. Some have suggested that, instead, families be limited to one son, that is, families may continue to have children until a son is born. Suppose such a policy were the case. Investigate the following questions using more than one strategy. (If you would like to read a complete article on teaching this lesson, see "Teaching Probability through Modeling Real Problems" by Clifford Konold (*Mathematics Teacher*, 86 (April 1994): 232–235).)

a. What would be the ratio of births of girls to births of boys?

b. What would be the average number of children in a family?

4. *Formal Conditional Probability*. Suppose you have six standard playing cards: ace of hearts, king of hearts, queen of hearts, ace of spades, king of spades, and ace of diamonds. The cards are well shuffled. Start each item below using the entire six-card deck.

a. What is the probability that you draw an ace?

b. Suppose you draw a king. What is the probability that it is also a heart?

c. Suppose you draw a heart. What is the probability that it is also a king?

TEACHING AND LEARNING HIGH SCHOOL MATHEMATICS © 2010 John Wiley & Sons, Inc.

d. You should have been able to solve problems 4a–c by examining the six cards and thinking about each case. The problem can also be solved using the formal definition of *conditional probability*:

If E and G are events from a set of outcomes S and if the probability of event G, symbolized P(G), is not equal to zero, then the conditional probability of E given that G has occurred is

$$P(E|G) = \frac{P(E \text{ and } G)}{P(G).}$$

Solve problems 4b and 4c again, this time using the formal definition of conditional probability. In problem 4b, for example, the problem is finding the conditional probability of drawing a heart given that a king is already drawn.

e. Find connections between your original solutions and your solutions in problem 4d using the formal definition of conditional probability.

5. *More on Conditional Probability*. Consider the deck of six cards: ace of hearts, king of hearts, queen of hearts, ace of spades, king of spades, and ace of diamonds.

a. Is $P(\text{heart} \mid \text{king}) = P(\text{king} \mid \text{heart})$? Explain.

b. Is $P(\text{ace} \mid \text{diamond}) = P(\text{diamond} \mid \text{ace})$? Explain.

c. Draw a conclusion from your results in problems 5a and 5b.

6. *Pascal and Chevalier de Méré*. Probability had some of its earliest roots in a problem posed by Chevalier de Méré (the gambler Antoine Gombaud) to the mathematician/philosopher Blaise Pascal. De Méré wanted to know which was more likely: to roll a single die four times and produce at least one ace (a face with one dot) or to roll a pair of dice 24 times and obtain at least one double ace. (Chevalier de Méré thought they were equally likely.) The problem is simplified by using the *complement*, that is, the probability that an event does *not* occur is 1 minus the probability that it does occur. Use this idea to solve the historical problem.

7. *The Birthday Problem*. In his book *Innumeracy*, Paulos (1988) argues that what people take as coincidence is just the natural result of the many possible experiences in our lives that have something in common. As an example, he cites the famous birthday problem, variations of which are investigated below.

a. Suppose there are 367 people in a room. Why may you be certain that there is at least one pair of people with the same birthday (ignore the year)?

b. Suppose there are two people in a room. What is the probability they do not share a birthday? Suppose there are three people, what is the probability they do not share a birthday?

c. Suppose there are 23 people in a room. Show that the probability that none of them share a birthday is less than 0.5. (Hint: Think about problem 7b, and use a calculator.)

d. Once on the *Tonight Show*, a guest tried to explain problem 7c to Johnny Carson, the host from 1962 to 1992. Johnny was incredulous. He attempted to test the idea by stating his birthday and asking whether any of the 120 people in the studio audience had the same birthday. No one did. Carson was not asking the appropriate question. Explain.

e. Suggest a systematic procedure to check whether any students in your class have the same birthday.

8. *Chuck-a Luck*. An old carnival game, Chuck-a Luck, has the following rules: The player chooses a number from 1 to 6. The operator rolls three fair dice. If the player's number comes up all three times, he wins $3. If his number comes up exactly twice, he wins $2. If his number comes up exactly once, he wins $1.

 © 2010 John Wiley & Sons, Inc.

However, if the number fails to appear at all, the player must pay the operator \$1. Is this game fair or unfair? Explain.

DEVELOPING MATHEMATICAL PEDAGOGY

Reflecting on Our Own Learning

9. Sometimes students engage in a mathematical learning activity but fail to understand the mathematical point.
   a. Recall an instance in your own learning history when this happened to you but you later understood the mathematical point. What facilitated your later understanding?
   b. What implications does your experience have on the importance of a teacher having a clear focus on the learning outcomes of a lesson and using those outcomes to frame key questions in the **Share and Summarize** portion of a lesson?

Preparing for Instruction

10. Return to the **Annotated Lesson Planning Guide** in the Unit Four Appendix.

   a. Combine the information from **Explore** items 3–5 into a **Lesson Planning Guide** for a lesson based on the student page Monty's Dilemma.
   b. Revise your solution to **Explore** item 4b to indicate learning outcomes for the lesson.
   c. What might have been the focus of a lesson previous to the lesson with Monty's Dilemma? What might be a subsequent lesson?
   d. Design a **Launch** for the student page.
   e. Refine the **Explore** and **Share and Summarize** portions from your class discussions. Recall that you have already drafted a **QRS Guide** for the student page.
   f. Complete the remaining elements for the lesson plan based on the Monty's Dilemma student page.
11. Below are several scenarios that might occur during a **Share and Summarize**. For each, comment on how you might address the situation.
   a. A small number of student responses are discussed but class time runs out before the major ones are presented.
   b. Some major misconception arises and no students can untangle it.
   c. An insightful solution is presented first and other students are deflated that their work looks so uninspired.
   d. The most important ideas of the lesson do not arise.
   e. Too many student solutions are alike.

12. Choose one of the student pages from earlier in this text and design a complete lesson plan for it using the **Annotated Lesson Planning Guide** to inform your work. Record your work on a **Lesson Planning Guide**.
13. Researchers believe that students overcome probabilistic misconceptions and learn to solve problems more effectively when a sequence like the one for Monty's Dilemma is used. Students are asked first to guess, then simulate by hand, then simulate a large number of times by pooling data or by computer, to solve the problem through informal analysis, and eventually, to solve analytically. What do you believe are some reasons for this point of view?

Analyzing Instructional Materials

14. Consider the following two student pages, Probabilities with OR and Conditional Probability and Independent Events.

TEACHING AND LEARNING HIGH SCHOOL MATHEMATICS © 2010 John Wiley & Sons, Inc.

a. For each student page, identify the learning outcomes.

b. How do Venn diagrams support student thinking?

c. Many educators believe it is effective to engage students with a concept and reserve the introduction of formal vocabulary for that concept until after students have built conceptual understanding for the relevant ideas. How is this principle used in these student pages?

d. Why might it be a bad idea to teach these two pages on the same day or consecutive days?

e. If you were planning a lesson for each page, what would you emphasize in the **Share and Summarize** portion?

f. For one of the student pages, complete a **Lesson Planning Guide** using the **Annotated Lesson Planning Guide** to inform your work.

Probabilities with OR

Recall that in logic and mathematics, the statement "Event *A* occurs or Event *B* occurs" is true when only *A* occurs, when only *B* occurs, or when **both** events *A* and *B* occur.

1. Consider the two events listed below. The events pertain to rolling a special six-sided die labeled: 1, 2, 3, 4, 4, 5.

Event *A*: You roll an even number.

Event *B*: You roll a number less than 3.

In the Venn diagram provided, indicate each of the six possible outcomes of the roll of one die in the appropriate region.

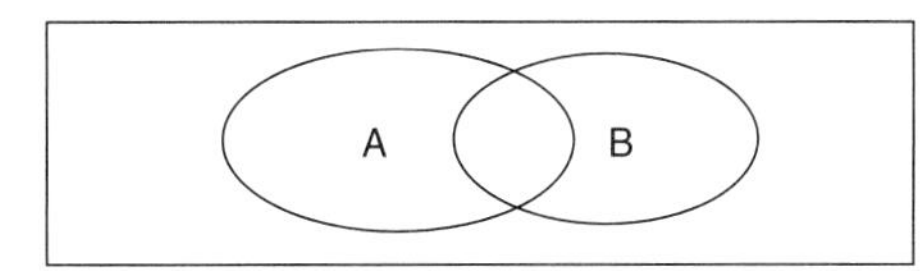

2. Use the Venn diagram to find each probability. The symbol *P*(*A*) means "the probability of event *A*."

a. *P*(*A*) **b.** *P*(*B*) **c.** *P*(*A* or *B*)

3. a. What did you need to do to find *P*(*A* or *B*)? Why did that make sense?

b. Write your process in problem 3a in words.

c. Symbolize your words in problem 3b.

4. a. Assuming the same die, create a Venn diagram to indicate Event *Q*, you roll a prime number, and Event *R*, you roll an odd number.

b. Find the probability that when you roll one die, the outcome is prime or odd. Use the Venn diagram to make a convincing case.

5. Suppose you roll the same special die. Find each probability.

a. *P*(4)

b. *P*(odd number)

c. *P*(4 or odd number)

d. *P*(4 and odd number)

6. a. Show a Venn diagram for problem 5c. What is special about this case?

b. Events that cannot occur at the same time are called **mutually exclusive**. Explain how the events in problem 5, "rolling a 4" and "rolling an odd number," relate to this definition.

c. Symbolize a formula for *P*(*A* or *B*) for mutually exclusive events *A* and *B*.

(*continued*)

TEACHING AND LEARNING HIGH SCHOOL MATHEMATICS © 2010 John Wiley & Sons, Inc.

7. How does the probability statement for mutually exclusive events (response to problem 6c) compare with the more general statement about the probability of event *A* or *B* (response to problem 3c)? Why?
8. Suppose you roll the same special die. Find each probability.
   - **a.** *P*(2) **b.** *P*(not 2)
   - **c.** *P*(2 or not 2) **d.** *P*(2 and not 2)
9. **a.** Show a Venn diagram for problem 8c. What is special about this case?
   - **b.** Two mutually exclusive events that account for all the possibilities in a situation are called **complementary events**. Explain how the events "rolling a 2" and "not rolling a 2" relate to this definition.
   - **c.** Symbolize a relationship between *P*(not *A*) and *P*(*A*).
10. Explain how to use the general formula for *P*(*A* or *B*) to find a formula for mutually exclusive events.
11. Explain how to use the general formula for *P*(*A* or *B*) to find a formula for complementary events.

Conditional Probability and Independent Events

Michelle's class was studying recycling. Part of their work included surveying themselves to learn who recycles at home. The table shows their findings.

	Recycles	Does not recycle	Total
Males	13	9	22
Females	8	6	14
Total	21	15	36

1. Imagine students are selected at random from Michelle's class. Use the table to find all of the probabilities indicated. Do not simplify fractions.
   - **a.** *P*(student is female and recycles)
   - **b.** *P*(student is female or recycles)
   - **c.** *P*(student is female given that a student recycles)
   - **d.** *P*(student recycles)
   - **e.** *P*(student recycles given that a student is female)
   - **f.** *P*(student is female)
2. Why are the denominators in 1c and 1e not 36?
3. How do you read the table differently to solve problems 1c and 1e compared to problems 1a, 1d, and 1f? What information do you need to use to solve problem 1b?
4. In problems 1c and 1e, you need to find a probability under restricted conditions. For example, in problem 1c, the condition is that the student recycles. This is called a **conditional probability**. In these cases, the number of possible outcomes is reduced to reflect the restrictions. To symbolize the conditional probability, *P*(student recycles given that a student is female), write *P*(student recycles | student is female). The vertical bar is read "given that." (*continued*)

TEACHING AND LEARNING HIGH SCHOOL MATHEMATICS © 2010 John Wiley & Sons, Inc.

a. Define the variables, R: a student recyles and F: a student is female. Symbolize the probabilities for problems 1c and 1e.

b. Students conjectured some relationships for problem 1. Which of the following relationships can be justified? Why?

i. $P(F|R) = \dfrac{P(F \text{ and } R)}{P(R)}$ **ii.** $P(F|R) = P(F) \bullet P(R)$

iii. $P(F \text{ and } R) = P(F) \bullet P(R)$ **iv.** $P(F \text{ and } R) = P(F) \bullet P(R|F)$

v. $P(F \text{ and } R) = P(R) \bullet P(F|R)$

c. Generalize the relationships you found in problem 4b for any events A and B. Symbolize the generalizations.

5. In some situations, the outcome of one event changes the probability of the outcome of another event. Suppose two blue and three red balls are placed in a jar. One ball is randomly chosen and its color is recorded. Then a second ball is chosen and its color is recorded. What is the probability of getting a blue ball on the first draw and a red ball on the second draw in each of the following situations?

a. The first ball is replaced before the second is drawn.

b. The first ball is not replaced before the second ball is drawn.

6. Use the situation described in problem 5 to find the probability that both balls drawn are blue.

a. Suppose the first ball is replaced before the second is drawn.

b. Suppose the first ball is not replaced before the second ball is drawn.

7. In which situations in problems 5 and 6 could you say that the results of the second ball *depend* on the results of the first ball?

8. Consider tossing two fair coins, one at a time.

a. There are four equally likely outcomes. List them.

b. Some people think there are three equally likely outcomes. What might they be thinking? How would you help them see why this is wrong?

c. What is the probability of getting a head on the second coin?

d. What is the probability of getting a head on the second coin, given that there was a head on the first coin?

e. How do your answers to problems 8c and 8d compare?

9. Two events are **independent** if and only if the occurrence of one event does not affect the probability of the occurrence of the other event. When two events are independent, you can multiply to find the probability that they both occur. That is, for independent events A and B, $P(A \text{ and } B) = P(A) \bullet P(B)$.

a. Are the events "female student" and "recycles" independent for the recycling data in problem 1? Explain.

b. Which events in problems 5, 6, or 8 are independent? Confirm using the probabilities as well as the context.

10. Give examples of events you think are independent. How can you test mathematically whether they are independent?

15. Consider the student page Dice and Drinks. Record your answers to problems 15a–15e on a **Lesson Planning Guide** using the **Annotated Lesson Planning Guide** to inform your decisions.

a. Identify the learning outcomes.

b. What might have been the focus of the lesson previous to this one?

TEACHING AND LEARNING HIGH SCHOOL MATHEMATICS © 2010 John Wiley & Sons, Inc.

c. Write another dice problem that would deepen or extend students' understanding. Provide a key giving satisfactory student responses.

d. Write another drinks problem that would deepen or extend students' understanding. Provide a key giving satisfactory student responses.

e. If you were planning a lesson for this page, what would you emphasize in the **Share and Summarize** portion?

Dice and Drinks

Consider the following set of equally likely outcomes for rolling two dice. Suppose the first die is red and the second is green and the two numbers are the result (red, green).

(1,1)	(1,2)	(1,3)	(1,4)	(1,5)	(1,6)
(2,1)	(2,2)	(2,3)	(2,4)	(2,5)	(2,6)
(3,1)	(3,2)	(3,3)	(3,4)	(3,5)	(3,6)
(4,1)	(4,2)	(4,3)	(4,4)	(4,5)	(4,6)
(5,1)	(5,2)	(5,3)	(5,4)	(5,5)	(5,6)
(6,1)	(6,2)	(6,3)	(6,4)	(6,5)	(6,6)

1. By inspecting the 36 equally likely possibilities, find each of the following probabilities.

a. *P*(at least one five) **b.** *P*(exactly one five)

c. *P*(exactly two fives) **d.** *P*(no fives)

2. Using your discoveries about probabilities and relationships from earlier student pages (**Probabilities with OR** and **Conditional Probabilities and Independent Events**), write each of the probabilities in terms of events that occur one die a time. Find the probability for each event and the related probability. Compare your results with those in problem 1.

a. *P*(at least one five) **b.** *P*(exactly one five)

c. *P*(exactly two fives) **d.** *P*(no fives)

3. a. In problem 2, when did it make sense to add (or add and subtract) probabilities? Why?

b. In problem 2, when did it make sense to multiply probabilities? Why?

4. Some special probability problems involve a situation with two mutually exclusive outcomes. Suppose, for example, that when a soda vending machine fills a cup, it overflows 10% of the time and is okay 90% of the time. Note: Each pouring from the machine is independent of previous pourings, that is, the machine has no "memory," so events are independent. What is the probability that each of the following happens?

a. In filling two cups, the machine never overflows.

b. In filling three cups, the machine never overflows.

c. In filling four cups, the machine never overflows. Why?

d. In filling four cups, the machine overflows exactly once. (Hint: What are all the possible ways this might happen?)

e. In filling four cups, the machine overflows exactly two times. (Hint: Think of all the possible pairs of cups that might overflow.)

(continued)

TEACHING AND LEARNING HIGH SCHOOL MATHEMATICS © 2010 John Wiley & Sons, Inc.

5. Brittany solved problem 4. Do you agree or disagree with her comments? Why?
   a. Brittany said, "I could just multiply in parts a, b, and c because each pouring is independent of the others."
   b. Brittany also said, "In part e, I could find the probability that a specific pair of cups overflows. There are six ways this can happen, so I could add that probability 6 times. Because the outcomes I was looking at are mutually exclusive, there was no intersection to subtract."
6. Write in your own words how to determine the probability that in five pourings, exactly two cups overflow. Explain why your work makes sense.

16. Consider the student page **What Is the Probability You Are Ill?**
   a. What is a benefit of using a two-way table as on the student page?
   b. What is a benefit of using an actual base (100,000 people) rather than a "whole" with decimals or percentages?
   c. How might you increase the level of challenge of this task?
   d. How might you decrease the level of challenge of this task?
   e. If you were planning a lesson for this student page, what would you emphasize in the **Share and Summarize** portion of the lesson?

   f. Complete a **Lesson Planning Guide** for the lesson in which this student page would be used. Use the **Annotated Lesson Planning Guide** to inform your work.

What Is the Probability You Are Ill?

Physicians use many tools to help diagnose various diseases. One caution with medical tests is that their reliability is limited. Sometimes, when a test suggests that a disease is present, the actual probability that the patient has the disease is surprisingly different from what someone might estimate without doing a thorough analysis. For this reason, doctors often use additional tools in their diagnosis. In this activity, you will investigate probabilities related to such tests.

1. Read, but do not solve the following problem. After reading, estimate the probability requested. Positive test results are interpreted as the patient having the disease.

> Suppose a certain disease occurs rarely in a population, 0.5% of the time. A test exists to detect the disease. The test gives an accurate positive result 94% of the time. The test may also give a false positive in 1% of the healthy people tested. Suppose you take the test and get a positive result. What is the probability that you actually have the disease?

2. To solve problems such as the medical test problem, a two-way table as shown can be helpful.

	Patient has disease	**Patient does not have disease**	**TOTAL**
Positive test result			
Negative test result			
TOTAL			

(continued)

TEACHING AND LEARNING HIGH SCHOOL MATHEMATICS © 2010 John Wiley & Sons, Inc.

To better understand the problem, use whole numbers, rather than decimals or percents. Suppose 100,000 people are being tested. Place the 100,000 as the grand total in the table.

a. If 0.5% of the population have the disease, write in the appropriate place in the table the number of people with and without the disease. Make any other deductions you can from the information you have so far.

b. If 94% of the people with the disease get an accurate positive result, calculate this number of people and place that value appropriately in the table. Continue to make deductions from the information you have so far.

c. If 1% of the healthy people get a false positive result, calculate this value, locate where it goes, and continue to make deductions.

d. Use the table to decide the probability that someone who tests positive actually has the disease.

e. Compare your findings from the analysis with your earlier estimate.

Analyzing Students' Thinking

17. Suppose someone tossed a penny nine times and obtained heads each time. Students have different ideas of what to expect next.

a. What reasons might be given that the next toss will likely be another head?

b. The Gambler's Fallacy would say that the coin is *due* for a tail. How would you help a student clarify why this is a misconceived notion?

Communicating Mathematically

18. Problems like **Monty's Dilemma** and **What Is the Probability You Are Ill?** are sometimes called *conditional probabilities*. In your work with logic in Unit One, you studied statements that were called *conditionals*.

a. Identify what is conditional about the probabilities in each student page.

b. What do these two uses of the term *conditional* have in common?

c. How might you help students become fluent with these new terms?

Reflecting on Professional Reading

19. Read "Orchestrating Discussions" by Margaret S. Smith, Elizabeth K. Hughes, Randi A. Engle, and Mary Kay Stein (*Mathematics Teaching in the Middle School*, 14 (May 2009): 548–556).

a. Explain each of the five practices.

b. What do the authors say about why the five practices they suggest are likely to help? What are your thoughts about the helpfulness of the five practices?

c. Suppose when working on a task several students have the same misconception. Some people argue that the misconception should be sequenced first in the **Summarize** portion. Others argue that it should be sequenced last. Give possible rationales for each perspective.

For problems 20–24, read "Data Analysis and Probability for Grades 9–12" (pp. 324–333) from *Principles and Standards for School Mathematics*.

20. **a.** The standard assumes students come to high school having had specific experiences in middle school with data analysis and probability. Identify those experiences.

TEACHING AND LEARNING HIGH SCHOOL MATHEMATICS © 2010 John Wiley & Sons, Inc.

b. Choose one of the middle school experiences and identify an activity or strategy you might use in your class if, in fact, students appear not to have studied that concept in middle school.

21. a. What are some questions students need to ask themselves as they design a survey?

    b. What are some difficulties students may experience in implementing a survey?

22. a. Draw two conclusions from the display of box plots on p. 328 in the reading.

    b. Recall a least squares regression line you produced in Lesson 3.3. Explain what *least squares* means in terms of Fig. 7.23 in the reading.

23. a. Refer to the distribution of Voters Supporting Mr. Blake on p. 330 in the reading. How likely is a sample of 20 voters that has 17 or more supporters for Mr. Blake?

    b. Complete: The distribution of voters supporting Mr. Blake suggests that in 95% of the samples the number of supporters is between ___ and ___.

24. a. Refer to the table at the bottom of p. 331 in the reading. How can you tell from this table that playing a musical instrument and playing on a sports team are not independent?

    b. Refer to Fig. 7.25 in the reading. Show in two different ways that $P(6|\text{even}) = 0.3$.

25. Read "Good Things Always Come in Threes: Three Cards, Three Prisoners, and Three Doors" by Laurie H. Rubel (*Mathematics Teacher*, 99 (February 2006): 401–405).

    a. How does the solution to the Three Cards problem using Bayes' theorem compare to your other approaches in earlier problems?

    b. Rubel researched the Three Cards problem with middle and high school boys. What was the nature of the misconceptions that were revealed?

    c. Try the Three Cards problem with some students. How did they think about the problem? What tasks or questions could you pose to move them toward a clearer understanding of the problem?

4.3 Blending Direct Instruction into a Lesson: Standard Deviation and Normal Distributions

In this textbook, and specifically in this unit, you have been introduced to inquiry-based teaching in which challenging tasks provide opportunities for students to grapple with and make sense of significant mathematics. You have been striving to plan lessons by working through a task and then anticipating student responses and appropriate support for students at various points in their solution or thinking processes. Typically, inquiry learning focuses on students' producing solutions that make sense to them using their strategies or representations. The summary of the lesson provides an opportunity for students, with teacher guidance, to look closely at various solutions and bring to the surface major ideas or processes. Rather than demonstrating standardized ways to solve a problem, teachers try to guide students from more naïve to more efficient approaches. Nevertheless, although inquiry learning limits the amount of time when teachers provide explanations, there are occasions when some topics need direct instruction. For example, when a new concept has been uncovered, the teacher might use direct instruction to introduce standard vocabulary. When a new tool is to be introduced, direct instruction might be needed to instruct students on its use. Also, when a particular procedure needs to be understood but is not likely to be developed by students, some aspects of direct instruction may be appropriate. So, in planning for inquiry-based instruction, teachers need to consider when and how to include brief segments of direct

TEACHING AND LEARNING HIGH SCHOOL MATHEMATICS © 2010 John Wiley & Sons, Inc.

instruction. This lesson will help you think about blending direct instruction into inquiry lessons that continue to develop ideas about statistics and probability.

In this lesson, you will:

- Investigate concepts related to standard deviation.
- Explore the normal distribution curve and its properties.
- Consider when and how to integrate brief segments of direct instruction into inquiry lessons.

Materials

- graphing calculators, one per student

Website Resources

- ***Computing Standard Deviation*** student page
- **Normal Distributions and z-Scores** student page

LAUNCH

1. In your schooling, you have likely experienced many mathematics lessons that have used direct instruction. Throughout this text, you generally experienced lessons using an inquiry approach with little or no direct instruction.
   a. Compare and contrast your own learning with these two instructional approaches.
   b. Find an example of a lesson in this text in which direct instruction is blended with an inquiry approach.
   c. What characteristics of direct instruction make it effective?
   d. What characteristics of direct instruction make it ineffective and problematic?

EXPLORE

2. Complete the student page ***Computing Standard Deviation.***
   a. Identify segments of direct instruction and what makes them effective.
   b. Identify any segments of inquiry on the student page.
   c. How does the student page attempt to engage students during the segments of direct instruction? during the segments of inquiry?

TEACHING AND LEARNING HIGH SCHOOL MATHEMATICS © 2010 John Wiley & Sons, Inc.

Computing Standard Deviation

Materials

- graphing calculators, one per person

Ms. Huns gave a mathematics test to two groups of students and recorded the following scores:

Group A: 78, 87, 92, 100, 85, 84, 81, 91, 82, 75, 80, 88, 89, 94, 99, 96, 84, 85, 89, 81

Group B: 75, 100, 100, 78, 79, 92, 95, 96, 83, 78, 88, 89, 90, 85, 85, 87, 90, 92, 79, 79, 93, 81, 79, 95

She was interested in comparing the performance of the two groups of students.

1. a. What strategies might Ms. Huns use to compare the scores from the two groups? Explain.

b. What comparisons can Ms. Huns make using the strategies you listed in problem 1a?

2. a. Find the mean, median, and mode of each group. What does each of these measures tell Ms. Huns about the two groups?

b. The **range** is the difference between the maximum and minimum scores. Find the range for each group of students. What does the range tell Ms. Huns about the two groups?

The **standard deviation** is a measure that describes the spread of a data set in relation to the mean. The smaller the standard deviation, the less variability there is in the data. As the word "deviation" suggests, the standard deviation uses the differences (deviations) between each value in a data set and the mean of that set. These deviations are then squared, summed, and averaged. That average is called the **variance** of the data. The square root of the variance is the standard deviation.

3. To find the standard deviation of the scores for the two groups, start by completing the following table. Use the means you computed in problem 2a. (Hint: Enter the data into your calculator in separate lists. Use the LIST capabilities to complete the table.)

Group A			Group B		
Score	**Score − Mean**	**(Score − Mean)^2**	**Score**	**Score − Mean**	**(Score − Mean)^2**
78			75		
87			100		
92			100		
100			78		
85			79		
84			92		
81			95		
91			96		
82			83		
75			78		
80			88		

(*continued*)

TEACHING AND LEARNING HIGH SCHOOL MATHEMATICS © 2010 John Wiley & Sons, Inc.

88			89		
89			90		
94			85		
99			85		
96			87		
84			90		
85			92		
89			79		
81			79		
			93		
			81		
			79		
			95		

4. **a.** For Group A, find the mean of the values in the column for (Score – Mean)^2. This is the variance for Group A.
   b. Take the square root of your result in problem 4a. This is the standard deviation for Group A.
   c. Repeat problems 4a and 4b for Group B.
5. **a.** Use your results in problem 4 to describe the relative performance of the two groups of students.
   b. Construct dot plots for each data set. How do the dot plots support your response to problem 5a?
6. Summarize the steps needed to find the standard deviation of a set of data values. Use your summary to write a formula that gives the standard deviation of a set of data.
7. Historically, attempts to determine deviations did not use the sum of the squares of the deviations, but instead summed the deviations from the mean. Why might these earlier approaches have been less effective than the current process for finding the standard deviation?

SHARE

3. Discuss as a class **Computing Standard Deviation**.
   a. What are the learning outcomes for this student page?
   b. What segments of direct instruction does it include?
   c. What segments of inquiry does the student page include?
   d. What features of the activity contribute to its effectiveness?
   e. How does the student page attempt to blend direct instruction and inquiry?

EXPLORE MORE

4. Complete the student page **Normal Distributions and *z*-Scores**.

TEACHING AND LEARNING HIGH SCHOOL MATHEMATICS © 2010 John Wiley & Sons, Inc.

Normal Distributions and z-Scores

The graph of the equation $y = \frac{1}{\sqrt{2\pi}} e^{-x^2/2}$ is called the **standard normal distribution curve**. The distribution represented by this equation is among the most important distributions in probability. Many natural phenomena, including many human characteristics, are described by a normal distribution.

1. a. Graph the equation $y = p(x) = \frac{1}{\sqrt{2\pi}} e^{-x^2/2}$.

b. Describe properties of this distribution, including its domain, range, maximum, symmetry, and any asymptotes. Based on the equation, explain any properties observed from the graph.

The area between the standard normal distribution curve and the horizontal axis is 1. Because of the importance of this distribution, probabilities related to the standard normal curve have been calculated. By using calculus, one can show that the following three benchmarks are true:

- Approximately 68% of the data lies within 1 standard deviation of the mean.
- Approximately 95% of the data lies within 2 standard deviations of the mean.
- Approximately 99.7% of the data lies within 3 standard deviations of the mean.

2. Use the benchmarks as needed to determine the probability that a data value meets each condition. Justify your answers.

a. The value is less than or equal to the mean.

b. The value is between 1 standard deviation below the mean and the mean.

c. The value is between 1 and 2 standard deviations above the mean.

d. The value is between 1 standard deviation below the mean and 2 standard deviations above the mean.

3. The mathematics portion of the SAT college entrance exam was originally designed to have a mean of 500 and a standard deviation of 100. A **z-score** is a measure that transforms a data value into a number that reflects its distance from the mean in terms of standard deviation units. That is, $z = 1$ indicates a data value is 1 standard deviation above the mean; $z = -2.5$ indicates a data value is 2.5 standard deviations below the mean.

a. Sketch the graph of a normal curve with mean 500 and standard deviation of 100. Call the function $y = S(x)$.

b. How are $y = S(x)$ and the standard normal distribution function $y = p(x)$ related?

c. Determine the values of A, B, C, and D so that $S(x) = A \bullet p(Bx + C) + D$. Verify your work and explain how you know it is correct.

4. Use the meaning of z-score and your work with transforming the function $y = p(x)$ to obtain the graph of $y = S(x)$ to find each of the following.

a. The z-score for a SAT mathematics score of 650.

b. The z-score for a SAT mathematics score of 300.

c. The SAT mathematics score corresponding to a z-score of -2.75.

d. The SAT mathematics score corresponding to a z-score of 1.8.

e. Interpret two of problems 2b–2d in terms of SAT mathematics scores.

5. Use your responses to problems 3–4 to write a formula to find a z-score for any data set with mean μ and standard deviation σ. Explain your formula and how you know it is correct.

TEACHING AND LEARNING HIGH SCHOOL MATHEMATICS © 2010 John Wiley & Sons, Inc.

5. **a.** What are the learning outcomes for the student page **Normal Distributions and z-Scores**?
   b. Identify elements of direct instruction on the student page.
   c. Identify any elements of inquiry on the student page.
   d. How do direct instruction and inquiry complement each other to address the learning outcomes on this student page?

SHARE AND SUMMARIZE

6. **a.** Share your responses to **Explore** item 5.
   b. Refer to the segments of direct instruction on the two student pages. What made the direct instruction segments effective? What might improve their effectiveness?
7. Even with direct instruction, teachers need to plan carefully a **Share and Summarize** to ensure students master the main concepts of the lesson. Identify several points for each student page you would want to make clear during a **Share and Summarize** segment of the lesson.
8. In the introduction to this lesson, several examples of direct instruction blended with inquiry learning were shared. What might be some additional instructional situations where direct instruction could support inquiry learning?

DEEPENING MATHEMATICAL UNDERSTANDING

1. *Revisit Changing Scores*. Refer to the **Changing Scores** student page in Lesson 4.1 **DMP** problem 11. Recall that Mrs. Wilson is considering two ways to transform a set of data: adding 10 points to each score or increasing each score by 10%.
   a. Find the standard deviation of the original set of data.
   b. Consider the situation in which Mrs. Wilson adds 10 points to each score. Find the standard deviation of the translated data. How does it compare to the standard deviation of the original data?
   c. Consider the situation in which Mrs. Wilson increases each score by 10%. Find the standard deviation of the transformed data. How does it compare to the standard deviation of the original data?
   d. Think about the meaning of standard deviation. Why do your answers to problems 1b and 1c make sense?
   e. Write a proof to show that your results in problems 1b–1c hold in general when k points are added to each score or each score is increased by m percent.
2. *Comparing Graphs of Normal Distributions*. The equation $y = e^{-x^2/2}$ is the parent function for normal distributions.
   a. Graph $y = e^{-x^2/2}$. Describe the properties of this graph.
   b. Compare the graph of the parent normal distribution to the graph of the standard normal distribution from the lesson. How are they alike? How are they different?
   c. Use transformations to find and describe the relationship between the parent normal distribution and the standard normal distribution.
3. *Calculus and the Normal Distribution*. Refer to the equation for the standard normal distribution, $y = p(x) = \frac{1}{\sqrt{2\pi}}e^{-x^2/2}$. Use calculus and numerical approximation tools to estimate the area between the standard normal distribution curve and the x-axis for x in each of the following intervals:

TEACHING AND LEARNING HIGH SCHOOL MATHEMATICS © 2010 John Wiley & Sons, Inc.

a. Between 0 and 1

b. Between –1 and 1

c. Between –2 and 2

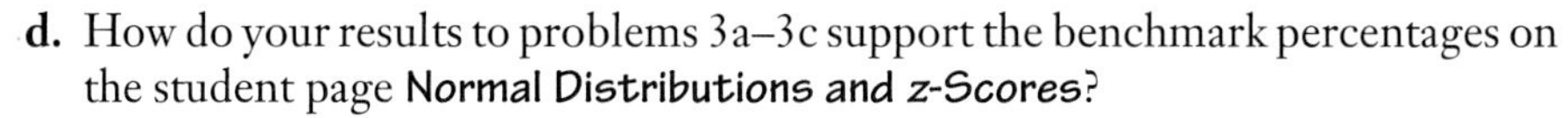

d. How do your results to problems 3a–3c support the benchmark percentages on the student page **Normal Distributions and *z*-Scores**?

4. *Population and Sample Standard Deviation*. Refer back to the test scores for Groups A and B on the student page **Computing Standard Deviation**.

a. Input the two data sets into a graphing calculator.

b. Use the statistics feature to calculate the descriptive statistics for the two groups.

5. *Bias and Variability*. Two major concerns researchers have when they sample from a population are bias and variability.

Bias is consistent repeated deviation of the sample statistic from the population value when many samples are taken.

Variability describes the spread of the values of the sample statistic when many samples are taken. Large variability means the results of sampling are not repeatable.

A good sampling method has both small bias and small variability.

In each target, imagine your goal is to hit the center of the target consistently. Each of the four examples illustrates a different combination of bias and variability. Match each diagram to its description.

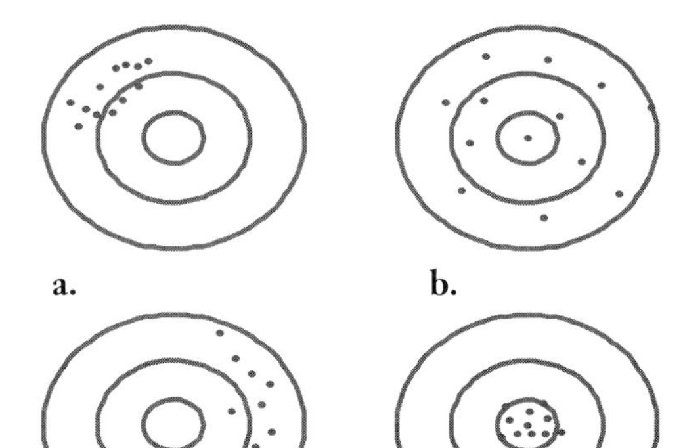

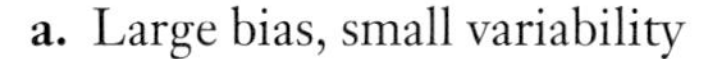

a. Large bias, small variability

b. Large bias, large variability

c. Small bias, large variability

d. Small bias, small variability

6. *Bias and Sampling*. Consider each of the following situations in which a sample is used to learn something about a population. Tell why each sample is potentially biased.

a. You want to see the strawberries in an enclosed package. Your sample consists of the berries on the top of the package.

b. You want to learn about the monthly food expenses of families in a city. You select a random sample of families from the telephone book.

c. You want to know how students at your school think about a topic. You select several friends to interview.

d. You want to know how the audience of a radio program thinks about a controversial topic. You invite people to call in to state their opinions.

7. *Biased Samples*. A *biased sample* systematically favors certain outcomes. Two types of sampling that are usually biased are convenience and volunteer. A sample of *convenience* is one that is readily available. A *volunteer* sample is one in which members of the population self-select. Which, if any, of the examples in problem 6 used convenience samples? Which, if any, use volunteer samples?

8. *Random Numbers from a Table*. A table of random digits is a long string of the digits from zero to nine meeting two criteria: (1) each entry in the table is equally likely to be any of the ten digits 0 through 9 and (2) the entries in the table are independent of each other. The second criterion means that knowledge of one part of the table gives no information about another part of the table. A sample of a portion of a random number table is given on page 333.

TEACHING AND LEARNING HIGH SCHOOL MATHEMATICS © 2010 John Wiley & Sons, Inc.

Portion of a Random Number Table

39975	06907	72905	91977	14342	36857	69578	40961	03969	61129	97336
16308	19885	04146	14513	06691	30168	25306	38005	00256	92420	82651

a. Consider the following list:
1357902468135790246813 57902468.

What about the list violates a criterion for a random number table?

b. One way to conceptualize generating a random number table is to imagine a bag filled with 10 congruent balls each labeled with one digit, 0 through 9. The balls are mixed well and one is selected, recorded, and returned to the bag. Then the balls are mixed again and the process is repeated.

i. How does this imaginary process meet the criteria for a random number table?

ii. Might this process produce five of the same number in a row?

iii. If no nines have appeared in 20 digits in a random number table, does that make it more likely for a nine to appear in the next few digits? Explain.

9. *Random Numbers from a Calculator*. A graphing calculator usually has a random number generator that generates numbers between 0 and 1. On a TI-83/84, the random number generator, `rand`, is accessed through the MATH menu and then the PRB (probability) menu. Find the `rand` command on your calculator, and press ENTER several times to produce several random numbers. Suppose you want to generate the six values on a standard die (1, 2, 3, 4, 5, 6). A formula you can use on your calculator is

```
int(6*rand(1)) + 1.
```

a. Try the formula a few times to see that it appears to randomly generate digits of a die.

b. Analyze the transformations the formula uses to produce the desired values. (Note: `rand(1)` produces one result at a time.) Why does the formula multiply `rand(1)` by 6? Why does it apply `int` (a function that rounds down to the nearest whole number)? Why does it add one?

c. Redesign the given function to produce random integer values from 1 through 8 such as an octahedral die might produce.

d. Redesign the given function to produce random integer values from 14 through 30 inclusive.

DEVELOPING MATHEMATICAL PEDAGOGY

Reflecting on Our Own Learning

10. It is possible that portions of your mathematics learning history included extended segments of direct instruction. What are some concerns or issues about a course almost entirely taught using the direct instruction format?

Preparing for Instruction

11. Choose one of the topics listed below.

- Learn which descriptive statistics are used together and for what reasons (i.e., mean and standard deviation and when they are appropriately used).
- Learn how to use dynamic statistical software or a graphing calculator to fit a nonlinear function to a set of data.

TEACHING AND LEARNING HIGH SCHOOL MATHEMATICS © 2010 John Wiley & Sons, Inc.

a. Prepare a student page that blends direct and inquiry-based instruction on the topic you selected.

b. Tell why you believe the direct instruction you designed was appropriate and effective.

c. Tell why you believe the inquiry instruction you designed was appropriate and effective.

Analyzing Instructional Materials

12. Look at high school textbooks for examples of effective direct instruction. What are qualities you think contribute to effective direct instruction?

13. Examine several high school algebra textbooks. How are statistics or probability integrated into the texts?

14. Complete the student page **Creating a Random Sample**.

    a. Identify segments of direct instruction.

    b. What helped make the direct instruction effective?

    c. What made the student page interactive?

    d. Anticipate student solutions (misconceived, partial, or satisfactory) to this activity. Record them in a **QRS Guide** along with appropriate teacher support.

Creating a Random Sample

Suppose you want to learn how many hours students in your class study. You will ask the question: How long did you study this past Monday evening? You decide to use a random sample. Suppose there are 35 people in your class and you will sample six of them. How do you do this?

STEP 1: Assign labels—give each member of the population a numerical label. Be sure each person has a different label. Use as few digits as possible. For example, for 35 persons you could assign the numbers 01 through 35. Assign one number to each name.

STEP 2: Use a random number table and read off the labels—start anywhere in the table (some people close their eyes and then point to the table). Then proceed systematically by rows or by columns. Every two digits produces a potential number for your sample. If the number is not in your set or repeats a number already selected, ignore it and go on to the next pair of numbers. (If you use a calculator, follow a comparable procedure. Use a formula to produce the range of numbers you want or just use the first two digits of the decimal number produced by the random number function.)

Example: Start with the fifth set of numbers in a portion of a random number table (as shown).

14342 36857 69578 40961 03969 61129 97336

These numbers produce person 14, 34, 23, 68 (omit), 57 (omit), 69 (omit), 57 (omit), 84 (omit), 09, 61 (omit), 03, 96 (omit), 96 (omit), 11. Six people have now been randomly selected: the ones labeled 14, 34, 23, 09, 03, and 11.

1. Make a team of eight students. Choose a random sample of three from your group. Describe your process.
2. Explain how to select a sample of 20 mice from a lab that has 135 mice.
3. When will selecting a random sample be easy to do? When will it be difficult? Why?
4. How does a random sample reduce bias?

TEACHING AND LEARNING HIGH SCHOOL MATHEMATICS © 2010 John Wiley & Sons, Inc.

15. Two ways to describe distributions are with graphs and with measures. The two student pages **Creating a Five-Number Summary and a Boxplot** and **Understanding Standard Deviations**, introduce some of these tools.

   a. For each student page, identify segments of direct instruction.

   b. For each student page, identify inquiry segments that engage students as they make meaning for themselves.

   c. How have the authors of the student pages tried to balance direct and inquiry-based instruction?

   d. Identify one or two questions you would pose in lesson summaries for each student page.

   e. For one of the student pages, write a complete lesson plan using the **Annotated Lesson Planning Guide** to inform your work. Be sure to complete a **QRS Guide** as part of your lesson plan.

Creating a Five-Number Summary and a Boxplot

Materials

- graphing calculators, one per person

Recall the data about the homerun statistics from some famous baseball players. How can their performances be compared?

Homerun Records for Some Major League Baseball Players

	Roger Maris	Mickey Mantle	Hank Aaron		Ty Cobb		Babe Ruth
1951		13		1905	1	1914	0
1952		23		1906	1	1915	4
1953		21		1907	5	1916	3
1954		27	13	1908	4	1917	2
1955		37	27	1909	9	1918	11
1956		52	26	1910	8	1919	29
1957	14	34	44	1911	8	1920	54
1958	28	42	30	1912	7	1921	59
1959	16	31	39	1913	4	1922	35
1960	39	40	40	1914	2	1923	41
1961	61	54	34	1915	3	1924	46
1962	33	30	45	1916	5	1925	25
1963	23	15	44	1917	6	1926	47
1964	26	35	24	1918	3	1927	60
1965	8	19	32	1919	1	1928	54
1966	13	23	44	1920	2	1929	46
1967	9	22	39	1921	12	1930	49
1968	5	18	29	1922	4	1931	46
1969			44	1923	6	1932	41
1970			38	1924	4	1933	34
1971			47	1925	12	1934	22
1972			34	1926	4	1935	6
1973			40	1927	5		
1974			20	1928	1		
1975			12				
1976			10				

Source: http://www.baseball-reference.com/leaders/HR_season.shtml, downloaded June 8, 2009.

(continued)

TEACHING AND LEARNING HIGH SCHOOL MATHEMATICS © 2010 John Wiley & Sons, Inc.

One tool is a boxplot. It shows five special values. First, put the data for each player in order from smallest to largest. Then find the center value, called the **median**. The Aaron data is used to illustrate the procedure:

10 12 13 20 24 26 27 29 30 32 34 34 38 39 39 40 40 44 44 44 44 45 47

There are 23 values, so the median is the twelfth value, 34. If the number of data values had been even, then the median would have been the average of the two most center values.

Next, find the median of the lower and upper halves of the set of data. These are called 1st (Q1) and 3rd (Q3) **quartiles** respectively, because, with the median, they divide the data into four sets, each with the same number of values. (The root of *quart* means *four*.)

1. Find the values for the 1st and 3rd quartiles.

The **five-number summary** of the distribution includes the 1st quartile, the median, the 3rd quartile, and the minimum and the maximum.

2. List the five-number summary for Hank Aaron in order from smallest to largest.

3. Find the five-number summary for Babe Ruth. (Hint: Organize the data first.)

4. Use the two five-number summaries to compare the Aaron and Ruth homerun records.

Your comparison in item 4 likely compared the players' median number of homeruns and found them to be fairly close. The difference between the maximum and minimum is called the **range** of a set of data. Another measure sometimes used is the difference between Q3 and Q1. This is called the **interquartile range** or **IQR**.

5. Find the IQR for each player.

The five-number summary of a distribution leads to a graph called the **boxplot**. On one grid, make a scale that will accommodate both sets of data. Use the five-number summary. Create a central box whose vertical edges are at the quartiles. Make a vertical line in the box to indicate the median. Make horizontal lines from the box to the minimum and maximum values; these are called whiskers. Here is an example for the Aaron data.

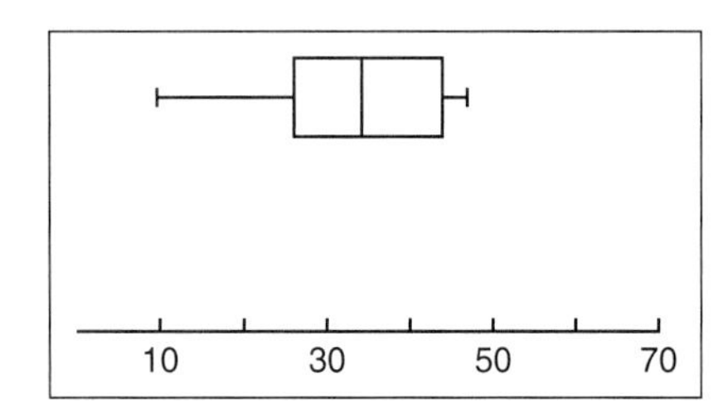

6. Verify that the five-number summary you listed in item 2 corresponds with the boxplot for Aaron.

7. Make another boxplot on the same grid for the Ruth homerun data.

8. Continue to make comparisons between the two players' homerun data. What does the pair of boxplots make clearer that you saw already? What does the pair of boxplots help you see beyond what you saw from the five-number summary?

9. A graphing calculator will also allow you to create boxplots. Use a calculator to make boxplots for the other baseball players in the table on p.335.

(continued)

TEACHING AND LEARNING HIGH SCHOOL MATHEMATICS © 2010 John Wiley & Sons, Inc.

On the TI-83/84, enter each player's data into a `LIST` accessed through the `STAT` menu, then `EDIT` command. Use `(2nd) STAT PLOT` to turn each plot `ON` and to select the boxplot icon from the graph choices (the fifth choice). A reasonable `WINDOW` is `Xmin = 0 to Xmax = 70`. Finally, press `GRAPH` to see up to three boxplots on the same screen. What comparisons can you make from these boxplots?

10. Some graphing calculators produce two different types of box plots, one showing extreme values called **outlier**s as separate points rather than simply on the whiskers of the graph. A point that is more than 1.5 times the IQR above the upper quartile (or below the lower quartile) is considered an outlier and is graphed as a separate point. Use your graphing calculator to produce a box plot with outliers for the case of Roger Maris. What does this tell you about his record?

Understanding Standard Deviation

Materials

- graphing calculators, one per person

Boxplots use a five-number summary of a data set to give a snapshot of a distribution. They can be helpful to compare distributions.

1. What do boxplots tell you about the consistency or reliability of a data set?

Even though the five-number summary and boxplots are helpful tools, they are not the most commonly used measures to study and compare distributions. The most common numerical summary measures of a distribution are its **mean** and **standard deviation**.

The mean, like the median, is a measure of the **center** of a data set. A center is a value around which all the other values are spread. The standard deviation, like the range or the interquartile range, is a measure of the **spread** of a data set. A spread provides a way to think about the variability of the data values in a set.

The standard deviation, σ, is the square root of the average of the squared deviations from the mean, $\sigma = \sqrt{\frac{\sum_{i=1}^{n}(x_i - \bar{x})^2}{n}}$.

You are likely wondering how someone invented a measure like standard deviation. Recall that the goal is to determine how distant or close the data values are from their mean. A deviation alone tells the distances of each value from the mean. But when the sum of the deviations is computed, the sum is zero. Another idea has been to use the absolute values of the deviations. Those would not sum to zero, but statisticians have found them less helpful for more sophisticated analyses. So instead, the squared deviations are averaged. Squaring, like taking an absolute value, eliminates the problem of negatives. Later the square root is taken so that the units in the numbers return to the original units.

A graphing calculator provides two measures for the standard deviation: s represents the standard deviation of a sample and the Greek letter sigma (σ) represents the standard deviation of a population. The minor differences in the values of s and (σ) occur when finding the average of the squared deviations; for the population, the divisor is n and for a sample the divisor is $n - 1$. Samples tend to vary less than populations, so dividing by $n - 1$ corrects for this difference.

Revisit the baseball homerun data from the student page *Creating a Five-Number Summary and a Boxplot.*

2. a. Use a graphing calculator with statistical features to enter the data for Aaron and find the standard deviation.

(continued)

TEACHING AND LEARNING HIGH SCHOOL MATHEMATICS © 2010 John Wiley & Sons, Inc.

b. Use a graphing calculator to find the standard deviations for the homeruns of the other players: Babe Ruth, Ty Cobb, Roger Maris, and Mickey Mantle.

3. Which player was the most consistent in hitting homeruns? How do you know?

4. Roger Maris's record is interesting, because his record of 61 homeruns in 1961 was so different from the rest of his career. Find the mean and standard deviation for his record, both with and without his 61 homeruns. How do the results compare? Why?

5. Consider the following three data sets, each with 13 data points and each with a mean of 3.

Set A: 1 2 2 2 3 3 3 3 4 4 4 4 4

Set B: 1 2 2 3 3 3 3 3 3 3 4 4 5

Set C: 2 2 2 2 2 2 3 4 4 4 4 4 4

a. Make a dot plot of each. From the dot plot, how does each set vary?

b. Complete the following table.

Measure of variability	**Set A**	**Set B**	**Set C**
Range			
IQR			
Standard deviation			

c. What are some possible benefits of graphing rather than just looking at measures to compare how sets of data vary?

This is a standard deviation contest. Choose six numbers from the whole numbers 0 to 20 inclusive with repeats allowed.

a. Choose a set of four numbers that have the smallest possible standard deviation.

b. Choose a set of four numbers that have the largest possible standard deviation.

c. Is there more than one possibility in either problem 6a or 6b? Explain.

Consider the data set *P*: 1, 2, 3, 4, 5, 6 and the data set *Q*: 1, 2, 3, 4, 5, 20. Find the median, mean, interquartile range, and standard deviation for each. Which of the four measures are more sensitive to outliers?

Item 5 is adapted from A.J. Nitko, 1983. *Educational Tests and Measurement: An Introduction,* cited in Aliaga, M. and B. Gunderson. *Interactive Statistics,* Second Edition. Pearson Education. Upper Saddle River, NJ: Pearson Education, 2003. Item 6 is adapted from Moore, David S. *Statistics: Concepts and Controversies.* New York: W.H. Freeman and Co, 2001.

16. Compare the student page **Understanding Standard Deviation** from **DMP** problem 15 with the student page **Computing Standard Deviation** from the lesson.

a. How are they alike in developing understanding of the concept as well as procedural facility to compute the standard deviation? How are they different?

b. Which one does a better job of blending direct instruction and inquiry learning? Justify your choice.

TEACHING AND LEARNING HIGH SCHOOL MATHEMATICS © 2010 John Wiley & Sons, Inc.

17. Consider the student page **Characteristics of the Normal Distribution.**

   a. What are the learning outcomes for this student page?

   b. How does the student page blend direct instruction with inquiry?

   c. If you were planning a **Share and Summarize** portion of the lesson, what critical questions would you include?

   d. Create two additional questions that would help students continue to develop or clarify this topic.

Characteristics of the Normal Distribution

Materials

- graphing calculators, one per person

Many distributions from biology produce a bell-shaped graph. The following table shows a frequency distribution of lengths of 1,000 ears of corn in inches, measured to the nearest half inch.

Lengths of ears (inches)	Frequency	Lengths of ears (inches)	Frequency
4.0	1	7.5	202
4.5	6	8.0	144
5.0	18	8.5	71
5.5	43	9.0	42
6.0	90	9.5	15
6.5	140	10.0	2
7.0	226		

1. Compute the median, mean, and standard deviation of these data. (On a TI-83/84, you can compute one-variable statistics for a set of data in frequency data form. Insert the inches into `L1` and the frequencies into `L2`. Then in the STAT menu, use the commands, `CALC`, `1-Var Stats`, `L1`, `L2`.)
2. Make a histogram of these data.
3. How do the mean and median of this data set compare?
4. On the horizontal axis of the histogram, mark points to locate the mean and 1, 2, and 3 standard deviations above the mean and 1, 2, and 3 standard deviations below the mean.
5. Find the percentage of the data that are
   a. Within one standard deviation of the mean.
   b. Within two standard deviations of the mean.
   c. Within three standard deviations of the mean.
6. Estimate a smooth curve that fits closely to the midpoints of the bars in your histogram. Describe its shape.

In many situations when large numbers of natural measures are collected, the resulting histogram of frequencies of the measures is close to a bell-shaped curve called a **normal curve**. A distribution with a histogram that follows a normal curve is called a **normal distribution**. In a normal distribution, the mean and median are equal and fall at the line of symmetry for the curve. Also, for a normal curve,

- About 68% of the data are within one standard deviation of the mean.

(continued)

TEACHING AND LEARNING HIGH SCHOOL MATHEMATICS © 2010 John Wiley & Sons, Inc.

- About 95% of the data are within two standard deviations of the mean.
- About 99.7% of the data are within three standard deviations of the mean.

7. How well does the distribution of corn measurements match that of the normal distribution?

8. Suppose a particular brand of candy bar is advertised to weigh 6.5 ounces. The actual weights vary. Assume the weights are normally distributed with a standard deviation of 0.15 ounce.

a. What percentage of candy bars likely weigh more than 6.5 ounces?

b. What percentage of candy bars likely weigh more than 6.65 ounces?

c. What percentage of candy bars likely weigh less than 6.8 ounces?

18. Compare the student pages **Characteristics of the Normal Distribution** from **DMP** problem 17 and **Normal Distributions and z-Scores** from the lesson. The two student pages repeat some of the same information, so teachers might not use both of them with the same group of students.

a. Comment on the mathematical sophistication students need for each student page. For what level course might you use each student page?

b. For each student page, create one or two additional questions you might ask students to assess their understanding of the underlying concepts.

Analyzing Students' Thinking

19. Producing a random sample is a careful and scientific process. Yet some students think random means haphazard. How might you help students better understand what random means?

20. Proportional reasoning is regularly identified as a critical strand for middle grades students so they are prepared for many high school mathematics topics. What role does proportional reasoning play in work in statistical reasoning?

Teaching for Equity

21. Read "Promoting Equity in Mathematics: One Teacher's Journey" by Alan D. Tennison (*Mathematics Teacher*, 101 (August 2007): 28–31).

a. What are the main premises of this article as they relate to equity?

b. How have the premises in the article been evident in your classroom observations?

Reflecting on Professional Reading

22. Read "The Histogram–Area Connection" by William Gratzer and James E. Carpenter (*Mathematics Teacher*, 102 (December 2008/January 2009): 336–340).

a. What do the authors mean by density histogram? Why do they believe these are important?

b. Compare the frequency histogram of snake length in Figure 1 in the reading with the density histogram in Figure 2. What misconceptions might arise when using Figure 1?

c. Learn how to produce, if possible, a density histogram on your calculator.

d. Many portions of this article are direct instruction. What contributes to the effectiveness of these segments?

TEACHING AND LEARNING HIGH SCHOOL MATHEMATICS © 2010 John Wiley & Sons, Inc.

4.4 Planning for Alternative Schedules: Statistical Decision Making

The time frames within which mathematics instruction occurs vary at different schools. Some high schools use *block schedules*. In one form of block scheduling, students complete what had formerly been a full year's course in one semester by meeting in 80–90 minute periods daily rather than in more typical 50–55 minute periods throughout the year. In other block schedules, students complete a course in a full year but meet in 80–90 minute periods every other day. Some schools have hybrid schedules, with classes meeting for longer periods of time on some days and more traditional or shorter lengths on other days. Still others have rotating schedules so that students might meet for a course every six school days out of seven. Although longer time periods create their own unique challenges in terms of planning, shorter periods every day create other challenges. Further, the inquiry teaching recommended by this textbook requires teachers to plan flexibly so that student thinking drives a lesson. Consequently, it is challenging to plan for even a week at a time, let alone for an entire unit.

Earlier in Unit Four, you planned lessons or major segments of lessons in detail. In this lesson, you will consider a sequence of lessons. You will have an opportunity to practice planning instructional time for different possible schedules.

This lesson includes a sequence of lessons to develop the statistics and probability ideas begun earlier in this unit. Assume that students enter these lessons having worked with probability, the multiplication counting principle, permutations, and combinations. (See student pages in Lesson 4.2 **DMP** problems 14 and 15 and in **DMP** problem 13 of this lesson for samples. These may need to be reviewed before engaging in this lesson.)

In this lesson, you will:

- Explore the mathematics of binomial experiments with simulation and with theory.
- Consider how statistical decision making works.
- Plan a portion of a unit, given different scheduling constraints.

Website Resources

- **A Case of Bias?** student page
- **Binomial Experiments** student page
- **Decision-Making with Probability** student page

LAUNCH

1. Your school is considering changing from a 55-minute daily class schedule (Schedule A in what follows) to a block schedule (Schedules B or C). You are on a committee to determine the viability of each possible plan. You and two colleagues represent the concerns of mathematics teachers. Together, you are looking at how a unit of instruction would play out with the original and alternative plans.

   Schedule A: Each course meets daily for 55 minutes.

   Schedule B: Each course meets on alternating days for 110 minutes.

   Schedule C: Each course meets Monday, Tuesday, Thursday, and Friday for 62 minutes and Wednesday for 27 minutes. (The half day on Wednesday provides opportunities for teachers to engage in professional development during the work day.)

TEACHING AND LEARNING HIGH SCHOOL MATHEMATICS © 2010 John Wiley & Sons, Inc.

a. What are some general ideas you have about how you might use the different class period lengths to best support student learning?

b. What might you think or do differently in each of the different time frames?

EXPLORE

2. Read the three student pages **A *Case of Bias?* Binomial Experiments**, and **Decision-Making with Probability**. Assume these sequential pages are part of an eleventh grade unit on statistical decision making.
   a. Identify the learning outcomes for each page.
   b. How might you **Launch** the lesson in which each student page is used?
   c. For each student page, identify the important ideas you want to ensure are highlighted in the **Share and Summarize** portion of the lesson.

SHARE

3. a. Share what teams decided were learning outcomes for the three student pages.
   b. For each student page, share how you might **Launch** a lesson in which the student page is used.
   c. For each student page, share the important ideas you want to ensure are highlighted in the **Share and Summarize** portion of the lesson.

TEACHING AND LEARNING HIGH SCHOOL MATHEMATICS © 2010 John Wiley & Sons, Inc.

A Case of Bias?

Materials

- graphing calculators, one per person

Solo Productions is hiring 10 new associates for multimedia jobs. Suppose a very large number of equally qualified people apply, 45% are women and 55% are men. Ultimately, of the 10 persons hired, eight are men. Some women wonder whether the process was biased. Investigate whether this outcome is likely to have happened by chance or whether bias might be involved.

1. What are likely numbers of women you would expect to be hired, if all the candidates were equally qualified? Why?
2. How likely do you think it would be for 0 women to be hired? 1 woman? 2 women?
3. Suppose the pool of candidates had 10% women and 90% men. In that case, how likely would you expect it to be that 0 women were hired? 1 woman? 2 women?

Imagine a simulation to investigate this problem. Because the candidates were equally qualified, suppose the results were strictly due to chance. You would like to look at the distribution (or graph) this situation produces. You might make a deck of 100 blank cards and label 55 *men* and 45 *women*, then repeatedly shuffle and select sets of 10. You would determine how many *men* were in each set of 10.

Rather than do the simulation manually, it is easier to use a calculator with a randomizing feature. On a TI-83/84 calculator, `randInt(1,100)` will produce a random integer between 1 and 100, inclusive. Suppose you designate numbers from 1 to 55 to represent men and numbers from 56 to 100 to represent women. `randInt(1,100,10)→ L1` will create a set of 10 *people* and store them in `L1`. You can then inspect the list and count how many are men (numbers 1–55).

4. With your partners, perform this simulation nine times. Record your findings.

Trial	1	2	3	4	5	6	7	8	9
Number of men									

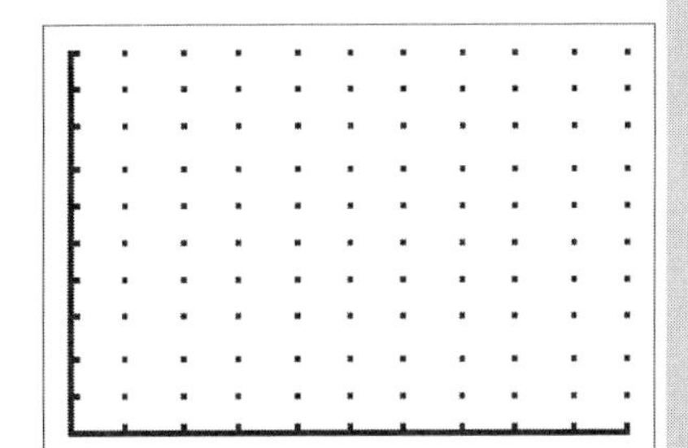

5. Pool your findings for the entire class. Create a frequency graph of the results. (The horizontal axis will be the number of men appearing in a trial and the vertical axis will be the frequency with which that number appeared.)
6. What is the general shape of the graph of the distribution?
7. What numbers of men occurred most often?
8. What numbers of men occurred least often?
9. Use the graph or data to estimate the probability of each event.
   - **a.** Exactly two of the people hired are women.
   - **b.** Exactly one of the people hired is a woman.
   - **c.** Exactly zero of the people hired are women.
   - **d.** Two or fewer of the people hired are women.
10. How do you use the graph to estimate the probability that two or fewer of the people hired are women? Why does this make sense?

(*continued*)

TEACHING AND LEARNING HIGH SCHOOL MATHEMATICS © 2010 John Wiley & Sons, Inc.

Binomial Experiments

Materials

- graphing calculators, one per person

The experiment described in the student page A *Case of Bias?* is a special type of situation called a **binomial experiment**. In a binomial experiment:

- There are exactly **two** outcomes for each trial (e.g., either a man or a woman is hired).
- The trials are independent (e.g., the selection of each candidate is not influenced by the selection of another).
- Each trial has the same probability of success (e.g., the probability that a man is hired is 0.55 for each trial).
- The experiment has a fixed number of trials (e.g., exactly 10 people were hired).

A binomial experiment can be analyzed with a simulation, as was done in the previous activity, or with probability theory as is done on this page.

1. Consider first a simpler case; suppose there were just five jobs to fill.

a. In how many ways can the five slots be filled with exactly two women? Use two different methods.

b. Jan was solving problem 1a. He claimed,

I can use a combination, $C(5, 2) = \frac{5 \bullet 4}{2!} = 10$. *Call the positions job 1, job 2, and so on to job 5. There are 5 possible jobs for the first woman. There are 4 possible jobs for the second woman. So there are 20 ways the five jobs could be filled by two women. But I need to divide by 2 because each pair of jobs comes up twice.*

Evaluate Jan's reasoning.

2. Now consider the original case where there are 10 jobs. Suppose exactly two of the 10 slots are filled with *W* (a woman), which is the same as exactly eight being filled with a man.

a. In how many ways might this be done?

__ __ __ __ __ __ __ __ __ __

b. How could you use the notation for combinations to symbolize the number of ways to select exactly two of the 10 slots?

3. Your result in problem 2 tells the possible ways that the 10 jobs could be filled with exactly eight men and two women. If you actually wrote down all the possibilities, it would be a very long list! Imagine just starting to write the first several cases.

Case 1: WWMMMMMMMM

Case 2: WMWMMMMMMM

Case 3: WMMWMMMMMM

Case 4: WMMMWMMMMM

. . .

Consider the probability of Case 1. Recall that the pool is large and all the candidates are equally qualified, so you may assume that filling one job has no effect on filling another. In other words, all the hirings are independent of one another. Recall that 55% of the applicants are men and 45% are women.

a. What is the probability that the first job is filled by a woman?

b. Show the calculation you would use to find the probability that the first job is filled by a woman and the second job is filled by a woman. Explain why your method is reasonable.

(continued)

TEACHING AND LEARNING HIGH SCHOOL MATHEMATICS © 2010 John Wiley & Sons, Inc.

c. Show the calculation you would use to find the probability that the first job is filled by a woman, the second job is filled by a woman, and the third job is filled by a man (as in Case 1).

d. Continue in the same way to find an expression to calculate the probability of exactly the situation of the 10 jobs filled in Case 1.

e. Use exponents to simplify the expression, but do not calculate.

4. In problem 3, you found an expression for the probability of exactly one case.

a. The probability that Case 1 occurs is the same as the probability for any of the other cases with exactly two women and eight men. Explain why.

b. In problem 2 you found that hiring two women and eight men for the 10 jobs could occur in many ways. (Let n represent the number of ways you found in problem 2.) You want the probability that in one case or another there were exactly two women and eight men hired. To find this, you need an expression that could be calculated to find the probability of Case 1 OR Case 2 OR Case 3 OR . . . Case n. Create such an expression.

5. Below is an incomplete formula for finding the probability that exactly two women (and eight men) occur in a set of 10 trials where the probability of a man on each trial is 0.55. Complete it and explain why it makes sense.

P (exactly 2 women in a set of 10 where probability of a woman is 0.45)

$$= C(\_\_\_, \_\_\_) \cdot (0.55)^8 (\_\_\_)^2.$$

6. Use the completed formula from problem 5 or an appropriate modification to calculate the following:

a. the probability that exactly two persons hired were women

b. the probability that exactly one person hired was a woman

c. the probability that no person hired was a woman

d. the probability that two or fewer people hired were women

7. In solving problem 6d, you used addition. Why is addition reasonable?

8. Compare your results from problem 6d with those of the graph the class generated in *A Case of Bias?* problem 9d.

a. How close was your estimate to the calculated probability?

b. Do you believe there was bias in the original situation? Explain.

TEACHING AND LEARNING HIGH SCHOOL MATHEMATICS © 2010 John Wiley & Sons, Inc.

Decision Making with Probability

One of the important uses of probability and statistics is to make decisions like, "How likely is it that an outcome is due to chance alone?" To make such a decision, researchers use these steps.

- Use a simulation or create a probability model to represent the situation.
- Assume the outcome is based solely on chance.
- Decide which probabilities to consider unusual if due to chance. Common values used are 0.01, 0.05, and 0.10. The value chosen is called a **significance level**. Use 0.05 unless specified otherwise.
- Use the simulation or a model to find probabilities.
- Compare the simulation or model with the actual results.
- Ask, "How does the probability from the simulation or the model compare to the significance level, for example, 0.05?"

If the probability is smaller, then the researchers suppose some other factors, not just chance, are involved. If the probability is larger, then the researchers decide the outcome could likely be due to chance.

In problems 1–3, use the 0.05 significance level.

1. What does your graph from the simulation in *A Case of Bias?* indicate about the possibility of bias in hiring?

2. What decision does your probability model from Binomial Experiments support about the possibility of bias in hiring?

3. What decision would the probability model from Binomial Experiments indicate if nine men were hired (and one woman)? Seven men (and three women)?

Statistical decision making never proves or disproves a claim. A statistical analysis helps you decide, at a specified level of probability (called the significance level), whether an outcome is likely due to chance or likely due to other circumstances.

Here is the logic behind statistical decision making:

If chance alone is operating, **then** a particular result is impossible.

The result happened.

Therefore, it is not the case that chance alone is operating.

The difficulty with this analysis is that "a particular result is impossible" is not exactly the case. Here is a better version:

If chance alone is operating, **then** a particular result is unlikely.

The result happened.

Therefore, it is unlikely that chance alone is operating.

4. Your friend's dad has a magic trick that allows him to toss a coin and call "heads" or "tails" with remarkable accuracy. He demonstrates that he can toss a coin and call it correctly 9 times out of 10.

a. Assume someone who is not a magician has a fair coin and tosses it. What is the probability that this person calls it correctly 9 or more times out of 10?

b. Write the logical argument that applies to this situation.

c. If you observed your friend's dad correctly call the coin 9 times out of 10, what scientific conclusion would you draw?

TEACHING AND LEARNING HIGH SCHOOL MATHEMATICS © 2010 John Wiley & Sons, Inc.

EXPLORE MORE

4. Estimate the time you would spend with students on each of the student pages in this lesson.
5. Your team will be assigned one of the three teaching schedules in the district. For your schedule, work out a time plan for teaching three lessons that use the student pages. Include time to support students as they re-establish concepts and processes they have already studied, engage with the student pages, and discuss homework. You may find the following *Block Scheduling Perspectives* helpful.

Block Scheduling Perspectives

- When a class meets for an extended period, activities must be rich and varied and employ strategies that help students engage, focus, and build understanding.
- How time is used depends on a lesson's learning outcomes, nature of the topic, activities involved, and the learning environment. One design will not be suitable for all lessons.
- A lesson should consist of at least three elements, including an introductory activity, an exploration, and a summary or assessment. Some suggestions follow:
  - Warm-up: Brief engaging activity with clear directions that is relevant to new or previous material, for students to begin as soon as they arrive. A warm-up that anticipates a problem that arises later in the lesson or looks back at earlier misconceptions makes the lesson more coherent.
  - Checking homework: Needs to be done efficiently. (Strategies are addressed in Unit Five).
  - Launch: An activity is motivated and introduced.
  - Explore: Students work in small groups on worthwhile mathematical tasks that can involve written materials, manipulatives, technology, etc.
  - Share: Students or teams share responses the teacher has selected and sequenced. The class may work together on a strategy.
  - Summarize: Teacher and students highlight the major ideas to be drawn from the activity.
  - Writing: Students individually record what they understand to a given point and what questions they have.
  - Brief quiz: A warm-up or closing activity assesses new ideas or maintains previously learned skills; it might include mental mathematics.

Adapted from Gilkey, Susan N. and Carla H. Hunt. *Teaching Mathematics in the Block*. Larchmont, NY: Eye on Education, 1998.

SHARE AND SUMMARIZE

6. Share with the class your team's thinking on **Explore More** questions 4 and 5.
7. Compare your thinking now with your response to **Launch** problem 1. What insights have you gained into planning a portion of a unit?

TEACHING AND LEARNING HIGH SCHOOL MATHEMATICS © 2010 John Wiley & Sons, Inc.

DEEPENING MATHEMATICAL UNDERSTANDING

1. *Generalization of Binomial Distribution*. Consider the formula in **Binomial Experiments** problem 5. Of the two outcomes (*man* or *woman*), consider one a *success* and the other a *failure*. Let p = probability of *success*. Let $q = 1 - p$ = probability of *failure*. Let n = number of trials.
   - **a.** Generalize the formula to find the probability of exactly r successes in n trials.
   - **b.** The generalization involves multiplying certain probabilities. Which ones are multiplied? What is the reasoning that allows this multiplication?
   - **c.** Generalize the formula to find the probability that **at least** r successes occur in n trials.
   - **d.** The generalization in problem 1c involves adding certain probabilities. Which ones are added? What is the reasoning that allows this addition?

2. *The Logic of Hypothesis Testing*. Consider the *if–then* reasoning discussed near the end of the student page **Decision-Making with Probability**.
   - **a.** In Unit One, you studied different forms of valid arguments. What form is being used on this student page?
   - **b.** A misconception made in statistics is to reason as follows:

     If A is true, **then** B happens.

     B happens.

     Therefore, A is true.

     Use a Venn diagram to show why this reasoning is invalid.
   - **c.** Compare the invalid argument in problem 2b with the following:

     If A is true, **then** B is unlikely to happen.

     B does not happen.

     Therefore, A is true.

     Tell why the argument above is invalid.

In problems 3–5, refer to Pascal's Triangle shown below. Pascal's Triangle arises in multiple contexts. In this lesson, the numbers occur in binomial experiments. The numbers also occur as coefficients of powers of binomials $(a + b)^n$. The source of both of these is combinations, $C(n, r)$ (from a set of n objects, choose r). Problems 3–5 help you explore why these same numbers occur in these seemingly different contexts.

Pascal's Triangle

Row 0   1
Row 1   1 1
Row 2   1 2 1
Row 3   1 3 3 1
Row 4   1 4 6 4 1
Row 5   1 5 10 10 5 1

3. *Connecting Combinations with Pascal's Triangle.*
   - **a.** Find the values for Row 6 of Pascal's Triangle.
   - **b.** What combinations problems do the values in Row 6 represent?
   - **c.** Where would you find $C(7, 3)$ in Pascal's Triangle?
   - **d.** A defining characteristic of Pascal's triangle is that each value is the sum of the two values above it in the previous row. This corresponds to the

TEACHING AND LEARNING HIGH SCHOOL MATHEMATICS © 2010 John Wiley & Sons, Inc.

relationship $C(n, r) + C(n, r + 1) = C(n + 1, r + 1)$. Verify this use of the variables.

e. Use the general formula for a combination, $C(n, r) = \frac{n!}{r!(n-r)!}$ and algebra to prove the equality in problem 3d.

4. *Connecting Coin Tosses with Pascal's Triangle.* Consider flipping a coin five times.

a. Use the criteria in **Binomial Experiments** to confirm this is a binomial experiment.

b. Imagine flipping a coin five times. How many possible outcomes are there?

c. How does the total number of outcomes for flipping a coin five times arise from Pascal's Triangle? Hint: How many ways are there to get exactly zero heads? one head? two heads? etc.

5. *Connecting Binomial Distributions with Pascal's Triangle.*

a. Explain how expanding $(0.55 + 0.45)^{10}$ (without calculating) is like finding the probabilities of various numbers of men getting jobs out of 10 new hires (as in the student page **A Case of Bias?**).

b. Explain in your own words why both binomial expansions in algebra and binomial experiments involve combinations.

6. *Comparing a Binomial with a Normal Distribution.* A graphing calculator can be used to produce a histogram of a binomial distribution and compare it to a normal distribution.

a. Do enough of the student page **Distributions on a Graphing Calculator** so that you are able to complete the last question.

b. Some statistics books caution that a normal distribution is not a reasonable approximation of a binomial function when $np < 10$. (The value n is the number of trials. The value p is the probability of "success.") Why would statisticians propose such a restriction?

Distributions on a Graphing Calculator

Materials

- graphing calculators, one per person

Recall the hiring scenario in **A Case of Bias?** and **Binomial Experiments.** You can use your graphing calculator to see the full probability distribution for this or other binomial situations. (Here are directions for a TI-83/84 for the hiring situation in **A Case of Bias?**).

- First enter into `LIST L1` the numbers 0 through 10. (Press `STAT`, `choose 1: Edit`, then scroll over to `L1` and enter the values 0, 1, . . . , 10.)
- Then clear `L2` where entries will shortly be made. (Scroll your cursor up to the name `L2` and press `CLEAR`, then press `ENTER`.) Press `QUIT` to return to the home screen.
- Press 2nd `DISTR` to locate the distribution menu (`DISTR` is written above the `VARS` key). Scroll and select `A:binompdf(` then press `ENTER`.
- `Binompdf` will allow you to generate all the values in a binomial probability distribution function (pdf). You enter two inputs: n (the number of trials) and p (the probability of success).
- For the hiring situation, enter 10, 0.55 and then close the parentheses.
- Press `STO` 2nd `L2` to store the values produced in `LIST L2`. Then press `ENTER`.
- Check that the values are stored by pressing `STAT`, then `Edit`, then scrolling to `L2`.
- For each value in `L1`, the value in `L2` gives the probability that exactly the given numbers of men were hired if the selections were totally random and there were 55 men and 45 women in the pool. Now graph these values.

(*continued*)

TEACHING AND LEARNING HIGH SCHOOL MATHEMATICS © 2010 John Wiley & Sons, Inc.

- To get ready to graph, go to the `Y=` screen and turn off all equations.
- Set up `STAT PLOT` to view a scatterplot with `L1` as the independent variable, `L2` as the dependent variable. Turn `PLOT ON`. Set the `WINDOW` with x in the interval [0,10] and y in the interval [–0.1, 0.3].
- Press `GRAPH`. You should see the values in the probability distribution. Press `TRACE` to examine the values. For each x number of people, the probability that exactly that number of men were hired is y.

1. What is the probability that exactly four men were hired?

2. What is the probability that exactly eight men were hired?

3. What is the probability that eight or nine or 10 men were hired?

4. At the 0.05 level of significance, would hiring eight men be considered possible due to chance alone?

A binomial distribution can be approximated by a normal distribution. You need to know the mean and the standard deviation to produce a normal distribution graph. In general, the mean of a binomial distribution is np and the standard deviation is $\sqrt{npq}$. (Recall that n is the number of trials, p is the probability of "success," and $q = 1 - p$ is the probability of "failure.") (See a more advanced text for a proof.)

5. According to the formula above, for the hiring case, what is the mean? Why does this make sense?

6. According to the formula above, for the hiring case, what is the standard deviation?

7. Enter the following function into the `Y=` menu: `y = normalpdf(`x, 5.5, 1.57). (On the TI-83/84, the `normalpdf` function is on the `DISTR` menu.) Press `GRAPH`. You should see a continuous normal distribution graph that fits the binomial plot. Describe the normal distribution graph.

8. Suppose the original pool of equally qualified applicants consisted of 30% men and 70% women, with seven men hired for 10 positions. Redo the analysis using the graphing calculator. Keep notes on the entries you make on the calculator and why you changed values as you did. Using the 0.05 level of significance, was there bias?

Note: For a normal distribution to be a reasonable estimate of a binomial distribution, the expected number of successes np and failures nq must each be at least 10.

7. *Clarifying What Is/Is Not Binomial.*

a. Suppose a pool of candidates for a job contains 13 men and 2 women.

i. What is the probability that the first person hired is a woman?

ii. Finding the probability that the second person being hired is a woman depends on whether a man or woman was the first person hired. Explain.

iii. Why is this situation not binomial?

b. Toss a coin until you produce a head. Consider the probability that a specific number of tosses are required to produce a head. Why is this situation not binomial?

8. *Comparing Simulated and Theoretical Results.* Dan doesn't want to study for his biology quiz. He knows it will have 10 true–false questions. He would like to score at least 80%. He figures his chance of getting any single item correct is 0.6. He figures he has a good chance if he guesses on each question.

a. Find a way to simulate Dan's testing situation 100 times. (Make up a key and use a random number table or generator or computer program to generate his answers, then check his paper.) From the simulation how likely is he to earn 80% or better?

TEACHING AND LEARNING HIGH SCHOOL MATHEMATICS © 2010 John Wiley & Sons, Inc.

b. Plot your results. What is Dan's most likely score?

c. State why Dan's situation can be viewed as a binomial experiment.

d. Using the binomial distribution, find the probability that Dan earns 80% or better.

e. How do the simulation and theoretical results compare?

DEVELOPING MATHEMATICAL PEDAGOGY

Preparing for Instruction

9. Select one of the following sets of student pages from earlier in this textbook. Imagine you have two 50-minute periods and one 25-minute period to help students engage with these materials using the **Launch**, **Explore**, and **Share and Summarize** format. Consider some homework that might be part of this segment of instruction. Draft a possible timetable for these lessons. Record your results in a **Lesson Planning Guide** using the **Annotated Lesson Planning Guide** to inform your work.

Topic	Student Pages (Lesson Number)
Venn Diagrams/Logic	Multiples (1.3) Baseball and Soccer (1.3)
Right Triangle Similarity	Similar Right Triangles (2.2) Exploring Relationships in Right Triangles (2.2)
Exponential Functions	Comparing Graphs of $y = B^x$ (3.2) Experimenting with Growth and Decay (3.2)

10. Study some high school texts that include the topics of the three student pages in this lesson. Identify or design what could be reasonable homework assignments for the lessons using the timetable your team had in the **Explore**.

11. Science classes often have laboratory lessons and sometimes have longer class periods even when other classes are shorter.

    a. Talk to a science teacher who uses laboratory experiments regularly to gather ideas for using longer class periods effectively.

    b. What might be some advantages and disadvantages to teaching mathematics as a laboratory class?

12. Students often wonder why 0.05 is commonly used as the *cut-off* between what is likely due to chance and what is not. The following activity can be used to help students get a sense of this measure (adapted from Burrill et al., 2003). Get two new decks of playing cards in sealed boxes. Carefully open the boxes and remove the cards, separating red from black. Put all the red cards, plus one red and one black joker into one box. Similarly, put all the black cards in the other box. Reseal the boxes, noting which has the red cards. Take the box of red cards to class and offer a prize to the first person who draws a black card. Break the seal in front of the class and remove the jokers to reinforce the idea that you have a sealed, intact, new deck. Have students individually draw cards, one at a time. After each draw, return the card to the deck and reshuffle, concealing the cards so colors do not show. It usually takes five students for the class to become suspicious.

    a. In a fair deck, what is the probability that all five cards drawn would be red?

    b. In a fair deck, what is the probability that the first four cards drawn would be red?

TEACHING AND LEARNING HIGH SCHOOL MATHEMATICS © 2010 John Wiley & Sons, Inc.

c. How do these probabilities relate to statistical decision making?

d. How might you do a similar trick with a two-headed coin from a magic store?

13. A major prerequisite for students to work with binomial experiments is their understanding of combinations (and, previous to that, permutations and the multiplication counting principle). Following are a sequence of three student pages (**Counts and Probabilities**, **Ice Cream Social**, and **The Clarinet Club**) that could provide this background.

a. What are the learning outcomes of each student page?

b. How does each student page engage students in determining the critical relationships for themselves?

c. For each student page, write at least two problems that would help students continue to build understanding of the ideas on that page.

d. What would you emphasize in the **Share and Summarize** portion of a lesson for each student page?

e. For one of the student pages, complete a **Lesson Planning Guide**, using the **Annotated Lesson Planning Guide** to inform your work. Include a completed **QRS Guide** for the student page as part of your lesson plan.

Counts and Probabilities

Gavin is going to an orientation for arts camp.

1. He is deciding what to wear. He has five favorite shirts (orange, royal blue, navy, lemon, and red) and three favorite slacks (khaki, black, blue jeans).
   a. Use a rectangular array to show the outfits he can make.
   b. Use a tree diagram to show all the possible outfits.
   c. How many outfits can he make?
   d. How does the information in the array connect with the information in the tree diagram?
   e. What operation could you use to find the number of possible outfits? Why does this make sense?
   f. Suppose Gavin gets a new favorite shirt and new slacks before he starts camp. Now how many outfits can he make?
   g. Complete: If one choice can be made in m ways and a second choice can be made in n ways, then there are ___ ways of making the first choice followed by the second choice.
2. Gavin will participate in painting, clarinet, and ceramics at the camp. At 8 a.m., there is painting; he can take watercolor, oil, or acrylics. At 10 a.m., there is clarinet; he can take jazz or classical. At 1:00, there is ceramics; he can take wheel, construction, or free form.
   a. Use a tree diagram to find the possible schedules Gavin might create.
   b. How many possible schedules can Gavin create?
   c. How could you use the number of choices for each activity to find the number of possible schedules? Why does this make sense?
   d. How is the problem like the outfits situation in problem 1? How is it different from the outfits situation?
3. At the camp orientation, campers take a survey about sports they might enjoy. There are five sports listed. For each sport, the camper may indicate a level of interest as "willing to participate" or "unwilling to participate." Use more than one way to find all the possible ways Gavin might respond to the survey.

(*continued*)

TEACHING AND LEARNING HIGH SCHOOL MATHEMATICS © 2010 John Wiley & Sons, Inc.

In problems 4 though 6, use the following information. As an icebreaker, the camp orientation involves a card game with a special deck. The deck consists of king, queen, and jack in each of four suits (spades, hearts, diamonds, and clubs).

4. How many cards are in the deck? How is finding this number like the previous problems?

5. Two cards are drawn from the deck, one at a time, without replacing the first.

a. How many possible outcomes (pairs of cards) are there? Why?

b. Suppose both cards happened to be kings. In how many ways might this occur? Why?

c. Use the results from problems 5a and 5b to find the probability that when two cards are drawn, both are kings.

6. Reconsider the problem of getting two kings (without replacing the first card).

a. What is the probability that the first card is a king?

b. Assuming the first card drawn is a king, what is the probability that the second card is also a king?

c. Multiply the probabilities and compare the result to what you found in problem 5c.

d. Recall that $P(A \text{ and } B) = P(A) \cdot P(B|A)$. How does this compare to your work in problem 6c?

e. Apply your thinking from problem 6d to another case: What is the probability that a king is drawn and then (without replacing) a queen is drawn. Solve in two ways and explain how they do the same job in two different ways.

7. Write a few sentences explaining what you understand about uses of multiplication from these activities.

Ice Cream Social

Gavin's arts camp orientation includes an ice cream social. There are five flavors: vanilla, chocolate, pistachio, mint, and cherry. Students can have a single-, double-, or triple-scoop cone or they can have a bowl.

1. Gavin decides to have a double cone with two different flavors. To his way of thinking (or tasting!), vanilla with chocolate on top is different from chocolate with vanilla on top. How many different double cones might he create? Solve in two different ways (e.g., lists, arrays, trees, arithmetic), and show that the methods agree and produce the same result.

2. Gavin's pal Valen will also have two different flavors. But he prefers a bowl.

a. How many different bowls are possible for Valen? Solve in two different ways and show that the two ways agree.

b. How is the number of Valen's bowls related to the number of Gavin's cones? Why?

3. Jackson, another friend, is a real ice cream lover. He wants a bowl with three different flavors.

(continued)

TEACHING AND LEARNING HIGH SCHOOL MATHEMATICS © 2010 John Wiley & Sons, Inc.

a. How many different bowls are possible for Jackson? Solve in two different ways and show that they agree.

b. How does the number of bowls for Valen compare with the number of bowls for Jackson? Why?

4. Carleigh, an even greater ice cream lover, decides she will have a triple cone with three different flavors. Like Gavin, the order in which the flavors are stacked matters to Carleigh.

a. How many different cones are possible for Carleigh? Solve in two different ways and show that the two ways give the same result.

b. How does Carleigh's number of cones with three scoops compare to Jackson's number of bowls with three scoops? Why?

5. Dave is an assistant at the soda fountain making the cones and bowls. Before the ice cream social began, he went to the freezer and only found three flavors. (Yikes!) The vanilla and chocolate were missing.

a. How many ways can Dave line up the three different tubs in a row from left to right? Solve in two ways.

b. Dave quickly ordered the vanilla and chocolate and had them rushed over to the social. First, he placed the vanilla tub. In how many ways can four different tubs of ice cream be arranged from left to right?

c. How can the results with three tubs be used to determine the results for four tubs?

d. Then he placed the chocolate tub. In how many ways can five different tubs be arranged in a row from left to right?

e. How can the results for four tubs be used to determine the results for five tubs?

f. Identify patterns in the numbers of arrangements of three tubs, four tubs, and five tubs.

6. Decide if the following statements are true or false. Explain your reasoning.

a. The number of ways to arrange five tubs equals $5 \cdot$ (number of ways to arrange four tubs).

b. The number of ways to arrange five tubs equals $5 \cdot 4 \cdot$ (number of ways to arrange three tubs).

c. The number of ways to arrange five tubs equals $5 \cdot 4 \cdot 3 \cdot$ (number of ways to arrange two tubs).

d. The number of ways to arrange five tubs equals $5 \cdot 4 \cdot 3 \cdot 2 \cdot$ (number of ways to arrange one tub).

7. Summarize what you understand about the different counting problems on this page.

TEACHING AND LEARNING HIGH SCHOOL MATHEMATICS © 2010 John Wiley & Sons, Inc.

The Clarinet Club

Once he is at Arts Camp, Gavin decides to join the Clarinet Club.

1. There are three officers of the Clarinet Club: president, vice president, and secretary. As they prepared for their photo for the camp newsletter, they wondered how many ways they might stand in a row. How many ways are there? Explain your thinking.

2. Right before the photo was taken, the camp director arrived and wanted to be included in the photo. Now how many possible ways are there for the four people to stand in a row? Explain your thinking.

3. Factorial notation is helpful in many counting problems. **Factorials** apply to non-negative integers; $n!$ represents the product of the given number and all the previous positive integers. (For later purposes, $1! = 1$ and $0! = 1$.) For example, six factorial is $6! = 6 \cdot 5 \cdot 4 \cdot 3 \cdot 2 \cdot 1 = 720$.

a. Identify recent problems you have done (on this page or previous ones) where factorial notation could be helpful. Rewrite your answers, where appropriate, with factorial notation.

b. Marcus claims the number of possible "line-ups" for a baseball team of nine players is 9! (which equals 362,880). Do you agree or disagree with Marcus? Why?

4. Evaluate each of the following expressions. Find an easy way that does not use a calculator. Explain your reasoning.

a. $\frac{7!}{5!}$ **b.** $\frac{10!}{7!}$ **c.** $\frac{(n+1)!}{n!}$

Finding all the ways that n different objects can be placed in a definite order is called a **permutation**. As the previous problems illustrate, you can choose the first person n ways, then the second person $(n - 1)$ ways, then the third person $(n - 2)$ ways, and so on down to 1. Consequently $n!$ represents the number of permutations or arrangements of n distinct objects. There are other counting problems where the order (first, second, third) does not matter.

5. The conductor is looking for volunteers for a clarinet trio. Five players volunteer. The players and conductor wonder how many ways there are to create the trio from the five players. Several players had different strategies for figuring this out. Some of them need assistance. Study what each student did.

If the person's strategy is well-conceived and complete, explain it.

If the person's strategy is well-conceived but incomplete, complete it.

If the person's strategy is not well conceived, correct it.

Dina: *I made a list. I called the five people A, B, C, D, and E. I tried to make an organized list. But I didn't find as many trios as other people did. I only found 8 trios.*

ABC	*ACD*	*ADE*	*BCD*	*BDE*
ABD	*ACE*		*BCE*	

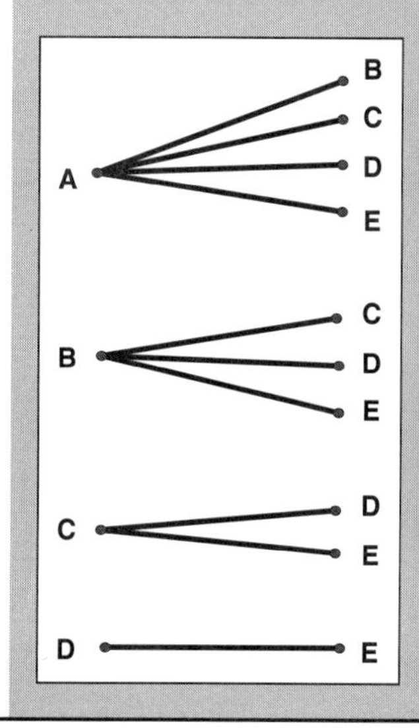

Effie: *I recalled another problem. If we choose three people from five, then two are left out. So, counting the number of pairs is the same as counting the number of triples. I used a tree diagram. I liked Dina's idea of using five letters for the five people. I found ten possible duos from the five players, so there are ten trios.*

Jason: *I thought I could use multiplication. There are 5 people who might be first, then 4 people who might be second, or 20 duos. Then there are 3 people who could be third. So there could be $20 \cdot 3 = 60$ trios. That's way more than the others got!*

(continued)

TEACHING AND LEARNING HIGH SCHOOL MATHEMATICS © 2010 John Wiley & Sons, Inc.

6. In working problem 5 (the trio problem), you may have found that Jason's work was incorrect but could be used as a basis for a correct solution. Actually Jason's strategy would work if the question were "With five clarinetists, in how many ways could three of them stand in a row?"

a. Explain why Jason's work fits the "stand in a row" question.

b. Explain why 5! (five factorial) itself does not work for the "stand in a row" question.

c. Jason's solution was $5 \cdot 4 \cdot 3$. Which of the following formulas produces the product $5 \cdot 4 \cdot 3$? Identify all that apply.

i. $\frac{5!}{3!}$ **ii.** $\frac{5!}{2!}$ **iii.** $(5 - 2)!$ **iv.** $\frac{5!}{(5-3)!}$

v. $5(5 - 1) \cdot \ldots \cdot (5 - 3 + 1)$

d. Suppose there were seven clarinetists and a photo was being taken with just four of them in a row from left to right. Write an expression to show how many different photos are possible. Explain.

7. Jason's solution to problem 5 was $5 \cdot 4 \cdot 3 = 60$, which represents the number of ways that three players could be chosen from five players and arranged in a row. The correct solution to problem 5 was 10, or $\frac{60}{6}$.

a. What does the factor of 6 mean in terms of the problem of selecting a trio? Why does it make sense to divide by 6?

b. Suppose there were nine clarinet players to volunteer for the trio. Now how many trios are possible? Solve in more than one way. Explain why each method makes sense.

c. Suppose there were nine clarinet players to volunteer; rather than a trio, a quartet was to be formed. How many quartets are possible? Solve in more than one way. Explain why each method makes sense.

8. Gavin was trying to summarize what he had learned about the trio and quartet problems and others like them. Read his summary. If you agree, then symbolize what he is saying. If you disagree, edit his statement to be correct.

The trios are not permutations because the order does not matter. But if I pretend at first that order does matter, then I can use that to begin to solve. So say there are 13 players and we want to make a quintet (five players). If order mattered, there would be 13 • 12 • 11 • 10 • 9 possibilities. But any time I pick the same 5 people, I could order them in 5! ways. So I need to divide by 5!. So if there are 13 volunteers and I choose 5, the number of possible quintets is $\frac{13 \cdot 12 \cdot 11 \cdot 10 \cdot 9}{5 \cdot 4 \cdot 3 \cdot 2 \cdot 1}$*. It's pretty cool that this fraction will simplify to a whole number.*

9. Problems like the duos and trios, where order does not matter, are called *combinations*. Gavin's summary in problem 8 generalizes. When there are n objects and r of them are selected, then the number of possibilities is $\frac{n(n-1)(n-2)\ldots(n-r+1)}{r!}$, symbolized as $C(n, r)$ and read as "n choose r" or "the combinations of n things taken r at a time."

a. Return to problem 5, determine the values for n and r, and confirm that the formula works.

b. Return to Gavin's example in problem 8 of quintets and confirm that the formula works.

c. Another form for the combinations formula can be derived by multiplying the fraction above by 1 in the form of $\frac{(n-r)!}{(n-r)!}$. Use this strategy to find another version of the formula for $C(n, r)$. (Hint: Try first with specific numbers.)

 © 2010 John Wiley & Sons, Inc.

14. Suppose you had two 110-minute periods to help students learn the ideas on the three student pages in problem 13.

   a. Design a timetable for these materials. Include problems from the pages or additional items you create that would make appropriate homework.

   b. Some teachers believe it is not a good idea to teach permutations and combinations too close together as students will confuse them. How might you address this concern in your timetable?

15. Find and read the book *Anno's Mysterious Multiplying Jar* by Masaichiro and Mitsumasa Anno (New York: Philomel Books, 1983), which introduces 10!.

   a. In what different ways do the authors strive to help readers understand factorials?

   b. When and how might you integrate the book into instruction using the student page **The Clarinet Club**?

   c. Read the book's Afterword. What other ideas does this stimulate for you about students' understanding of factorials?

Reflecting on Professional Reading

16. Read "Is Central Park Warming?" by Christine A. Franklin and Madhuri S. Mulekar (*Mathematics Teacher*, 99 (May 2006): 600–605).

   a. How is the initial graph of the trend data of Central Park's mean annual temperatures revealing?

   b. What is the statistical question that is the centerpiece of the investigation?

   c. Explain the logic of the investigation.

   d. Learn to simulate a coin toss on a calculator or computer available to you. Run it and find, for your data, the number of times in 100 sets of 30 coin tosses that there are 23 or more heads. What p value is suggested by your data for this outcome?

Preparing to Discuss Observations

17. In preparation for Lesson 4.5, review your notes from your observations in high school mathematics classes. Also, review the articles you read in preparation for the observations. Be prepared to discuss your findings with your peers.

   a. Identify the big issues in the Teaching Principle from *Principles and Standards for School Mathematics*. What elements of the Teaching Principle were evident in the classes you observed? What elements were lacking?

   b. Comment on how the teacher addressed students' emergent understandings of the mathematics, including resolving misconceptions or responding to students' incorrect responses.

   c. Consider the tasks used during the lessons. In what ways did they intrigue you? In what ways did they appear to intrigue or challenge the students? In what ways did students find the tasks or the lessons intellectually challenging?

   d. Reflect on the teachers' actions. How did the teacher help maintain the intellectual challenge for students if they struggled, or how did the teacher increase accessibility when necessary?

   e. Describe how multiple perspectives were integrated into classroom instruction or how you could modify tasks so they could be approached from multiple perspectives. If you designed a new task that addresses the same mathematical concepts using multiple perspectives, share the task.

   f. How did the teacher help students understand and summarize the important mathematical ideas inherent in the tasks?

18. What aspects of teaching that you observed incorporated the motivational factors outlined by Middleton and Spanias in the Observation?

TEACHING AND LEARNING HIGH SCHOOL MATHEMATICS © 2010 John Wiley & Sons, Inc.

Preparing to Discuss Listening to Students

For problems 19 and 20 and in preparation for Lesson 4.5, review your notes from your interviewing of students. Recall your reading of the article "Research on Students' Understandings of Probability" by J. Michael Shaughnessy (in Jeremy Kilpatrick, W. Gary Martin, and Deborah Schifter (Eds.), *A Research Companion to Principles and Standards for School Mathematics*, pp. 216–226. Reston, VA: NCTM, 2003).

19. **a.** What are the major misconceptions about probability that Shaughnessy identifies?

b. Refer to the student page **Thinking About Probability**. Be prepared to share your **QRS Guide** for the page, including updated student responses based on the interviews.

20. Summarize the responses of the students you interviewed to the student page **Thinking About Probability**.

a. What misconceptions did your interviewees have?

b. What partial or fragile conceptions did they have?

c. What robust conceptions did they have?

d. What similarities or differences did you find between your students' responses and those in the Shaughnessy article?

4.5 Summarizing Classroom Observations and Listening to Students

At the beginning of this unit, you were asked to observe mathematics classrooms with a focus on teaching. During your observations, you were asked to focus on the types of tasks teachers use and how they engage students in those tasks in ways that maintain the mathematical challenge.

In addition, you were asked to interview several students as they worked on a series of tasks, **Thinking about Probability**. What difficulties or misconceptions did students have? What might have accounted for these misconceptions, and what might teachers do to overcome them?

In this lesson, you will:

- Compare the results of your observations with those of your peers.
- Reflect on what your observations suggest about teaching and the extent to which your classroom observations and experiences embody the vision of *Principles and Standards for School Mathematics*.
- Compare the findings from your student interviews with those of your peers and think about what they suggest about student thinking related to probability.

LAUNCH

1. Share your overall impressions about the classes you observed. What more are you able to listen for and attend to now than in earlier observations?

EXPLORE RESPONSES TO CLASSROOM OBSERVATIONS

2. Review the reading "The Teaching Principle" from *Principles and Standards for School Mathematics* (pp. 16–19). The principle highlights instructional challenges in which teachers must balance their planned lessons with the unexpected instructional opportunities that arise on a regular basis.

 © 2010 John Wiley & Sons, Inc.

a. Identify a time in your classroom observations when you realized the instructor was encountering unexpected discoveries or difficulties. What did this incident help you think about related to your future teaching?

b. This textbook emphasizes lesson planning that focuses on anticipating student responses and thinking in advance about teacher's support for those responses. Even with this effort, surprises may result. Some first-year teachers claim that surprises happen daily! Identify incidents from your classroom observations or experiences while listening to students that you did not expect or that caught you by surprise. In what ways do these incidents prepare you to anticipate student responses in a comparable lesson in your own classroom?

3. Consider the tasks used during the lessons you observed.

   a. In what ways did they intrigue you? In what ways did they appear to intrigue or challenge the students? In what ways did students find the tasks or the lesson intellectually challenging?

   b. Reflect on the teacher's actions. How did the teacher help maintain the intellectual challenge for students if they struggled, or how did the teacher increase accessibility when necessary?

   c. Describe how multiple perspectives were integrated into classroom instruction or how you could modify tasks so they could be approached from multiple perspectives. If you designed a new task that addresses the same mathematical concepts using multiple perspectives, share the task.

   d. How did the teacher help students understand and summarize the important mathematical ideas inherent in the tasks?

4. What aspects of teaching that you observed incorporated the motivational factors outlined by Middleton and Spanias in the observation?

SHARE

5. The Teaching Principle highlights the importance of a classroom environment that challenges students while being supportive, as discussed in Unit One. What are elements that help develop and maintain such an environment? Consider what the teacher says, how he or she maintains the mathematical challenge, what is done to encourage students, use of space and time, tone of voice, and other contributing factors.

EXPLORE RESPONSES TO LISTENING TO STUDENTS

6. Misconceptions regarding probability are pervasive, even among mathematically educated adults. You, too, might have incorrectly solved some of the items on **Thinking about Probability**. Share your initial responses when you solved the tasks. What makes probabilistic reasoning so challenging?

7. Having read the Shaughnessy article "Research on Students' Understandings of Probability," you now have researchers' ideas about the sources of several misconceptions.

   a. Share student responses from the interview. Compare your student responses to those of the researchers reported in the Shaughnessy article.

   b. Use the Shaughnessy article to clarify with one another the researchers' interpretations of the sources for student errors on comparable items.

8. Discuss with your partners Shaughnessy's suggestions for teaching probability.

TEACHING AND LEARNING HIGH SCHOOL MATHEMATICS © 2010 John Wiley & Sons, Inc.

SHARE AND SUMMARIZE

9. As a class, discuss one or two of the most interesting misconceptions the interviews revealed. What teaching strategies suggested by Shaughnessy might help prevent these misconceptions?
10. Compare and critique your **QRS Guide** with those of your peers. Revise your **QRS Guide** to include student responses that you did not previously consider.

DEVELOPING MATHEMATICAL PEDAGOGY

Reflecting on Our Own Learning

1. The Teaching Principle emphasizes that teaching requires knowing and understanding mathematics, students as learners, and pedagogical strategies. How do you see yourself making progress in these three areas?
2. During this lesson you compared and critiqued your **QRS Guide** with your peers. What have you learned throughout this textbook about writing appropriate Teacher Support questions?

Debating Issues

3. Suppose your high school mathematics department is debating whether to create a new course in data analysis and probability that any student might elect that is not an advanced placement course. Some teachers believe such a course is necessary for informed citizens. Some teachers believe it is sufficient if data analysis and probability are integrated into the existing algebra and geometry courses. Some teachers believe no specific work in statistics or probability is needed at the high school level. State your position on this issue and justify your position.

Analyzing Students' Thinking

4. Following are several misconceptions researchers have found related to students' thinking in probability, some of which you likely found in your interviews. For each statement, (i) describe the misconception and (ii) speculate about the source of the misconception.

   a. A bag is presented to a child who is told it holds two red and one black marble. The child is asked, "Which is more likely, that we pull out a red marble or that we pull out a black marble? Or, do you think that these two things have the same chance of happening?" The child says,

   "They are equally likely."

   [cited in Shaughnessy, 2003, p. 218].

   b. Reconsider the same situation as in problem 4a. The student says,

   "Anything can happen."

   c. Students were presented with a jar containing 100 colored chips: 50 red, 30 blue, and 20 yellow. Researchers first asked students how many reds they thought would be pulled out in a handful of 10 chips. Then they asked the students what would happen if the experiment were repeated six times, each time pulling out 10 chips, recording the colors, then replacing and remixing the chips. What would the number of reds be? One response was

   "1, 3, 5, 7, 9, 10."

   d. Reconsider the same situation as in problem 4c. The student says,

   "It could be any number of reds."

TEACHING AND LEARNING HIGH SCHOOL MATHEMATICS © 2010 John Wiley & Sons, Inc.

Reflecting on Professional Reading

For problems 5 through 8, read "It's a Home Run! Using Mathematical Discourse to Support the Learning of Statistics" by Kathleen S. Himmelberger and Daniel L. Schwartz (*Mathematics Teacher*, 101 (November 2007): 250–256).

5. **a.** What makes the activity productive for discourse?

b. Identify at least three different strategies the students used.

6. **a.** What was the teacher's role while students worked on the task?

b. In oral language, people frequently use pronouns (e.g., it, this, that) whose reference may be clear in the conversation but is less clear in a transcript. Identify some examples of unclearly referenced pronouns in the dialogue on p. 252 in the article. What might be more careful ways of speaking?

c. What are some valuable features of the teacher's role in the dialogue on p. 253 in the article?

d. How is student discourse valuable to a teacher?

e. Give examples of the teacher's decision making during the lesson.

7. How did the teachers take advantage of the varied timetable at the school for planning the sequence of activities in this lesson?

8. What brief segment of direct instruction did the teacher insert into the lesson? Why?

TEACHING AND LEARNING HIGH SCHOOL MATHEMATICS © 2010 John Wiley & Sons, Inc.

SYNTHESIZING UNIT FOUR

Mathematics Strands: Data Analysis and Probability

Lesson planning is a critical aspect of teachers' work. Effective mathematics instruction requires careful planning, especially when instruction is focused toward inquiry learning. The manner in which a teacher launches a lesson often determines students' interest and motivation to engage in the lesson. The nature of the exploration—the central part of a lesson—determines both the content that students explore and the depth to which they explore it; hence, the tasks teachers choose are critical in order for students to engage in cognitively challenging mathematics. Finally, how the teacher facilitates the summary of students' explorations influences whether students take from the lesson the important mathematics they need to learn. Without appropriate summary, inquiry lessons can leave students unclear about the essential concepts of the lesson.

In addition, high schools around the country have varied formats for scheduling. Instructional planning for classes that meet every day for 45 to 55 minutes is likely different from instructional planning for classes that meet 80 to 90 minutes every day for a semester or every other day for an entire school year. Thinking about the demands of time and the needs of adolescents requires considerable planning on the part of the teacher.

Today's population is bombarded with data. Statistics are, or should be, used to inform decisions for a person, a school, a city, a country, or the world. The interplay of data analysis and probability provides a basis on which reasonable decisions can be made in the face of variability.

Having arrived at the end of this unit, it is time to reflect on your understanding of the lessons learned throughout the unit. The following items provide an overview of the unit. You should be able to respond meaningfully to each item, providing support for your answers with specifics drawn from the various lessons, discussions, readings, homework problems, or observations in high school classes.

Website Resources

- **Where Do I Stand?** survey

MATHEMATICAL LEARNINGS

1. Given a set of data for some context, determine different ways to display the data that are appropriate. Be prepared to discuss how different displays do or do not highlight variability in the data.
2. Understand the importance of recognizing variability in situations involving chance.
3. **a.** Identify different real-world situations in which the median would be the preferred measure of center. Repeat for the mean and then for the mode.

   b. Given a data set, determine which set of measures are most appropriate to analyze the data (median, quartiles, and interquartile range versus mean and standard deviation). Compare sets of data using one of these analyses and an appropriate graph.
4. **a.** Determine the conditional probability of an event, given that another event has occurred, $P(A|B)$.

   b. Determine the probability of the union of two events, $P(A \text{ or } B)$.

   c. Determine the probability of the intersection of two events, $P(A \text{ and } B)$.
5. Explain one or more strategies for simulating a random sample.
6. Identify benchmarks of a normal distribution.

TEACHING AND LEARNING HIGH SCHOOL MATHEMATICS © 2010 John Wiley & Sons, Inc.

7. Compute standard deviation for a sample or population and use the standard deviation to describe a data set.
8. Determine whether a situation can be modeled by a binomial experiment. If the situation is binomial, use a binomial distribution to determine the probability that a given event occurs.
9. Distinguish situations that involve combinations from situations that involve permutations. Be able to determine the number of combinations or permutations for a given situation.

PEDAGOGICAL LEARNINGS

10. Articulate the major elements of the Teaching Principle from *Principles and Standards for School Mathematics.*
11. Highlight some of the elements of effective motivation in mathematics classes.
12. Identify some of the common misconceptions students have about basic concepts related to probability.
13. Plan instruction for data analysis using the sequence:
    - guess a solution,
    - simulate physically,
    - simulate repeatedly or with technology,
    - analyze a pooled or large set of data,
    - solve informally,
    - solve analytically (once prerequisite knowledge is in place).
14. **a.** Identify the essential elements of the *Thinking Through a Lesson Protocol.*
    b. What are five practices that are important in order to orchestrate productive mathematical discussions?
15. Generate some or all of a lesson plan for a given lesson.
16. Compare direct instruction and inquiry instruction. What similarities and differences do they have? When and how might direct instruction be blended into inquiry learning?
17. Identify some of the issues and challenges of instructional planning based on different class schedule formats. Discuss possible variations in your planning based on different time formats and justify your decisions.
18. Revisit your responses to the survey **Where Do I Stand?** for items 1, 2, 3, 6, 8, 11, 14, and 17. How have your responses changed, depend or grown?
19. Revisit your responses to item 24 on the survey **Where Do I Stand?** Are there any additional challenges you expect to face as a high school mathematics teacher?

TEACHING AND LEARNING HIGH SCHOOL MATHEMATICS © 2010 John Wiley & Sons, Inc.

UNIT FOUR INVESTIGATION: BUILD YOUR OWN DREAM TEAM

In this investigation, you will continue to investigate the Dream Team Mania computer game from the Unit Four Team Builder. For the investigation, you are designing your own personal dream team for your favorite sport and pitting your team against other players' dream teams. Start by selecting a sport. Then design your dream team, one player for each position that would be on the playing surface at one time. Enter your sport and players into an electronic file. Take turns to "earn" players for your team. On each of your turns, the computer randomly selects one of the players for your team. Each player on your defined list is equally likely to be generated on each turn, so the computer may generate a duplicate of a player you already have. Just continue taking turns until you have earned all your players. The first person to complete his or her team wins the first game of the season. When you have a full team, you can enter them into competition in that sport with teams that other players have created.

1. Identify a sport, other than hockey, that you want to investigate. Determine the number of players on the playing surface at any one point in time. Your sports team should have at least five players; choose a sport for which the number of players is different from the number in hockey. Identify your favorite players in this sport, or create a set of players with names of your choice. Here are some possibilities:

   - Baseball: 9
   - Football: 11
   - Basketball: 5
   - Arena football: 11
   - Soccer: 11
   - Field hockey: 5
   - Men's squash: 5

2. Use your responses from the Unit Four Team Builder to help you answer the following.
   a. Estimate the number of turns you would need to obtain a full set of players for your team. Explain your reasoning for this estimate.
   b. Identify at least one technique from the Team Builder that can be used to simulate the number of turns you need. Based on your work in the Team Builder, refine or modify a plan to simulate the first round of the computer game where you win a full set of players for your team (e.g., use a random number table, generate random numbers on a graphing calculator, use *Fathom*™). Identify any assumptions you make.
3. Implement your simulation plan for the dream team you selected. With your group, collect results for at least 100 trials.
   a. Detail your implementation steps. Explain any difficulties that arose. Explain any strategies you used to simplify or automate the work.
   b. Show your data collection records.
   c. Make a graph of the number of turns you needed during each trial to earn a full set of players for your dream team.
   d. Describe the graph (shape, range, anomalies, intervals of likely outcomes, quartiles, etc.).
   e. What was the mean number of turns you needed to earn a full team?
   f. Do the values center on the mean that you found? Should they? Why or why not?
4. Explore the stages in data collection.
   a. Suppose you have all but one of the players. What is the expected number of turns you need to get the last player on your team?

TEACHING AND LEARNING HIGH SCHOOL MATHEMATICS © 2010 John Wiley & Sons, Inc.

b. Suppose you have all but two of the players. What is the expected number of turns you need to get these last two players for your team?

c. Refer back to the beginning of your work. Design a way to keep track for several trials of the number of turns you need before you obtained the first player, second player, third player, etc.

d. What patterns do you notice in the mean number of turns you need to earn the next player for your team?

5. Explore the problem theoretically.

   a. If you already have one player, what is the probability that you will earn a different player the next time?

   b. If you already have one player, what number of turns do you need to earn the next new player?

   c. Continue with this reasoning to determine a theoretical way to calculate the mean number of turns you need to collect a complete team. Justify your reasoning.

   d. If possible, generalize your results to earn teams of different sizes. Justify the reasoning behind your generalization.

   e. If you can generalize the results, compare what you found to other functions or mathematical topics you have studied.

6. Refer back to your work in the investigation.

   a. What do you understand about the mathematics underlying this problem?

   b. What questions do you still have?

   c. What learning outcomes do you see in this investigation?

   d. Identify some anticipated responses you would expect from high school students at critical junctures in this investigation.

   e. This investigation used a specific format recommended for statistics and probability work with students.

      - Estimate an answer.
      - Design a simulation. Distinguish between a trial and a "successful" trial.
      - Carry out the simulation many times, possibly using technology. Keep track of the outcomes.
      - Produce a graph from the simulations and describe the variation and the patterns.
      - Solve the problem informally.
      - Consider the problem theoretically once students have sufficient mathematics background.

      What benefits do you see in these steps and their sequence?

   f. Discuss other reflections or ideas this investigation brings to mind.

TEACHING AND LEARNING HIGH SCHOOL MATHEMATICS © 2010 John Wiley & Sons, Inc.

UNIT *five*

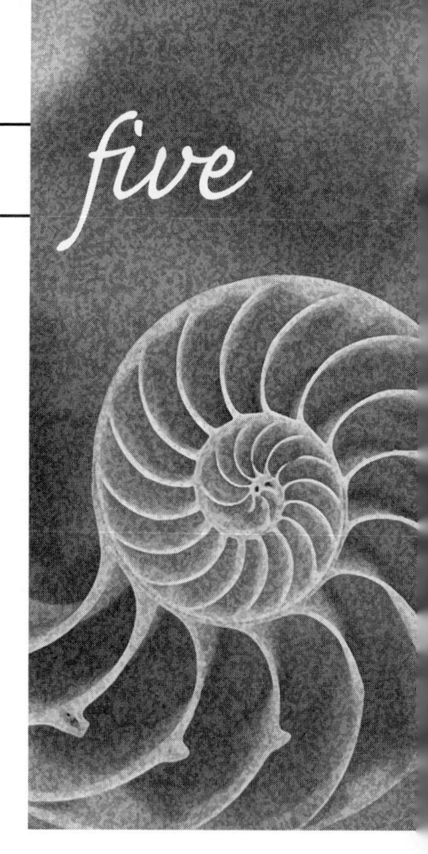

ASSESSMENT OF STUDENTS' LEARNING

Mathematics Strand: Precalculus

Assessment is an integral part of classroom instruction. For students and adults alike, what is assessed reflects what is valued. Hence, it is critical that instruction and assessment are aligned if they are to work together to support learning. If instruction emphasizes multiple approaches or the use of technology, then assessment must reflect such emphasis. In addition, students will likely focus primarily on those aspects of instruction that are most prominent in their daily interactions with mathematics classes and also those that most influence their grades. So if classroom instruction and exams focus on conceptual understanding but daily homework assignments focus on procedural understanding, students' time working on homework can affect their view of the real intent of the course. For fairness and effectiveness it is important for instruction and assessment to be aligned.

As indicated in Unit Two, the *assessment-centered lens* for teaching and learning focuses on providing frequent opportunities for assessment so that teachers and students make thinking visible to guide further instruction (Donovan and Bransford, 2005). Assessment should be an open process, so students know from the beginning of a unit what will be expected of them. Frequent informal assessments can help students learn to monitor their own progress (Stiggins, 2005).

Although teachers and students often equate assessment with grades, assessment and evaluation actually are two different terms. According to *Assessment Standards for School Mathematics* (National Council of Teachers of Mathematics, 1995), *assessment* involves collecting information about students' mathematical knowledge and using that knowledge to make inferences, such as for future instruction. In contrast, *evaluation* involves assigning value to students' work for judgment purposes, such as to assign grades. Evaluation is one purpose of assessment, but it is not the only purpose; regular assessment that helps teachers and students understand what students know without assigning a value (i.e., a grade) to the assessment should be a natural part of classroom instruction.

In *Assessment Standards*, four phases of assessment are identified: planning the assessment, gathering evidence, interpreting the evidence, and using the results. These four phases are interconnected; that is, they inform each other in a constant intertwining cycle. Through interpreting the evidence, teachers use the results to modify and plan subsequent instruction. Throughout this unit, you will consider a range of tools and approaches that can be used for assessment, and sometimes for evaluation.

The mathematical focus of this unit, precalculus, provides many challenges for thinking about assessment. With the advent of powerful graphing calculators that include computer algebra systems, many of the paper-and-pencil skills that were previously emphasized in precalculus and calculus can now be handled efficiently with technology. Hence, instruction and assessment can focus on important conceptual understandings and applications of the concepts. Determining what paper-and-pencil skills need to be mastered and what paper trail students need to provide when using technology are issues that the mathematics education community will likely continue to face for some time.

The mathematicians who gave us the foundations of calculus could not have dreamed of the readily available technology that permits the symbolic, visual, and tabular representations of important concepts to be handled simultaneously. Since the time of the Greeks, mathematicians have been interested in problems related to determining rate of change and optimization of functions and finding the area between two curves, both important problems in calculus that require solid understanding of functions from precalculus. The ancients had solved such problems in specific cases. But the founders of calculus, Isaac Newton and Gottfried Leibniz, were able to generalize a process that could be used with any function.

In the 17th century, René Descartes (1596–1650) and Pierre de Fermat (1601–1665) provided two different approaches to introduce analytic geometry. Fermat began with equations and then described the curve. In contrast, Descartes described curves geometrically and then derived their equations. Fermat developed techniques that could be used to find tangents to a curve; Descartes was able to find a normal to a curve at any point. Both ideas could be applied to a range of new curves not necessarily known to the Greeks.

The introduction of many new curves through the study of analytic geometry also led to attempts to find the area bounded by these curves. The ancient Greek and Islamic mathematicians knew how to find areas bounded by particular curves, generally breaking up the region into many small regions whose area could be found. Thus, this method of *indivisibles* was similar in many respects to our present-day methods of bounding the area under a curve between a series of inscribed and circumscribed rectangles or trapezoids.

When Newton and Leibniz began their process of bringing together ideas about areas and tangents that eventually became what is known as calculus, they were building on the work of mathematicians from many earlier centuries. Newton and Leibniz approached the calculus from two different perspectives; "Newton's approach was through the ideas of velocity and distance while Leibniz's was through differences and sums" (Katz, 1998, p. 531). One of the interesting controversies in mathematics revolves around the claims of which of these two geniuses should be credited with the discovery of calculus. Their approaches were different and the notation of Leibniz was easier to use. Because of the dispute, English mathematicians sided with Newton while those on the European continent sided with Liebniz. The continental mathematicians continued to make progress in analysis while those in England failed to avail themselves of the progress made throughout the 18th century.

This unit explores issues related to assessment while investigating mathematical ideas that form the foundation for the derivative and the integral. With technology, many of these foundational concepts of calculus can be explored easily and informally in precalculus courses. Throughout your work on assessment in this unit, you will also revisit many of the student pages and concepts you explored earlier in this text and consider how the assessment ideas discussed in this chapter might be applied to those concepts.

TEACHING AND LEARNING HIGH SCHOOL MATHEMATICS © 2010 John Wiley & Sons, Inc.

Unit Five Team Builder: Conic Conundrums

In *Conic Conundrums*, teams choose a conundrum to solve from those that follow. Use only the materials provided. Teammates will name their group based on the conic conundrum they attempt. All teams complete problems 1–4 below.

1. Consider lines m and n that intersect in point O forming an angle that measures less than 90°. What geometric shape is created by rotating line m around line n in three-dimensional space?
2. **a.** What figures are possible if the shape in item 1 is intersected by a plane?
   b. Describe how the plane must intersect the shape in item 1 to get each figure you determined in item 2a.
3. Solve a conundrum on one of the following student pages.
4. Create a name for your team based on the conic section you explored in this team builder and the letters of your team members' names.

Materials

- newsprint paper, one large sheet per team (paper with large grid is best)
- markers, two per team
- string, 12 inches per **Conundrum I** team and several feet per **Conundrum II** team
- pushpins, two per team
- large paperclips, one box per **Conundrum III** team

Website Resources

- **Conundrum I** student page
- **Conundrum II** student page
- **Conundrum III** student page

For further work with conic sections, see the Unit Five Investigation at the end of the unit.

TEACHING AND LEARNING HIGH SCHOOL MATHEMATICS © 2010 John Wiley & Sons, Inc.

Conundrum I

1. On a large sheet of newsprint, draw two points, *P* and *Q*, so they are no more than 10 inches apart and are somewhere near the center of the paper.
2. Cut a string at least 2 inches longer than the distance between *P* and *Q*. Tie a small knot in both ends of the string.
3. Choose two team members; each team member uses a pushpin to hold an end of the string, one at point *P* and the other at point *Q*. (Put something under the paper to prevent damage to the desk below.)
4. A third team member pulls the string taut with a marker and uses the marker to draw a curve while the endpoints stay fixed.
5. Determine the name of the resulting figure. How do you know?
6. How must a plane intersect a double cone to obtain this figure?

Materials

- newsprint paper, one large sheet per team (paper with large grid is best)
- one marker per team
- string, 12 inches per team
- pushpins, two per team

Conundrum II

1. On a large sheet of newsprint, draw point *P* somewhere near the center of the paper.
2. Draw a straight line, *m*, across the paper so that the distance, *d*, between *P* and *m* (measured along a perpendicular to *m*) is no more than 10 inches.
3. Cut strings of several different lengths. The first string should be length *d*. All others must be longer than *d*. Mark the midpoints of each of the strings.
4. Use the strings to find points that are the same distance from *P* and *m* (once again, the distance from *m* must be measured along a perpendicular line). Use the marker placed at the midpoint of each string to mark points on the curve.
5. Determine the name of the resulting figure. How do you know?
6. How must a plane intersect a double cone to obtain this figure?

Materials

- newsprint paper, one large sheet per team (paper with large grid is best)
- markers, two per team
- string, several per team
- pushpins, two per team

Conundrum III

1. On a large sheet of newsprint, draw two points, *P* and *Q*, so they are approximately 10 inches apart and are somewhere near the center of the paper.
2. Using large paperclips, make two chains so one chain is two paperclips longer than the other and the combined lengths are at least the distance between points *P* and *Q*.
3. Choose two team members; each team member uses a pushpin to hold an end of a paperclip chain. One member holds the end of one chain at point *P*. The other member holds the end of the other chain at point *Q*.
4. The two remaining team members will use these chains to create a curve. Pulling their respective chains taut, these two team members mark all of the points of intersection of the two chains, then exchange chains and mark all the new points of intersection.
5. Each of the team members in item 4 adds another paperclip to their respective chains and repeats item 4.
6. Repeat item 5 three more times. For each step, how are the chain lengths related?
7. Determine the name of the resulting figure. How do you know?
8. How must a plane intersect a double cone to obtain this figure?

Materials

- newsprint paper, one large sheet per team (paper with large grid is best)
- markers, two per team
- pushpins, two per team
- large paperclips, one box per team

TEACHING AND LEARNING HIGH SCHOOL MATHEMATICS © 2010 John Wiley & Sons, Inc.

Preparing to Observe Mathematics Classrooms: Focus on Assessment

The assessments teachers use help students understand what concepts and procedures they are expected to master. As indicated in the unit introduction, assessment and evaluation are not synonymous, even though people often use them interchangeably. Teachers frequently use informal assessments throughout a unit to gauge how well students understand the mathematical concepts studied; homework assignments are one form of informal assessment as students engage with problems and tasks that practice and extend concepts first investigated in class. More formal assessments at the end of a unit, often in the form of tests at the high school level, provide evaluative information to students and teachers about the extent to which the mathematics content of the unit has been mastered.

The Assessment Principle, the focus of this unit's observation, states that assessment needs to support students' learning of mathematics and provide helpful information about that learning to both students and their teachers. Assessment needs to help guide and support student learning and not be just something that is done *to* students at the end of a unit of study. Too often, assessment in mathematics classes is done only for summative purposes, to assign a grade. Formative assessment, to inform how learning is progressing so that instruction can be modified or particular misconceptions can be addressed, is essential. Unfortunately, though formative assessment is needed more, it often is used less than summative assessment.

Throughout Units One, Two, Three, and Four, you explored teaching and learning from a constructivist perspective in which students are encouraged to build their own mathematical understanding through rich activities that teachers facilitate. You used Stages of Questioning to help students think through a problem-solving task and you completed **QRS Guides** to record anticipated student responses and related teacher support. You encouraged the development of proof and reasoning skills. You considered multiple representations for concepts. You designed lessons that focus on students' thinking. Instruction and assessment must be aligned (aimed at the same goals) to work together to enhance mathematics learning.

The purpose of the observation in this unit is to look carefully at issues related to assessment. In what ways do assessment practices align with instruction in the classroom? In what ways do assessment practices align with your own philosophy of mathematics teaching and learning? Is there attention to formative as well as summative assessment?

Website Resources

- "Improving Classroom Tests as a Means of Improving Assessment" by Denisse R. Thompson, Charlene E. Beckmann, and Sharon L. Senk. (*Mathematics Teacher*, 90 (January 1997): 58–64).
- **Template for Classifying Test Items**

1. *Prepare for the Observation*. In preparation for the observation, read "The Assessment Principle" (pp. 22–24) from *Principles and Standards for School Mathematics*.
   - **a.** List several of the informal assessment strategies mentioned in the discussion of the principle.
   - **b.** The discussion mentions the importance of helping students recognize the difference between an excellent response and one that is mediocre. What are some benefits of having students analyze responses and see models across a range of levels (e.g., from an excellent response to a poor response)? What are

TEACHING AND LEARNING HIGH SCHOOL MATHEMATICS © 2010 John Wiley & Sons, Inc.

some issues that teachers will likely need to consider as students critique responses of their peers? (Note: For privacy and legal reasons, students should never be asked to assign grades to other students' written responses when the author can be identified.)

2. *Conduct the Observation.* If possible, observe a class you visited for a previous unit. Observe at least two full class periods in the same class. Complete Observations 5.1 and 5.2 using the guidelines below; if possible, complete observation 5.3.
3. *Observation 5.1: Focus on Informal Assessment.* As you observe, make note of the mathematical intent of the lessons and keep track of the informal assessments the teacher uses to gauge student understanding of the lesson.
   - **a.** What types of questions did the teacher ask throughout the lessons to assess students' understanding? How did the teacher use responses to the questions to adjust instruction?
   - **b.** In what ways did the teacher use observations? questioning of individuals or small groups of students?
   - **c.** Were students expected to write about their experiences in journals or via writing prompts? Give examples of prompts or journal assignments.
4. *Observation 5.2: Focus on Formal Assessment.* Note the mathematical intent of the lessons. Take enough notes so you can compare an assessment instrument with the content of the lessons.
   - **a.** From the teacher, request a copy of a quiz or test that assesses students' knowledge of the mathematics in the lessons you observed. (Depending on the proximity of your observation to the formal assessment, you might have to return or use email to get a copy of a quiz or test.)
   - **b.** Compare the items on the quiz or test to the notes you took during the lessons. (Confine your comparison to only those test or quiz items that relate to the mathematical instruction you observed.) What similarities and differences do you find? Are the assessment items at the same level as those used during instruction?
   - **c.** Read "Improving Classroom Tests as a Means of Improving Assessment" by Denisse R. Thompson, Charlene E. Beckmann, and Sharon L. Senk. (*Mathematics Teacher*, 90 (January 1997): 58–64). (See the course website for a copy of the article and the Unit Five Appendix for the **Template for Classifying Test Items**.) In the article, the authors provide a set of criteria for analyzing test items. Code all the items on the classroom test according to the criteria provided by Thompson, Beckmann, and Senk. What are the strengths of the test or quiz? In what areas might the test or quiz be improved?
5. *Observation 5.3: Focus on Grading.* If possible, review a set of homework papers or grade a set of quiz or test papers pertaining to the lessons you observed in Observations 5.1 and 5.2.
   - **a.** What misconceptions, if any, did students have?
   - **b.** Distinguish students' levels of understanding for selected items and comment on students' opportunities to learn the mathematics these items addressed.
   - **c.** What issues arose as you were grading students' papers?
   - **d.** What did you find interesting about students' mathematical work?
6. *Interview a Teacher.* Interview a teacher about his or her grading practices, preferably related to one of the classes you observed in this course.
   - **a.** Have the teacher describe his or her overall grading plan for the course. What is his or her philosophy? How are students' grades determined? Does the teacher assign different weights to formative assessments than to summative assessments?

TEACHING AND LEARNING HIGH SCHOOL MATHEMATICS © 2010 John Wiley & Sons, Inc.

b. How does the teacher develop a quiz or test? How does the teacher determine the important content emphases of these assessments? How does the teacher assess the process standards (e.g., communication, representation, reasoning and proof, connections, problem solving)?

c. Is the teacher's grading plan consistent across courses or levels? If the plan varies by course or level, what factors influence the differences?

d. What is the most challenging aspect of grading for the teacher?

7. *Summarize the Observations.* Start by providing a context for the classes; state the level of the class and the mathematical intent of the lessons observed.

a. Refer to the questions for Observations 5.1 and 5.2. Summarize your responses to the questions. Justify any conclusions with specific instances from the readings or the class observations.

b. Summarize your results from grading the assessment in Observation 5.3 if you had this opportunity.

c. Summarize your interview with the teacher from item 6. What aspects of the teacher's grading plan would you like to emulate in your own classroom? What aspects, if any, would you want to modify?

d. Based on your observations from all five units and your interviews with teachers, outline an overall grading plan for one course that you are likely to teach in high school.

Be prepared to discuss your observations and reflections in the last lesson of this unit.

LISTENING TO STUDENTS REASON ABOUT PRECALCULUS

The National Assessment of Educational Progress (NAEP) is often called the Nation's Report Card. It is a test mandated by Congress in the 1960s to assess students' knowledge in a wide range of subject areas. Although the test originally assessed students at particular ages, it now assesses students at grades 4, 8, and 12. Mathematics often is assessed every two to four years. The test uses a stratified random sample of students from across the United States.

The NAEP provides a website tool that enables users to find released items administered on prior NAEP assessments, together with information about the achievement of students on the item. In this **Listening to Students**, you will have an opportunity to investigate this NAEP tool to select items appropriate for a group of high school students with whom you have been working.

Go to the NAEP website at http://nces.ed.gov/nationsreportcard/ITMRLS/. You might want to bookmark this website for future reference.

Website Resources

- **QRS Guide**

1. Find a link for **state profiles**. When you click on this link, you will see a map of the United States. Click on your state to obtain a profile of how students in your state perform on the NAEP at the eighth grade level. (Information does not seem to be available at the state level for 12th grade students.) What do the results suggest about the level of mathematical proficiency of students in your state as they prepare to enter high school?

2. Click on the **Questions Tool** to link to released items for previous NAEP administrations. Choose the **Main NAEP** assessment. You will need to select **Mathematics** as the subject and then a specific grade. Choose the most recent year available of those highlighted. Other selection criteria may be displayed, and

TEACHING AND LEARNING HIGH SCHOOL MATHEMATICS © 2010 John Wiley & Sons, Inc.

you should make appropriate selections; if selecting item type, be sure to select **all** so that you obtain constructed-response items in addition to multiple-choice items. Show the results for your grade and subject. A list of items will appear indicating the grade levels at which the item was administered, the year, and a description of the item. (Note: 8 (4, 12) means the item was administered at grade 4 and 12 as well as grade 8.)

a. Select at least 10 items of interest whose description suggests content in the late high school or precalculus curriculum (including work with functions). Check the selected items to put them in a folder. Notice that you can navigate to another page of item descriptions to obtain items 21–40, 41–60, etc. Save each of your selected items.

b. After you identify the items of interest, select **View Print Folder**. You can now select from options to save or print the following: the items, scoring rubrics for constructed-response items, student responses, and performance data. Be sure to obtain student responses and performance data about your items. Click **Assemble Document**.

c. Review the items, sample responses if available, and performance data. Use your knowledge of the background of your high school students to select 5 to 10 of these items to administer to your students. Assemble the items in an order that seems reasonable and in which the items progress from easy to more difficult.

3. Identify two to three high school students who are willing to solve the NAEP items you selected and explain their thinking to you.

a. Prior to administering the items, identify potential misconceptions that you anticipate students might have. Complete a **QRS Guide** with your responses and those that you expect students to make at three levels (misconceived, partial, and satisfactory). Also provide appropriate teacher support.

b. Predict how well you expect your students to perform on these items. Justify your prediction based on your knowledge of their mathematics background.

c. Administer the items to students. As they work, have them explain their reasoning about the items both orally and in writing. Make note of their comments.

d. Compare your students' results to the national results. What similarities and differences did you find?

e. What unanticipated misconceptions did students have? Add these to the **QRS Guide** with appropriate teacher support.

f. Note any results that vary significantly from those reported for the NAEP item.

Be prepared to share your insights and your students' results during the last lesson of this unit.

5.1 Daily Assessments: Limits

A variety of assessments take place every day, both inside class and beyond. The role of questioning between students or students and teacher is one of the most important aspects of daily assessment because it provides immediate clues about how a student is thinking or what misconceptions a student may harbor. Other forms of daily assessment include homework, short writing assignments, and journaling. The latter forms can be used in or out of class.

In order for students to develop competence with mathematics, they need opportunities to explore concepts and to practice procedures in more depth than is often possible during normal class time. Assigning homework, short out-of-class

TEACHING AND LEARNING HIGH SCHOOL MATHEMATICS © 2010 John Wiley & Sons, Inc.

writing assignments, or journal entries are ways to assist students as they deepen their understanding and make connections between the mathematics they are learning and their prior knowledge. Students should view daily assignments as an essential part of learning, an opportunity to identify any confusions, begin to make connections among concepts, and develop procedural fluency.

The challenge for teachers is how to manage homework so that it does not become a paper burden, particularly for teachers who have five classes with an average of 30 students per class. One hundred fifty papers a night can be overwhelming even for the most organized teacher!

This lesson addresses the mathematical concept of limits. How might limits be intuitively addressed to pave the way for more formal approaches in a later calculus course? To ground the discussions about homework, a precalculus **Sample Unit Outline: Introducing Derivatives** is provided in the Unit Five Appendix.

In this lesson, you will:

- Explore the concept of limits and think about function concepts that underpin later work with derivatives.
- Consider a precalculus unit outline as a basis for thinking about homework issues.
- Compare and contrast different techniques for managing and assigning daily work.
- Consider factors that might influence daily assignments.
- Consider various forms of daily assignments, including homework, short writing assignments, and journal entries.

Website Resources

- **Sample Unit Outline: Introducing Derivatives**

LAUNCH

For questions 1–3, think back to your own experiences in high school mathematics classes.

1. What were your teachers' or schools' expectations for homework in terms of frequency and the amount of time they expected you to spend on mathematics each night?
2. **a.** What types of daily assignments were you asked to complete?
   b. Were daily assignments always directly from a text or did the teacher give additional directions to fit the needs of the students in your class?
   c. Were any of the assignments of the variety, "Find an example in your daily life of a use of the mathematics we are studying"?
   d. Were there any homework problems that previewed the discussion for the next lesson?
3. Was homework a part of your grade for a given marking period? If so, what percentage of your marking period grade came from homework?

EXPLORE

4. Study the precalculus **Sample Unit Outline: Introducing Derivatives**. Keep this outline in mind as an example while you think about different issues related to homework.
5. Following are four clusters of issues related to daily assessments (**Assigning Homework**, **Checking Homework**, **Grading Homework**, or **Alternatives to Homework from Textbooks**). Each team should explore the issues within one

TEACHING AND LEARNING HIGH SCHOOL MATHEMATICS © 2010 John Wiley & Sons, Inc.

cluster and present the issues to other teams. Each cluster should be addressed by at least one team.

Cluster A: Assigning Homework

High school students often have many claims on their time, including extracurricular activities, a heavy academic load (e.g., AP or honors classes), or part-time jobs. So, teachers need to be cognizant of these time pressures as they assign homework. At the same time, students need to realize that homework is a critical component of the learning process.

I. Below are several ways that teachers assign homework. Compare and contrast the different approaches by considering advantages and disadvantages of each. Which approaches might you consider using in your own classroom? Why? What concerns do you have about an approach?

- **a.** *Given daily at the end of class.* Teachers give students the homework assignment each day, adjusting the assignment based on the flow of instruction on a given day.
- **b.** *Weekly assignment.* Teachers give students homework for the week on Monday by section, either by providing the assignment written on a weekly sheet, by posting the week's assignment on a board in the classroom, or by posting the homework on the school's website.
- **c.** *Unit assignment.* At the beginning of a unit, teachers give students all the homework assignments for that unit, with assignments listed by section of the textbook. Students are periodically reminded which sections of homework should be completed with due dates.
- **d.** *Assignment for a marking period.* At the beginning of a new marking period, teachers give students all the homework assignments for the entire marking period by section in the text.

II. Many teachers have Web pages for class, with students and parents able to go to the website to access assignments, support, or other information related to class. How might such a website be used with the different approaches to assigning homework that were outlined in question I? What concerns do you have about homework Web pages?

III. One rationale teachers use for giving students homework assignment sheets for a week or a unit is that students are able to "keep up" if they are sick or absent from school for some reason (e.g., school-related field trips or athletic events). Why might it be particularly important to help students keep up in mathematics?

IV. In addition to homework problems on a given lesson, some teachers like to have students read the lesson for the next class as part of each day's homework assignment. What might be some advantages and disadvantages of having students preread a lesson before it is discussed in class?

V. Mathematics teachers often like to have some time in class, even 10–15 minutes, for students to begin work on their homework assignment. What are some reasons other than "giving students time" that teachers would build homework time into their lesson time? What are disadvantages of this practice?

Cluster B: Checking Homework

I. Homework is an opportunity for students to assess their own understanding of the mathematics concepts and procedures discussed in class. So, there needs to be some opportunity for students to determine whether or not their homework is correct. Discuss the pros and cons of the following mechanisms for checking homework.

- **a.** *Students check their own work.* The teacher displays the answers via an overhead or a computer projection device and students check their own work, making note in red of any incorrect solutions.

TEACHING AND LEARNING HIGH SCHOOL MATHEMATICS © 2010 John Wiley & Sons, Inc.

b. *Students check work in peer groups.* Students gather in small groups and discuss any problems where they have questions. They determine the correctness of their solutions together.

c. *Teacher checks students' work.* The teacher checks each student's work and marks incorrect answers or gives other feedback.

II. After students have had an opportunity to check answers, they need an opportunity to discuss problems that gave them difficulty. Outline some of the advantages and disadvantages of the following methods of addressing homework.

a. *Students list difficult problems and the teacher works them.* As students enter class or after they have checked answers, they make a list of those problems that were incorrect. The teacher then works those problems for the class, asking students for input as appropriate.

b. *Students work through difficulties in a group.* After students have checked answers, they work with peers in small groups to discuss those problems that posed difficulties for members of the group. Only if no member of the group is able to explain the problem does the teacher work the problem.

c. *Students work and explain problems for the class.* From a list of problems generated by students or the teacher, student volunteers present a problem on the board and then explain it to the rest of the class.

d. *Teacher identifies those problems likely to be difficult for many students.* The teacher predetermines problems that he or she believes will pose difficulties for a significant number of students and works those problems or has students work those problems on the board.

e. *All homework problems are worked out in detail in class.* The teacher or other students work out all homework problems on the board for the entire class.

f. *A solutions manual is provided.* Detailed written solutions are provided for all problems. Students who do not understand the written solution are encouraged to seek additional help outside of class.

Cluster C: Grading Homework

I. Determining how to grade homework without it becoming a paperwork burden is a challenge for mathematics teachers. Discuss the pros and cons of each of the following methods for grading homework.

a. *Effort grades.* Homework is graded based on a reasonable effort on the problems. Some teachers circulate around the room as students check their answers and note whether or not students made a reasonable attempt on all or the majority of problems. Grades are denoted as check ✓ (or –) in a grade book or teachers place their name stamp or initials on each paper if the student made a good effort.

b. *Teacher collects and grades every assignment.* The teacher collects and grades every assignment, marking problems as correct or incorrect. A grade is recorded based on the percentage correct or the number of points correct.

c. *Teacher collects and grades selected problems.* In general, homework is graded in terms of effort. However, once or twice per week a selected set of 4–5 problems are graded for correctness.

d. *Homework notebook quiz.* In general, homework is graded in terms of effort. However, once each week or twice per unit students have a homework quiz. Using only the homework solutions in their notebook, students record the responses for selected problems from multiple lessons. Students are not permitted to use their textbook while taking the homework quiz.

e. *Homework collected and marked once per unit.* Homework is corrected in class and questions are answered daily. Students are encouraged to correct any problems they got wrong. Students submit homework for a grade only on the day of the chapter test, and submit all homework for the chapter on that day.

TEACHING AND LEARNING HIGH SCHOOL MATHEMATICS © 2010 John Wiley & Sons, Inc.

f. *Homework collected once a week.* Papers are collected weekly and checked to see if students have worked assigned problems and made corrections based on class discussions.

II. Some high school teachers choose not to include homework as part of students' grades, particularly at the Algebra II and higher levels. Instead, they want students to realize that homework is essential in order to be successful. Comment on the pros and cons of not checking to see if high school students complete homework.

III. Depending on personal philosophies and school policies, homework may account for up to 25% of a grade per marking period.

a. Within your group, provide an argument for teachers who want homework to count for 0% of a marking period grade.

b. Within your group, provide an argument for teachers who want homework to account for 25% of a marking period grade.

Cluster D: Alternatives to Homework from Textbooks

I. Discuss the pros and cons of each of the following alternatives to traditional textbook homework assignments.

a. *Journal writing.* In place of solving textbook problems, students write their own interpretation of the meaning of a concept with an example and a nonexample. Or they write about what they do and do not yet understand about a concept or process being studied.

b. *Determining contexts.* Students find and analyze examples of the concept they are studying in class in their daily lives. For example, students notice that cell phone billing plans are examples of piecewise linear functions. They create a table of values, graph the functions, and compare pricing plans.

c. *Other representations.* Students create a Frayer Model or Link Sheet (see Lesson 2.2) for the current topic to replace two to three regular homework problems. Or they make a concept map (see Lesson 2.5 **DMP** problem 18) showing key ideas and their relationships for current content.

d. *Write quiz questions.* Students pose and solve three to four problems or questions that illustrate their estimation of what is important in a unit and their understanding of those topics.

e. *Choose your own.* Students are responsible for selecting for themselves problems to solve from the textbook that demonstrate what they consider the most challenging problems they can do.

f. *Forward homework.* A homework assignment has students do in a more primitive way something they will learn to do more efficiently in the next day's lesson. For example, students might graph related polynomial functions and then use them the next day to see how transformations could more efficiently be used to predict where the graphs fall. Or homework might bring to the fore needed prerequisites for the next day's lesson.

g. *Teach another.* The daily homework assignment is to teach what was learned in class to a friend or family member. Alternatively, students write a letter to another student explaining the lesson they missed.

h. *Creative writing.* Students create a poem, rap, song, short story, or script that demonstrates their understanding of critical ideas in the day's lesson.

II. Writing good journal prompts requires some thought. Teachers want items to be open enough for students to have to explain their thinking, but focused enough that the response provides effective information. Revisit the **Sample Unit Outline: Introducing Derivatives**. Write journal prompts for three of the learning activities described in the unit outline.

TEACHING AND LEARNING HIGH SCHOOL MATHEMATICS © 2010 John Wiley & Sons, Inc.

III. Initially, students and teachers find it difficult to determine real contexts related to the mathematics being studied. Decide on at least two real contexts that teachers might expect students to consider as they find examples from their daily lives (as in item Ib in this cluster) while working through the unit shown in the **Sample Unit Outline: Introducing Derivatives**. Suggest some resources that students might consult. Note that newspapers or the Internet are too broad to be helpful. Teachers will need to suggest specific sections of the newspaper or specific websites.

SHARE AND SUMMARIZE

6. Share the discussion of the issues in each cluster with the entire class.

7. What additional issues related to homework would you raise from the discussion or your own experience? Can these issues be addressed by one or more techniques discussed in one of the clusters? If not, what suggestions do you have to address your concerns?

8. How do daily assessments impact the day-to-day decision making teachers must do for effective teaching?

DEEPENING MATHEMATICAL UNDERSTANDING

1. *Behavior at a Point.* In Lesson 3.6, you investigated rational functions through graphing the numerator function and the denominator function, then finding the quotient function by dividing values from the graphs of the numerator and denominator functions for several values of x.

a. Complete the student page **Behavior at a Point.**

b. Quotient functions behave differently under various conditions where the denominator function has a value of zero.

i. How might the graph of a quotient function $\left(f(x) = \dfrac{g(x)}{h(x)}\right)$ appear over small intervals that include an x-value for which $g(x) \neq 0$ and $h(x) = 0$? Give different examples from those provided in the student page and explain each.

ii. How might the graph of a quotient function appear over small intervals that include an x-value where $g(x) = 0$ and $h(x) = 0$? Give different examples from those provided in the student page and explain each.

c. Limits are used to describe the behavior of a function when x approaches a particular value. For example, if a vertical asymptote occurs at $x = a$, where $h(a) = 0$, the mathematical notation $\lim_{x \to a^+} f(x) = \infty$ indicates that the function $y = f(x)$ increases without bound as x approaches a from values greater than a (or from the right). The notation $\lim_{x \to a^-} f(x) = -\infty$ indicates that the function decreases without bound as x approaches a from values less than a (or from the left). Revisit the student page **Behavior at a Point** and use limit notation to indicate the behavior of each function at points where the function has a distinct endpoint, jump, hole, or asymptote.

TEACHING AND LEARNING HIGH SCHOOL MATHEMATICS © 2010 John Wiley & Sons, Inc.

Behavior at a Point

Study the four sets of graphs each containing $y = f(x)$, $y = g(x)$, and $y = h(x)$ where $f(x) = \frac{g(x)}{h(x)}$. Notice that in each case, $g(0) = h(0) = 0$.

1. From the graphs of $y = g(x)$ and $y = h(x)$, why do you think the graph of $y = f(x)$ appears as it does for x near 0?

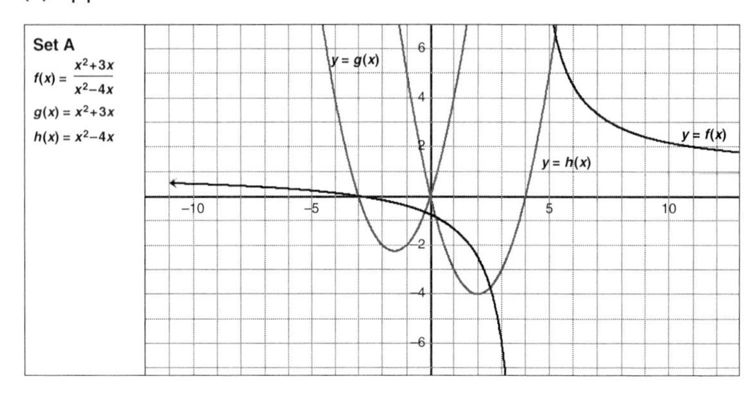

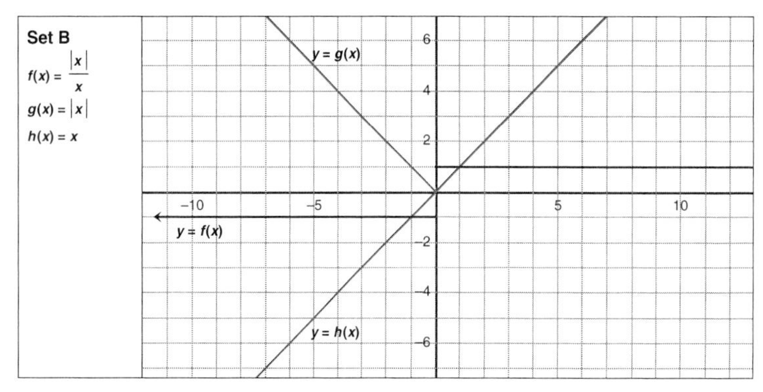

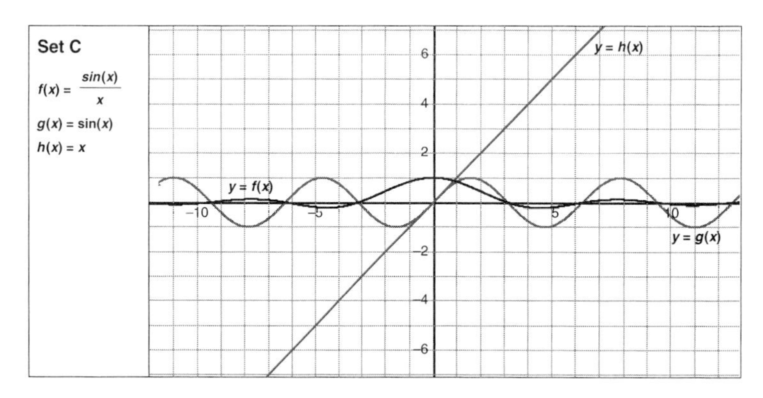

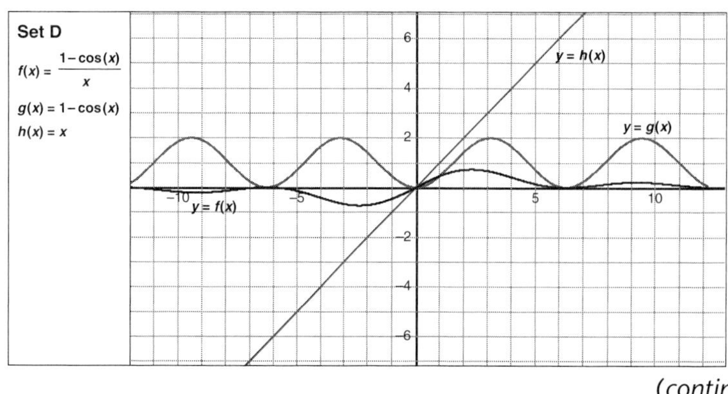

(*continued*)

Materials

- graphing calculators, one per person

TEACHING AND LEARNING HIGH SCHOOL MATHEMATICS © 2010 John Wiley & Sons, Inc.

2. For each set of graphs, plot the graphs of $y = g(x)$ and $y = h(x)$ on the viewing window $[-1, 1]$ by $[-1, 1]$.
   a. Identify which of the graphs is $y = g(x)$ and which is $y = h(x)$.
   b. How do these graphs appear on this window?
   c. From this view, what additional insights do you have about why the graph of $y = f(x)$ appears as it does for x near 0?
3. For each set of graphs, determine values of $x = a$ where $g(a) = 0$ and $h(a) \neq 0$.
   a. Describe the behavior of $y = f(x)$ for x near a.
   b. Explain why the graph of $y = f(x)$ appears as it does for x near a.
4. For each set of graphs, change the denominator function to $h(x) = x^2 + x$.
   a. Describe any changes to the graph $y = f(x)$ at $x = 0$. Why does the graph of $y = f(x)$ behave as it does for x near 0?
   b. Determine values of $x = a$ where $g(a) \neq 0$ and $h(a) = 0$.
      i. Describe the behavior of $y = f(x)$ for x near a.
      ii. Explain why the graph of $y = f(x)$ appears as it does for x near a.
5. Consider the graph of $f(x) = \frac{g(x)}{h(x)}$ over small intervals that include an x-value where the denominator function equals zero.
   a. Suppose $g(x) \neq 0$ and $h(x) = 0$. How might the graph of $y = f(x)$ appear? List different cases, give an example for each case and explain how you know the quotient function will behave as you say it does.
   b. Suppose $g(x) = 0$ and $h(x) = 0$.
      i. What different behaviors might the graph of $y = f(x)$ exhibit over such intervals?
      ii. List different cases, give an example for each case, and explain how you know the quotient function will behave as you say it does over such intervals.

Student page is inspired by Beckmann and Sundstrom (1992) and Beckmann (1993).

2. *Behavior at the Ends.* In Lesson 3.6, you worked with rational functions and determined how the function behaves as x increases without bound or decreases without bound (that is, as $|x| \to \infty$). Behavior of a function as x approaches ∞ or as x approaches $-\infty$ is called *end behavior*.
   a. When the numerator function is of higher degree than the denominator function, describe the end behavior of the quotient function. Note that there are two cases to address.
   b. List three rational functions $y = f(x)$ for which $\lim_{x\to\infty} f(x) = k$ for constant k. Explain why each function behaves as you say it does for large values of x.
   c. List three rational functions $y = f(x)$ for which $\lim_{x\to\infty} f(x) = 0$. Explain why each function behaves as you say it does for large values of x.
   d. List three rational functions $y = f(x)$ for which $\lim_{x\to\infty} f(x) = \infty$ and $\lim_{x\to-\infty} f(x) = \infty$. Explain why each function behaves as you say it does for large values of $|x|$.
   e. List three rational functions $y = f(x)$ for which $\lim_{x\to\infty} f(x) = \infty$ and $\lim_{x\to-\infty} f(x) = -\infty$. Explain why each function behaves as you say it does for large values of $|x|$.
   f. Revisit problems 2b–2e. Find functions that are not rational that satisfy the limits in each problem. Explain your work.

3. *Slopes of Secant Lines.* The **Sample Unit Outline: Introducing Derivatives,** expects students to make sense of the expression below.

$$\frac{\textit{change in } y}{\textit{change in } x} \text{ or } \frac{f(b) - f(a)}{b - a}$$

a. Sketch a graph with labels to illustrate what the expression represents.

b. Say in words what the expression means.

c. How does this expression relate to $\dfrac{f(c+h)-f(c)}{h}$?

4. *Exploring Secant Lines for Trigonometric Functions.*

   a. Sketch the graph of $f(x) = 0.04(\cos(\text{abs}(25x)) + x$ for x in $[-10, 10]$ and y in $[-10, 10]$. Describe the graph. (Use radian mode.)

   b. Zoom in on the graph several times until the x-interval is no more than 0.01 units wide. Describe the graph on this window.

   c. Is there any interval containing the graph of $y = f(x)$ for which the graph cannot be made straight over high enough magnification? Why or why not?

   d. What does your response to problem 4c tell you about secant lines to this function over small intervals of x?

   e. In problem 4a, replace cosine with sine. Repeat problems 4a–4d for this function.

5. *More Secant Lines for Trigonometric Functions.* Repeat problem 4 for several other functions. How well can the instantaneous rate of change of most functions be estimated? Why do you think so? Justify your point of view with graphical and symbolic representations.

6. *Polynomial Functions and Their Rate of Change Graphs.* Sketch an approximate graph of the instantaneous rate of change function for $f(x) = x^2$ by completing the following on a graphing calculator:

   i. Enter `Y1=X^2.`

   ii. Enter `Y2=(Y1(X+.001)-Y1(X))/.001.`

   a. Explain how Y_2 is related to an instantaneous rate of change function.

   b. Sketch and describe the graph of the rate of change function for $y = x^2$. Does it seem reasonable? Why or why not?

   c. Replace x^2 with x^3. Repeat problem 6b. What function appears to be graphed as the rate of change of $y = x^3$? Does this seem reasonable?

   d. Continue to increase the power of the polynomial by one. What relationship appears to exist between a polynomial function and its rate of change function?

DEVELOPING MATHEMATICAL PEDAGOGY

Preparing for Instruction

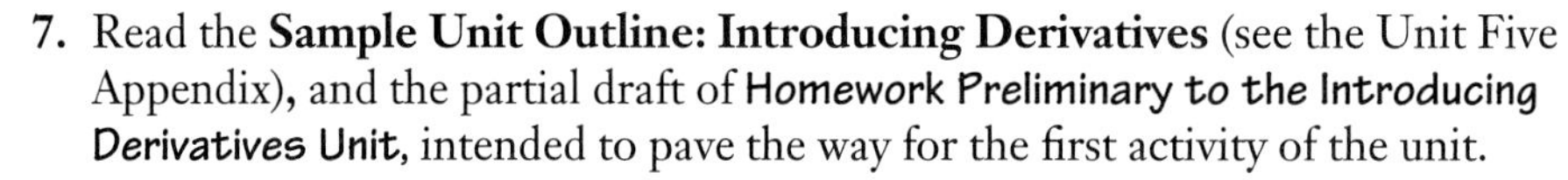

7. Read the **Sample Unit Outline: Introducing Derivatives** (see the Unit Five Appendix), and the partial draft of **Homework Preliminary to the Introducing Derivatives Unit,** intended to pave the way for the first activity of the unit.

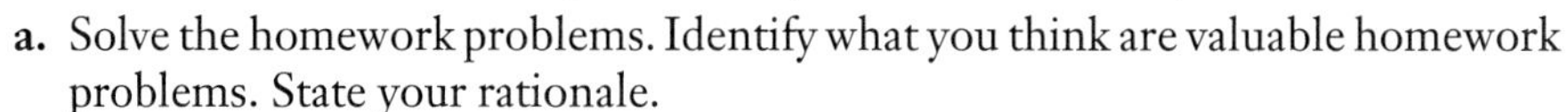

   a. Solve the homework problems. Identify what you think are valuable homework problems. State your rationale.

   b. Identify items you think need improvement. State your rationale.

   c. Rewrite one of the items needing improvement so that it focuses on conceptual understanding.

   d. Design two homework questions for learning activity 1 on the outline. Ensure that at least one of these questions focuses on conceptual understanding.

   e. Design another homework question that focuses on conceptual understanding and uses technology.

   f. Rewrite one of the problems to increase its accessibility for a student who needs more support.

TEACHING AND LEARNING HIGH SCHOOL MATHEMATICS © 2010 John Wiley & Sons, Inc.

Homework Preliminary to the Introducing Derivatives Unit

[To be critiqued]

1. Define each of the following:

a. function

b. argument

c. range

2. As she grew, Nesrin recorded her height at different ages. Here are her heights on several birthdays:

Age	10	11	12	13	14	15	16	17
cm	135	138	142	148	155	161	164	167

If a = Nesrin's age and h = Nesrin's height, then $h = f(a)$ represents her height as a function of her age.

a. Which is the dependent variable? Why?

b. Complete $f(13)$ =______. Explain in words what your sentence says.

c. Is $f(13 + 2) = f(13) + 2$? Is $f(13 + 2) = f(13) + f(2)$? Why or why not?

d. In which year did Nesrin grow fastest? How do you know?

e. How did Nesrin's rate of growth compare in the interval from ages 11–13 versus from ages 14–17? What do you need to think about to make this comparison?

3. Draw a graph for each of the following situations.

a. Xu is learning to play a video game. He records his best score each day. At first his "bests" improve quickly. Later he continues to improve, but more slowly.

b. The population of a city was steady for many years. Then it declined sharply and later continued to decline, but more slowly. Finally it began a slow recovery.

4. Consider the function g with the table of values shown.

x	-6	-3	0	3	6	9	12
$y = g(x)$	35	12	0	6	30	72	132

a. Find $g(-3)$. **b.** Find $g(6)$. **c.** Find $g(9)$.

d. Find $2 \cdot g(-3)$. **e.** Find $g(2 \cdot (-3))$.

f. Find $g\left(\frac{12}{2}\right)$. **g.** Find $\frac{g(12)}{2}$.

8. Study the **Sample Unit Outline: Introducing Derivatives**. Some teachers are concerned that once students learn procedures (for example, rules for finding the derivative of a polynomial function), they forget the concept underlying the procedure.

a. How does the sample unit attempt to avoid this problem?

b. Give another example of the same phenomenon in algebra or geometry. Explain how you would plan in order to reduce the likelihood of students forgetting underlying concepts once procedures are introduced.

TEACHING AND LEARNING HIGH SCHOOL MATHEMATICS © 2010 John Wiley & Sons, Inc.

9. Revisit the student page **Behavior at a Point** from **DMU** problem 1 in this lesson.
   a. Determine the learning outcomes for the student page.
   b. Complete a **QRS Guide** for the student page. Anticipate student responses at three levels for each major problem.
   c. Write teacher support questions for each student response and add these to the **QRS Guide**.
   d. Decide on a cooperative learning (CL) strategy that best fits use of the student page with a class of students. Why do you think this CL strategy is appropriate? Indicate the CL strategy and the learning outcomes at the top of the **QRS Guide**.
   e. Write three homework problems that reinforce or extend what students will learn while working on this student page. One of the problems must be in context. Another should be a short writing assignment or journal prompt.

   f. Try the activity with one or more high school students. Record any additional student responses you did not anticipate and related teacher support in the **QRS Guide**. What more did you learn by using the student page and **QRS Guide** with students?

   g. Complete a **Lesson Planning Guide** using the **Annotated Lesson Planning Guide** to assist you for a lesson containing the student page **Behavior at a Point**.

10. When students are disruptive, some teachers are tempted to give them seatwork or extra homework in order to keep them occupied until the end of class.
   a. Identify some negative effects that arise from such a practice.
   b. One teacher gives each student a disk at the beginning of each class. If a student is disruptive, he picks up the student's disk and puts it in a jar on the teacher's desk. If at the end of class fewer than 5 disks are picked up, students in the class earn 5 minutes of in-class homework time to be used on Friday. What are benefits of this practice? What are disadvantages?
   c. Suggest alternatives that will encourage good behavior in class.

Making Historical Connections

11. Early mathematician such as Nicole Oresme, Johannes Kepler, and Galileo explored mathematical concepts that foreshadowed the development of calculus, such as infinitestimals, indivisibles, and uniform velocity. Investigate one or more of these ideas and other work by these mathematicians that paved the way for the developments of Leibniz and Newton.

Reflecting on Professional Reading

12. Beckmann, Senk, and Thompson (1999), in "Assessing Students' Understanding of Functions in a Graphing Calculator Environment" (*School Science and Mathematics*, 99 (December 1999): 451–456), describe assessment in a graphing calculator environment. The authors describe items in terms of the potential for technology use: inactive (calculator is not helpful), neutral (calculator can be used but problem can be easily solved without it), or active (calculator is essential).
   a. Consider the following three graphing items. Match the items with the potential for technology use. Justify your answer.
      i. Graph $y = 5(x + 4)^2 - 3$.
      ii. Graph $y = 0.3(0.4x + 1.5)^2 - 2.9$.
      iii. Graph $y = ax^2$ for $a > 0$.
   b. The authors suggest the following strategies to modify traditional items to make them more appropriate for a calculator environment:

TEACHING AND LEARNING HIGH SCHOOL MATHEMATICS © 2010 John Wiley & Sons, Inc.

- Have students justify their answers.
- Have students analyze tables and graphs that a calculator might have produced without providing students the algebraic representation.
- Have students explore problems set in a real context.

From a precalculus textbook, find three traditional sample items. Modify each using one or more of the strategies suggested in the article. In each case, show both the original and the modified item and explain your modification process.

13. In preparation for Lesson 5.2, read "Using Rubrics in High School Mathematics Courses" by Denisse R. Thompson and Sharon L. Senk (*Mathematics Teacher*, 91 (December 1998): 786–793). (See the article on the course website.) Reading the article will provide background information needed during the **Launch** and **Explore** portions of Lesson 5.2. During Lesson 5.2, you should be able to discuss the following.

a. What is a rubric?

b. What is the difference between an analytical rubric and a holistic rubric?

c. On a rubric with a scale from 0 to 4, how do the authors decide whether a paper is a 3 or a 4? 2 or 3? 1 or 2?

5.2 Rubrics: Rates of Change

When students explain their reasoning on challenging mathematical tasks, their thinking becomes visible. Teachers gain insight into students' thoughts and can use that insight to enhance their instruction, addressing misconceptions in thinking or extending understandings in new directions. However, high school teachers also have to assign grades to students, and so the issue arises about how such open-ended work, or constructed responses, can be assessed in ways that are fair and equitable to students as well as informative to teachers.

Rubrics provide one means to assess extended responses. According to Thompson and Senk (1998), "a *rubric* is a set of guidelines for evaluating students' responses to one or more tasks" (p. 786). High school teachers long have had students *show their work* and often have given partial credit to students who make progress on a problem but have errors in their work. So, why worry about rubrics?

Partial credit typically is awarded on a problem by problem basis, so that 3 points may mean something very different from one problem to another. A rubric is a general standard; when applied consistently from problem to problem, teachers are able to compare a student's performance across problems. Suppose a teacher is using a rubric that ranges from 0 to 4. On five problems, one student gets 2, 2, 2, 2, 2 and another student gets 4, 0, 4, 0, 2. Both students have 10 points, but their pattern of performance is quite different. The first student makes some progress on all five problems but is not able to bring any of them to satisfactory completion. The second student solves two of the problems completely, makes some but incomplete progress on another problem, and makes no progress at all on two of the problems. Although receiving the same number of points overall, these two students likely need different instruction to help them move forward.

In this lesson, a holistic rubric is used, that is, a rubric that is applied to a response in its entirety. In the **DMP** questions, other types of rubrics are explored. The use of performance assessment as part of a state's accountability measures for students and schools has increased interest in the use of rubrics with constructed-response tasks. You are encouraged to become conversant with the rubrics used in your state, if such rubrics exist, and compare their use in your state to the discussion of rubrics in this lesson.

TEACHING AND LEARNING HIGH SCHOOL MATHEMATICS © 2010 John Wiley & Sons, Inc.

In this lesson, you will:

- Consider issues that arise when using a rubric.
- Explore similarities and differences between a general rubric and an item-specific rubric.
- Use a rubric to score student responses to problems involving rates of change.
- Identify features of tasks that make them appropriate for scoring with rubrics.

Website Resources

- *Rate of Change from Two Points of View* student page
- **QRS Guide**

LAUNCH

A *general rubric* is an outline or framework that describes expected standards of performance across various tasks.

Consider the following general rubric, adapted by Thompson and Senk (1998) from the work of Malone and others (1980). This general rubric has been used to grade hundreds of student papers from prealgebra through precalculus. The categories should look quite familiar after your work with **QRS Guides** throughout this textbook.

A General Rubric	
Satisfactory (Successful) Responses	
4	The response is complete and correct for the task. The response might be considered a model response.
3	The response is almost complete and correct for the task. There may be minor errors that do not detract from the understanding expressed about the problem. Any errors are considered clerical, rather than conceptual, in nature (e.g., minor arithmetic mistake for an advanced student, minor error in notation).
Partial or Misconceived (Unsuccessful) Responses	
2	The response indicates major progress in the proper direction. The student might complete about half the problem and stop. There is a chain of reasoning. However, the response terminates before completion. (*Partial*) OR The response indicates major progress toward an acceptable solution with a chain of reasoning. However, at some point the response indicates the student made a major conceptual error. (*Misconceived*)
1	The response indicates some limited entry into the problem. The student demonstrates some initial understanding about the problem but reaches an early impasse. The response indicates no chain of reasoning. (*Partial and/or Misconceived*)
0	The response has nothing mathematically meaningful or correct in terms of the problem.

1. What are some pros and cons of making a general rubric available to students?
2. Discuss the rubric as a class. What would you consider to be the rubric's strengths? What would you consider to be its weaknesses? Considering the work you have done in anticipating satisfactory, partial, and misconceived student responses as you prepared **QRS Guides,** do you believe you could apply this rubric consistently?

TEACHING AND LEARNING HIGH SCHOOL MATHEMATICS © 2010 John Wiley & Sons, Inc.

EXPLORE

Some states use only a general holistic rubric in scoring extended performance tasks. *Anchor papers* (i.e., sample papers at different score levels) are used to help scorers understand the levels of performance needed for a particular score. Anchor papers are shared and discussed so that all scorers understand how to apply the rubric consistently.

3. Refer to the student page **Rate of Change from Two Points of View**. Work through the problems on the student page, anticipating and recording potential satisfactory, partial, and misconceived student responses in a **QRS Guide**.

Rate of Change from Two Points of View

On his travels to school one morning, Dorian kept track of his time (in minutes) and distance traveled from home (in miles) at several points along his route. The data and related graph are provided.

Time	Distance	Time	Distance	Time	Distance
0.0	0.0	10.0	4.6	20.0	10.1
0.5	0.1	10.5	4.8	20.5	10.4
1.0	0.4	11.0	5.0	21.0	10.8
1.5	0.7	11.5	5.2	21.5	11.3
2.0	1.0	12.0	5.5	22.0	11.4
2.5	1.3	12.5	5.7	22.5	11.8
3.0	1.6	13.0	5.8	23.0	12.2
3.5	1.9	13.5	6.0	23.5	12.7
4.0	2.1	14.0	6.3	24.0	12.9
4.5	2.4	14.5	6.4	24.5	13.1
5.0	2.7	15.0	6.4	25.0	13.4
5.5	3.0	15.5	6.5	25.5	13.7
6.0	3.2	16.0	6.7	26.0	14.0
6.5	3.4	16.5	6.8	26.5	14.3
7.0	3.7	17.0	7.1	27.0	14.7
7.5	4.0	17.5	7.7	27.5	14.9
8.0	4.0	18.0	8.2	28.0	15.1
8.5	4.2	18.5	8.8	28.5	15.3
9.0	4.3	19.0	9.4	29.0	15.6
9.5	4.5	19.5	9.9	29.5	15.9
				30.0	16.1

(*continued*)

TEACHING AND LEARNING HIGH SCHOOL MATHEMATICS © 2010 John Wiley & Sons, Inc.

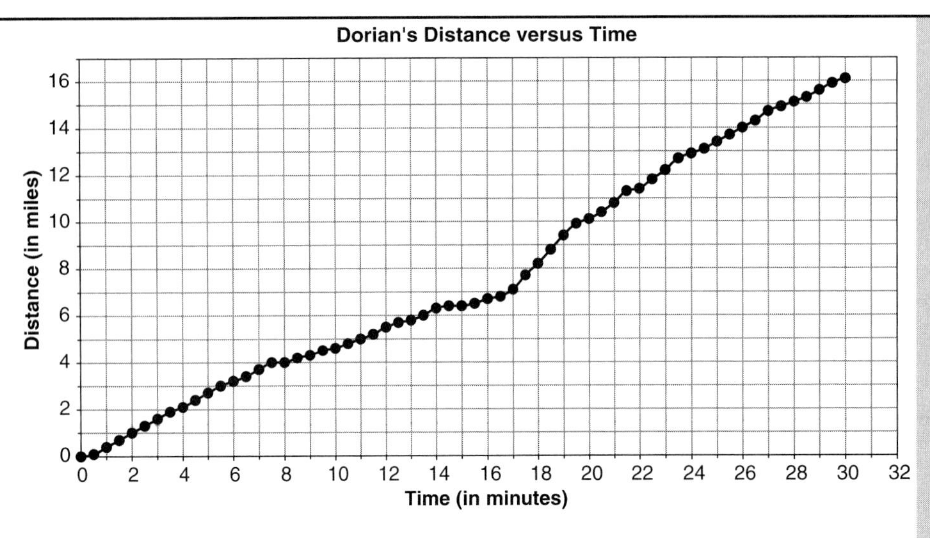

1. **a.** Describe Dorian's ride to school. Indicate where he seems to be traveling slowly, traveling quickly, stopped, etc. Describe on what sorts of roads you think he might be traveling during various parts of his travels. Say why you think so.
   b. From the table, what is Dorian's average velocity?
   c. How is Dorian's average velocity related to the graph of his travels to school? Show the relationship on the graph.
   d. During what time intervals does Dorian appear to be traveling fastest? How do you know? Give evidence from both the table and the graph.
2. Suppose Dorian travels only on city streets (no highways).
   a. Over what interval of time would a police officer be interested in Dorian's travels? Indicate such an interval on the graph.
   b. Estimate Dorian's velocity during the interval indicated in problem 2a.
   c. How is Dorian's velocity over the interval indicated in problem 2a related to the graph of his travels to school? Show the relationship on the graph.
3. In the graph at right, $t = a$ indicates the initial time and $t = b$ indicates the final time of travel of a commuter. Assume distance is a function f of time t, $d = f(t)$.

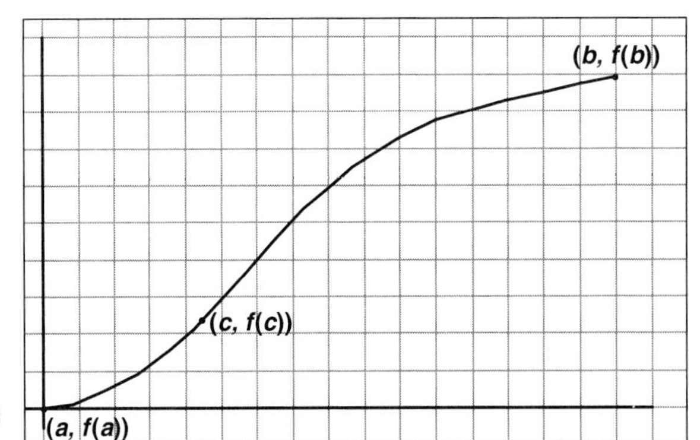

   a. What do $f(a)$ and $f(b)$ represent?
   b. Write an expression for the average rate of change (velocity) of the commuter over the interval $[a, b]$.
   c. What does this quantity represent in terms of the graph of the function? Explain.
4. A police officer is interested in a different point of view than a commuter. Suppose a police officer begins to observe the commuter at time $t = c$.
   a. Suppose the police officer determines the commuter's velocity for h seconds. Symbolize the beginning and ending times of the interval.

(*continued*)

TEACHING AND LEARNING HIGH SCHOOL MATHEMATICS © 2010 John Wiley & Sons, Inc.

b. Symbolize the beginning and ending distances over the time interval.

c. Write an expression for the rate of change (velocity) of the commuter over the time interval in problem 4a.

d. The police officer is most interested in detecting velocity before a commuter has the opportunity to change speeds. To make the police officer's velocity reading favorable to his point of view, how large should h be?

e. Indicate a point on the graph in problem 3 that corresponds with a time h seconds after time c.

   i. Sketch a line segment connecting the two points on the graph. Such a line is called a **secant line**.

   ii. How does the velocity from the police officer's point of view relate to this line? Explain.

   iii. What happens to the secant line of the graph of $d = f(t)$ as h is allowed to decrease to 0?

f. Discuss how limits might be used to obtain an accurate velocity reading.

5. Summarize your findings. Compare the commuter's point of view of average rate of change to the police officer's point of view of instantaneous rate of change (rate of change as the time interval decreases to zero). Discuss symbolic representations and their relationship to graphical representations of each type of rate of change.

Student page inspired by Beckmann and Sundstrom (1992) and Beckmann (1993).

It often is helpful to create an *item-specific rubric* for each item to be scored. The item-specific rubric applies the general rubric to the specific context of the problem and outlines the types of thinking about the problem that result in each score level.

Consider the following item-specific rubric for problem 1 on the student page.

Item-Specific Rubric for Dorian's Travel Problem 1	
Satisfactory (Successful) Responses	
4	• Student response indicates a thorough understanding of the relationship of the slope to the velocity. • The response indicates that the steeper the graph, the faster the speed; the student hypothesizes that faster speeds occur on interstates or major highways and slower speeds occur on city streets. • Stops may be associated with red lights or stop signs. • The student accurately finds the average velocity (0.54 mi/min or 32.2 mi/hr) and connects this to the slope of the segment joining the initial point and the endpoint of the graph and draws an appropriate segment. • The student is able to connect the fastest speed to the steepest segments for a given interval (the interval from 17 to 19.5 min or subparts of this interval), and backs up the conclusion with appropriate references to the graph and to the table.
3	• Student response indicates a solid understanding of the relationship of the slope to the velocity but students might omit units in their response.

(*continued*)

TEACHING AND LEARNING HIGH SCHOOL MATHEMATICS © 2010 John Wiley & Sons, Inc.

3	*(continued)* • The response indicates that the steeper the graph, the faster the speed; some questions are incompletely or inappropriately addressed, for example, the student fails to hypothesize the type of roads on which travel occurs, or the causes for flat segments of the graph (stops). • The student accurately finds the average velocity and connects this to the slope of the segment joining the initial point and the endpoint of the graph and draws an appropriate segment. • The student recognizes that the fastest speed relates to the steepest segments for a given interval but makes some minor arithmetic or copying errors in backing up conclusions from the graph and/or table of data values.
Partial or Misconceived (Unsuccessful) Responses	
2	• Student response indicates basic understanding of the relationship of slope to speed, indicating that steeper slopes indicate faster speeds. • The student is able to determine the average velocity from a computational perspective but is not able to connect the average velocity to the graph. • The student has difficulty determining the intervals in which Dorian is traveling the fastest, basing decisions solely on a visual inspection of the graph without supporting the decision with appropriate data from the table or vice versa.
1	• Student response indicates only a minimal understanding of the graph and/or the situation. For instance, the student indicates that Dorian gets farther from home the longer he travels and that he travels at different speeds. • The student seems unable to connect the speed to the slope of the corresponding segments; it is not clear if the student is able to determine average velocity, or if so, to connect it to the context of the problem.
0	• There is nothing mathematically correct that relates to the problem.

4. With your team, use the item-specific rubric to score the following sample student responses. Be prepared to justify your score to members of other teams.

a. *The longer Dorian travels, the farther he gets away from home. From 7.5 minutes to 8.0 minutes, Dorian is stopped because his distance doesn't change. He is also stopped from 14.5 to 15 minutes. The total distance he travels is 16.1 miles in 30 minutes. So, if I use $d = rt$ and change 30 minutes to 0.5 hour, his average speed is $30 \div 0.5 = 32.2$ miles per hour.*

b. *Dorian travels a total of 16.1 miles in 30 minutes or 0.5 hour. When I look at the graph, I see that different segments have different steepnesses. I know that the slope is the change in y over the change in x, so this is change in distance divided by change in time which is speed. So, steeper parts of the graph mean faster speeds. When the slope is 0 and the graph is horizontal, Dorian is standing still. I think he would travel fast on an interstate and travel a lot slower on a city street. He probably had to sit still when he was at a red light or a stop sign when there was a lot of traffic. The average velocity would be*

$$\frac{16.1 - 0.0 \text{ miles}}{30.0 - 0.0 \text{ minutes}} = \frac{16.1 \text{ miles}}{30 \text{ minutes}} = \frac{16.1 \text{ miles}}{0.5 \text{ hours}} = 32.2 \text{ miles per hour}$$

The average velocity is the slope of the segment drawn from the starting point to the ending point and is the speed at which he would be traveling if he could travel at a constant speed with no stops or slow downs. I think he is going fastest

TEACHING AND LEARNING HIGH SCHOOL MATHEMATICS © 2010 John Wiley & Sons, Inc.

from 17.0 to 17.5 minutes, from 18.0 to 18.5 minutes, and from 18.5 to 19.0 minutes because in each of these intervals he travels 0.6 miles. He doesn't travel that far in any other half-minute interval.

c. *Dorian travels 16.1 miles in half an hour, so his average speed would be 16.1 ÷ 0.5 or 32.2 miles per hour. He travels at different speeds at different times on his trip. I know he is traveling at different speeds because the slope of the different segments is different from interval to interval. I know that slope is change in y over change in x so this is distance over time and that is the same as the rate or speed because d/t = r. A larger slope number means a faster speed, a smaller slope number means a slower speed, a 0 for slope means no speed or standing still. Dorian is traveling fastest in the interval from 18.0 to 19.0 minutes because he travels 1.2 miles in 1 minute or 72 miles per hour.*

d. *Dorian travels 30 miles to get to school. He probably had to go on some city roads. There's not enough information to know how fast he's going.*

e. *I know that distance divided by time is speed. I also know that the slope is the change in distance (y) over the change in time (x); so the slope is the same as the speed. Different parts of the graph have different slopes because the segments slant differently. The steeper the slant, the greater the slope and the greater the speed. The slope from the beginning to the end is* $\frac{16.1 - 0}{30 - 0} = 0.54$*, so Dorian is going really slow. I know from the table that he didn't move from 7.5 to 8.0, so he had to be stopped then. When I checked all the differences in the distances, I found three places where the distance went up by 0.6 from one time to the next, so these were the times when he traveled the fastest—from 17.0 to 17.5, 18.0 to 18.5, and 18.5 to 19.0.*

5. Think about the rubric in relation to the sample responses you scored.

a. What changes, if any, would you recommend in the rubric to clarify its use?

b. Although item-specific rubrics are helpful, it is important to build them from a general rubric that can be used when a response does not fit neatly into the categories specified by the item-specific rubric. Were there any responses in item 4 for which you needed to refer to the general rubric because the item-specific rubric did not cover the response completely? Explain.

6. Among your team, form two pairs. One pair should develop an item-specific rubric for item 3 on the student page. The other pair should develop an item-specific rubric for item 4 on the student page. After creating your rubric, share and critique it with the other pair.

SHARE AND SUMMARIZE

7. As a class, tally the scores provided by each team to the sample responses in item 4.

a. How consistent were the scores assigned by different teams? What reasons were given for differences in scoring?

b. When large numbers of papers are going to be scored with a rubric, it is common to have, as mentioned earlier, a set of anchor papers that represent typical responses at each of the score levels. All scorers read these responses and practice applying the rubric to the response; discrepancies are discussed and the rubric is modified or clarified as needed. Having scored the responses, which student responses would you use as anchor papers? How might a set of anchor papers have helped you to apply the rubric? What would you like to add to the item-specific rubric to help you grade additional work on this item?

TEACHING AND LEARNING HIGH SCHOOL MATHEMATICS © 2010 John Wiley & Sons, Inc.

c. Based on your responses to items 7a and 7b, comment on how rubrics can be used to grade open-response problems in an objective manner.

8. Share the rubrics created for items 3 and 4 on the student page from the different teams. Look for similarities and differences in the rubrics. Modify the rubrics as appropriate based on the differences in the rubrics.

Typically, rubrics can only be applied to tasks that meet certain conditions:

- The task is open with multiple answers.
- If the task has only one correct answer, there are multiple ways to obtain the answer.
- The task requires students to explain their thinking.

If a task has only a single or small number of answers or ways to obtain an answer, it is not usually rich enough to warrant the use of a rubric.

9. Consider the following items related to rate of change from an Algebra I course through precalculus. Which would be appropriate for scoring with a rubric? Justify your answer.

a. Find the rate of change from $(-2, 5)$ to $(6, 8)$.

b. A 7-pound turkey requires 3 hours to cook. A 14-pound turkey requires 5 hours. A 22-pound turkey requires 6 hours. Find the rate of change in cooking time from a 7-pound to a 14-pound turkey and from a 14-pound to a 22-pound turkey. Interpret the rate of change in the context of the problem.

c. True or false: The rate of change for an interval is the same as the slope of the graph over that interval.

d. Compare the rate of change over an interval to the instantaneous rate of change at a point. How are they related to each other? Represent any relationships symbolically, if possible.

DEEPENING MATHEMATICAL UNDERSTANDING

1. *Revisiting Exploring Quadratic Functions.* Revisit the student page **Exploring Quadratic Functions** from Lesson 3.5 **DMU** problem 1.

a. Use the table you created in item 2a on the student page to make a conjecture about the rate of change for the function $f(x) = x^2$ between $(x, f(x))$ and $(x - 1, f(x - 1))$.

b. Prove your conjecture from problem 1a.

2. *Cubics and Rates of Change.* Refer to the function $f(x) = x^3$.

a. Find the rate of change between $(x, f(x))$ and $(x + \Delta x, f(x + \Delta x))$ when $x = 2$ and $\Delta x = 2,\ 1,\ 0.5,$ and 0.25.

b. What happens to the rate of change as Δx approaches 0?

c. How does your result from problem 2b relate to the derivative?

3. *Rational Reciprocal Functions.* Consider the function $g(x) = \frac{1}{x}$ for $x \neq 0$.

a. Find the rate of change between $(x, g(x))$ and $(x + \Delta x, g(x + \Delta x))$ when $x = 1$ and $\Delta x = 2,\ 1,\ 0.5,$ and 0.25.

b. What happens to the rate of change as Δx approaches 0?

c. Relate your result in problem 3b to the derivative of function g at $x = 1$.

4. *Connecting Rates of Change and Graphical Representations.* Sketch a graph of a function f and label x-values a, b, c, d, and e such that the following conditions are true.

a. The average rate of change between points $(a, f(a))$ and $(b, f(b))$ is positive.

b. The average rate of change between points $(a, f(a))$ and $(c, f(c))$ is negative.

c. The average rate of change between points $(a, f(a))$ and $(d, f(d))$ is 0.

d. The average rate of change between points $(a, f(a))$ and $(e, f(e))$ is 4.

TEACHING AND LEARNING HIGH SCHOOL MATHEMATICS © 2010 John Wiley & Sons, Inc.

5. *Acceleration*. Refer back to the student page **Rate of Change from Two Points of View**.
   a. Find the rate of change (velocity) for each half-minute in the interval from $t = 0$ minutes to $t = 10$ minutes. Record your results in a table.
   b. Use your results from problem 5a to find the rate of change of velocity in each half-minute interval. What does this number represent?
6. *Rate of Change and Exponential Functions*. Consider the exponential function $f(x) = e^x$.
   a. Find the rate of change between $(x, f(x))$ and $(x + \Delta x, f(x + \Delta x))$ when $x = 1$ and $\Delta x = 1, 0.5, 0.25, 0.1,$ and 0.01.
   b. Use a graphing tool to graph secant lines containing the points $(x, f(x))$ and $(x + \Delta x, f(x + \Delta x))$ for $x = 1$ and $\Delta x = 1, 0.5, 0.25, 0.1,$ and 0.01.
   c. How do the secant lines change as Δx approaches 0? How does each secant line compare to the graph of the original function over the interval $[x, x + \Delta x]$ as Δx approaches 0?
7. *Teacher Certification Exam Enrollments*. The table at right gives the number of first-time examinees for two of the teacher certification exams in Florida by years.

Year	Number of Examinees for Elementary Education Exam	Number of Examinees for 6–12 Mathematics Exam
2004	9,043	1648
2005	10,071	1473
2006	12,725	1469
2007	12,969	1868

Source: www.fldoe.org/asp/ftce/pdf/firsttime-ftce-examinees.pdf (downloaded July 30, 2009)

   a. Between which two consecutive years is the rate of change in first-time examinees for the elementary education exam the greatest?
   b. Find the rate of change in first-time examinees for the elementary education exam from 2004 to 2007.
   c. Compare the rate of change from 2004 to 2005 with the rate of change from 2006 to 2007 for the first-time examinees for the 6–12 mathematics exam.
   d. Find out if your state has an exam required of individuals seeking teacher licensure. If so, locate enrollment numbers for the past few years. What is the rate of change in the number of examinees over this period of time?

8. *Revisiting On a Roll*. Revisit the student page **On a Roll**, from the **Explore** of Lesson 3.3. A possible distance versus time graph for the experiment is shown at right.

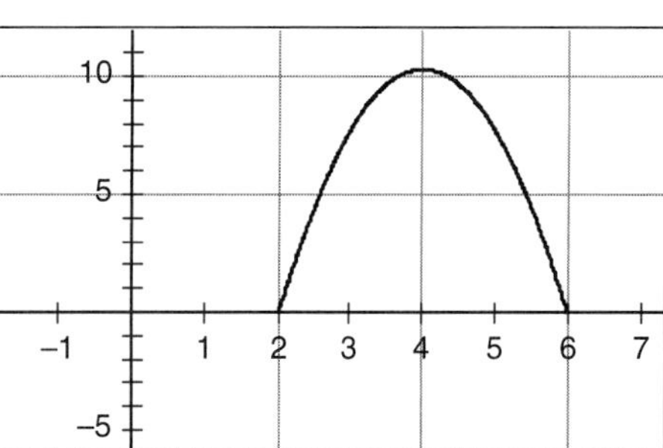

   a. Over which intervals is the distance versus time graph increasing? What does this tell you about the velocity versus time graph?
   b. Over which intervals is the distance versus time graph decreasing? What does this tell you about the velocity versus time graph?
   c. When is the velocity zero? How do you know?
   d. Use your answers to problems 8a–8c and what you have learned about estimating rate of change over very small intervals of time to sketch the graph of velocity versus time for this function. Show your work for at least five points.
   e. Complete the sentence: For a quadratic function, the derivative function is a ___________ function because _______________. Explain your answers using your work in problems 8a–d and another representation.
9. *Falling Bodies*. A rock is dropped from the edge of a cliff 400 feet high. Its height, h in feet, at any time, t in seconds, is given by the equation $h = -16t^2 + 400$.
   a. Find the velocity of the rock 1 second after being dropped.
   b. Find the velocity of the rock when it hits the ground.

TEACHING AND LEARNING HIGH SCHOOL MATHEMATICS © 2010 John Wiley & Sons, Inc.

DEVELOPING MATHEMATICAL PEDAGOGY

Preparing for Instruction

10. Research your state's assessment program at the high school level. Does your state use performance tasks? If so, how are they scored? If a general rubric is used, compare your state's rubric to the one provided in this lesson.

11. Connecting rubrics to grades is an issue that teachers need to consider. A score of 3 on a 0 to 4 rubric should not be considered 75 percent and assigned a grade of C; likewise, a score of 2, although unsuccessful, is not equivalent to 50 percent and a grade of F.

- **a.** Determine appropriate letter grades to connect with each rubric score. Justify your answer.
- **b.** Discuss issues that arise when grading an exam where some items are scored with a holistic rubric and others are scored using an analytic scale (such as 2 points for a correct definition and 1 point for a graph). What concerns do you have about combining scores from these scoring methods?

12. *Analytic rubrics* often are used to score a response at several phases of the problem-solving process, such as interpreting the problem, planning a solution, and carrying out the plan. Use the Internet to research the difference between an analytic and a holistic rubric. What are similarities and differences? Which do you prefer? Why?

13. Select one of the student pages identified below.

Student Page	Location
How Many Oranges?	Lesson 1.1
Similar Right Triangles	Lesson 2.2
The Geometry of Functions	Lesson 3.4
What Is the Probability You Are Ill?	Lesson 4.2

- **a.** Determine the learning outcomes for the student page.
- **b.** Complete a **QRS Guide** for the student page. Anticipate student responses at three levels for each major problem. Using the general rubric from the **Launch** of Lesson 5.2, indicate the score students would earn for each level of response.
- **c.** Write teacher support questions for each student response and add these to the **QRS Guide**.
- **d.** Write two homework problems requiring the use of a rubric that either reinforce or extend what students are expected to learn through the student page. Write an item-specific rubric for each homework problem.

- **e.** Try the activity and homework with one or more high school students. Record any additional student responses you did not anticipate and related teacher support on the **QRS Guide**. Grade the homework using the rubric. Revise each rubric as needed.

Focusing on Assessment

14. Revisit the rubric you created for **Explore** problem 6. Choose item 3 or 4 on the student page Rate of Change from Two Points of View. Write at least one student response for each score of the rubric you created.

TEACHING AND LEARNING HIGH SCHOOL MATHEMATICS © 2010 John Wiley & Sons, Inc.

15. Refer back to one of the interviews you conducted with students from Units One through Four. As you interviewed students, you collected students' responses on a number of tasks. Collaborate with one or more peers so that you have a collection of sample student responses.
    a. Create a rubric that could be used to score these responses.
    b. How did having student work enhance your creation of the rubric?
16. a. Find a task in a precalculus textbook that is suitable for a rubric assessment. Draft a rubric for the task.
    b. Try the task in problem 16a with at least a dozen students who have some background on the topic. Use their responses to refine your drafted rubric.
    c. Evaluate the papers with your rubric. Then share with colleagues some of the responses that were more difficult to evaluate. Describe important points you discussed.
    d. What did you learn from the processes of task selection, rubric drafting, and rubric refining?
17. Refer back to the sample items in problem 9 of the **Share and Summarize** portion of Lesson 5.2. For any items you considered inappropriate for scoring with a rubric, suggest a possible modification to the item so that it would be appropriate for use with a rubric.

Reflecting on Professional Readings

18. Read "Using Rubrics in High School Mathematics Courses" by Denisse R. Thompson and Sharon L. Senk. (*Mathematics Teacher*, 91 (December 1998): 786–793), which you may have read in **DMP** question 13 in Lesson 5.1. (See the article on the course website.)
    a. What is a rubric?
    b. Distinguish an analytic rubric from a holistic rubric.
    c. On a rubric with a scale from 0 to 4, how do the authors decide whether a paper is a 3 or a 4? 2 or 3? 1 or 2?
    d. The authors identify several issues to consider when using rubrics. Summarize the issues. Which issues were of concern for you?
19. Read "EMRF: Everyday Rubric Grading" by Rodney Y. Stutzman and Kimberly H. Race. (*Mathematics Teacher*, 97 (January 2004): 34–39).
    a. What were the authors' goals in designing their EMRF system?
    b. The authors begin rubric development at the M level. What are they looking for there?
    c. What does R represent to the authors?
    d. What benefits do the authors cite for sharing a task's rubric with students when papers are returned?
    e. What are the authors' rules for revisions?
    f. How do the authors convert EMRF grades to ABCDE?
    g. What is your current thinking about an EMRF grading system?

20. (In preparation for Lesson 5.3) Read "Improving Classroom Tests as a Means of Improving Assessment" by Denisse R. Thompson, Charlene E. Beckmann, and Sharon L. Senk. (*Mathematics Teacher*, 90 (January 1997): 58–64). (See the article on the course website.) During discussion of the lesson, be prepared to classify test items according to the categories in the table in the article.

TEACHING AND LEARNING HIGH SCHOOL MATHEMATICS © 2010 John Wiley & Sons, Inc.

5.3 Designing and Aligning Tests with Instruction: Derivatives

Throughout this text, you have engaged in and focused on developing inquiry lessons for use in your own classroom. A major part of inquiry lessons is the nature of the tasks and questions you use with students. In particular, you have generated questions that require students to explain their thinking and justify their results in some way (e.g., using graphs, tables, other visual representations, logic, descriptive language, proof). You have also considered the importance of the **Share and Summarize** portion of a lesson to ensure that students master the major learning outcomes of the lesson.

Equally important are the summative assessments teachers use to gauge student learning and determine grades. These summative assessments, often in the form of tests at the high school level, send important messages to students about what is really valued. Instruction and assessment should be aligned so they support each other. When aligned, results on assessments should be informative to students to help them learn what has been mastered and what still requires work. Assessments should also help teachers determine what concepts need to be strengthened before students can progress mathematically, how effective the instruction was, and what modifications need to be made the next time a lesson or unit is taught.

Assessment that is not aligned with instruction can have the effect of undermining instructional goals. For instance, teachers might encourage students to explain and justify their thinking, but if assessment then focuses only on procedural skills or factual knowledge, students will quickly learn that explaining and justifying are not all that important. Otherwise, those processes would be part of their grade!

Two types of alignment need to be considered when developing summative tests: *content alignment* and *alignment with instructional approaches*. Content alignment means the content being assessed corresponds to content students have had an opportunity to learn. Important concepts should be included on assessments and concepts that students have not had an opportunity to learn should not be included. Alignment with instructional approaches means that assessments emphasize what instruction has emphasized. In classrooms where the use of technology, real-world applications, explanations, multiple solutions, and representations are part of instruction, they need to be part of assessment.

Although tests are likely to be a part of the high school curriculum in most situations for the foreseeable future, tests are not the only form of summative assessment. Some important features of assessment are not well addressed in a timed, on-demand format such as a test. In this lesson, the focus is on tests and quizzes; other forms of assessments are discussed elsewhere in this unit.

In this lesson, you will:

- Identify factors that influence the development of unit tests.
- Analyze and modify test items to make them more appropriate and useful to inform teachers about their students' thinking.
- Consider strategies for helping students review for tests.

Website Resources

- **Sample Unit Outline: Introducing Derivatives**
- Sample Test on Derivatives
- **Template for Classifying Test Items**

LAUNCH

1. Review the **Sample Unit Outline: Introducing Derivatives,** found in the Unit Five Appendix. What are some issues you would want to consider when developing a test for this unit?

TEACHING AND LEARNING HIGH SCHOOL MATHEMATICS © 2010 John Wiley & Sons, Inc.

EXPLORE

Assessments should reflect important content as well as mathematical processes. In the Unit Observation, you read an article by Thompson, Beckmann, and Senk (1997) in which categories were described to assess the nature of items on an assessment. The authors analyzed tests to determine the extent to which skills, real contexts, open-ended problems, representations, reasoning, and technology were emphasized.

2. Critique the categories identified by Thompson, Beckmann, and Senk.
   - **a.** Which ones would you view as particularly important for high school students?
   - **b.** What changes, if any, would you make to this list?
3. Consider the **Sample Test on Derivatives.**
   - **a.** Classify each of the items on the following characteristics: skills, real-world contexts, open-ended problems, representations, reasoning, and use of technology. A **Template for Classifying Test Items** is provided in the Unit Five Appendix.
   - **b.** Based on your responses in item 3a, discuss the test's strengths and weaknesses.
   - **c.** What items, if any, would you delete? What additional items would you add?
   - **d.** What items do you anticipate are most likely to create difficulties for students?
4. Analyze the **Sample Test on Derivatives** in terms of use of technology.
   - **a.** What items would you change if students are allowed to use graphing calculators while taking the exam?
   - **b.** What items would you change if students are allowed to use computer algebra systems while taking the exam?
   - **c.** Change one of the items identified in items 4a or 4b so that the item is appropriate to include on a test and also requires high cognitive demand, assuming students are using the technology indicated.

TEACHING AND LEARNING HIGH SCHOOL MATHEMATICS © 2010 John Wiley & Sons, Inc.

Sample Test on Derivatives

[*To be critiqued*]

1. Cassie was 3 feet 2 inches at age 3, and 5 feet 8 inches at age 15. Find the average rate of change in her growth.

2. Consider the following statement: If the average rate of change between $(a, f(a))$ and $(b, f(b))$ for $a < b$ is negative, then $f(b) < f(a)$.

  a. Is the statement true or false? Explain your answer.

  b. Sketch a graph to illustrate your answer in question 2a.

3. a. Sketch a graph of distance from home (in miles) versus time (in hours) for the following situation:

  Jack leaves home and drives at 50 miles per hour for 2 hours. He stops for lunch for 1 hour. When he resumes his trip, he continues driving away from home for 3.5 hours at an average speed of 65 miles per hour.

  b. What is his average speed for the entire trip?

  c. Jack leaves from his final location and drives straight home at a constant speed. Modify your graph to show Jack's trip home. How can you determine his speed for this portion of the trip from the graph?

4. At right is a graph of $f(x) = x^2 + 3x - 2$.

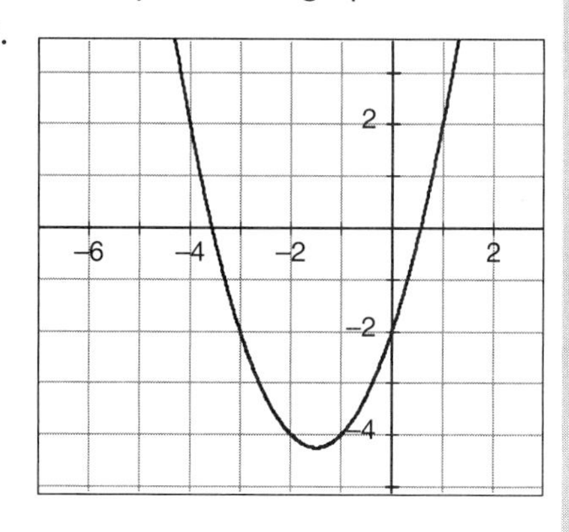

  a. Find a symbolic description for the average rate of change of f from $x = 1$ to $x = 1 + \Delta x$.

  b. Find the instantaneous rate of change of f at $x = 1$.

  c. Estimate $f'(-3)$. Explain how you obtained your estimate.

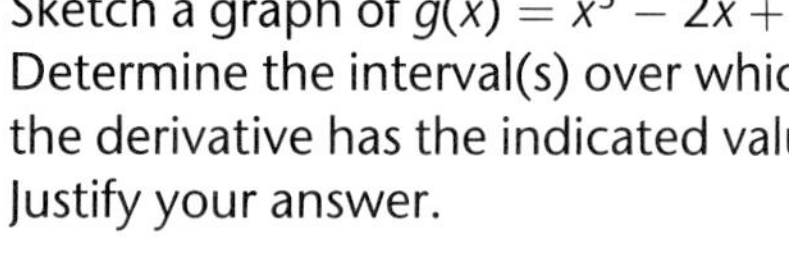

5. Sketch a graph of $g(x) = x^3 - 2x + 1$. Determine the interval(s) over which the derivative has the indicated value. Justify your answer.

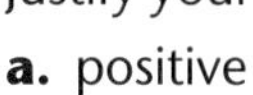

  a. positive

  b. negative

  c. 0

SHARE

5. Compare your perceptions of the test with your team and with other members of the class. What learning outcomes are not yet assessed with this test?
6. What other aspects of students' understanding about derivatives and limits might you still need to assess in some manner other than on a timed test?

EXPLORE MORE

After students have studied the various lessons in the **Sample Unit Outline: Introducing Derivatives**, they likely need some opportunities to consolidate their learning prior to the test. That is, teachers generally need to provide some type of unit review to help students solidify their thinking, determine what knowledge is still fragile, and understand how the current mathematical learnings relate to each other.

TEACHING AND LEARNING HIGH SCHOOL MATHEMATICS © 2010 John Wiley & Sons, Inc.

7. Teachers use a variety of strategies to help students review. Compare and contrast the following review strategies for high school students. Which methods would you be interested in using in your own classroom? Why?
   - Students take a practice test the night before the actual test.
   - Students take a practice test two or three days before the actual test. Based on the results, they choose specific review problems to address in class.
   - In small groups, students work through a large collection of teacher-created review problems during class, seeking help from each other and the teacher as needed.
   - Students work in groups to write sample test questions. The teacher pools these together to develop a review for the class.
   - Students are selected at random to go to the board to work on practice problems while the rest of the class works the same problems at their seats.
   - Students play adaptations of games like *Jeopardy* to help them review. Questions can come from supplements, homework, teacher-made reviews, or student-written questions.
8. Some teachers dislike practice tests. They think students learn how to solve only the practice problems and fail to understand important concepts. When actual test items vary from the practice test, students complain that the test is unfair.
   a. Comment on this perspective.
   b. What approaches could be used to avoid students misusing a practice test?

SHARE AND SUMMARIZE

9. Share your team's responses to questions 7 and 8 with those of other teams.

DEEPENING MATHEMATICAL UNDERSTANDING

1. *Derivatives and Graphing Calculators.* Some graphing calculators, such as the TI-83/84, enable students to investigate the derivative of a function without having a rule for the derivative. The `nDeriv` function is located in the `MATH` menu. Explore these ideas in the following questions.
   a. Let `Y1=X^3−2X+1`. `nDeriv(`stands for the numerical derivative and is accessed through the `MATH` menu. Let `Y2=nDeriv(Y1,X,X)`; the notation means you are estimating the numerical derivative of `Y1` with respect to `X` at each value of `X`. Graph `Y1` and `Y2` and explain what you see.
   (Use the 2nd `VARS` menu under function to enter `Y1` into `Y2`. You can now change the function in `Y1` and graph its derivative without making any further changes.) Use an appropriate viewing window to distinguish between the two graphs.
   b. Find an equation for the derivative of $f(x) = x^3 - 2x + 1$. Graph it. How does the graphical result in problem 1a compare with the graph of your equation for the derivative?
   c. Now make `Y1=e^(X)`. Leave `Y2=nDeriv(Y1,X,X)`. Graph the results. Turn off the graph of `Y1` and regraph `Y2`. Describe what happens. What does this suggest is a rule for the derivative of $f(x) = e^x$?
2. *Derivatives and Computer Algebra Systems.* Some graphing calculators with a computer algebra system (CAS) enable students to find a symbolic rule for the derivative. Explore these ideas in the following questions.
   a. On the TI-89, access the `F3 Calc` menu to find `d(` for the derivative. Use the syntax: `d(ln(x),x)` to find the derivative of $f(x) = \ln x$ with respect to x. What is displayed when you press `ENTER`?

TEACHING AND LEARNING HIGH SCHOOL MATHEMATICS © 2010 John Wiley & Sons, Inc.

b. Use the appropriate syntax for your calculator to find a symbolic rule for the derivatives of the following functions:

i. $g(x) = \ln(x + 1)$ **ii.** $h(x) = \ln(x + 2)$ **iii.** $k(x) = \ln(x - 5)$

c. Based on your results in problems 2a and 2b, conjecture a rule for the derivative of function f with $f(x) = \ln(x + k)$ for $x > k$.

d. Why must $x > k$? Explore using the preceding examples.

e. Study the graphs of $g(x) = \ln(x + 1)$ and its derivative. Explain why that derivative makes sense for the natural logarithm function.

3. *Derivatives and Hours of Daylight.* People who live a distance north of the equator are aware that the days are considerably shorter in winter and longer in summer, with equinoxes (equality of daylight and darkness) on the first days of fall and spring. Not so obvious to everyone is that the rate at which days lengthen or shorten varies, too.

   a. In `Y=`, enter `Y1 = sin(X)+2`. Be sure your calculator is in radian mode. Use an appropriate window. Graph. Interpret the graph in terms of the daylight scenario. What do the axes represent? What special points can you determine? Explain your thinking.

   b. In `Y=`, enter `Y2 = nDeriv(Y1, X, X)`. Graph. What function appears?

   c. Interpret the relationship between the graphs of `Y1` and `Y2`. When are the lengths of days increasing rapidly? Increasing slowly? Decreasing rapidly? Decreasing slowly? To what times of year do these changes correspond?

   d. Some people say that for those who live in the northern hemisphere, "Once people get through the period from Thanksgiving to Valentine's Day, things are looking brighter." Use each graph to interpret this statement.

   e. On the Internet, find data showing the number of minutes of sunlight for your latitude. Determine A, B, C, and D in $f(x) = A \sin(Bx + C) + D$ to find an equation that gives the number of minutes of sunlight per day for your latitude based on day of the year. Find the derivative of this equation and determine dates for the longest and shortest days of the year (in terms of sunlight). Also determine numbers of minutes of sunlight on each day. When are the numbers of minutes of sunlight increasing most rapidly?

DEVELOPING MATHEMATICAL PEDAGOGY

Preparing for Instruction

4. Refer back to **DMU** problem 1 and **DMU** problem 2 in which technology was used to explore derivatives. In these problems, technology was used to explore the derivatives of functions that students might not be able to study symbolically.

   a. What might be some of the advantages of using technology to explore derivatives in this way?

   b. What might be some of the disadvantages of using technology in this way?

   c. Consider again the function in **DMU** problem 1, $y = x^3 - 2x + 1$.

      i. Sketch the graph by hand on a coordinate grid for x in $[-1, 4]$.

      ii. Draw the tangent lines to the curve for at least 10 points in the x-interval $[0, 3]$.

      iii. Estimate the slope of each of the tangent lines you drew in problem 4c.ii.

      iv. Plot the points (x, slope of tangent line at x) for each of the tangent lines you drew in problem 4c.iii.

   d. Compare your understanding of the derivative from your work in problem 4c and your work in **DMU** problem 1.

TEACHING AND LEARNING HIGH SCHOOL MATHEMATICS © 2010 John Wiley & Sons, Inc.

e. Under what circumstances might you give the tasks in problem 4c and the tasks in problem 1 on a summative assessment for students?

5. Design a student page in which students use technology to conjecture that the derivative of a quadratic function is always a linear function. Be sure to incorporate questions that help students make conceptual connections between these ideas. Include at least two modifications of the activity that increase the accessibility of the task for students who need additional support.

6. Refer to the student page you created in **DMP** problem 5.

a. Complete a **QRS Guide** for the student page and recommended modifications. Anticipate student responses at three levels for each major problem.

b. Write teacher support questions for each student response and add these to the **QRS Guide**. Also include tone-setting notes, cautious points, and timing for the lesson.

c. Decide on a cooperative learning (CL) strategy that best fits use of the student page with a class of high school students. Indicate the CL strategy at the top of the **QRS Guide**.

d. Use the **Annotated Lesson Planning Guide** in the Unit Four Appendix to guide your completion of a **Lesson Planning Guide** for your student page. Indicate the learning outcomes on the **Lesson Planning Guide**. Also complete the sections for Materials, Use of Space, Launch, and Share and Summarize.

e. Try the activity with one or more high school students. Record any additional student responses you did not anticipate and related teacher support in the **QRS Guide**. What more did you learn by using the student page, given the work you did in the **Lesson Planning Guide** and related **QRS Guide**?

Using Internet Resources

7. Look up the game *Jeopardy* on the Internet. Study the rules. There are several sites available, including some with blank templates to allow you to include your own questions. Create a *Jeopardy* game to help students review for a test on derivatives. What categories will you use?

Focusing on Assessment

8. Write assessments to accompany the student page you created in **DMP** problem 5, as follows:

a. Write three homework problems to reinforce or extend what students will learn through the lesson in which the student page will be used. Include contexts and short writing assignments for two of the problems. One of the problems should require use of a rubric for grading.

b. Write one test problem related to your student page that assumes the use of technology. Write an item-specific rubric to grade the test problem.

In problems 9–15, refer to the **Sample Test on Derivatives** in the **Explore** of this lesson.

9. **a.** Write a key for the test. For some problems, there might be alternative correct answers; in such cases, provide a sample response and some indication about the range of responses that are reasonable to accept.

b. What paper trail must students using graphing calculators leave? What if students are using CAS? Write directions to tell students what you expect.

c. Write a rubric for scoring item 2 on the sample test.

10. Item 1 on the sample test focuses on skills. What additional questions might teachers ask about this item to obtain better insight into what students understand about rate of change?

TEACHING AND LEARNING HIGH SCHOOL MATHEMATICS © 2010 John Wiley & Sons, Inc.

11. In a good test, the first few items should be ones that the teacher expects most students are able to do.

    a. What often happens to students when the first items on a test are particularly difficult?

    b. Critique the first few items of the sample test. Are these items appropriate? If not, revise the first few items on the test, suggest replacement items, or remove inappropriate items.

12. The sequence of problems in a test is also an important consideration. Although it is important that the first few items be accessible to all students, it is also important that not all the most challenging problems are at the end of the test when students might be tired or pressed for time.

    a. Which items do you think are the easiest? Which do you perceive to be the hardest?

    b. Comment on the sequence of the items.

13. An important consideration when scoring a test is the distribution of points. Although the current test has 5 questions, they are not of equal difficulty. Assume the test is worth 100 points. Provide a distribution of points by item and a rationale for your decisions.

14. Create at least one additional item set in a real-world context that would be appropriate for this test. Identify any sources used.

15. The time it takes for students to complete a test is something that teachers need to consider carefully when writing a test. Sometimes teachers take a test themselves and multiply by a factor of 3 or 4 or more to determine the amount of time students are likely to need. Based on your time to create a key for the test, how much time would you expect high school students to need for this test?

16. When classes meet on a block schedule, teachers often face dilemmas about how much of a 90-minute period should be available for the test. Outline some advantages and disadvantages of each of the following teaching and testing strategies:

    - Review for the test and give the test on the same day.
    - Review for the test for the first half of the block, teach a new lesson during the second half of the block, and give the test during the first half of the block during the next time the class meets.
    - Teach a new lesson for the first half of the block and then test during the second half of the block.
    - Test for the first half of the block and then teach a new lesson during the second half of the block.

17. If students work in groups on a regular basis, it is natural to consider whether part of their assessments should occur in groups.

    a. Discuss advantages and disadvantages of each of the following testing scenarios.

        - The test consists of two parts. On one part, students work individually. On a second part, with more complex tasks, students work in a group. All students receive the same score on the group portion. The final score is the sum of the individual score and the group score.
        - Students are given a few minutes to discuss test problems with members of their group but are not permitted to write anything down. When time is called, students must complete the test on their own.
        - The entire test is completed as a group and all students receive the same score.
        - Students complete the test in pairs chosen randomly or by the teacher on the day of the test. Both students in the pair receive the same score.
        - The test is given individually and group members earn their own scores.

TEACHING AND LEARNING HIGH SCHOOL MATHEMATICS © 2010 John Wiley & Sons, Inc.

b. What other approaches might be considered to assess some aspects of the unit in a group context?

18. Research your state's high stakes accountability measures for high school students, if such a measure exists. For what high school courses is this test likely to impact your instruction? What are the stakes for students and schools in terms of performance on this test?

19. Locate a teacher's edition (TE) for a high school textbook for one of the courses you have been observing throughout this course. Within the TE or in a supplement, locate a test for one unit, preferably for a unit being taught while you were observing. Critique the test using the categories in **Explore** item 3a and record your categorizations on the **Template for Classifying Test Items**. What changes would you make and why?

Teaching for Equity

20. In a given class, a teacher has three students who are English Language Learners and two additional students with learning disabilities. The teacher believes these students are able to take the same test as the rest of the class but that some accommodations are likely needed. Often, accommodations are specified in individualized education plans (IEPs) for students with special needs.

a. Outline some advantages and disadvantages of each of the following possible accommodations to be made for the five students.

- Students are given additional time to complete the test.
- Students are permitted to use their notes during the test.
- Students are permitted to use an English-native language dictionary during the test.
- Students are allowed to have the questions read to them.
- Students are allowed to take the test orally with a teacher.
- Students are allowed to listen to a recording of the test.

b. Some students in the class believe it is unfair for only some students to have accommodations, such as being able to use notes. How would you handle such student concerns?

c. If possible, contact a counselor in the school where you are observing. Interview the counselor about school policy concerning accommodations for students with special needs.

Debating Issues

21. High school teachers are sometimes conflicted on whether students should have access to technology (i.e., graphing calculators or computers) when taking a test. Some believe students should take tests with no technology available, sometimes for issues of equity. Others believe technology should be used all the time. Still others believe that a test should have two parts—one part with technology and one part without. Discuss your position on this issue. What arguments would you make to support your position?

Reflecting on Professional Reading

22. In Lesson 5.2, **DMP** problem 20, you were asked to read "Improving Classroom Tests as a Means of Improving Assessment" by Denisse R. Thompson, Charlene E. Beckmann, and Sharon L. Senk. (*Mathematics Teacher*, 90 (January 1997): 58–64). Based on that reading, answer the following. (Find the article on the course website.)

 © 2010 John Wiley & Sons, Inc.

a. The authors discuss several ways to modify items. Take one of the items from the **Sample Test on Derivatives**. Modify it in at least two different ways as described in the article.

b. The authors mention several cautions when modifying items to make them more open or to require students to explain their thinking. Identify some of the cautions. What other cautions might you identify?

5.4 Alternative Assessments: Accumulations

Tests and quizzes are common tools teachers often use to gauge student achievement in mathematics classes. These measures have the potential to provide teachers with information on specific items that are similar or related to work done recently in class. But tests and quizzes have limitations. For one, not all students perform their best under conditions where time to respond might be limited. For another, mathematics education recommendations encourage more extended thinking, more communication, more reasoning and justification; tests and quizzes might not be the best venue for students to show what they can do with respect to these broader processes. Also, college applications and job interviews ask students to share significant accomplishments. Students are expected to synthesize, apply, and share publicly their skills in reading, writing, using technology, solving problems, and reasoning. Moreover, learning activities that are substantial and significant form memorable life experiences.

This lesson provides an opportunity for you to consider assessment tools beyond tests and quizzes, such as projects, investigations, and portfolios. These alternative assessments provide challenges to students and teachers as well as many benefits.

This lesson continues to address mathematics at the precalculus level. It refers to the **Sample Unit Outline: Introducing Derivatives** (see Unit Five Appendix) and, in the **DMU** problems, addresses finding areas and accumulations that underlie the calculus concept of integral.

In this lesson, you will:

- Consider several alternative assessments, their possible benefits, and implementation issues.
- Continue to think about the precalculus curriculum, in particular in relation to assessment.
- Learn how the concept of integral can be foreshadowed.

Website Resources

- **Sample Unit Outline: Introducing Derivatives**
- **Sample Confidential Group Evaluation Form**
- **Sample Peer Evaluation for Oral Presentations**

LAUNCH

1. Reflect on your own mathematics studies in high school and college, including this course. What experiences did you have with investigations, projects, reports, problems of the week, portfolios, self-assessments, or other assessments besides quizzes and tests? If you experienced alternative assessments, what benefits did you find in them?
2. What are some limitations of assessment plans that include only homework, quizzes, tests, and possibly, participation?

TEACHING AND LEARNING HIGH SCHOOL MATHEMATICS © 2010 John Wiley & Sons, Inc.

EXPLORE

Review the **Sample Unit Outline: Introducing Derivatives**. Keep this unit plan in mind as an example while you think about alternative assessments.

3. Following are three clusters of issues related to alternative assessments (**Projects**, **Investigations**, and **Portfolios**). Explore the issues within one cluster. Be prepared to present the issues to other teams and conduct a discussion. Each cluster should be addressed by at least one team.

Cluster A: Projects

Projects are extended assignments generally requiring a product illustrating a significant application of mathematical principles and processes studied in class. (Contrast projects with investigations in Cluster B, which center on mathematics not studied in class.) These assignments are expected to be done outside of class and take two or more weeks to complete. Some class time may be used for progress checks. In addition to written components, projects often include visual aids, technology integration, and an oral component. Some textbooks provide students with scaffolding for a project throughout a unit; the project is introduced at the outset of the unit and one or two problems in the homework for each lesson provide connections to the newly studied mathematics. These homework problems support students in the next few steps of completing their project. When projects are complete, some classes hold a Math Fair or other event for public display or presentation of student work. A project might be assigned once a year, once a term, once a marking period, or for each chapter or unit.

I. Review the **Sample Unit Outline: Introducing Derivatives**, including the project described. Identify examples you think would be appropriate for the project.

II. Create a rubric to use for the project in the unit outline. Be sure to include oral, written, and visual criteria.

III. When students work in groups outside of class, the teacher can evaluate the mathematical work submitted by the group but is unable to assess how well students work together. It is helpful in such cases to ask students to complete a confidential group evaluation form to evaluate their own contributions as well as to evaluate the contributions of other members of their groups. (The Unit Five Appendix contains a **Sample Confidential Group Evaluation Form.**) Read this evaluation form. Comment on how you might use students' responses from this form.

IV. Some teachers have found it beneficial when students do oral reports to have the rest of the class (a) take notes so they can learn from others or (b) complete a peer review form. For each of these two cases, draft a strategy for how you would orchestrate this plan in your own classroom. For example, what would you include on a peer review form? How would you use these reviews? How would you assess listeners' learning from presenters? (The Unit Five Appendix contains a **Sample Peer Evaluation for Oral Presentations**; if needed, find and modify this rubric for the project in the unit outline.)

V. Some schools encourage students to present projects at parent nights or Math Fairs.

a. What are some benefits of such student presentations?

b. What would you do to prepare students for such an event?

Cluster B: Investigations

In this course, each unit provides an opportunity to complete in-depth investigations in which you can explore independently or with a partner some significant

TEACHING AND LEARNING HIGH SCHOOL MATHEMATICS © 2010 John Wiley & Sons, Inc.

mathematics that is not otherwise studied formally in the course. High school students, too, can be asked to work on investigations of mathematics beyond classroom lessons. (Contrast investigations with projects in Cluster A, which apply mathematics studied in class.) Ideally, the investigation would involve mathematics new to the student. Investigations can also be used to encourage students to discover links between diverse areas of mathematics (the Unit Two Investigation on transformations in this book likely could serve this role for some individuals) or to delve more deeply into an area of mathematics that students have studied (such as the function families in the Unit Three Investigation). An investigation might be assigned once a year, once a term, once a marking period, or for each chapter. In some schools, problems or topics are suggested and students define their own areas of research within them. Students may submit a proposal or interim progress report so the instructor can give feedback as they go. Investigations may lead to science/engineering fair submissions and to scholarship prizes. Some textbooks and websites provide students and teachers with ideas and references for exploration.

I. Propose a mathematical question at the precalculus level you have studied or are curious about that might be explored by high school students in an investigation. State the topic and at least three significant questions related to the topic that students could explore. (Consult the unit investigations in this textbook for examples.)

II. Consider the following roles teachers may play in supporting student investigations. Choose two or three of these roles and explain why they are supportive or how they might be more supportive.

- Motivating students to engage in an extended investigation
- Providing support in understanding or defining the problem and ways to proceed
- Providing materials or guidance in finding resources
- Planning a multi-week timetable that coordinates work on the investigation with other related studies
- Explaining the expectations for the investigation and the assessment criteria
- Suggesting possible starting points
- Helping students develop a workable timetable for themselves
- Reviewing early drafts of student work
- Helping students self-assess their progress and achievement

III. Consider the following elements teachers might require for an investigation report. Decide which of the elements are essential and comment on why you feel these are valuable.

- Title page, including a statement of the problem, questions investigated, overview of what was done, and summary of what was learned
- Details of the investigation and what was learned
- A story of the thinking and problem solving the students did, including false starts and how they were recognized, tools and problem-solving strategies attempted, and insights or relationships discovered through the process
- Justification of several of the conjectures or patterns generated
- Summary of the mathematical methods used and examples
- References, including acknowledgments of help given or received from people, books, the Internet, and so on
- Appendices, for example, extensive data sets, spreadsheets, and repetitive calculations
- Self-evaluation of what was learned, how well partners collaborated, and other aspects of the investigation
- Questions or further conjectures for related future research

TEACHING AND LEARNING HIGH SCHOOL MATHEMATICS © 2010 John Wiley & Sons, Inc.

IV. For your suggested investigation from item I in this cluster, create a guide that allocates 100 points across the various elements chosen from among the bullets in item III. Give reasons for your allocations.

V. Choose one of the more substantive questions from your investigation. What would you like to see from students? Create a rubric for the question.

VI. What support from colleagues or administrators would help you as a teacher to orchestrate investigations in your classroom? Think about teachers in other fields as well as mathematics teachers, counselors, curriculum leaders, and principals.

Cluster C: Portfolios

There is a long history among artists, architects, journalists, photographers, and others to collect their best work into a portfolio to encapsulate their achievements. Some schools today have modified this concept as a part of educational assessment. Having a good variety of things for students to select for a portfolio requires that the school program at least occasionally include alternative assessments—extended problem solutions, investigations, projects, reports, journals, computer/calculator explorations, and other formats besides tests and quizzes. Students may select, or teachers with students may jointly select, samples of work that illustrate particular attributes, for example, "where I made the most progress," "how I used technology meaningfully," "how I learned well with others," "where I need to continue to study," "evidence of my persistence," and so on. Students also share why they selected each piece. Although portfolios may not be a strategy used by a novice teacher, portfolios are a strategy that new teachers should look forward to using as their careers progress.

I. Review the **Sample Unit Outline: Introducing Derivatives**, including the project described. Identify opportunities within the unit for activities students might later opt to include in a portfolio.

II. The introduction to this cluster provided several possible categories for student self-selection of items they might include in a portfolio. Identify at least two other possible categories that would help you and the student get a better sense of his or her achievement.

III. What are benefits of students self-selecting items and justifying the choice for each of the required portfolio categories?

IV. For parent–teacher conference nights, some schools do "student-led conferences." Students prepare their portfolios in advance of the conferences, then present their work to their parents while the teacher listens.

- **a.** If you were planning for your grade 12 students to do student-led conferences, what specific sorts of items would you require them to include in their portfolio?
- **b.** What types of items would you recommend students include in their portfolios?
- **c.** How might you help students rehearse for this parent presentation?
- **d.** What parent questions would you anticipate and how would you plan to address them?

V. Some schools have students create electronic portfolios or e-folios.

- **a.** What are some items that are appropriate to include in an e-folio?
- **b.** What are some items that might be more difficult to include in an e-folio? Suggest, if possible, how to link or include these items into an e-folio.
- **c.** What are some items that are easier to include in an e-folio than in a paper-based portfolio?

VI. High school students often need support in staying organized, especially if part of their evaluation ultimately rests on having saved materials throughout the term to use as a basis for portfolio selections.

TEACHING AND LEARNING HIGH SCHOOL MATHEMATICS © 2010 John Wiley & Sons, Inc.

a. How would you support students in keeping their materials organized?

b. How would you suggest their portfolios be structured for the term or marking period? Think about loose-leaf binders with dividers, folders within folders, or electronic files (in the case of an e-folio). Think about what would make these easy for you and your students to manage and evaluate.

c. How would you introduce portfolios to students?

VII. Design a grading rubric for a portfolio for a term of high school precalculus.

SHARE AND SUMMARIZE

4. Share selections from each team's responses to their cluster of tasks.
5. Choose one or two of the following topics. Discuss the topic in-depth with your team.
   a. What are some benefits of having students engage in mathematics that goes beyond solving ordinary textbook problems? Keep in mind the five process standards from *PSSM*: problem solving, reasoning and proof, communication, connections, and representation.
   b. What are benefits of assessment plans that include a variety of assessment formats?
   c. What are some benefits of incorporating students' self-assessments into an assessment plan? What are some concerns you have about self-assessment?
   d. When and how might you provide an opportunity for students to revise and improve their work for one or more of these alternative assessments? How is a revision option related to creation of equitable classrooms (high expectations with support for all students)?
   e. What are some benefits of having clear, well-defined expectations and scoring guides provided to students at the outset of a major assignment?
   f. The Internet may enhance student learning or it may undermine students' opportunities to think for themselves. How will you plan alternative assessments so that the Internet becomes a research tool and not a source for ready-made solutions?

DEEPENING MATHEMATICAL UNDERSTANDING

1. *Connecting Distance and Velocity*. Complete the student page **Determining Distance from Velocity.**
   a. How can the student page help build understanding for the relationship between derivatives and integrals?
   b. How does the student page help build understanding for the integral as area under a curve?
   c. How does the student page help build understanding for the integral as accumulation?

TEACHING AND LEARNING HIGH SCHOOL MATHEMATICS © 2010 John Wiley & Sons, Inc.

Determining Distance from Velocity

1. Ping is training for a 100-mile bike ride. She is able to comfortably ride at a consistent speed of 18.5 miles per hour on flat roads. She is gradually increasing the amount of time that she can keep up the desired speed. She does not begin timing herself until she is traveling at the desired speed.

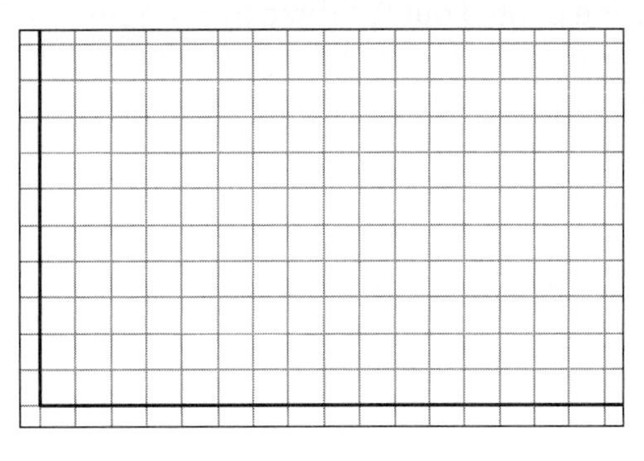

a. Once she is up to speed, how far does she ride in 1 hour? 2 hours? 3.5 hours? *h* hours? How do you know?

b. Write an equation indicating Ping's velocity at any time during her timed ride.

c. Sketch a graph of her velocity function on the grid.

d. Find a relationship between Ping's velocity versus time graph and the distance she travels for a fixed time $t = h$. Explain what you find.

2. Suppose Ping is able to maintain a speed of 20 miles per hour for 2 hours and then immediately slows to 15 miles per hour for a third hour.

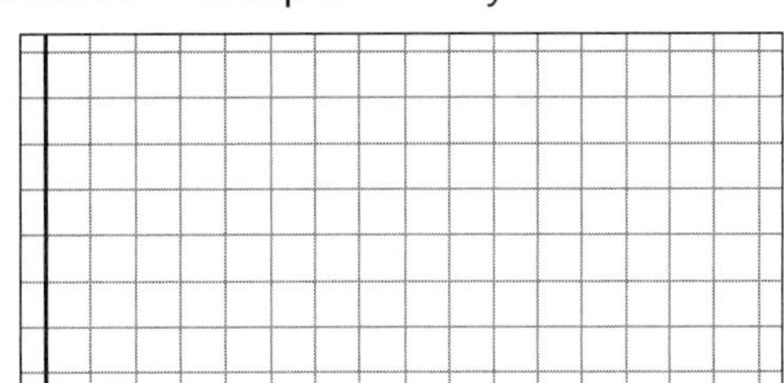

a. Sketch a graph of her velocity function for this scenario.

b. Write an equation indicating Ping's velocity at any time during her ride. (Use a piecewise-defined function.)

c. How far does Ping ride in these three hours?

d. Can the total distance Ping rides in three hours be found using just the graph? Explain.

e. Describe Ping's velocity at exactly 2 hours into her timed ride. Is it possible for her velocity to drop immediately from 20 to 15 miles per hour? Explain. If not possible, sketch a more reasonable graph to model the situation.

3. Suppose Ping maintains a speed of 20 miles per hour, gradually slows to 15 miles per hour, and then rides at 15 mph for the rest of the third hour as shown in the graph at right.

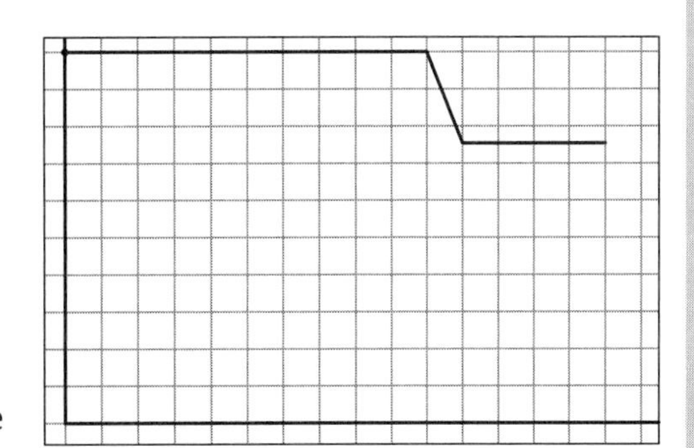

a. How can the total distance Ping rides in three hours be found from the graph? Suggest at least two methods you might use.

b. Use one of your methods from problem 3a to estimate how far Ping rides in these three hours.

(*continued*)

TEACHING AND LEARNING HIGH SCHOOL MATHEMATICS © 2010 John Wiley & Sons, Inc.

4. On very windy or rainy days, Ping rides a stationary bike to keep up with her training. The stationary bike has the advantage that it records Ping's speed over time. She printed out the three training sessions labeled A, B, and C. Gridlines indicate time increments of 20 minutes (horizontal axis) and speed increments of 2 miles per hour (vertical axis).
   a. Estimate the distance Ping rode during each session.
   b. Explain your process.
   c. How can distance be determined from velocity when the velocity function is given graphically?
   d. For the graph of session C, what techniques did you consider in order to estimate the distance traveled from this velocity graph?
   e. Determine an underestimate and overestimate for the distance for the graph of session C. Describe your process.
   f. How might you make your estimate even better?

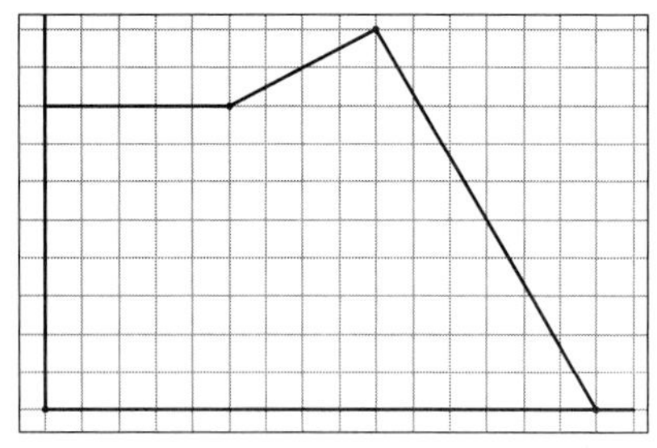

Session A

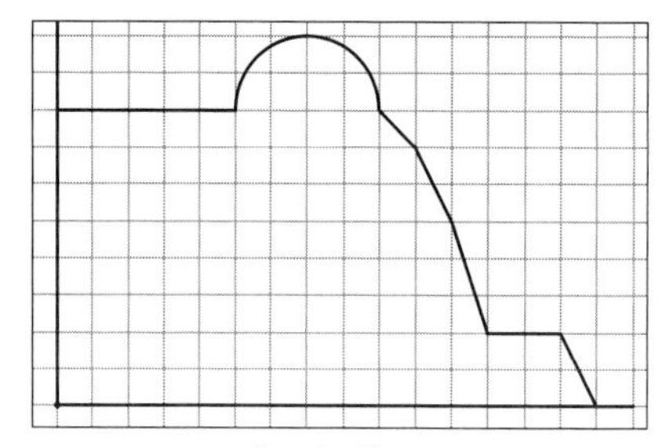

Session B

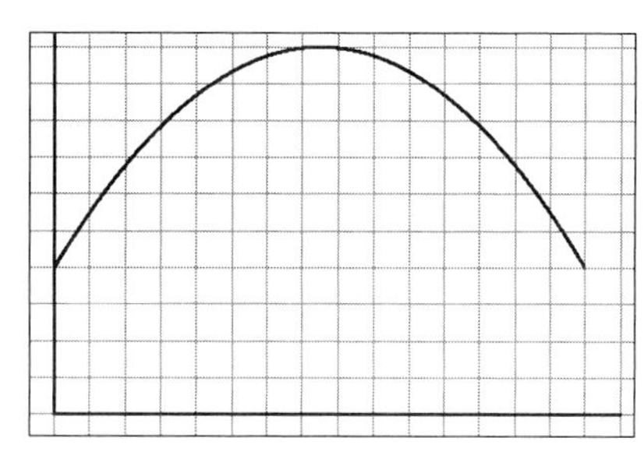

Session C

5. Sketch the graph of the velocity function, $y = x$, for x in [0, 5]. Keep the y-interval [0, 5] throughout problem 5.
   a. If $y = x$ is a velocity function, how can you find the distance traveled for x in [0, 5]?
   b. Sketch $y = x$ for x in [3, 4]. Describe the graph over this interval. How can you find the distance traveled for x in [3, 4]?
   c. Sketch $y = x$ for x in [3, 3.1]. Describe the graph over this interval. How can you find the distance traveled for x in [3, 3.1]?
   d. Sketch $y = x$ for x in [3, 3.01]. Describe the graph over this interval. How can you find the distance traveled for x in [3, 3.01]?
   e. Convince yourself with other functions that over small intervals the graphs behave as you experienced in problems 5a through 5d as the interval width for x decreases and the interval height for y remains unchanged.
   f. Generalize your findings from problem 5 to suggest a method for finding distance from velocity for any function $y = f(x)$.

Student page inspired by Beckmann and Sundstrom (1992) and Beckmann (1993).

TEACHING AND LEARNING HIGH SCHOOL MATHEMATICS © 2010 John Wiley & Sons, Inc.

2. *Rectangles versus Trapezoids*. Graph the function f defined by $f(x) = x^2 + 3$ for x in the interval $[-5, 5]$. Find the area under the curve between $x = 0$ and $x = 5$ using the following methods.

   a. Create a set of rectangles that lie completely under the curve so the width of each rectangle is 0.5 and the height is the value of the function at the left endpoint of the interval. Find the sum of the areas of the 10 rectangles.

   b. Create a set of rectangles that lie completely above the curve so the width of each rectangle is 0.5 and the height is the value of the function at the right endpoint of the interval. Find the sum of the areas of the 10 rectangles.

   c. Create a set of trapezoids to estimate the area under the curve. Let the parallel sides of each trapezoid be 0.5 units apart and determine the lengths of the two parallel bases from the function values associated with the x-values for each endpoint. Find the sum of the areas of the 10 trapezoids.

   d. Compare the areas from problems 2a, 2b, and 2c. What happens to the areas as the width of the interval decreases to 0?

   e. How does your result in problem 2d compare to what you know about integrals?

3. *Determining the Value of an Integral without Calculus*. The notation $\int_a^b f(x)dx$ symbolizes the area under the function f from $x = a$ to $x = b$. Depending on the function, you can often evaluate this integral without using rules of calculus.

   a. Consider the function f defined by $f(x) = \sqrt{16 - x^2}$. Sketch the graph of the function; describe the graph. From the graph of $y = f(x)$, evaluate $\int_0^4 \sqrt{16 - x^2}dx$ using what you know about the area of the region the integral represents. Explain your thinking.

   b. Show how to evaluate $\int_{-2}^{3} (4x - 5)dx$ without using calculus by focusing on the meaning of this notation.

4. *A Sum of Integrals*.

   a. Without using calculus techniques, show that
   $\int_0^5 (2x + 4)dx = \int_0^2 (2x + 4)dx + \int_2^5 (2x + 4)dx$.

   b. Will the result in problem 4a generalize to the area under any continuous function f for x in the interval $[a, b]$? Provide a rationale based on the meaning of this notation.

   c. Write a general rule for any continuous function based on your work in problem 4b.

DEVELOPING MATHEMATICAL PEDAGOGY

Analyzing Instructional Materials

5. Locate a teacher's edition (TE) of a high school mathematics text. Examine closely one unit and whatever general introduction the TE offers related to assessment.

   a. Find examples of alternative assessments.

   b. Comment on how often the authors expect alternative assessments to be used.

   c. Choose an alternative assessment that you think is good and state why you think so. If none exist, describe how you would modify an existing assessment or create an alternative assessment. Explain the thinking behind your work.

Debating Issues

6. For a period beginning in 2001 with the No Child Left Behind legislation, states were required to test students at grades 3–8 to be eligible for federal

TEACHING AND LEARNING HIGH SCHOOL MATHEMATICS © 2010 John Wiley & Sons, Inc.

education funds. Some states extended this external assessment by mandating statewide high school completion exams. These steps caused considerable controversy. Some argued that schools should be accountable and external tests are a means to provide that accountability. Others argued that the tests were not well aligned with district goals, students did not "buy in," and time was lost from more appropriate instruction. Write at least two other arguments for and two against external testing as a requirement for school funding.

Focusing on Assessment

7. Glaser and Silver (1994, cited in Kilpatrick and Silver (2000)) argued that changes in assessment practices could offer opportunities for improved student learning. As assessment and instruction become better aligned, as criteria for judging student performance become clearer to the students themselves, as students become better able to judge for themselves how well they are meeting specified criteria, the more independent and self-directed students can become in their learning.

  a. If possible, identify an experience in your life where your own assessment of your learning enhanced your achievement.

  b. In essence, Glaser and Silver are arguing for assessment practices to be open so assessment is not a mystery to students. How do students benefit when expectations for performance are clear? What difficulties arise when expectations are not clear?

  c. There often exists a tension between providing clear expectations while leaving a task open and providing so much guidance that students are allowed very little room for creativity or flexibility in approaching and solving a task, thereby reducing the cognitive demand of the task. What are some ways teachers can address this tension?

8. Teachers often discover in the course of grading complex student work like projects or investigations that there were other criteria they wished they had included earlier in the description of the assignment and the scoring guide.

  a. How might you keep track from term to term of improvements you want to make in assignment descriptions, rubrics, and point allocations?

  b. Suppose you found a major omission in your planned point allocation after you had collected student work. What might you do before you return the work to your students, assuming the point allocation had been provided to students as they prepared their work?

9. Return to the clusters in this lesson. Choose one cluster different from the one your group explored in class.

  a. What interests you about this assessment type?

  b. Respond to two items in that cluster.

  c. Find Internet resources that appear to be of high quality related to that assessment alternative. In your opinion, what makes the resources good quality?

10. Return to the items in **Share and Summarize** item 5. Respond to one of the bulleted items different from the ones discussed in class.

Using Internet Resources

11. Find two project ideas from the Internet you think would be of interest to students in the schools where you have been observing.

  a. Describe why you think the project would interest students.

  b. Evaluate the projects. Identify the learning outcomes for each project.

TEACHING AND LEARNING HIGH SCHOOL MATHEMATICS © 2010 John Wiley & Sons, Inc.

Preparing for Instruction

12. Throughout this unit, several student pages have been included in the **DMU/ DMP** questions accompanying the lessons. While students are building conceptual development for the underpinnings of calculus in class, they can be working on a project or investigation outside of class.
    a. Suggest a project or investigation students could complete to reinforce or extend their growing understanding of one or more of these concepts outside of class.
    b. Create a rubric for the project or investigation.
13. One use for integrals is to find the volume of a solid of revolution.
    a. Physical models often are useful to help students recognize that the solid is being sliced into many cylinders (think of cardboard disks under take-out pizzas) whose volumes are then summed. What other physical models might be used to illustrate this phenomenon?
    b. What solid arises from revolving $y = 5$ about the x-axis from $x = 2$ to $x = 6$? Find the volume using a known formula.
    c. Repeat problem 13b to find the volume of the solid of revolution from $x = 2$ to $x = 5$ when $y = 5 - x$ is rotated about the x-axis.
    d. Repeat problem 13b for the solid created when revolving the region bounded by $x = 0$, $x = 4$, and the function $f(x) = \sqrt{16 - x^2}$ about the x-axis.
    e. When you worked to find areas under curves (for example, **DMU** problem 1), you found that rectangles, if they were narrow enough, could be used to find a good estimate. Imagine revolving one of the narrow rectangles about the x-axis. What solid results? How can you find its volume?
    f. Revisit problems 13b–d. Use your results from problem 13e and calculus to find the volume of each of the solids. Compare the calculus results with those obtained from known formulas.
    g. Write a problem that requires revolving a region bounded by the x-axis about the x-axis that cannot be solved with known formulas.
    h. Help students extend what they learned through problems 13b–g to find a volume formula when any region bounded by the x-axis is revolved about the x-axis.
    i. Combine problems 13b–h into a student page. Give the student page a title.
    j. Write a lesson plan to use with this student page as its core. Think about problem 13a as you design the launch for the lesson.
14. Consider the student page earlier in this lesson, **Determining Distance from Velocity**.

    a. Write a lesson plan for the student page. Begin by completing a **QRS Guide** for the student page, then completing the rest of the elements on the **Lesson Planning Guide**.
    b. Write appropriate homework items to accompany the student page. Make certain at least one item is in a context different from that on the student page. Include one short writing assignment. Write a rubric for the short writing assignment.

    c. Try the modified activity with one or more high school students. Record any additional student responses you did not anticipate and related teacher support in the **QRS Guide**. What more did you learn by using the student page, **QRS Guide**, and lesson plan with students?

Teaching for Equity

15. Locate the article "Lessons Learned from the 'Five Men Crew': Teaching Culturally Relevant Mathematics" by Lillie R. Albert. (In *Changing the Faces of*

© 2010 John Wiley & Sons, Inc.

Mathematics: Perspectives on African Americans, edited by Marilyn E. Strutchens, Martin L. Johnson, William F. Tate (pp. 81–88). Reston, VA: National Council of Teachers of Mathematics, 2000).

a. How did issues of equity arise within this teacher's assessment strategies?

b. What other issues related to alternative assessments did the author face?

Reflecting on Professional Reading

16. Read "A, E, I, O, U, and Always Y: A Simple Technique for Improving Communication and Assessment in the Mathematics Classroom" by Lorna Thomas Vazquez (*Mathematics Teacher*, 102 (August 2008): 16–23).

a. Describe what the author means by A, E, I, O, U, and Y.

b. Review the author's rubric. Compare and contrast it with the rubric in Lesson 5.2.

c. Which rubric is more appropriate for projects and bigger assessments? Why do you think so?

17. Read "The Power of Investigative Calculus Projects" by John Robert Perrin and Robert J. Quinn (*Mathematics Teacher*, 101 (May 2008): 640–646). The authors discuss a number of issues and challenges when using investigative projects. Compare the issues and challenges they raise to those discussed in this lesson in the Cluster on Investigations. What more did you learn from reading this article? What concerns do you have about using investigations in your own classes? What other challenges or issues do you think might arise when using investigations with students in classes below the level of calculus?

Preparing to Discuss Observations

18. In preparation for Lesson 5.5, review your notes from your observations in high school mathematics classes. Also, review the articles you read in preparation for the observations. Be prepared to discuss your findings with your peers.

a. What types of questions did the teacher ask throughout the lesson to assess students' understanding? How did the teacher use these responses to adjust instruction?

b. In what ways were observations used? questioning of individuals or small groups of students?

c. Were students expected to write about their experiences in journals or via writing prompts? If so, provide an example.

d. Refer to your coding of a classroom test using the categories described by Thompson, Beckmann, and Senk in "Improving Classroom Tests as a Means of Improving Assessment." What similarities and differences did you observe between items on the test or quiz and the instruction? Are the assessment items at the same level as those used during instruction?

e. What strengths or areas of improvement did you note for your test?

f. Be prepared to discuss your findings from interviewing a teacher about his or her grading practices.

Preparing to Discuss Listening to Students

19. Refer to your notes from **Listening to Students Reason about Precalculus**. Be prepared to share some NAEP items for which your students' results were substantially different from those indicated on the website or for which their responses surprised you. Why do you think you obtained these results?

TEACHING AND LEARNING HIGH SCHOOL MATHEMATICS © 2010 John Wiley & Sons, Inc.

5.5 Summarizing Classroom Observations and Listening to Students

At the beginning of this unit, you were asked to observe in at least two class periods and to focus your observations on issues related to informal as well as formal assessment. As part of your work with formal assessments, you were to code a test using a given set of criteria based on a particular reading. If possible, you were to review a set of homework papers or grade a set of quiz or test papers and to interview a teacher about his or her grading practices.

In addition, as part of **Listening to Students**, you were to explore the questions tool from the National Assessment of Educational Progress (NAEP) website, select a set of items to use with your high school students, and compare their performance to that of a national sample. The NAEP website is a valuable tool to use in assessing your students' knowledge against a benchmark, particularly if you are the only teacher in a school teaching a particular course. In such situations, you lack other classes to provide comparisons when your students struggle with some concepts or find other concepts particularly easy to understand. Comparing your students' performance on particular content to that of a national sample can be helpful to determine whether instruction needs major changes or whether the concept is just one that often is difficult for students.

In this lesson, you will:

- Compare the results of your observations with those of your peers.
- Reflect on what your observations suggest about teachers' assessment practices.
- Compare the performance of local students on NAEP items to that of a national sample of students.

LAUNCH

1. Reflect on your own assessment experiences in mathematics classes, at either the high school or college levels.
   - **a.** Describe, if possible, an assessment practice that supported your learning.
   - **b.** What assessment practices have you particularly liked? Why?
   - **c.** What assessment practices have you felt needed improvement? Why?

EXPLORE RESPONSES TO CLASSROOM OBSERVATIONS

2. Review your notes from Observation 5.1, which focused on informal assessment.
   - **a.** What types of questions did the teacher ask throughout the lesson to assess students' understanding? How did the teacher use responses to the questions to adjust instruction?
   - **b.** In what ways did the teacher use observations? questioning of individuals or small groups of students?
   - **c.** Were students expected to write about their experiences in journals or via writing prompts? If so, provide an example.
3. Review your notes from Observation 5.2, which focused on formal assessment.
   - **a.** What similarities and differences did you observe between items on the test or quiz and the instruction? Are the assessment items at the same level as those used during instruction?
   - **b.** Share your observations from the coding of your test with your group members. What are strengths of the various tests or quizzes? In what areas might the various tests or quizzes be improved?

TEACHING AND LEARNING HIGH SCHOOL MATHEMATICS © 2010 John Wiley & Sons, Inc.

4. Summarize Observation 5.3, the interviews with teachers about their grading practices. What similarities and differences did you find among the grading plans of the various teachers? Which of these practices might you consider using in your own class? Which practices would you want to avoid? Why?

SHARE

5. Share your team's responses to item 2 with the entire class. What aspects of informal assessment seem to be common across a range of teachers? What aspects seem to be unique and used by only a single teacher?
6. Share your team's responses to item 3 with the entire class. What strengths and/or weaknesses are evident across many tests administered by high school teachers?
7. Compare and contrast the grading plans of the various teachers who were interviewed. Discuss your perceptions of the pros and cons of their practices. Share any practices that you would like to adopt and explain why.

EXPLORE RESPONSES TO LISTENING TO STUDENTS

8. Share a few NAEP items for which the results surprised you for some reason or were very different from the national sample. Why do you think you obtained the results you did? What did you learn about student thinking on some items?

SHARE AND SUMMARIZE

9. As a class, discuss how the NAEP assessment tool might be used as part of classroom instruction. What sample items did you find that assessed conceptual understanding? What items assessed procedural understanding?
10. What were the most interesting or important insights you gained about student understanding from your interviews?

DEEPENING MATHEMATICAL PEDAGOGY

Analyzing Instructional Materials

1. Locate the teacher's edition of at least one current high school mathematics textbook and the assessment resources that accompany it. Choose a topic of interest.

   a. Evaluate the quality of the assessments that accompany the textbook for the topic chosen. Use the criteria provided in the article by Thompson, Beckmann, and Senk (see the **Template for Classifying Test Items** in the Unit Five Appendix) or a set of criteria that you develop. Discuss the strengths and weaknesses of the assessments that accompany the textbook. How will you need to supplement the assessment resources? Why do you think so?
   b. Most teachers' editions have the student pages embedded within them. Review the student text. Determine the ways in which the student text provides opportunities for students to monitor their progress as they proceed through a lesson or a chapter.

Debating Issues

2. High-stakes state assessments, whether required for graduation or just as an accountability measure, need to be addressed by teachers. Some argue that teachers need to spend special preparation time to review for these assessments. Others argue that good mathematics instruction will take care of these

TEACHING AND LEARNING HIGH SCHOOL MATHEMATICS © 2010 John Wiley & Sons, Inc.

assessments with only minimal supplemental review. Choose one side of this issue and provide several points you would use to make your case to the school board.

3. Many standardized tests and college placement tests are multiple-choice. Consequently, some would argue that mathematics teachers should use multiple-choice items on tests so that students become familiar with this format. Many teachers would argue that they prefer to see students' work to determine how they need to modify instruction. Take a position on this issue and provide at least two other points in the debate for the side you choose.

Focusing on Assessment

4. Some school districts mandate that teachers in all disciplines give a writing grade every marking period. If you were teaching in such a district, identify one or more writing assessments you would use with your students in the course you observed.

Teaching for Equity

5. Identify some considerations related to assessment that support establishing or maintaining equity in a mathematics classroom. What are some ways in which assessments may create inequity?

Reflecting on Professional Reading

6. Read "The Myth of Objectivity in Mathematics Assessment," by Lew Romagnano. (*Mathematics Teacher*, 94 (January 2001): 31–37).
   a. Romagnano suggests that *validity* and *reliability* are more useful characteristics to judge an assessment than *objectivity*. What do each of these three terms mean?
   b. How would you score the work in Figure 1 in the article? Why? Why might other teachers score it differently?
   c. Romagnano analyzes an AP item in detail. What is his point?
   d. Romagnano also explains how variability is overlooked in reporting SAT scores. What are the implications of this omission?
   e. What does Romagnano suggest to make assessments more consistent and useful?
   f. Romagnano argues that objectivity in assessment is a myth and consequently should not be used as a basis for rejecting alternative assessments. Where do you stand on this issue?

TEACHING AND LEARNING HIGH SCHOOL MATHEMATICS © 2010 John Wiley & Sons, Inc.

SYNTHESIZING UNIT FIVE

Mathematics Strand: Precalculus

Assessment is a critical part of the teaching and learning process. Formative assessments during instruction help teachers and students monitor learning as it occurs. Formative assessments might take the form of warm-up questions at the beginning of a lesson, questions asked during the lesson, homework, or noticing puzzled looks on students' faces. Summative assessments at the close of a unit of study help teachers and students know what has been mastered and what still needs study in order for mathematical learning to progress. Assessment and instruction should be related in a continuous cycle in which instruction informs assessment, assessment informs instruction, and so on.

Homework is a daily form of assessment that is crucial to provide opportunities for students to engage with mathematics in independent and group settings. Homework needs to be meaningful and tied to the instructional outcomes of a lesson. Tests tend to be regular fixtures of high school mathematics classes, although some teachers also use alternative forms of assessment (e.g., projects, investigations, portfolios) to engage students in important mathematics content and processes not assessed well in a timed format. On tasks or problems that involve multiple answers or multiple paths to an answer, rubrics provide a mechanism for teachers to score papers in an equitable manner.

Precalculus topics provide high school students on the cusp of post-secondary education an opportunity to deepen their understandings of function concepts and to engage informally in exploration of three of the basic concepts of calculus—the limit, the derivative, and the integral. Students have been studying slopes and rates of change since beginning algebra; in calculus, slopes and rates of change are related to the concept of the derivative. The integral is related to accumulation and area under a curve. Helping students extend their understanding of slope and rates of change to build conceptual understanding for derivatives and to extend their understanding of accumulations and area to build conceptual understanding of integrals should be an important part of the precalculus curriculum.

Having arrived at the end of this unit, it is time to reflect on your understanding of the lessons learned. The following items provide an overview of the unit. You should be able to respond meaningfully to each item, providing support for your answers with specifics drawn from the various lessons, discussions, readings, homework problems, or observations in high school classes.

Website Resources

- *Where Do I Stand?* survey

MATHEMATICAL LEARNINGS

1. Given a rational function,
   a. Describe and explain the behavior of the function as x approaches a specific value.
   b. Describe and explain the behavior of the function as $x \to \infty$ or $x \to -\infty$.
   c. Relate the behavior of the function symbolically to its graphical and tabular representations.
2. Given a real-world context with data values,
   a. Determine the average rate of change between two values.
   b. Interpret the average rate of change in context.

TEACHING AND LEARNING HIGH SCHOOL MATHEMATICS © 2010 John Wiley & Sons, Inc.

3. Given a graph,
   a. Determine the instantaneous rate of change at a point.
   b. Explain how the instantaneous rate of change relates to the average rate of change and to the derivative.
4. Given a graph of velocity versus time,
   a. Determine the distance traveled in a specific period of time.
   b. Explain why the sums of areas of rectangles or trapezoids under a curve give reasonable estimates for the integral of a function over an interval.

PEDAGOGICAL LEARNINGS

5. Design some tasks that help students develop informal understandings of the derivative and integral.
6. Compare and contrast formative and summative assessments. Identify some possible assessment strategies for each.
7. a. Identify several different approaches for assigning homework at the high school level.
   b. Discuss the pros and cons of each approach to assigning homework.
   c. Identify different ways that teachers can grade or manage homework to avoid being overwhelmed with papers while still helping students learn.
8. Describe the differences between
   a. Holistic and analytic rubrics
   b. General and item-specific rubrics
   c. Rubrics and partial credit
9. a. Given a rubric for a particular task, apply the rubric to score a set of responses.
   b. Given an appropriate task, develop a rubric to be used in scoring that task.
10. a. Describe some characteristics to consider when developing a test for a unit.
    b. Given a test, critique the test and offer specific recommendations to improve the test.
    c. Suggest different testing strategies for students accustomed to working with peers to solve mathematics problems.
11. Identify various alternative assessment strategies. Discuss some advantages and disadvantages of each strategy, including their relationship to classroom equity.
12. Revisit your responses to the survey **Where Do I Stand?** for items 1, 3, 6, 9, 10, 12, 15, and 20. How have your responses changed, deepened, or grown?
13. Revisit your responses to item 24 on the survey **Where Do I Stand?** Are there any additional challenges you expect to face as a high school mathematics teacher?

TEACHING AND LEARNING HIGH SCHOOL MATHEMATICS © 2010 John Wiley & Sons, Inc.

UNIT FIVE INVESTIGATION: CONIC SECTIONS

The investigation for Unit Five requires you to explore conic sections in several different ways. You will find examples of conic sections in the real world, analyze conic sections informally, and investigate their reflective properties. You will also investigate the parametric and polar forms of each conic section. Finally, you will find your own example of the use of each conic section.

Following are questions to guide your investigation. Choose some that interest you and take them as far as you can. Explain your solutions in coherent sentences. Include diagrams to make your arguments clear. Provide complete citations for all resources used to complete this investigation (including names of persons you consulted). Attach copies of examples of conic sections from resources, such as newspapers, photographs, or the Internet.

Website Resources

- **Exploring Parabolas** student page
- **Exploring Ellipses and Circles** student page
- **Exploring Hyperbolas** student page

1. If you have not already done so, complete problems 1–2 in the Unit Five Team Builder. Describe how each of the conic sections arises from cutting a double cone by a plane. Describe the symmetries of each conic section and how these arise from the double cone and planar intersection.
2. Complete the student pages **Exploring Parabolas**, **Exploring Ellipses and Circles**, and **Exploring Hyperbolas**. Summarize what you learn about each of the conic sections in a table, concept map, Frayer Model, or Link Sheet. (See Lesson 2.2 for descriptions of a Frayer Model, or Link Sheet, and Lesson 2.5 **DMP** problem 18 for a description of a concept map.)
3. In general, conic sections are not functions. For several values of x, there can be more than one value of y associated with it. Parametric equations allow both coordinates, x and y, to be defined as functions of a parameter, t, where t is allowed to vary over a specified interval. Polar coordinates provide a coordinate system that depends on a distance from a central point, r, and the amount of rotation, θ, from an initial direction (in degrees or radians). Air traffic controllers use polar coordinates to map the flights of inbound and outbound air traffic from an airport.

   a. A point P in the polar coordinate system is indicated by the coordinates (r, θ). The same point in a rectangular coordinate system is represented by (x, y). Use the figure at right to label the corresponding parts of the right triangle indicated.

   $P = (r, \theta) = (x, y)$

   b. Let $t = \theta$ and sketch the graph of the parametric curve $(\cos(t),\ \sin(t))$ for t in the interval $[0, 2\pi]$. (Set the mode on the graphing calculator to Parametric. In the function list, define $x(t) = \cos(t)$ and $y(t) = \sin(t)$.) Use a viewing rectangle of $[-5, 5]$ by $[-3, 3]$.

      i. Describe the graph. Why does the graph appear as it does?

      ii. How can you make the radius longer? Shorter? Explain.

      iii. What happens when you restrict t to the interval $[0, \pi]$? Why does this happen?

   c. Graph the parametric curves $(A\cos(t),\ B\sin(t))$ for different values of $A \neq B$ for t in the interval $[0, 2\pi]$.

      i. Create a table listing the values of A and B and a brief description of the graph. Why does the graph appear as it does?

TEACHING AND LEARNING HIGH SCHOOL MATHEMATICS © 2010 John Wiley & Sons, Inc.

ii. How can you make the major axis longer? shorter? Explain.

iii. What happens when you restrict t to the interval $[0, \pi]$? Why does this happen?

d. Graph the parametric curves $(A\sec(t), B\tan(t))$ for different values of $A \neq B$ for t in the interval $[0, 2\pi]$.

i. Create a table listing the values of A and B, equations of asymptotes, and a brief description of the graph. Why does the graph appear as it does?

ii. How can you make the major axis longer? shorter? Explain.

iii. What happens when you restrict t to the interval $[0, \pi]$? Why does this happen?

e. For each of the standard conic sections listed below, suggest a parameterization to graph each using trigonometric functions. Comment on any restrictions to the domain or range of each of the conic sections when parameterized using trigonometric functions for x and y.

i. Circle, $x^2 + y^2 = r^2$

ii. Ellipse, $\frac{x^2}{a^2} + \frac{y^2}{b^2} = 1$

iii. Hyperbolas, $\frac{x^2}{a^2} - \frac{y^2}{b^2} = 1$ and $\frac{y^2}{a^2} - \frac{x^2}{b^2} = 1$

f. How can the parameterizations of the conic sections in problem 3e be adjusted to graph the following general conic sections? Explain.

i. Circle, $(x - h)^2 + (y - k)^2 = r^2$

ii. Ellipse, $\frac{(x-h)^2}{a^2} + \frac{(y-k)^2}{b^2} = 1$

iii. Hyperbolas, $\frac{(x-h)^2}{a^2} - \frac{(y-k)^2}{b^2} = 1$ and $\frac{(y-k)^2}{a^2} - \frac{(x-h)^2}{b^2} = 1$

g. What benefits arise from using parametric equations to help define conic sections as functions? What limitations occur when trigonometric functions are used to define the parametric equations? Explain.

TEACHING AND LEARNING HIGH SCHOOL MATHEMATICS © 2010 John Wiley & Sons, Inc.

Exploring Parabolas

Definition: A **parabola** is the set of all points in the plane equidistant from a fixed point *F* (the **focus**) and a fixed line *l* (the **directrix**).

1. a. Use the definition and the template below to sketch a parabola using the focus *F* and line 1 as the directrix. Repeat for lines 2 and 3 obtaining two more parabolas.

i. Describe your process in determining points on each parabola.

ii. Compare the parabolas. What do you observe?

b. Using only a compass and a straightedge, how can you construct a parabola from its definition? Construct a parabola using the method you describe.

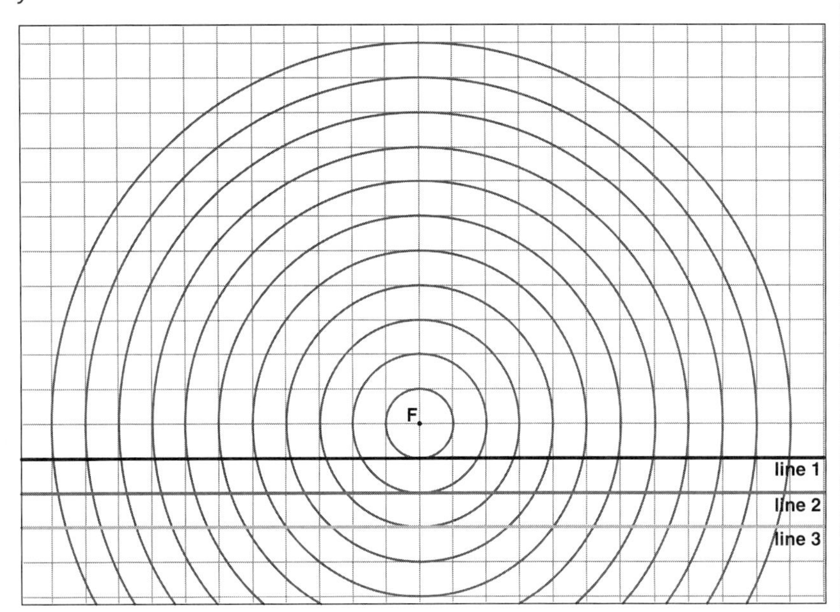

2. a. From the definition of a parabola, derive the equation for a parabola in standard position (vertex at the origin). Use the sketch at right. Show your work.

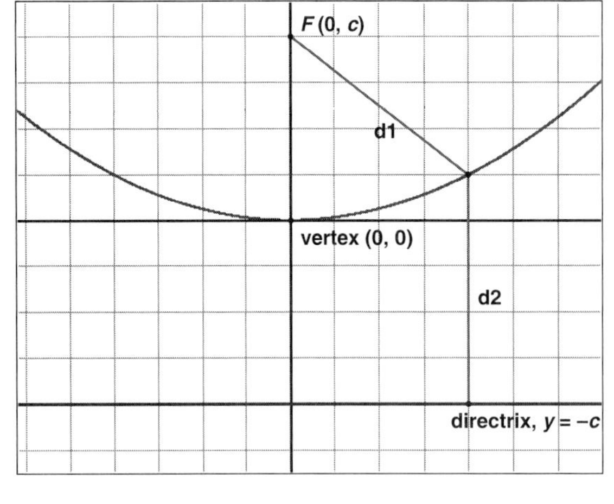

b. Determine the coordinates of the focus and the equation of the directrix for the parabola, $y = x^2$. Explain your work in terms of the derivation in problem 2a.

c. Suppose a parabola in standard position has a vertex at the origin and focus on the positive *x*-axis. What is the standard form of such a parabola? Label the focus, directrix, vertex, and axis of symmetry.

3. a. What is the equation of a parabola with directrix parallel to the *x*-axis whose vertex is the point (h, k)? Explain in terms of transformations of functions.

(continued)

TEACHING AND LEARNING HIGH SCHOOL MATHEMATICS © 2010 John Wiley & Sons, Inc.

b. Determine the focus, directrix, and axis of symmetry for the parabola in problem 3a.

c. Determine the vertex, focus, directrix, and axis of symmetry for each of the following parabolas:

Parabola	Vertex	Focus	Directrix	Axis of Symmetry
$y = x^2 + 2$				
$y = (x + 2)^2$				
$y = 3x^2$				
$y = 3x^2 + 2$				
$y = 3(x - 1)^2 + 2$				

4. Parabolas have the reflective property that a ray traveling into the parabola on a path parallel to the axis of symmetry bounces off the parabola and is directed through its focus.

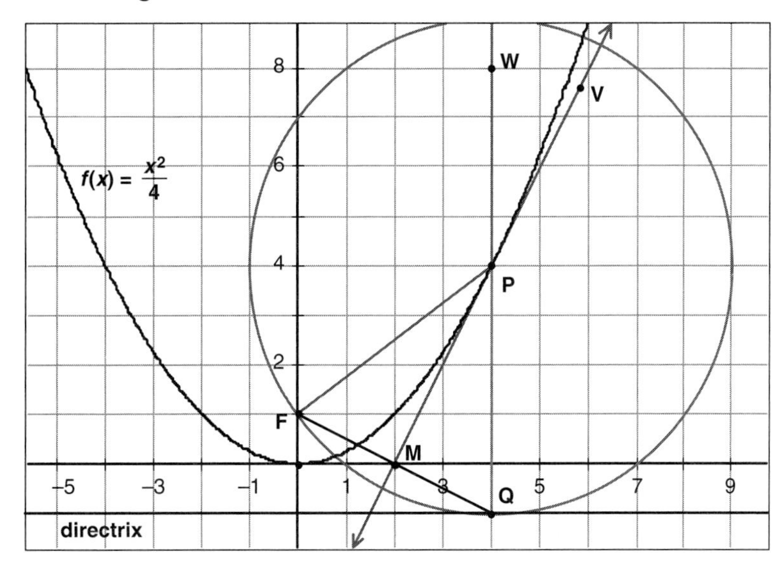

a. Study the figure above. Point F is the focus. The directrix is labeled. Line PM is the angle bisector of $\angle QPF$. Solve and explain the following:

i. How are the lengths PF and PQ related?

ii. How are $\angle QPM$ and $\angle FPM$ related?

iii. How are $\angle VPW$ and $\angle QPM$ related?

iv. How are $\angle FPM$ and $\angle VPW$ related?

v. How does line MP appear to be related to the parabola?

b. Generalize the work above to all points on the parabola.

c. Search the Internet to find at least two uses of the reflective properties of parabolas in real life. In each case, explain how the reflective properties of the parabola are helpful for the application.

TEACHING AND LEARNING HIGH SCHOOL MATHEMATICS © 2010 John Wiley & Sons, Inc.

Exploring Ellipses and Circles

Definition: An **ellipse** is the set of all points in the plane, the sum of whose distances from two fixed points (the **foci**) is a constant.

1. a. Use the definition and the template below to sketch an ellipse for which the sum of the distances between a point on the ellipse and two fixed points (foci) is 8. Repeat for a sum of distances of 10. Describe your process in determining points on the ellipse.

b. Using only a compass and a straightedge, how can you construct an ellipse from its definition? Construct an ellipse using the method you describe.

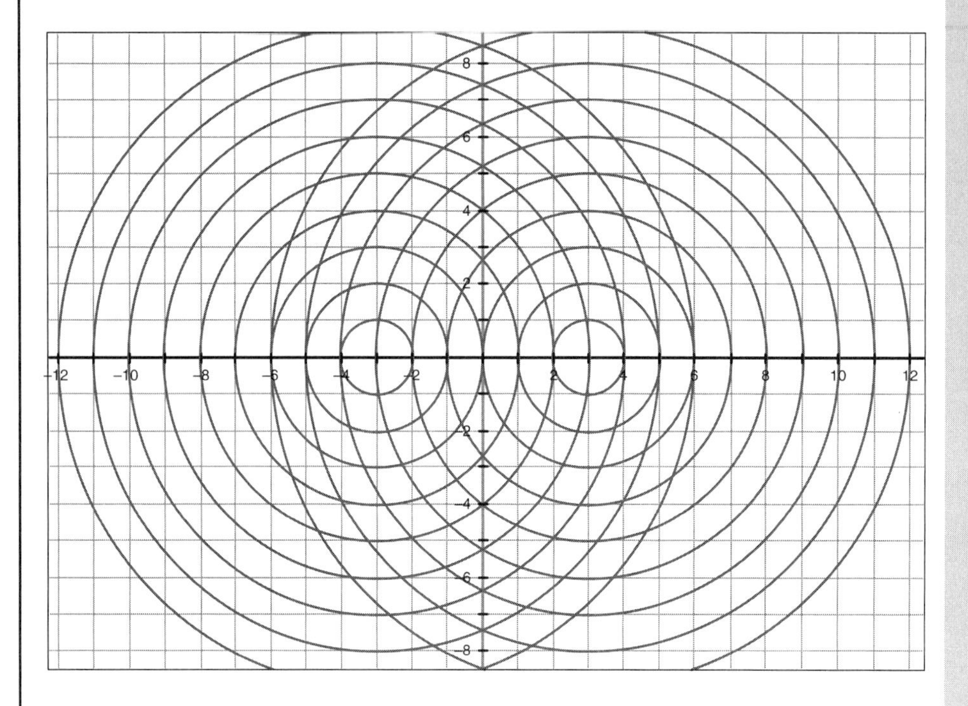

2. For each of the following equations, plot each ellipse. Determine the x and y intercepts, the major axis (the axis along which the ellipse is widest), the minor axis, the center of the ellipse, and the foci.

a. $\frac{x^2}{4}+\frac{y^2}{9}=1$ **b.** $\frac{x^2}{9}+\frac{y^2}{4}=1$ **c.** $\frac{x^2}{16}+\frac{y^2}{9}=1$

3. For an ellipse in standard position, centered at the origin, let the coordinates of the foci be $(c, 0)$ and $(-c, 0)$. The equation of an ellipse in standard position is:

$$\frac{x^2}{a^2}+\frac{y^2}{b^2}=1$$

a. What are the x- and y- intercepts of the ellipse?

b. i. What is the sum of the distances between two fixed points (the foci) for this ellipse in standard position?

ii. Use the definition of an ellipse to determine the relationship between a, b, and c. Explain your work.

c. Complete the table on the following page. Constant sum refers to the sum of the distances between a point on the ellipse and the foci. Summarize your work.

(*continued*)

TEACHING AND LEARNING HIGH SCHOOL MATHEMATICS © 2010 John Wiley & Sons, Inc.

Problem	Equation	Foci	Major Axis	Minor Axis	x-Intercepts	y-Intercepts	Constant Sum
1a							8
1a							10
2a							
2b							
2c							
3							

4. The **eccentricity** e of an ellipse is defined as the ratio $e = \frac{c}{a}$ where $2c$ is the focal length (distance between foci) and $2a$ is the length of the major axis (the distance between x- or y-intercepts, whichever is greater, when the ellipse is in standard position).

a. What bounds are there for values of e for every ellipse? Explain.

b. Describe the shape of an ellipse when the eccentricity approaches its lower bound.

c. Describe the shape of an ellipse when the eccentricity approaches its upper bound.

d. Is a circle an ellipse? Explain.

5. a. Use your previous experience with transformations of functions to determine an equation of an ellipse with center (h, k), major axis length $2a$, and minor axis length $2b$. State how you know your work is correct.

b. Sketch the graph and determine an equation of an ellipse with foci (2, 3) and (6, 3) with major axis length 8 and minor axis length 6. Explain your work.

c. In light of your response to problem 4d, what is a general equation of a circle?

6. Ellipses have the reflective property that a light or other ray sent from one focus bounces off the ellipse at any point and is reflected through the other focus.

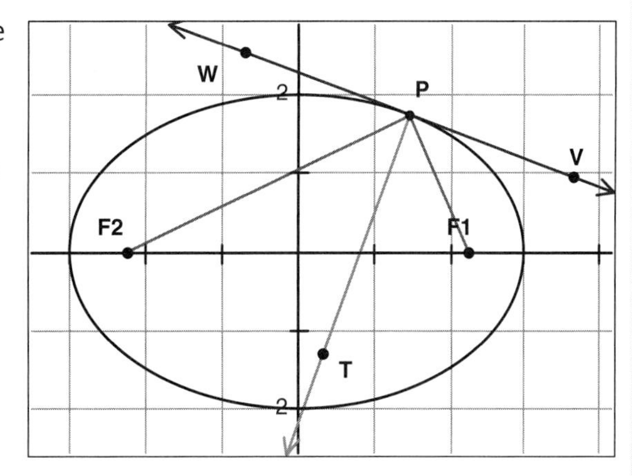

a. Study the figure at right. Ray PT is the angle bisector of $\angle F_1PF_2$. Line WV is perpendicular to ray PT at point P.

i. How are $\angle WPF_2$ and $\angle VPF_1$ related?

ii. How does line WV appear to be related to the ellipse?

b. Generalize your findings above to all points on the ellipse.

c. Search the Internet to find at least two uses of the reflective properties of ellipses in real life. Explain how the reflective properties of the ellipse are necessary in each case.

 © 2010 John Wiley & Sons, Inc.

Exploring Hyperbolas

Definition: A **hyperbola** is the set of all points in the plane, the differences of whose distances from two fixed points (the **foci**) is a constant.

1. a. Use the template below and a constant difference of 2 to sketch a hyperbola using the definition. Repeat for a constant difference of 5. Describe your process in creating each hyperbola.

b. Using only a compass and a straightedge, how can you construct a hyperbola from its definition? Construct a hyperbola using the method you describe.

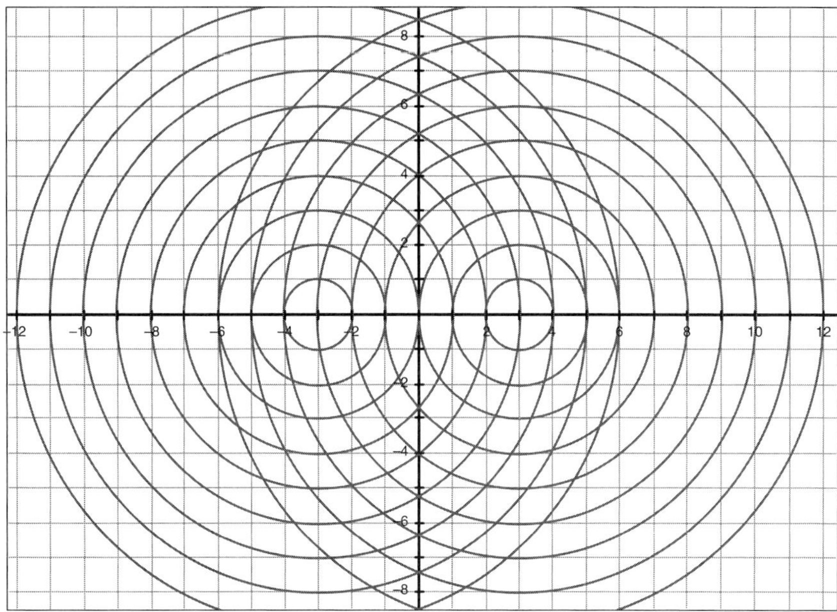

2. For each of the following equations, plot each hyperbola. Determine the x- or y-intercepts (whichever is appropriate), the center of the hyperbola, the foci, and the asymptotes. Compare the results. How can you tell when the foci will be on the x-axis?

a. $\frac{x^2}{4} - \frac{y^2}{9} = 1$ **b.** $\frac{x^2}{9} - \frac{y^2}{4} = 1$ **c.** $\frac{x^2}{16} - \frac{y^2}{9} = 1$

3. For a hyperbola in standard position, centered at the origin, let the coordinates of the foci be $(c, 0)$ and $(-c, 0)$. The equation of a hyperbola in standard position is:

$$\frac{x^2}{a^2} - \frac{y^2}{b^2} = 1 \text{ with } b = \sqrt{c^2 - a^2}$$

a. What are the x- or y-intercepts of the hyperbola? How can you tell which intercepts will be on the hyperbola by looking at its equation?

b. Use the standard form of the equation to describe the shape and position of the hyperbola for which

i. $a < b$ **ii.** $b < a$

c. What is the constant difference between the distances from a point (x, y) on the hyperbola in standard position and its foci? Explain.

d. Estimate the asymptotes for the hyperbola in standard position. Provide a justification for why your estimates are reasonable in terms of the equation of the hyperbola.

e. Repeat items 3b–3d for a hyperbola with equation $\frac{y^2}{a^2} - \frac{x^2}{b^2} = 1$.

(continued)

TEACHING AND LEARNING HIGH SCHOOL MATHEMATICS © 2010 John Wiley & Sons, Inc.

Problem	Equation	Foci	Major Axis	*x*- or *y*- Intercepts	Asymptotes	Constant Difference
1a						2
1a						5
2a						
2b						
2c						
3						
3e						

f. Complete the table on the following page. Constant difference refers to the difference between the distances from a point on the hyperbola and each foci. Summarize your work.

4. The **eccentricity** e of a hyperbola is defined as the ratio $e = \frac{c}{a}$ where $2c$ is the focal length (distance between foci) and $2a$ is the length of the major axis (the distance between *x*- or *y*- intercepts when the hyperbola is in standard position).

a. What bounds are there for values of e for every hyperbola? Explain.

b. Describe the shape of a hyperbola when the eccentricity approaches its lower bound.

c. Describe the shape of a hyperbola when the eccentricity is large.

5. a. Consider a hyperbola in standard position with center at the origin, distance between *x*-intercepts of $2a$, and focal length $2c$. Use your previous experience with transformations of functions to determine an equation for the hyperbola if the center has been translated to (h, k). State how you know your work is correct.

b. Determine equations for the asymptotes of the hyperbola in problem 5a. Explain your work.

c. Sketch the graph and determine an equation of a hyperbola with foci (2, 3) and (6, 3) and a constant difference $d = 1.5$ between a point and each of the foci on the hyperbola. Explain your work.

6. Hyperbolas have the reflective property that a light or other ray directed toward one focus bounces off the hyperbola at a point P and is reflected through the other focus.

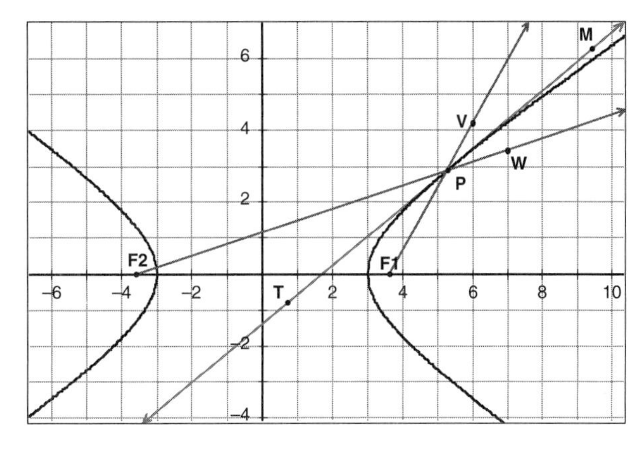

(continued)

TEACHING AND LEARNING HIGH SCHOOL MATHEMATICS © 2010 John Wiley & Sons, Inc.

a. Study the figure. Line *PT* is the angle bisector of $\angle F_1PF_2$. Solve and explain the following:

i. How are $\angle TPF_2$ and $\angle TPF_1$ related?

ii. How are $\angle TPF_1$ and $\angle MPV$ related?

iii. How are $\angle TPF_2$ and $\angle MPV$ related?

iv. How are $\angle TPF_1$ and $\angle MPW$ related?

v. How does line *TP* appear to be related to the hyperbola?

vi. Does it matter if the ray originates inside the hyperbola or outside of it?

b. Generalize your findings above to all points on the hyperbola.

c. Search the Internet to find at least one use of the reflective properties of hyperbolas in real life. Explain how the reflective properties of the hyperbola are helpful for the application.

7. The graph of $y = \frac{1}{x}$ is a hyperbola. Determine the axes of symmetry, the asymptotes, and foci and show that the geometric definition of the hyperbola is satisfied.

TEACHING AND LEARNING HIGH SCHOOL MATHEMATICS © 2010 John Wiley & Sons, Inc.

UNIT six

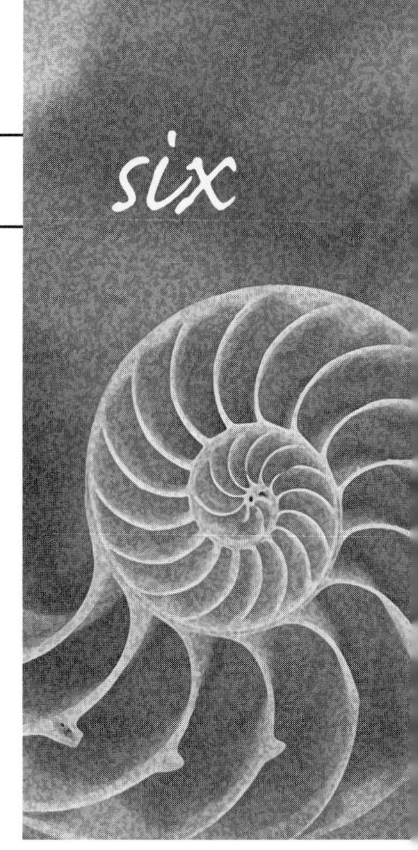

COLLABORATING WITH EDUCATIONAL PARTNERS

Just as successful students benefit from mathematical collaboration, successful educators benefit from professional collaboration. In this unit, you will consider a number of educational venues where collaboration plays a role: selecting and evaluating curriculum materials, considering issues of articulation between institutions from which your students arrive (middle schools) or to which they will go (e.g., colleges or work force), and addressing controversial mathematics education issues.

This unit does not focus on a particular strand of mathematics; nor is there a team builder or observation connected to the unit. Rather, this unit addresses a breadth of curriculum and related issues.

Listening to Educational Partners about Issues in Mathematics Education

In place of a classroom observation and interview with students in this unit you will revisit the principles of *Principles and Standards for School Mathematics* (*PSSM*) (NCTM, 2000) and think about them in light of current issues. You will interview educators, identify a spectrum of views, and prepare to share a variety of perspectives in Lesson 6.4.

1. Reread the sections on the six principles in *PSSM* (NCTM, 2000). Reread your observation reports from the earlier units. Which principle seems most challenging to implement? Why? Use details from your observations, your own and colleagues' learning experiences, and other readings.
2. Choose one issue from the following list (or your class may identify its own issue) that will be the focus of your interview.
   - Uses of technology in mathematics teaching and learning—scientific and graphing calculators, computer algebra systems (CAS), interactive geometry software, the Internet, or other computer software
   - Placement of students in courses—tracking (homogeneous grouping) versus nontracking (heterogeneous grouping)

- Block scheduling—the use of longer class periods, semester courses, and other alternative schedules
- External testing—national, state, district, or other external tests (e.g., SAT, ACT) and how they are designed and implemented, as well as how results are reported and used
- National and international research on high school student achievement (e.g., Programme for International Student Assessment (PISA) from the Organization for Economic Cooperation and Development (OECD), Trends in International Mathematics and Science Study (TIMSS), or other international sources of data) and what they suggest for high school programs

Identify at least two people to interview from the following choices: practicing high school teachers, school administrators, parents of high school students, counselors, or others involved in high school education. If possible, find two people with differing roles and viewpoints on the topic. Make an appointment to talk to these individuals to learn what they think about the issue you chose. Tell the person you are interviewing that you want to understand a variety of perspectives on the particular issue.

3. **a.** For your issue, ask your interviewee some open and unbiased questions similar to the following.
   - Tell me about discussions or experiences you have had related to the topic of ___.
   - What is your thinking about what works well regarding ___?
   - What are some reasons for your position?
   - What are some other positions on this subject, and how do you respond to them?

   b. Include at least one other question of your own that aligns with your thinking derived from one of the *PSSM* principles.
4. Write about your own views on the interview issue. Organize your findings and categorize different viewpoints so they can be shared with your peers.

6.1 Evaluating Curriculum Materials

The word *curriculum* comes from a root meaning *to run*, or *move swiftly*. From the same root, we get the words *current*, like the flow of a river and *course*, like a racecourse, or a course of study. The idea of a *course* suggests the curriculum should start somewhere, have a destination, and a plan for getting there. In education, the word *curriculum* may refer to an element as small as a learning activity, to increasingly larger elements such as a lesson, to a unit, a course, a sequence of courses, up to an entire educational program (Usiskin, 2003). A curriculum centers on goals or learning outcomes to be achieved by learners. These may be skills, concepts, processes, attitudes, habits of mind, or other learning outcomes. The major tools for bringing these goals to fruition are the teaching materials and their related instructional and assessment programs. In the past, curriculum materials were solely textbooks. In today's world, curriculum materials include software, manipulatives, supplements, and supports (e.g., teacher's editions) that help teachers coordinate and make the best use of these tools.

Evaluating, selecting, and implementing new curricular materials are monumental tasks. Effective processes involve teams of teachers and curriculum leaders over extended periods, sometimes years. This lesson looks ahead to a point in a teacher's career when he or she is serving on such a team. How does a textbook evaluation process work? What criteria should be used to evaluate new texts or related curriculum materials?

TEACHING AND LEARNING HIGH SCHOOL MATHEMATICS © 2010 John Wiley & Sons, Inc.

In this lesson, you will:

- Consider criteria for evaluating curriculum materials.
- Use established criteria to evaluate high school mathematics materials.

Materials

- a variety of high school algebra textbooks or integrated textbooks that include Algebra I concepts and procedures
- teacher's guides for the student textbooks to be examined
- curriculum standards and textbook alignment documents for your state

LAUNCH

Assume you have been asked to serve on a curriculum materials review committee. Your team will ultimately recommend materials for your school district's Algebra I program. At the first meeting, the chair asks committee members to share what they have learned from reading a number of professional resources, such as books and articles on teaching, equity, and young adolescents.

1. From your readings in *Principles and Standards for School Mathematics* and other course resources, identify learning goals and other important qualities you would envision for the school's algebra program.

EXPLORE

Evaluating several textbooks, even if they are all for one course such as Algebra I, can be a daunting task. Prior to looking at the materials, a team needs to decide on the criteria they will use to evaluate the books. They also need to decide on which particular part or parts of the course they want to study in detail.

2. When evaluating textbooks, some teachers like to look in detail at the part of the curriculum corresponding to what they are currently teaching. What are advantages and disadvantages of this approach?
3. Another approach when evaluating textbooks is to study in detail a part of the curriculum teachers have found challenging to learn or to teach. What are advantages and disadvantages of this approach?
4. Although single units might be evaluated in their entirety, in the case of an entire book or series of books it is wise to sample selected units or strands of instruction. These selections may act as indicators for the larger product. For an Algebra I course, which topics or sequences of topics do you believe are of high importance to be examined in a textbook evaluation?
5. Another important resource for determining which topics should be studied in a textbook evaluation is the curriculum standards document for your state. Some states also provide textbook alignment documents. (Find these resources online, typically from your state's department of education.) How does the list of important topics align with those you determined in item 4?

The American Association for the Advancement of Science (AAAS), as part of its Project 2061, developed criteria for evaluating mathematics materials at the algebra level. (For more information, go to http://www.project2061.org/publications/textbook/algebra/ summary/critdet.htm. Look under the navigation bar to find other links.) The criteria provided by AAAS are built from recommendations of the NCTM and the AAAS's *Benchmarks for Science Literacy* (1993). The AAAS analyses used selected *idea sets* and instructional criteria; an idea set is a major conceptual area in the curriculum.

The idea sets for algebra that AAAS examined were (1) functions, their representations, and uses as models; (2) variables as representations of quantities; and

TEACHING AND LEARNING HIGH SCHOOL MATHEMATICS © 2010 John Wiley & Sons, Inc.

(3) operating with symbols and equations. Activities, examples, explanations, exercises, and assessments are studied within these three idea sets to determine (a) if they address the standard substantively, (b) if they are at the appropriate level of sophistication, and (c) if they address all parts of the standard.

The activities identified by the content analysis are then further scrutinized for quality of instruction based on qualities identified in research and best practices to characterize effective mathematics instruction.

6. Read the following summaries of the *AAAS Categories for Evaluating Instructional Materials*. What seem to be some of the underlying principles of these criteria? How is equity embedded?

AAAS Categories for Evaluating Instructional Materials

Category I: Identifying a Sense of Purpose

Is the purpose of the lesson or unit clear and motivating to students? Does the lesson or unit have an overarching problem used as motivation for study? Does the lesson or unit flow in a logical sequence that builds understanding?

Category II: Building on Student Ideas about Mathematics

Do the materials identify and ensure that students have the needed prerequisite knowledge for a lesson or unit? Do the materials help teachers probe students' thinking, including misconceptions, and offer support to use that information to enhance student learning?

Category III: Engaging Students in Mathematics

Do the materials provide opportunities for students to engage with mathematics in multiple contexts?

Category IV: Developing Mathematical Ideas

Do the materials help students see the importance and validity of mathematical ideas? How is vocabulary introduced and used to facilitate communication (e.g., always provided or introduced only as necessary)? Do the materials make connections among different concepts explicit? Do the materials help students reason mathematically by modeling complex processes, such as developing generalizations or proofs? Do students have sufficient opportunities to practice working with important concepts?

Category V: Promoting Student Thinking about Mathematics

Do the materials expect students to explain their thinking and justify their reasoning to peers and the teacher in order to obtain feedback? Do the materials sequence tasks and questions to build reasoning about important mathematical ideas? Are students expected to monitor their own progress?

Category VI: Assessing Student Progress in Mathematics

Are assessments and learning outcomes aligned? Are students expected to apply new learning to something unfamiliar in ways that require more than just memorization? Do embedded assessments provide suggestions for teachers on how to use them to rescale activities?

TEACHING AND LEARNING HIGH SCHOOL MATHEMATICS © 2010 John Wiley & Sons, Inc.

Category VII: Enhancing the Mathematics Learning Environment

Do the materials help teachers enhance their own understanding of mathematics and its applications? Do the materials help teachers build a classroom environment that welcomes and expects student curiosity, creativity, and questioning and that has high expectations for all students?

Adapted from Project 2016 (Association for the Advancement of Science. www.project2061.org/publications/textbook/algebra/summary/critdet.htm, downloaded July 29, 2009).

7. **a.** As a class, select one of the three AAAS idea sets for algebra (e.g., variables as representations of quantities) and one of the preceding categories.
   b. Select an algebra or integrated mathematics textbook. Find where it addresses the content area you selected.
   c. Read the lessons and the related teacher's edition material.
   d. Evaluate how the materials meet the criteria for that idea set, according to your interpretation of the criteria.
   e. Cite evidence to support your evaluation.
   f. How does the AAAS category you used embed issues of equity (high expectations and support for all learners)?

SHARE AND SUMMARIZE

8. Share with the class:
   a. Your understanding of the criteria for the category your team studied.
   b. Your evaluation of the textbook your group examined against the criteria for that category.

DEVELOPING MATHEMATICAL PEDAGOGY

Analyzing Instructional Materials

1. Using the same idea set that your group studied in the **Explore**, repeat your analysis of an algebra or integrated mathematics textbook using one of the other AAAS categories for instructional criteria. Visit the AAAS website and read more specifically to understand the criteria for that category (www.project2061.org/publications/textbook/algebra/summary/critdet.htm).
   a. Identify your category and describe what AAAS intends for an evaluator to examine.
   b. How well does the book you examined meet the criteria? Give specifics.
   c. How does the experience of thinking in detail about instructional criteria and looking closely at instructional materials help you think more clearly about supporting student learning?
   d. How does the AAAS category you used embed issues of equity?
2. Many novice teachers find instructional resources on the Internet. The quality of such materials varies. For the following items, select a student page from this textbook.
   a. Identify the learning outcomes of the student page you selected.
   b. Search the Internet for at least two other instructional materials with the same learning outcomes as the student page. Find, if possible, activities that take distinctly different approaches. Provide copies and cite the websites for the two

TEACHING AND LEARNING HIGH SCHOOL MATHEMATICS © 2010 John Wiley & Sons, Inc.

resources. Summarize each of the three activities (student page and two others). Explain distinctions between them.

c. For each of the three activities, identify what would precede and follow in a curriculum sequence (e.g., for the *Continuity* portion of the **Lesson Planning Guide** in the Unit Four Appendix).

d. Evaluate each of the three items against AAAS Category IV: Developing Mathematical Ideas.

e. What cautions would you suggest to others who are looking for Internet teaching materials?

3. In "The Case of the University of Chicago School Mathematics Project—Secondary Component," Zalman Usiskin (2007) describes the thinking behind the development of high school curriculum. (In *Perspectives on the Design and Development of School Mathematics Curricula*, edited by Christian R. Hirsch, pp. 173–182. Reston, VA: National Council of Teachers of Mathematics, 2007).

   a. Usiskin identifies the following principles related to textbook features:
   - Students should be active learners.
   - Students need to learn to read mathematics.
   - Mastery learning is effective but difficult to implement.
   - Content not assessed is typically not taught well.
   - Long-term projects can capture students' creativity.

   Identify one principle about which you have strong feelings (pro or con). State your position, and give your rationale.

   b. Usiskin also argues for integrating mathematics topics, such as statistics with algebra; transformations with geometry and functions; applications; and technology. What experiences have you had with studying these topics in an integrated manner (beyond those in this textbook)? How did the textbooks used in the classes you observed integrate these topics?

   c. Usiskin (2007) describes four dimensions of understanding: skills or procedures, properties or mathematical principles, applications or uses of mathematics, and representations or visual displays of content. How were these different dimensions of understanding evident in the curriculum materials used in the classes you observed?

Reflecting on Professional Reading

For questions 4–8, read "Selecting High-Quality Mathematics Textbooks" by James E. Tarr, Barbara J. Reys, David D. Barker, and Rick Billstein (*Mathematics Teaching in the Middle School*, 12 (August 2006): 50–54).

4. a. What evidence do the authors provide that textbooks play a central role in classrooms?

   b. Why are so many textbooks so long in number of pages, but short in depth?

5. The article is written about middle school texts but the principles and criteria are general and may be applied to high school texts. Identify one key topic that spans several high school grades, for example, functional relationships. Examine a series of three or more high school mathematics books from one publisher and trace the development of the topic across courses using the authors' *quality considerations for mathematics content.*

6. Read the authors' criteria for *quality considerations for the instructional focus* of mathematics textbooks. Study one unit in a high school textbook and evaluate how well it meets the criteria.

7. a. What sort of information do the authors recommend be provided to teachers to support their teaching?

TEACHING AND LEARNING HIGH SCHOOL MATHEMATICS © 2010 John Wiley & Sons, Inc.

b. Read the authors' criteria for *quality considerations for teacher support* in mathematics textbooks. Study one unit in a high school textbook teacher's edition and evaluate how well it meets the criteria.

8. Compare the criteria for evaluating curriculum materials outlined in this article with those of AAAS.

6.2 Coordinating Curricula beyond the Classroom

Early career teachers usually are fully occupied with day-to-day planning, understanding their students, teaching, assessing, grading, communicating with parents, and classroom management. But as teachers grow in their profession, they begin to think more about a bigger picture of students' mathematical education. Questions arise. What did students learn in middle school and how does our high school program build from that program? How does our program align with students' future needs? How does mathematics learning relate to student learning in science, social studies, or other disciplines? In this lesson, you will have opportunities to consider some of these issues of larger scale curriculum planning and collegial collaboration.

In this lesson, you will:

- Consider mathematics education issues beyond the classroom.

LAUNCH

1. As a team, answer one of the following questions.
   - How is high school mathematics related to other subjects students are studying (e.g., chemistry, economics, art, composition, vocational classes)?
   - What middle grades concepts and processes are critical to high school mathematics achievement?
   - How is studying mathematics in college different from studying it in high school?

EXPLORE

Each team should select one of the following clusters of questions (**Cross-Curricular Considerations, Middle School–High School Articulation, College and the Work Force: Beyond High School**) related to mathematics education issues beyond the classroom. Be prepared to share your responses to the questions.

Cluster A: Cross-Curricular Considerations

One of the major goals of mathematics learning is applications. Opportunities for many applications exist very naturally within the other high school courses students are taking. Imagine you are collaborating with a teacher in another subject. You have one or two weeks when your students can work in both classes on a joint project.

I. Identify a topic on which you could join forces with a colleague in another discipline. Some ideas are listed at the end of this cluster, but you are encouraged to think of your own. Think about the different courses high school students take. Be sure to consider vocational as well as college preparatory courses. Explore Internet resources to identify or develop ideas.

II. What might be the products of this project?

TEACHING AND LEARNING HIGH SCHOOL MATHEMATICS © 2010 John Wiley & Sons, Inc.

III. What might be the mathematical learning outcomes?

IV. What might students be learning in the other subject(s)?

V. Draft a plan for one to two weeks of instruction in each class.

VI. What do you see as benefits of such a joint project?

VII. One concern often raised about integrating mathematics and other subjects is that integration results in mathematics getting too little attention. What would you do in your plan to reduce this concern?

Starter ideas for cross-curricular projects

Earth Science—Use logarithms for carbon dating.

Biology—Use predator–prey models to investigate populations sharing an environment.

Chemistry—Use number theory and algebra to balance chemical equations.

Physics—Understand the inverse square law, where it applies, and why.

Economics—Use spreadsheets to study effects of different inflation/deflation factors.

Government—Use graph theory to represent relationships among government agencies.

History—Research the introduction in western Europe of Hindu-Arabic numerals, their role, and related controversies.

Composition—Write data-based research reports.

World languages—Study how sets of related mathematics terms (e.g, quadrilaterals) are expressed in other languages and what is revealed.

Graphic design—Learn how animation software uses geometric transformations.

Computer Science—Study the mathematics of bin-packing. Evaluate different algorithms for their efficiency.

Design Science—Learn the geometry of arches.

Cosmetology—Use ratio and proportions to calculate dye formulas.

Art/archeology—Study symmetry patterns in different cultures.

Music—Learn the mathematics behind some musical instruments (e.g., guitar or piano).

See Dalton (1987) and McConnell (1995) for more ideas.

Cluster B: Middle School–High School Articulation

School districts that have a record of high student achievement usually have strong articulations between different school levels, for example, middle school and high school, so that transitions are well planned and implemented. Imagine you are the mathematics representative from your high school to such a middle/high school articulation committee.

I. What goals would you like to result from your work on the articulation committee? Think of both short-term goals (within one or two semesters) and long-term goals (within two to four years).

II. Explore Internet resources to find articles or references to support your committee's work.

III. What questions would you like to ask the middle school mathematics teachers?

IV. What would you like to share about your high school program?

V. What activities might you propose so teachers from different levels better understand one another?

TEACHING AND LEARNING HIGH SCHOOL MATHEMATICS © 2010 John Wiley & Sons, Inc.

VI. What steps would you suggest the committee take to work toward the goals identified in item I?

VII. What challenges would you anticipate in your committee work? How would you address them?

Cluster C: College and the Work Force—Mathematics beyond High School

Students leave your high school and enter a variety of different programs: college, the military, vocational programs, and others. They might also go directly into the work force. Complete item I and one other item.

I. Choose a couple of school districts (at least one out of your state) where high school teaching jobs are expected to be plentiful in the next year or so. Learn as much as you can about where the students from those districts go after high school.

II. Some schools teach advanced placement (AP) calculus, computer science, or statistics. Learn more about AP courses. One starting site is http://collegeboard.com.

- **a.** Learn about the curriculum for AP calculus, computer science, or statistics.
- **b.** What qualifications do teachers need to teach these courses?
- **c.** What courses must students take to prepare for these courses (and at what grade levels)?
- **d.** How do students fare after taking these courses? Do they continue with calculus? Are they successful in sequential courses at the colleges they attend?

III. Some schools teach international baccalaureate (IB) courses, designed for international recognition. Learn more about IB programs. One site to visit is www.ibo.org.

- **a.** Learn about the high school mathematics curriculum for IB.
- **b.** What qualifications do teachers need to teach these courses?
- **c.** What courses must students take to prepare for these courses (and at what grade levels)?

IV. Some of your students may pursue careers in the military. Identify one branch of the military (army, navy, air force, marines, national guard, coast guard).

- **a.** What educational opportunities are there for members of this branch?
- **b.** Choose one career path within that service that includes continued education. What mathematics is involved? How might that mathematics be seeded in high school?
- **c.** Suppose a student has studied trigonometry. What career opportunities does this level of mathematics afford within the branch you are researching?

V. Some of your students may pursue careers with on-the-job training (e.g., building trades apprenticeships such as electricians, carpenters, plumbers).

- **a.** Choose one field and research its educational prerequisites.
- **b.** What entrance exams do such programs include?
- **c.** What information would be helpful for students interested in these careers to know about the mathematics entailed?

VI. Some students may pursue associate's degrees at local community colleges.

- **a.** Identify a community college in the vicinity of the school district you are studying. What programs at that college have good job prospects?
- **b.** Choose two career paths. What mathematics is involved in each?

VII. Imagine you are planning an articulation meeting with mathematics faculty and admissions representatives from a local college or university.

TEACHING AND LEARNING HIGH SCHOOL MATHEMATICS © 2010 John Wiley & Sons, Inc.

a. What questions would you have for these colleagues?

b. If you have read the article about college placement (see **DMP problem** 7), then use that to help you formulate other discussion points.

c. What are some activities the participants of the meeting might consider for building better communication between the institutions?

SHARE AND SUMMARIZE

Teams should take turns sharing their responses to the selected cluster. During or after each presentation, other students should raise questions and contribute additional ideas.

DEVELOPING MATHEMATICAL PEDAGOGY

Reflecting on Our Own Learning

1. Recall your own experiences studying mathematics at middle school, high school, and college or university levels.
   - a. Comment on the transitions. Describe any similarities or differences. Consider aspects such as how you perceived mathematics, how it was presented, studied, or assessed, what materials were used, or your expectations of the courses.
   - b. What did or might have supported a smooth transition from middle school to high school?
   - c. What did or might have supported a smooth transition from high school to college or university?
2. Interview family members, friends, or acquaintances about the uses of mathematics in their jobs. What do they use? What did they use in earlier years? What was helpful from their education? What would have been more helpful in their mathematics education? What implications can you find for your teaching career?

Preparing for Instruction

3. Choose one of the ideas for cross-curriculum projects at the end of Cluster A or one your team generated in class. Research the topic. Outline a three-day series of lessons that address that topic. Identify the mathematical and other learning outcomes of the lessons. Include one student page that could guide part of the sequence. Tell how you would assess student learning from that sequence.
4. Many high schools have a service learning requirement for high school graduation. Identify a service learning project that could incorporate mathematics. Draft some tasks you could have students do in association with the service learning project you envision.

Focusing on Assessment

5. College-bound students typically take the Scholastic Achievement Test (SAT) and/or the American College Test (ACT). Find samples on the Internet of mathematics portions of each of these exams.
   - a. Find a high school topic (e.g., rate of change, exponential functions, circle theorems) for which you can identify items from each exam. Compare the items.
   - b. Some educators have commented that the SAT exam uses more reasoning and problem-solving while the ACT uses more skills. Cite evidence to support or refute these comments.

TEACHING AND LEARNING HIGH SCHOOL MATHEMATICS © 2010 John Wiley & Sons, Inc.

Reflecting on Professional Reading

6. Read "Facing Facts: Achieving Balance in High School Mathematics" by Lynn Arthur Steen (*Mathematics Teacher*, 100 (Special Issue: *100 Years of Mathematics Teacher*, December 2006): 86–95).
   - **a.** What was the impact of Sputnik on mathematics education? What was the reaction to "new math"?
   - **b.** How were mathematics education reform efforts of the 1980s different from those of the 1960s and 1970s?
   - **c.** What does Steen find about U.S. student achievement from the NAEP trends data?
   - **d.** What are the two different goals Steen identifies in today's schools?
   - **e.** Figure 3 lists many commonly made statements about what is wrong with today's schools. Choose two about which you have factual knowledge. What evidence can you use to support one and refute the other?
   - **f.** State in your own words Steen's three recommendations.
   - **g.** What is your opinion of Steen's recommendations?
7. Read "Placement Tests: The Shaky Bridge Connecting School and College Mathematics" by Sheldon P. Gordon (*Mathematics Teacher*, 100 (October 2006): 174–178).
   - **a.** Gordon is a college professor. How does he perceive the intentions of the NCTM standards at the high school level?
   - **b.** Identify several issues related to placement tests that cause distress for students, parents, and teachers at both high school and college levels.
   - **c.** In analyzing the dynamics of placement testing, Gordon speaks about the perspectives held by college administrators that differ from the needs or interests of others involved in placement testing. Give details of some of these differences.
   - **d.** What suggestions does Gordon make for improving this situation?
   - **e.** What do you think would be helpful to bridge the current rift between high school and college expectations?

6.3 Continued Professional Development

Principles and Standards for School Mathematics (NCTM, 2000) opens with its vision for school mathematics. The *Standards* authors recognize the vision as "highly ambitious" and enormously challenging, but view its accomplishment as essential. *Teaching and Learning High School Mathematics* has attempted to provide research, readings, foundations, experiences, and strategies to support teachers in reaching that vision. But the reality is that many schools need support in working toward that vision. What can teachers do? How can teachers collaborate with colleagues at their school to move toward stronger educational visions? These questions are the focus of this lesson.

In this lesson, you will:

- Find ways to incorporate many aspects of *Standards*-based education in your own classroom regardless of outside limitations.
- Consider ways to support your school in bringing *Standards*-based education to all students, thus promoting equity.
- Identify ways to continue to grow professionally.

TEACHING AND LEARNING HIGH SCHOOL MATHEMATICS © 2010 John Wiley & Sons, Inc.

LAUNCH

What are the attributes of *Standards*-based mathematics education? Develop your list from recalling *PSSM* principles and standards, readings, textbook activities, interviews with high school students, and values that have evolved throughout the use of this textbook. When your team has what it believes to be a good list, reread the vision on page 3 of *PSSM* and compare your view with that of the NCTM.

EXPLORE

Address one of the following five scenarios (**In Your Classroom**, **Parents Back to School Night**, **New Teacher Working with Other Colleagues**, **The Experienced Teacher Trying to Grow**, or **The Standards-Based Setting**), with each team choosing one scenario. Each team should prepare to share its ideas with other teams. After you have worked individually and together, read the additional questions for each scenario provided at the end of this lesson. Use them to help you continue the discussion and think creatively.

Scenario 1: In Your Classroom

You are teaching at a school where there are few physical materials or calculators. The textbook for your Algebra I class is traditional. The book starts with some contexts but quickly shows students one way to solve each type of problem, then has them practice. The state framework is based on an earlier version of the NCTM *Standards*, and your book is loosely correlated to it. The teacher next door was on the textbook selection committee; while the innovative books the committee examined had lots of good problems, the teacher thought they were too challenging for the students at this school.

What are some strategies you can use within this setting to help your colleagues begin to develop the vision of the NCTM for *all* students?

Scenario 2: Parents Back to School Night

Parents at your high school are concerned about student progress and generally unaware of any national educational recommendations. Often parents are not able to attend meetings. They work. They need baby sitters for younger children. They are tired. Nevertheless, the school holds an annual Parents Back to School Night. The idea is to give parents a sense of what and how their students are learning. It is not a time for individual conferences.

a. How might the school organize this event (e.g., parents following their child's schedule, parents interacting at booths in a gym)?

b. What will be your mathematical goals for your interactions with parents? What will you want to emphasize? What activities might you provide? What special features might you include?

c. How will you evaluate your portion of the event?

Scenario 3: The New Teacher Working with Other Colleagues

This term you have begun teaching at a new high school (new to you!). You feel strongly about your values, but you want to be cautious initially as you meet and get to know other teachers and administrators in the building. You are concerned that some people might not share your ideas. What are reasonable and helpful steps you can take to learn what others are doing and engage them in professional conversations without creating difficulties?

TEACHING AND LEARNING HIGH SCHOOL MATHEMATICS © 2010 John Wiley & Sons, Inc.

Scenario 4: The Experienced Teacher Trying to Grow

You are in a high school where you can work toward the *PSSM* vision to some extent but generally in limited ways. You would like to implement more aspects of the vision. You have been at this school for three years. The location, salary, benefits, and schedule are good for you. The most forward-looking teachers you have met are one science teacher and one English teacher. What can you do to help yourself and your mathematics department move toward more *Standards*-based education?

Scenario 5: The Standards-Based Setting

You are teaching in a high school that is working hard to implement the NCTM *Principles and Standards* while also satisfying state requirements for mathematics. The school supplies a classroom set of graphing calculators for every mathematics teacher and has additional physical materials available for check-out. Your colleagues are in general agreement with the *PSSM* and each one works to implement the standards in his or her classroom. What can you and your colleagues do to stay informed and make decisions about further implementation as new technologies and materials become available? How might colleagues assist each other?

SHARE AND SUMMARIZE

Present the scenarios and ideas generated by your team. As your team reports, invite other teams to share more ideas about your topic.

DEVELOPING MATHEMATICAL PEDAGOGY

Analyzing Instructional Activities

1. Go to the website of the National Council of Teachers of Mathematics, www.nctm.org.
   a. Find *Figure This! Math Challenges for Families*, a resource for Internet-based problem solving (www.figurethis.org/index.html). Although this resource was designed for families of middle school students, the ideas are applicable across a range of grade levels. How might you use this resource (i) in your class, (ii) for a Family Math Night, or (iii) for helping other colleagues think differently about teaching mathematics?
   b. Find other electronic resources on the NCTM's website, such as the electronic examples *Illuminations* or *Student Math Notes*. Evaluate at least one of these resources. How might you use these resources in your classroom or school?
2. Use links from the NCTM's website or Math Forum (www.mathforum.org) to find at least two additional resources, one for teaching and one for professional development. Evaluate these resources. How might you use the teaching resource in your classroom? How might you benefit from the professional development resource?

Developing Professionally

3. Many formats exist for professional development, some with outside leaders or consultants, some with internal leadership. Examples include university courses; departmental or school committees; professional reading groups; practice-based professional development; case study analysis groups; affiliation with local, state, or national professional organizations; attending and presenting at professional conferences; coordinating enrichment activities for students outside of school;

TEACHING AND LEARNING HIGH SCHOOL MATHEMATICS © 2010 John Wiley & Sons, Inc.

and lesson study. Identify one of these formats and learn more about it. If possible, talk to a teacher who has been involved with that format. Prepare for your classmates a five-minute presentation with an informative handout that gives highlights and sources for the format you studied.

4. Teachers can assist each other in improving their teaching skills through working together to plan lessons, observing each other teach and discussing the results, and revising lessons and questioning techniques based on the observations. Faculty members at Clemson University have devised a rubric to assist teachers in learning to teach by inquiry. (J.C. Marshall, B. Horton, J. Smart, and D. Llewellyn. (2008). EQUIP: Electronic Quality of Inquiry Protocol: EQUIP.pdf file, Clemson University's Inquiry in Motion Institute, www.clemson.edu/iim). Retrieve the EQUIP.pdf file from their website under the **Research and Evaluation** tab.
   - **a.** Read the rubrics for Instructional, Discourse, Assessment, and Curriculum Factors. What benefits do you think might be derived from using the rubrics for evaluating your own instruction?
   - **b.** What benefits might be derived from working with a colleague to conduct evaluations of each other?
   - **c.** Choose one of the four Factor rubrics. Think about how you expect to teach. Evaluate yourself.
   - **d.** Choose one of the lesson plans you wrote during this course. Evaluate the lesson based on the rubric you chose in problem 4c.

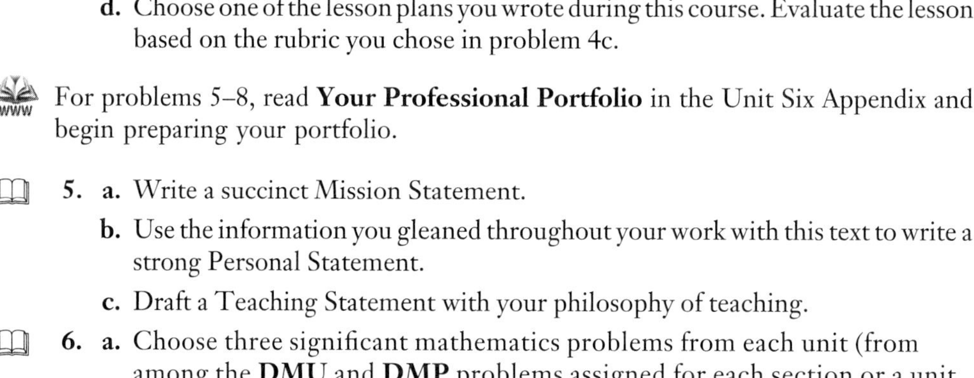

For problems 5–8, read **Your Professional Portfolio** in the Unit Six Appendix and begin preparing your portfolio.

5. **a.** Write a succinct Mission Statement.
   - **b.** Use the information you gleaned throughout your work with this text to write a strong Personal Statement.
   - **c.** Draft a Teaching Statement with your philosophy of teaching.

6. **a.** Choose three significant mathematics problems from each unit (from among the **DMU** and **DMP** problems assigned for each section or a unit investigation) to include in your portfolio as *Examples of Personal Deep Understanding of Mathematics*.
   - **b.** Write a page describing why you included the items you did, indicating how they give a picture of your deep mathematical understanding.

7. **a.** For the portfolio section on *Examples of Exemplary Teaching Materials*, provide a collection of teaching materials which you prepared. For example, you might include exemplary **QRS** segments or complete lesson plans, function family examples and activities, concept maps, student pages, software programs, or games. Choose three or four of your best examples. These items can be ones you completed individually or with team members.
   - **b.** Write a descriptive page to insert at the beginning of this part of your professional portfolio that indicates what can be found in the section and why you have included the particular samples.

8. **a.** Provide a collection of assessment items and student work you prepared or collected as you used this text. Choose your three or four favorites for the section on *Examples of Assessments and Student Work*. These items can be ones you completed individually or with team members.
   - **b.** Consider the high school students you interviewed at different points during your work with this textbook. Assemble their work by student. Write about the growth in each student's mathematical understanding, ability to reason and write, and their growth in mathematical power, that is, their confidence in solving challenging problems in mathematics.

TEACHING AND LEARNING HIGH SCHOOL MATHEMATICS © 2010 John Wiley & Sons, Inc.

c. Write a descriptive page to insert at the beginning of this part of your professional portfolio that indicates what can be found in the section and why you have included the particular samples.

Debating Issues

9. Many parents are concerned that the changing mathematics curriculum does not provide their children with the basics needed to succeed. Read the article "Parrot Math" by Thomas C. O'Brien in the February 1999 issue of *Phi Delta Kappan* (available online at http://findarticles.com/mi_6952/is_6_80/ai_n28757692/?tag=content;col1, downloaded July 29, 2009). Although the article focuses on arithmetic, similar comments often are heard about the learning of algebra. Identify several points in the article you might use in a discussion with parents about a hands-on, activity-based, and technology-based approach to Algebra I or Algebra II. Supplement your list with what you have learned through this textbook.

Reflecting on Professional Readings

10. Reread Chapter 1, "A Vision for School Mathematics," from *Principles and Standards for School Mathematics* (pp. 3–8). (Reston, VA: National Council of Teachers of Mathematics). What strikes you differently now than when you read about the vision in Unit 1?

11. Skim the other readings you have done in *Principles and Standards for School Mathematics*.

- **a.** What parts have been most helpful to you? Why?
- **b.** How might this document be of assistance to you in working with other teachers, administrators, and parents?

For problems 12–15, read "Participation in Career-Long Professional Growth," Standard 5 of Standards for the Education and Continued Professional Growth of Teachers of Mathematics. (In *Mathematics Teaching Today*, edited by Tami S. Martin (pp. 157–170). Reston, VA: National Council of Teachers of Mathematics, 2007).

12. **a.** Many new teachers believe their work is individual and private. How does this chapter help you think differently about this idea?

- **b.** The authors identify many formats for teachers' continuing professional growth. Identify at least five different formats. Identify one that sounds particularly interesting to you and say why it interests you.

13. **a.** Solve the condominium problem in a way different from that explained in the vignette.

- **b.** What major point is the facilitator in the condominium vignette trying to make?

14. **a.** In the redesigning instruction vignette, how did the principal support the teacher?

- **b.** In the redesigning instruction vignette, how did the middle school teacher support the elementary school teacher? How does this vignette suggest that you, as a high school teacher, could support a middle school teacher?

15. **a.** How are teachers in the technology vignette connecting their knowledge of technology to teaching?

- **b.** What principles that you have been studying in this course do you see in the vignettes?

Preparing to Discuss Interviews

16. In preparation for Lesson 6.4, review your notes from your readings of the principles in *PSSM* and from your interviews with educational partners. Be prepared to discuss your interviewees' responses with your peers.

TEACHING AND LEARNING HIGH SCHOOL MATHEMATICS © 2010 John Wiley & Sons, Inc.

a. Identify your topic and the positions (jobs/roles) of your interviewees. State the question(s) you posed to the interviewees.

b. What perspectives and reasons did the interviewees share about the mathematics education issue you chose for the interview?

c. How did your interviewees respond to other perspectives on this issue that were different from their own?

Questions to Guide Your Thinking while Discussing the Scenarios

Scenario 1: In Your Classroom

- Even with traditional textbooks, what can you do to get students to think, not just blindly use algorithms?
- What can you do to increase the number of approaches students use for solving problems, doing mental math, and representing mathematics?
- What materials would you consider so valuable that you would find a way to make them or obtain funding for them?
- What can you do to promote a supportive, collaborative classroom environment?
- How can you set high expectations for all students?
- What are strategies for supporting all mathematics learners to set high expectations and to achieve them?
- How can you use assessment strategically or creatively?
- What decisions about course content can you make based on your state's standards? Must lessons be taught in the sequence found in the course textbook? Must every section in the course textbook be taught?

Scenario 2: Parents Back to School Night

- How can you involve your high school students in planning and participating in the event?
- How might you use activities from the curriculum?
- What mathematics activities might you provide for younger children, perhaps under high school student supervision, so parents can participate without finding baby sitters?
- What students' work from your classes might be appropriate to display?
- Think about possibly contrasting a concept taught in a traditional way and in a more meaningful way. What would you select, and how would you engage parents to help them see the benefits to student understanding?
- What can you include in the Parents Back to School Night that would help parents better support their children outside of class?
- What might you include to help parents better understand the value of *Standards*-based education?

Scenario 3: The New Teacher Working with Other Colleagues

- What informal opportunities are there to get to know people?
- What are good communication skills to use with others who may not agree with you?

TEACHING AND LEARNING HIGH SCHOOL MATHEMATICS © 2010 John Wiley & Sons, Inc.

- How might you use the resources of your college or your college instructors to help you?
- What case studies, journal articles, videos, current events related to education, or other sources can you use for discussion starters?
- What are some modest goals you might set for your first year in the school?

Scenario 4: The Experienced Teacher Trying to Grow

- How can you use local, state, or national professional organizations to support you? What roles might you play in these to expand your professional network, learn more about what is happening at other schools, and increase access to information?
- How can you share more widely the positive steps you have already taken?
- How can colleges or universities in your region support you?
- What school or district committees can you join?
- What might you say to your principal or local curriculum supervisor?
- What steps might you take with the one most supportive colleague whom you have identified?
- How can you use your rapport with the science teacher to plan a joint *Standards*-based unit?
- How can you use your rapport with the English teacher to enhance your teaching?

Scenario 5: The Standards-Based Setting

- Though it seems as though you have the best of all worlds, how will you work with colleagues to continue to grow so your teaching does not stagnate over time?
- How will you stay informed of innovations in teaching and learning?
- Where will you find information?
- How will you and colleagues share information?
- When a colleague learns something of interest to other members of the mathematics department, how will the rest of the department learn enough to be able to make the decision of whether or not to implement the new idea or tool?

6.4 Summarizing Interviews on Educational Issues

Your professional philosophy—your beliefs about mathematics and effective mathematics teaching—has been evolving throughout your study of this textbook and will continue to develop over your career. You will find, however, that at any point in time, wherever you stand on certain issues, others in the educational community may not agree with you. These others may be teachers of mathematics, teachers of other subjects, parents, counselors, administrators, or members of the community at large. As you communicate with these other educational partners, it is important that you do so respectfully, clearly, and effectively. By the same token, you must listen carefully, respectfully, and effectively to their positions. Learning occurs through thinking more carefully about issues and through thoughtful interchange of ideas.

TEACHING AND LEARNING HIGH SCHOOL MATHEMATICS © 2010 John Wiley & Sons, Inc.

In this lesson, you will:

- Share perspectives on mathematics education issues.
- Identify what appears to be a spectrum of views on some issues.
- Continue to rethink your own positions on one or more issues.

LAUNCH

In preparation for this lesson, you should have interviewed at least two persons interested in high school mathematics education on at least one issue. As a group, decide on one or more issues you will discuss today.

1. Write personally for a few minutes about your own views on the selected issues. Be sure to include the reasoning behind your position.
2. Share with group members your notes from your interviews on the same issue. Make a record of all the views collected. Include reasons for each perspective. Put each issue on its own sheet.

EXPLORE

Assign each team one of the issues. Some teams may have the same topic. Give interview notes and personal opinions from **Launch** item 2 for each topic to the appropriate team. Be sure to include reasons.

3. Using the sheets prepared in **Launch** item 2, identify as a team a spectrum of positions on your topic. Include extreme and moderate views. As necessary, create additional views so you have at least three and preferably four or five ways of thinking about the topic. Prepare a means to share your spectrum with the entire group.

SHARE AND SUMMARIZE

4. As a team, present the spectrum of positions and rationales your team has created. Ask for responses from other teams. What do they think and why?
5. Reflect on your personal experience interviewing and considering a spectrum of views on mathematics education issues.
   a. What did you learn by thinking about opinions different from your own?
   b. What strategies support respectful conversation about difficult subjects?

DEVELOPING MATHEMATICAL PEDAGOGY

Preparing for Instruction

1. One issue that concerns some parents and members of the community is the use of various forms of technology in schools. For example, use of computer algebra systems (CAS) often raises particular concerns.
   a. Identify at least three high school topics better developed with CAS. Tell how CAS could be used to develop one of those topics.
   b. Identify at least three high school objectives for which you would not want students to use CAS initially. Tell how you would develop the topic without CAS.
   c. Identify at least one high school topic accessible to students with CAS that could not be understood meaningfully otherwise.
   d. Identify at least one high school topic appropriately developed using CAS and tested without it after students have developed deep understanding.

TEACHING AND LEARNING HIGH SCHOOL MATHEMATICS © 2010 John Wiley & Sons, Inc.

e. What are some issues that might arise with CAS-capable technology that might not arise with non–CAS-capable technology?

Developing Professionally

2. Learn about professional organizations, such as one of the organizations below. Some have student memberships of which you might want to take advantage now, if you are eligible. What are their missions? What services do they provide?
   - National Council of Teachers of Mathematics
   - Your state, provincial, or local council of teachers of mathematics
   - National Education Association
   - American Federation of Teachers

Teaching for Equity

3. Read "The Role of Mathematics Instruction in Building a Just and Diverse Democracy" by Deborah Loewenberg Ball, Imani Masters Gaffney, and Hyman Bass. (*The Mathematics Educator*, 15 (2005): 2–6).
   a. The authors believe teachers need to listen closely to students. How is this connected to justice and democracy?
   b. What are some of the authors' concerns about contexts used in textbooks?
   c. What are some of the authors' concerns about teachers giving praise to students?
   d. The authors suggest that mathematics can support a just and diverse democracy by developing tools for social change and as a setting for cultural knowledge. Explain each of these.
   e. Describe the authors' views about how the work of doing mathematics in classrooms provides opportunities for building a just democracy.
   f. What are the authors' ideas about how instruction may be designed to specifically develop practices in students that help them understand and live in a democracy?

TEACHING AND LEARNING HIGH SCHOOL MATHEMATICS © 2010 John Wiley & Sons, Inc.

Synthesizing Unit Six

Collaborating with Educational Partners

This unit discusses various aspects of collaboration with educational partners. Teachers' input is important in selection decisions about the curriculum materials used in mathematics classrooms. Thinking ahead of time about what one wants to see in such materials prepares teachers to make curricular decisions.

In addition, high school teachers need to consider ways of coordinating their mathematics curricula with other educational stakeholders, including teachers at the high school in different disciplines, with middle school mathematics teachers who provide the background for high school courses, with college or university mathematics departments to which students enroll, and with external agencies such as the College Board, which may administer Advanced Placement courses. The challenge for teachers is to find ways to coordinate with these other partners while maintaining a high level of meaning and mathematical challenge for students in their own classes.

At some point in their careers, many teachers find themselves in schools and environments that need support in addressing the reforms in mathematics education advocated by the NCTM *Standards*. Thus, this unit engaged teachers in thinking about actions they can take in such environments as they collaborate with colleagues to move toward a *Standards*-based perspective.

In the final lesson of this unit, teachers summarized information about educational issues obtained from interviews with a variety of educational partners, including administrators, other teachers, guidance counselors, and parents. All parties need to collaborate to support mathematical achievement by all students.

At the close of this unit, you should be able to answer questions like the following based on your engagement with the lessons, readings, and interviews.

Website Resources

- **Where Do I Stand?** survey

Pedagogical Learnings

1. Given a current issue pertaining to high school mathematics teaching or learning (e.g., the use of graphing calculators, CAS), identify a spectrum of positions surrounding the topic. Formulate and justify your own position on the issue.
2. Identify several important qualities you would want to see in mathematics materials adopted for use in your school or district. Provide a rationale for why these qualities are important.
3. Given a topic of study from the high school mathematics curriculum, be able to provide ideas about how this area of study can influence the study of at least one other discipline (e.g., science, social studies, language arts).
4. Suppose you are teaching in a high school with a traditional textbook that emphasizes procedures rather than concepts and problem solving. For a particular lesson, outline some actions you could take to transform the lesson into one with a more *Standards*-based approach.
5. Identify some actions teachers can take collaboratively within a high school mathematics department to inform other educational partners about what is important for effective mathematics learning.
6. Revisit your responses to the survey **Where Do I Stand?** for items 7, 9, 10, 13, 15, 16, 18, 19, and 24.
   a. How have your responses changed, deepened, or grown?
   b. How prepared do you feel for the challenges you have identified throughout this text?

TEACHING AND LEARNING HIGH SCHOOL MATHEMATICS © 2010 John Wiley & Sons, Inc.

APPENDIX: UNIT ONE

Contents

- Blank **Question Response Support (QRS) Guide**
- **Stages of Questioning**
- **Attribute Pieces for Colorful Shapes**
- **Question Response Support (QRS) Guide for Colorful Shapes**
- **Question Response Support (QRS) Guide for How Many Oranges?**

All resources in this appendix are also on the course website.

Question Response Support (QRS) Guide

Questions or Learning Activities	Students' Responses (Expected or Observed)	Teacher Support (Using Stages of Questioning)
	Misconceived	
	Partial	
	Satisfactory	
	Misconceived	
	Partial	
	Satisfactory	

TEACHING AND LEARNING HIGH SCHOOL MATHEMATICS © 2010 John Wiley & Sons, Inc.

Stages of Questioning

Questions are tools teachers can use to understand and support students' mathematical thinking. Teachers need to allow students the opportunity to think independently or in groups, asking questions only when students are not able to proceed. Questions should be phrased to be "just enough" to help students progress; they often are rhetorical and intended to help students think about alternatives. It is advisable to allow students the opportunity to pursue their own solution processes before asking them to consider other routes. Guidelines are provided below.

Stage 1: Asking Non-Directive Questions

Students need time to understand the problem and get started; the teacher should intervene as little as possible. Non-directive questions, such as the following, are used only when it is evident that a student has trouble getting started.

- What do you think the problem is asking you to do?
- What are you thinking? Please explain.

Stage 2: Suggesting General Problem-Solving Strategies

At times, a student is able to explain the problem but still struggles to begin. Questions at this stage may help a student think of ways to get started on the problem by giving him or her a range of strategies to consider. The teacher's tone of voice indicates that these strategies might or might not work. Teachers should be as non-directive as possible.

- Might drawing a picture help?
- Might a table help to solve this problem?
- Might any materials be helpful to model the problem?
- Have you solved similar problems that might help?
- Can you work backwards or simplify the problem?
- What must happen in the problem?
- What cannot happen in the problem?

Stage 3: Suggesting Specific Problem-Solving Strategies

At times, students understand the problem and have initiated a strategy, but are not able to use the work they have generated to bring the problem to closure. They need suggestions for more specific problem-solving strategies through questioning. Once again, the teacher should be as non-directive as possible, asking questions that suggest possibilities that might or might not help them move forward.

- What patterns do you see in the numbers? in the table? in the pictures?
- Is there a rule that describes the relationship in the table of values you created?
- Are the graph and the story related?

Stage 4: Asking for Metacognition

At this stage, the teacher helps the student think metacognitively about the problem, that is, to think about their thinking.

- How did you think about the problem?
- How would you convince someone who has never seen the problem that your approach is correct?
- How is this problem like others you have solved? How is it different?

The teacher's goal is to help students generate a coherent verbal explanation of the problem with representations as appropriate. Teachers might have a list of strategies posted in the classroom that students are reminded to consult when solving problems. In time, students begin to think about the list automatically and to ask themselves and each other such questions.

Table inspired by Goldin, Gerald A. "Observing Mathematical Problem Solving through Task-Based Interviews." In Anne R. Teppo (Ed.), *Qualitative Research Methods in Mathematics Education* (pp. 40–62). Reston, VA: National Council of Teachers of Mathematics, 1998.

TEACHING AND LEARNING HIGH SCHOOL MATHEMATICS © 2010 John Wiley & Sons, Inc.

Attribute Pieces for Colorful Shapes

 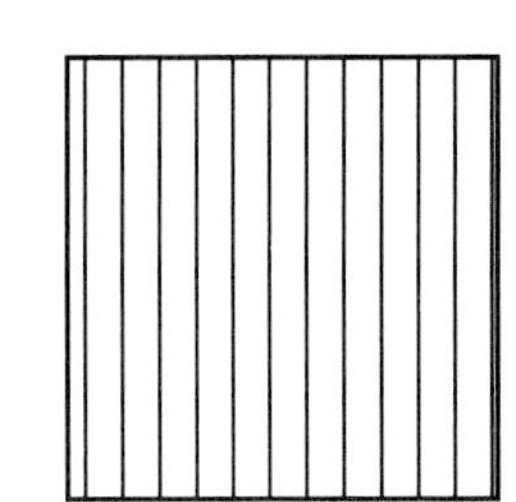 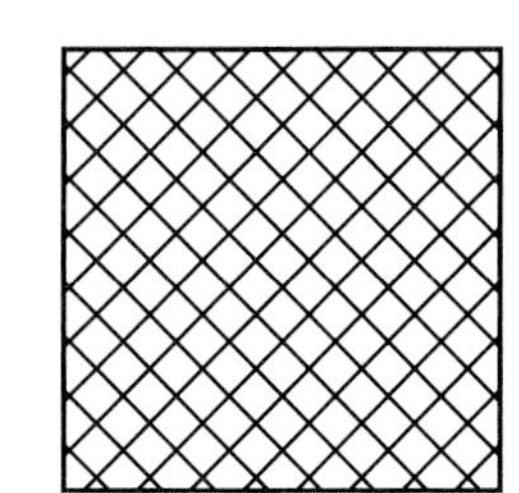 

 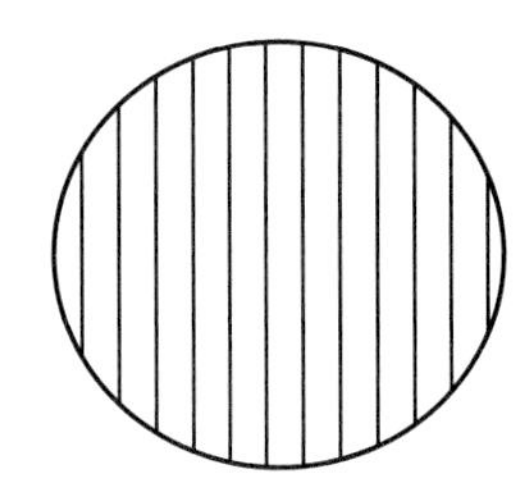 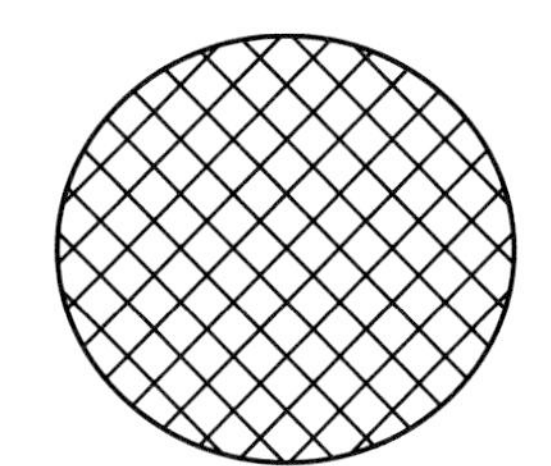  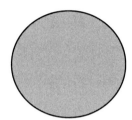

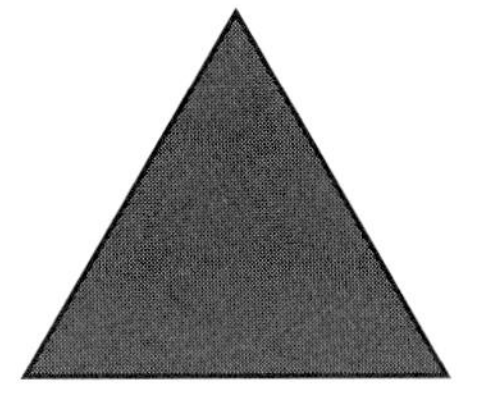 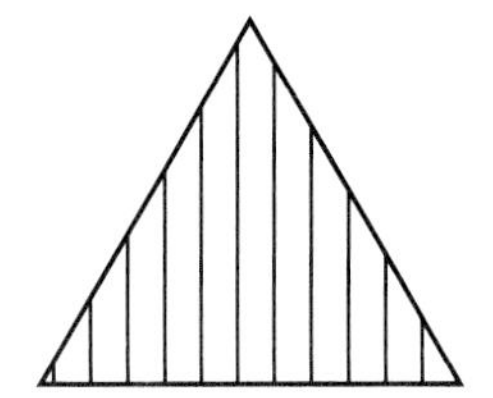 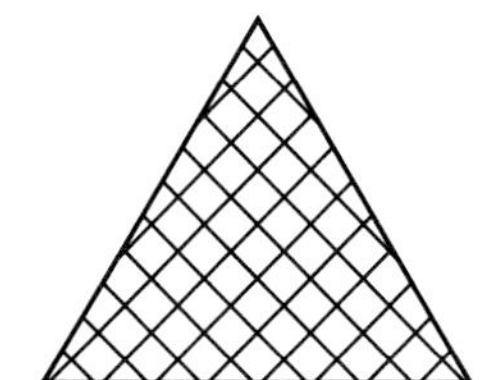 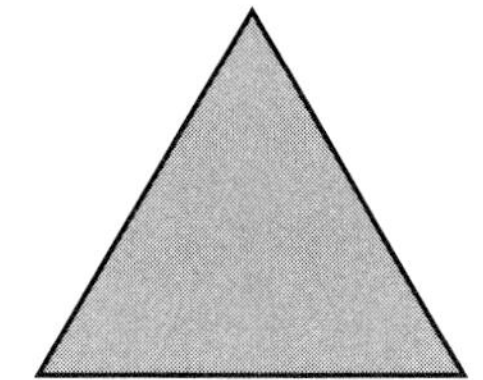

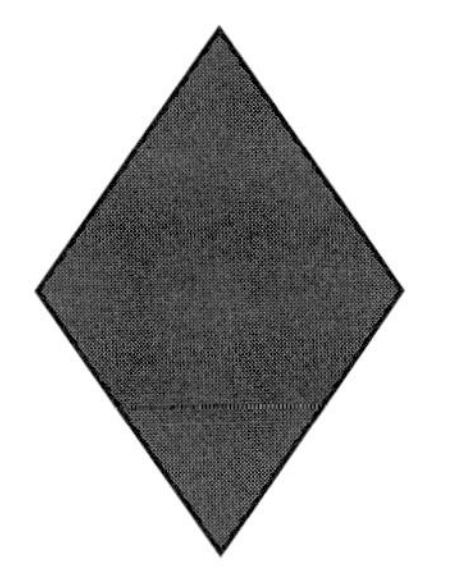 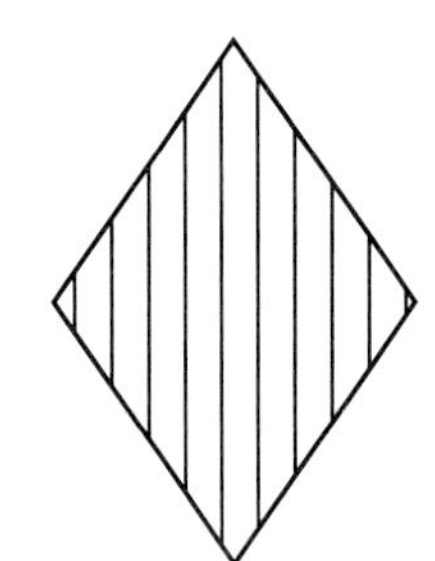 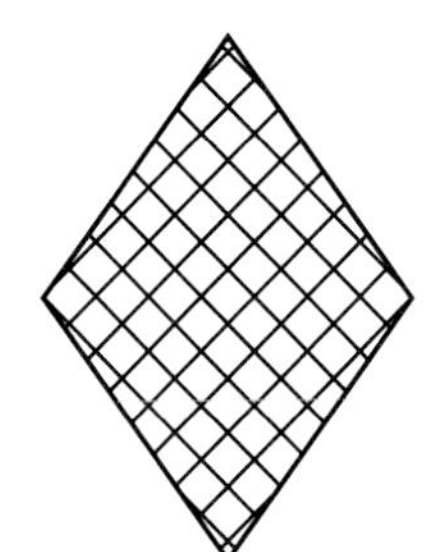 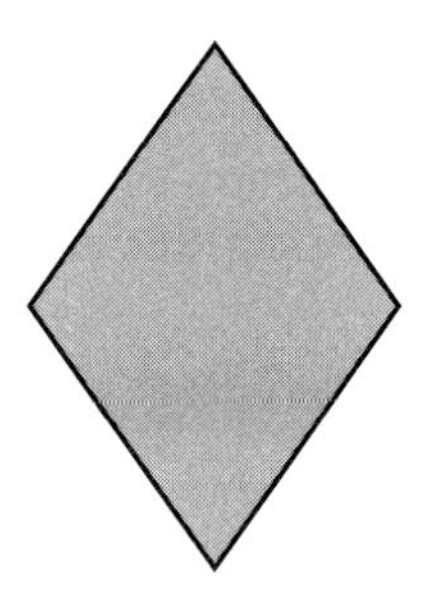

TEACHING AND LEARNING HIGH SCHOOL MATHEMATICS © 2010 John Wiley & Sons, Inc.

Question Response Support (QRS) Guide for Colorful Shapes

Materials: *Colorful Shapes* student page and 16 attribute pieces

<table>
<tr><th>Questions or Learning Activities</th><th>Students' Responses
(Expected or Observed)</th><th>Teacher Support
(Using Stages of Questioning)</th></tr>
<tr><td rowspan="3">Use the 4 by 4 grid in Colorful Shapes and one set of small attribute pieces. Place each piece in the grid so that each color and shape appears only once in any row, column, or main diagonal.</td><td>Misconceived: A student places two pieces of the same color or shape in the same row.</td><td>• Please reread the directions.
• Does your placement follow the conditions of the puzzle? Why or why not?</td></tr>
<tr><td>Partial: Student starts by placing all pieces of one color on the grid, with no two pieces in the same row, column, or main diagonal. Student continues placing each color, being careful not to violate the rules of the puzzle. Student tries several different arrangements but gets stuck each time.</td><td>• Show me what you have tried.
• Are there any other ways you can start the puzzle?</td></tr>
<tr><td>Satisfactory: Student starts by selecting 4 pieces of different colors (A, B, C, D) and shapes (1, 2, 3, 4). Student places each piece in a different row of the first column as shown.
<table>
<tr><td>A1</td><td></td><td></td><td></td></tr>
<tr><td>B2</td><td></td><td></td><td></td></tr>
<tr><td>C3</td><td></td><td></td><td></td></tr>
<tr><td>D4</td><td></td><td></td><td></td></tr>
</table>
Student notices that the other two corners must be B3 and C2 so places these pieces. Student fills in the remaining pieces of each main diagonal, noticing that these are entirely determined, then finishes the puzzle with a correct solution.
<table>
<tr><td>A1</td><td>C4</td><td>D2</td><td>B3</td></tr>
<tr><td>B2</td><td>D3</td><td>C1</td><td>A4</td></tr>
<tr><td>C3</td><td>A2</td><td>B4</td><td>D1</td></tr>
<tr><td>D4</td><td>B1</td><td>A3</td><td>C2</td></tr>
</table></td><td>• How did you decide on your approach to this problem? Was this the first approach you tried?
• Are there any other possible solutions to the problem?
• Could you have started the problem in a different way? If so, which ways are productive and which ways don't seem to work?</td></tr>
</table>

TEACHING AND LEARNING HIGH SCHOOL MATHEMATICS © 2010 John Wiley & Sons, Inc.

Question Response Support (QRS) Guide for How Many Oranges?

Questions or Learning Activities	Students' Responses (Expected or Observed)	Teacher Support (Using Stages of Questioning)
See How Many Oranges? student page. **Three friends picked oranges, then fell asleep. Melanie woke first and decided to leave. She took 1 orange, then one-third of what was left. Ben woke next. He took 2 oranges, then one-third of what was left. Colleen woke last. She took 1 orange and a third of what remained.** 1. What was the total number of oranges the friends picked? 2. How many oranges were left behind when Colleen departed? 3. Is more than one answer possible for the total number of oranges? If so, what other numbers are possible? Explain.	*Misconceived*	
	Partial	
	Satisfactory	

TEACHING AND LEARNING HIGH SCHOOL MATHEMATICS © 2010 John Wiley & Sons, Inc.

Contents

Except as noted, all resources in this appendix are also on the course website.

van Hiele Levels of Geometric Thought

Level 1: Visualization

At this level, students view geometric objects in their entirety, rather than considering attributes. For instance, a figure has a given shape based on its physical appearance. Students are able to reproduce figures and learn vocabulary.

Level 2: Analysis

Students begin to consider the attributes and characteristics of figures. Through examples, students are able to make generalizations for classes of figures. However, students have difficulty with interrelationships between figures, do not understand definitions, and do not appreciate proof.

Level 3: Informal Deduction

Students are able to deduce properties of various figures, deal with issues related to class inclusion, and understand definitions. Although students can follow formal proofs, they have difficulty constructing proofs on their own.

Level 4: Deduction

Students understand geometric structure, recognize the importance of proof, and understand the roles of definitions, theorems, and axioms. Students are able to create their own proofs and recognize that proofs may be completed in more than one way. Students rely on models to make sense of geometric concepts and proofs.

Level 5: Rigor

Students are able to understand axiomatic systems and study non-Euclidean geometries. They no longer need to rely on a particular model to understand and create proofs.

Adapted from Crowley, Mary L. "The van Hiele Model of the Development of Geometric Thought." In Mary Montgomery Lindquist and Albert P. Shulte (Eds.), *Learning and Teaching Geometry, K–12* (pp. 1–16). Reston, VA: National Council of Teachers of Mathematics, 1987 and Groth, Randall E. "Linking Theory and Practice in Teaching Geometry." *Mathematics Teacher*, 99 (August 2005): 27–30.

TEACHING AND LEARNING HIGH SCHOOL MATHEMATICS © 2010 John Wiley & Sons, Inc.

Levels of Demand

Lower-level demands (memorization):

- Involve either reproducing previously learned facts, rules, formulas, or definitions or committing facts, rules, formulas, or definitions to memory.
- Cannot be solved using procedures because a procedure does not exist or because the time frame in which the task is being completed is too short to use a procedure.
- Are not ambiguous. Such tasks involve the exact reproduction of previously seen materials, and what is to be produced is clearly and directly stated.
- Have no connection to the concepts or meaning that underlie the facts, rules, formulas, or definitions being learned or reproduced.

Lower-level demands (procedures without connections):

- Are algorithmic. Use of the procedure either is specifically called for or is evident from prior instruction, experience, or placement of the task.
- Require limited cognitive demand for successful completion. Little ambiguity exists about what needs to be done and how to do it.
- Have no connection to the concepts or meaning that underlie the procedure being used.
- Are focused on producing correct answers instead of on developing mathematical understanding.
- Require no explanations or explanations that focus solely on describing the procedure that was used.

Higher-level demands (procedures with connections):

- Focus students' attention on the use of procedures for the purpose of developing deeper levels of understanding of mathematical concepts and ideas.
- Suggest explicitly or implicitly pathways to follow that are broad general procedures that have close connections to underlying conceptual ideas as opposed to narrow algorithms that are opaque with respect to underlying concepts.
- Usually are represented in multiple ways, such as visual diagrams, manipulatives, symbols, and problem situations. Making connections among multiple representations helps develop meaning.
- Require some degree of cognitive effort. Although general procedures may be followed, they cannot be followed mindlessly. Students need to engage with conceptual ideas that underlie the procedures to complete the task successfully and develop understanding.

Higher-level demands (doing mathematics):

- Require complex and nonalgorithmic thinking—a predictable, well-rehearsed approach or pathway is not explicitly suggested by the task, task instructions, or a worked-out example.
- Require students to explore and understand the nature of mathematical concepts, processes, or relationships.
- Demand self-monitoring or self-regulation of one's own cognitive processes.
- Require students to access relevant knowledge and experiences and make appropriate use of them in working through the task.
- Require students to analyze the tasks and actively examine task constraints that may limit possible solution strategies and solutions.
- Require considerable cognitive effort and may involve some level of anxiety for the student because of the unpredictable nature of the solution process required.

These characteristics are derived from the work of Doyle on academic tasks (1988) and Resnick on high-level thinking skills (1987), the *Professional Standards for Teaching Mathematics* (NCTM 1991), and the examination and categorization of hundreds of tasks used in QUASAR classrooms (Stein, Grover, and Henningsen 1996; Stein, Lane, and Silver 1996).

From Smith, Margaret Schwan and Mary Kay Stein. "Selecting and Creating Mathematical Tasks: From Research to Practice." *Mathematics Teaching in the Middle School*, 3 (February 1998): 344–350. Reprinted with permission from *Mathematics Teaching in the Middle School*, ©1998 by the National Council of Teachers of Mathematics.

TEACHING AND LEARNING HIGH SCHOOL MATHEMATICS © 2010 John Wiley & Sons, Inc.

Link Sheet

Concept:	
Mathematics (Symbolic) Example	Everyday Example
Diagram/Picture/Graph	My Explanation

TEACHING AND LEARNING HIGH SCHOOL MATHEMATICS © 2010 John Wiley & Sons, Inc.

Frayer Model

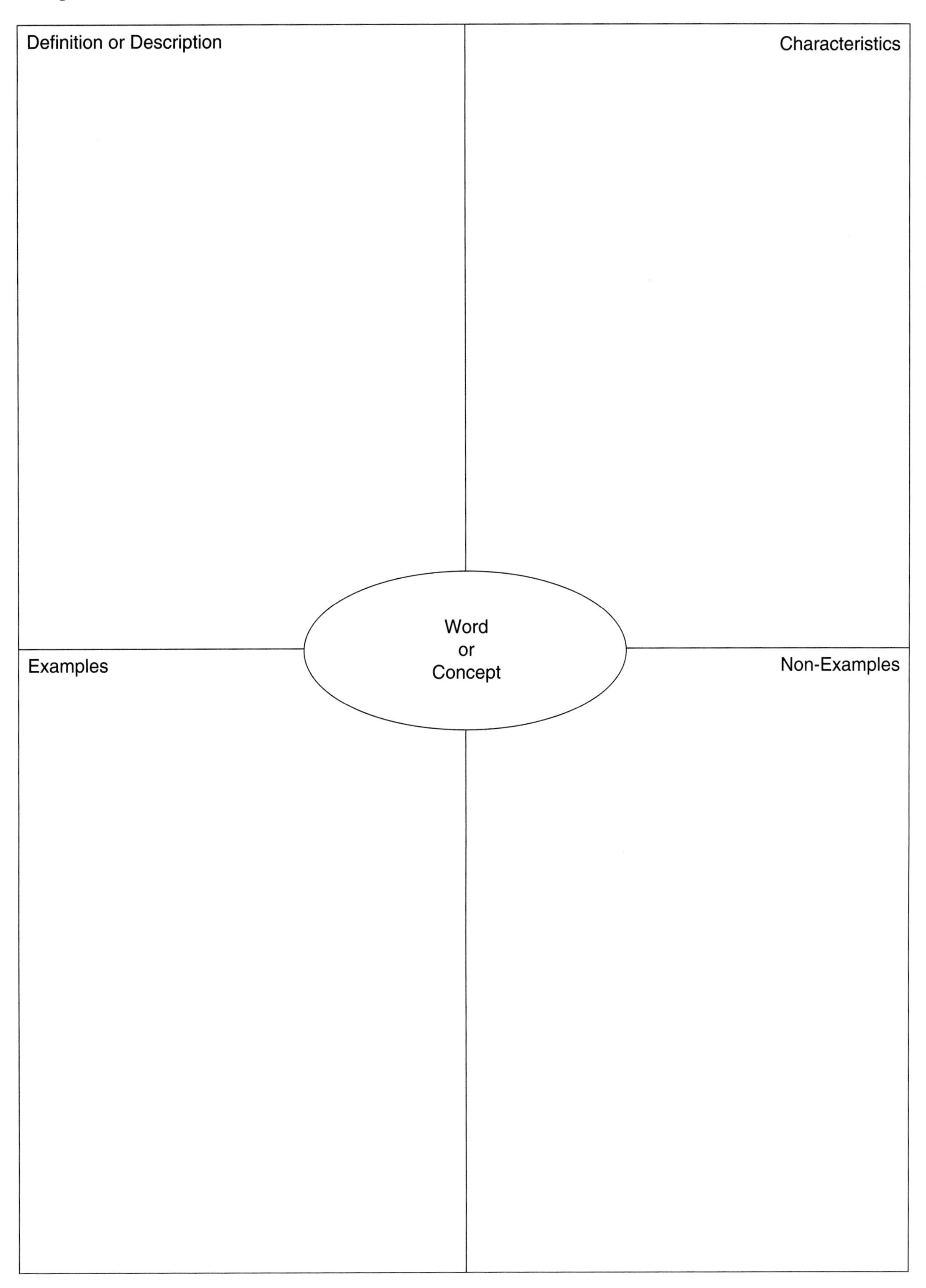

TEACHING AND LEARNING HIGH SCHOOL MATHEMATICS © 2010 John Wiley & Sons, Inc.

Circular Geoboard

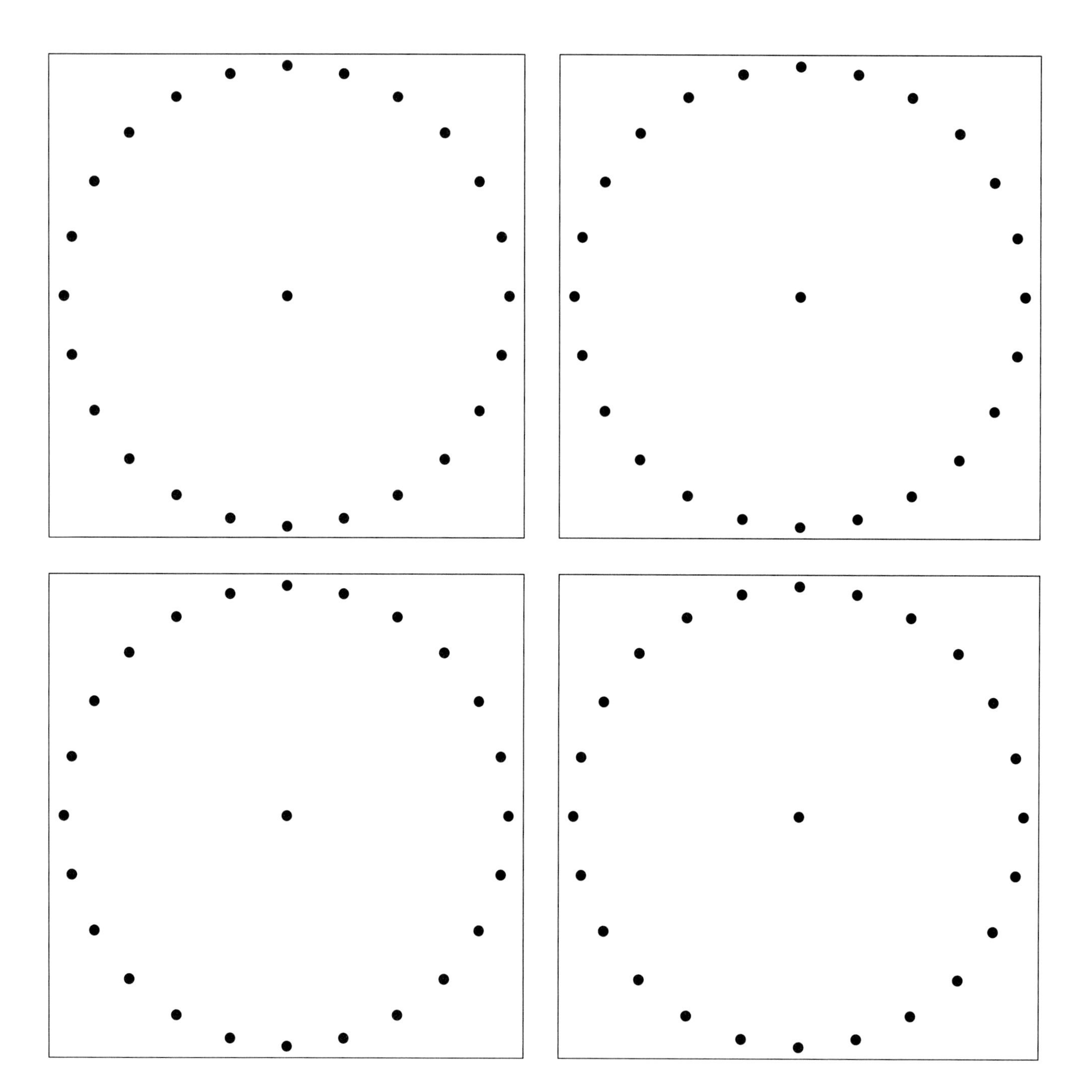

TEACHING AND LEARNING HIGH SCHOOL MATHEMATICS © 2010 John Wiley & Sons, Inc.

Square Geoboard

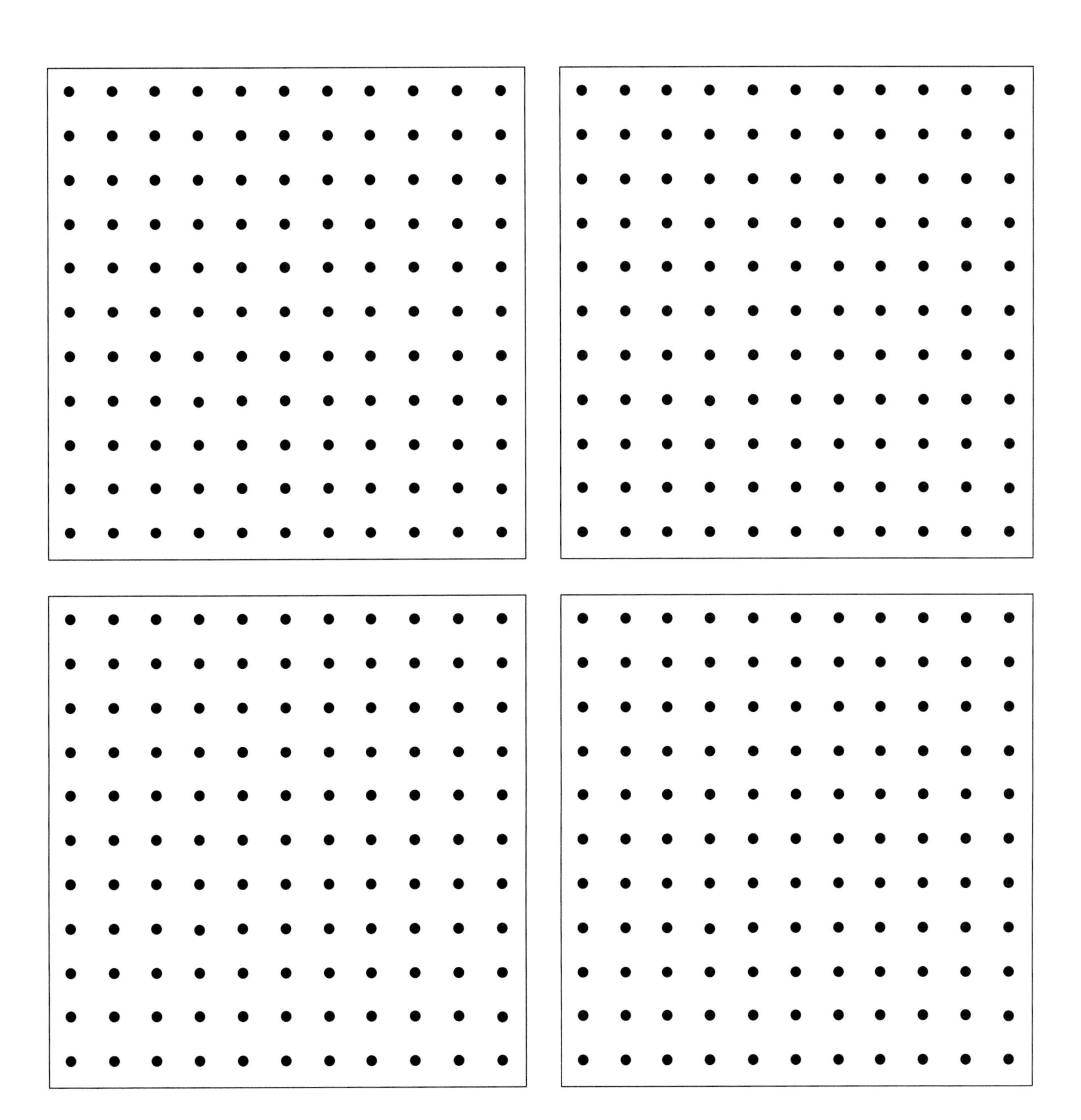

TEACHING AND LEARNING HIGH SCHOOL MATHEMATICS © 2010 John Wiley & Sons, Inc.

Cards for What Am I?

Quadrilateral Cards

Trapezoid	Isosceles Trapezoid	Kite	Rhombus
Parallelogram	Rectangle	Square	

Property Cards

At least one pair of parallel sides	Two distinct pairs of adjacent congruent sides	Equilateral
Two pairs of parallel sides	At least one pair of consecutive supplementary angles	Equiangular
At least one pair of opposite congruent sides	At least one pair of opposite congruent angles	Congruent base angles
Two pairs of opposite congruent sides	Two pairs of opposite congruent angles	Two distinct pairs of consecutive supplementary angles

TEACHING AND LEARNING HIGH SCHOOL MATHEMATICS © 2010 John Wiley & Sons, Inc.

QRS Guide for What Am I?

Questions or Learning Activities	Students' Responses (Expected or Observed)	Teacher Support (Using Stages of Questioning)
1. Play the game **What Am I?** until all of the **Quadrilateral Cards** have been used.	*Misconceived*: Students are randomly guessing quadrilateral names.	• Please read the rules again. • While students play the game, monitor teams to see that they are playing correctly.
	Partial: Student gives too many clues to help team members guess the quadrilateral.	•
	Satisfactory:	•
2. Use the results of the game to define each of the quadrilaterals based on relationships among their sides or angles. Record your work. • A trapezoid is . . .	*Misconceived*:	
	Partial: A trapezoid has at least one pair of parallel sides.	• Draw a hexagon with one pair of parallel sides. • Is this a trapezoid? • What is your definition missing? 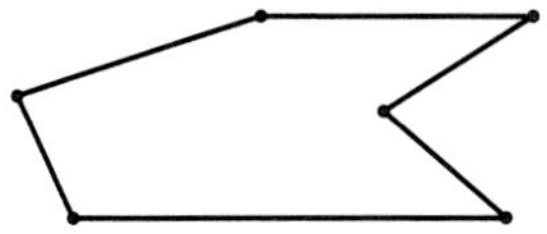
	Satisfactory: A trapezoid is a quadrilateral with a pair of parallel sides.	
• An isosceles trapezoid is . . .	*Misconceived*:	
	Partial: An isosceles trapezoid is a quadrilateral with a pair of congruent base angles.	• Draw a right trapezoid that is not a rectangle. • Is this an isosceles trapezoid? Why or why not? • What do you mean by base angles? 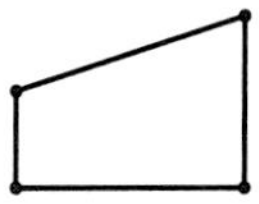
	Partial: An isosceles trapezoid is a quadrilateral with congruent base angles.	• Draw a general quadrilateral with a pair of congruent base angles whose opposite sides are not parallel. • Is this an isosceles trapezoid? • Draw a square. Is this an isosceles trapezoid? How do you know? 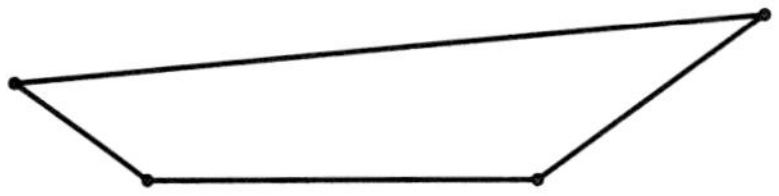
	Satisfactory: An isosceles trapezoid is a trapezoid with congruent base angles.	

(continued)

TEACHING AND LEARNING HIGH SCHOOL MATHEMATICS © 2010 John Wiley & Sons, Inc.

Questions or Learning Activities	Students' Responses (Expected or Observed)	Teacher Support (Using Stages of Questioning)
• A kite is . . .	*Misconceived*:	
	Partial:	
	Satisfactory:	
• A parallelogram is . . .	*Misconceived*:	
	Partial:	
	Satisfactory:	
• A rectangle is . . .	*Misconceived*:	
	Partial:	
	Satisfactory:	
• A rhombus is . . .	*Misconceived*:	
	Partial:	
	Satisfactory:	
• A square is . . .	*Misconceived*:	
	Partial:	
	Satisfactory:	

TEACHING AND LEARNING HIGH SCHOOL MATHEMATICS © 2010 John Wiley & Sons, Inc.

QRS Guide for Exploring the Diagonals of Quadrilaterals

Questions or Learning Activities	Students' Responses (Expected or Observed)	Teacher Support (Using Stages of Questioning)
1. Experiment with the diagonals of quadrilaterals by positioning the diagonals and connecting their endpoints. Try at least three different arrangements for each relationship listed. Determine which quadrilaterals must go in which cells.	*Misconceived*: For *diagonals bisect each other* several students also assume diagonals are perpendicular.	• What is meant by "bisect each other"? • Show me with the strips a pair of diagonals that bisect each other.
	Partial:	
	Satisfactory:	
• Keep track of students' responses. Ask questions of students as needed based on the difficulties shown at right. See answer key.	*Misconceived*: For *diagonals bisect each other*, student has listed "kite" under "Quadrilaterals whose Diagonals Might or Might Not Be Congruent."	• Tell me how you decided on this quadrilateral. • Does it fit both categories, diagonals might or might not be congruent and diagonals bisect each other?
	Partial: For *diagonals bisect each other*, student has listed "parallelogram" under Quadrilaterals whose "Diagonals Might or Might Not Be Congruent."	• How do you know this quadrilateral works? • What other quadrilaterals have you tried? How do you know they don't fit this category?
	Satisfactory: For *diagonals bisect each other*, student has listed "square" and "rectangle" under "Quadrilaterals whose Diagonals Are Congruent."	• How do you know this quadrilateral works?
2. Characterize the relationship between the diagonals for each of the following quadrilaterals. Write a definition for each of the quadrilaterals based on the relationships between their diagonals. A trapezoid is . . .	*Misconceived*:	
	Partial:	
	Satisfactory:	
An isosceles trapezoid is . . .	*Misconceived*:	
	Partial:	
	Satisfactory:	
A kite is . . .	*Misconceived*:	
	Partial:	
	Satisfactory:	
A parallelogram is . . .	*Misconceived*:	

(continued)

TEACHING AND LEARNING HIGH SCHOOL MATHEMATICS © 2010 John Wiley & Sons, Inc.

Questions or Learning Activities	Students' Responses (Expected or Observed)	Teacher Support (Using Stages of Questioning)
	Partial:	
	Satisfactory:	
A rectangle is . . .	*Misconceived*:	
	Partial:	
	Satisfactory:	
A rhombus is . . .	*Misconceived*:	
	Partial:	
	Satisfactory: a quadrilateral whose diagonals are perpendicular bisectors of each other.	
A square is . . .	*Misconceived*:	
	Partial:	
	Satisfactory:	
3a. From the relationships between the diagonals of the quadrilaterals, which categories of quadrilaterals belong to other categories? Explain briefly. **b.** Create a diagram showing a hierarchy of quadrilaterals.	*Misconceived*:	
	Partial:	
	Satisfactory:	

TEACHING AND LEARNING HIGH SCHOOL MATHEMATICS © 2010 John Wiley & Sons, Inc.

Sample Student Responses for Sprinkling around Obstacles

Sprinkling around Obstacles—Abe's Solution

1. The circle represents the tree. Construct the center of the circle. Label the center *T*. Show and explain your process and describe how you know it works.

   I measured the diameter and found another diameter of the same length. Their intersection is the center of the circle.

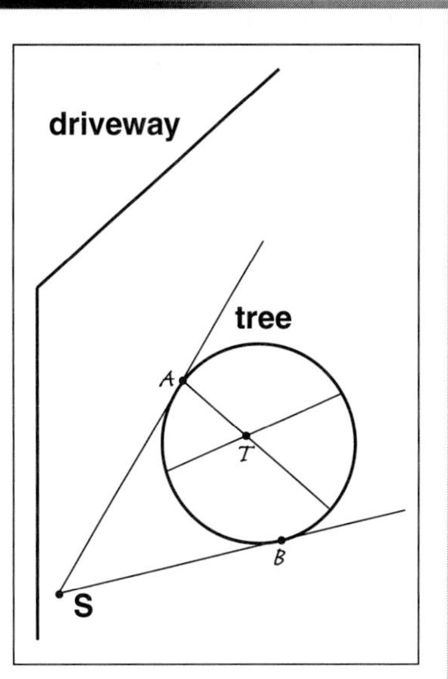

2. Construct tangent lines to the circle from the sprinkler head. Label the points of tangency *A* and *B*, respectively. Show and explain your process for finding the points of tangency and describe how you know your process works.

   Tangent lines are shown in the drawing.

3. Relative to the context of the problem, what do the tangent lines represent?

   Tangent lines show where the water hits the tree.

Sprinkling around Obstacles—Bev's Solution

1. The circle represents the tree. Construct the center of the circle. Label the center *T*. Show and explain your process and describe how you know it works.

   Construct a tangent line to the circle. The tangent line is perpendicular to the circle at the point of tangency. Do this again. The perpendicular lines intersect at the center of the circle.

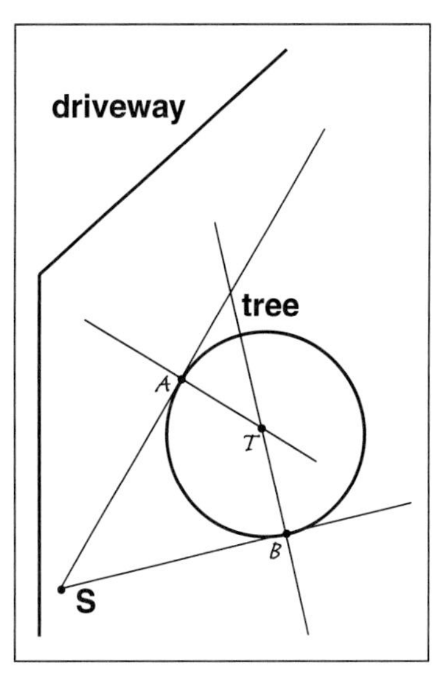

2. Construct tangent lines to the circle from the sprinkler head. Label the points of tangency *A* and *B*, respectively. Show and explain your process for finding the points of tangency and describe how you know your process works.

   Construct SA and SB infinitely past the tree to see where the sprinkler can't reach.

3. Relative to the context of the problem, what do the tangent lines represent?

   The part behind the tree between the points of tangency is where the sprinkler can't reach.

TEACHING AND LEARNING HIGH SCHOOL MATHEMATICS © 2010 John Wiley & Sons, Inc.

Sprinkling around Obstacles—Colin's Solution

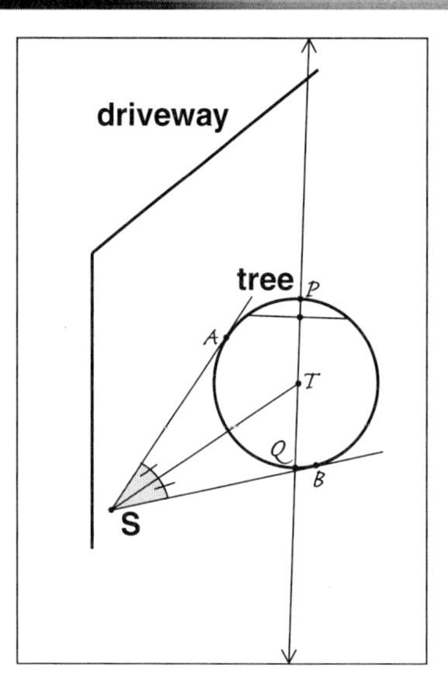

1. The circle represents the tree. Construct the center of the circle. Label the center *T*. Show and explain your process and describe how you know it works.

   To find the center, we draw a chord and construct its perpendicular bisector. There are two intersections of the perpendicular bisector with the circle. Call them P and Q. The midpoint of segment PQ is the center of the circle.

2. Construct tangent lines to the circle from the sprinkler head. Label the points of tangency *A* and *B*, respectively. Show and explain your process for finding the points of tangency and describe how you know your process works.

3. Relative to the context of the problem, what do the tangent lines represent?

   The part of the yard that isn't watered behind the tree lies between the tangent lines and the side of the tree away from the sprinkler.

Sprinkling around Obstacles—Darma's Solution

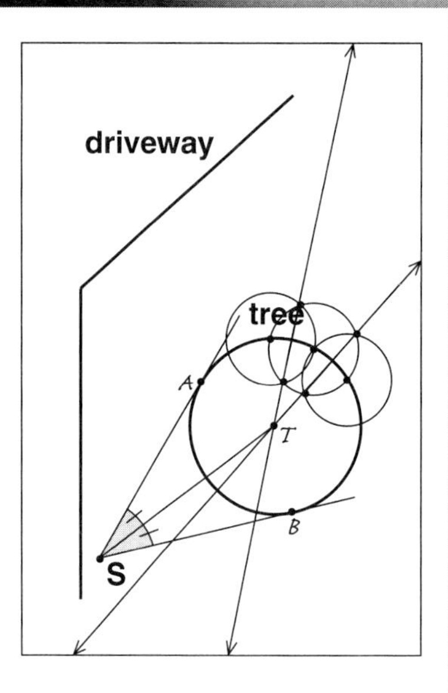

1. The circle represents the tree. Construct the center of the circle. Label the center *T*. Show and explain your process and describe how you know it works.

   Construct 3 circles, all with the same radius (see my construction). Find their intersection points and draw lines through each pair. The intersection of the two lines is the center of the circle.

2. Construct tangent lines to the circle from the sprinkler head. Label the points of tangency *A* and *B*,respectively. Show and explain your process for finding the points of tangency and describe how you know your process works.

   I know that tangent lines intersect the tree at 90 degree angles to the radius of a circle. Draw two lines tangent to the tree through point S.

3. Relative to the context of the problem, what do the tangent lines represent?

   The tree blocks the water between the tangent lines.

TEACHING AND LEARNING HIGH SCHOOL MATHEMATICS © 2010 John Wiley & Sons, Inc.

Sprinkling around Obstacles—Everest's Solution

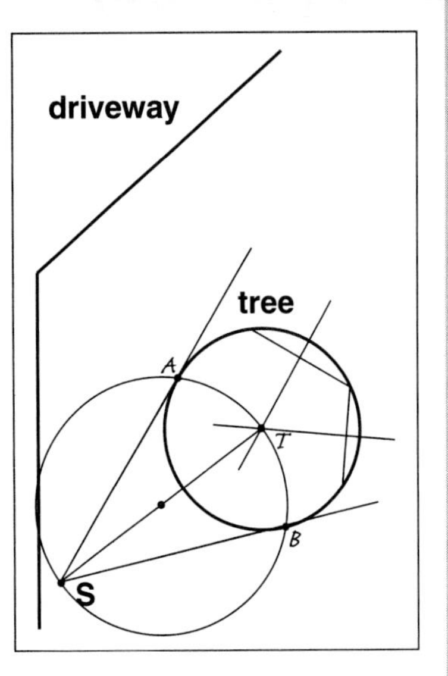

1. The circle represents the tree. Construct the center of the circle. Label the center *T*. Show and explain your process and describe how you know it works.

To construct the center, we first draw 2 chords and then determine the midpoints of the chords. We then draw a perpendicular to each chord at the midpoint. The intersection of these perpendiculars is the center.

2. Construct tangent lines to the circle from the sprinkler head. Label the points of tangency *A* and *B*, respectively. Show and explain your process for finding the points of tangency and describe how you know your process works.

We first create the segment TS and then construct the midpoint on the segment. Then we construct a circle using the midpoint as the center and the distance to S as the radius. The intersection of this new circle and the tree circle is the point of tangency for the line from S to the original circle.

3. Relative to the context of the problem, what do the tangent lines represent?

The tangent lines show the boundary of where the water can spray beyond the tangent points on the tree. Water can't reach the area behind the tree between the tangent lines, unless it is windy!

TEACHING AND LEARNING HIGH SCHOOL MATHEMATICS © 2010 John Wiley & Sons, Inc.

Sample Student Responses for Sprinkling around Obstacles When Using *The Geometer's Sketchpad*™

Basil: *I drew tangent lines and found the intersections with the circle. When I did this at first, I had 2 intersection points for each "tangent" so I moved each tangent line until the intersection was one point.*

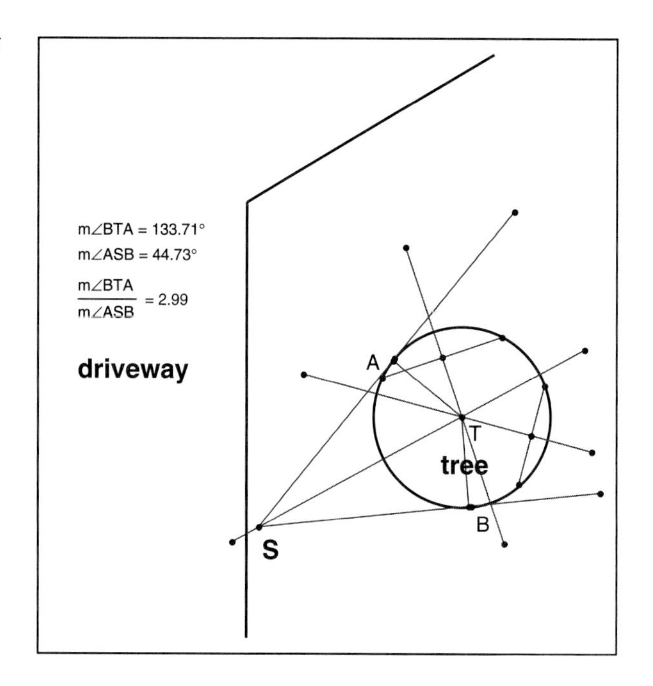

Camellia: *I constructed a radius and the perpendicular to the radius at the endpoint of the radius on the circle. I moved the endpoint on the radius until the perpendicular line contained point S. I did the same thing for a radius on the other side of the tree. Then I measured the lengths between S and the points of tangency and found that these lengths are just about the same. I assume the difference is because of the accuracy of the program.*

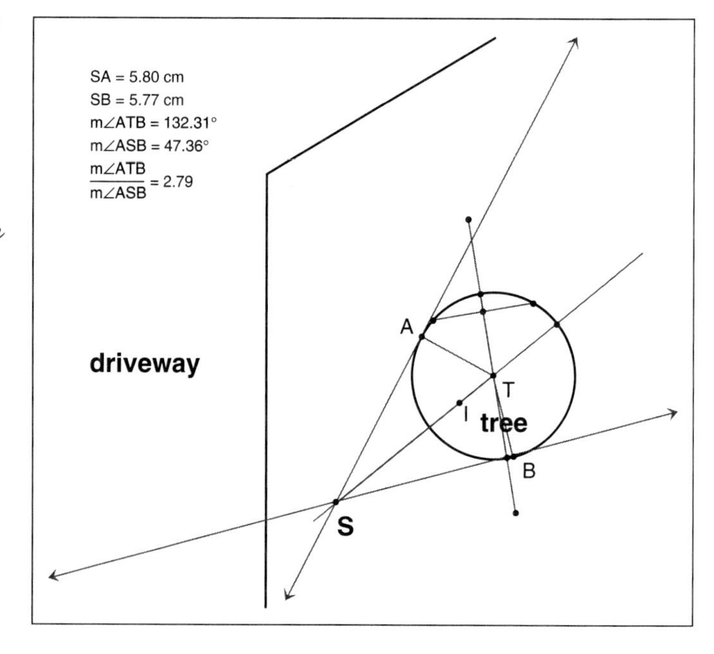

TEACHING AND LEARNING HIGH SCHOOL MATHEMATICS © 2010 John Wiley & Sons, Inc.

Dill: I know that the tangent line is perpendicular to a radius, so I have a right triangle and can use the Pythagorean Theorem to find the length of the side from S to the point of tangency. I then constructed a circle with center S whose radius is the desired length. The intersections of those circles give the points of tangency on both sides of the tree.

Because we have a right triangle, the acute angles are ALWAYS complementary. The lengths of segments AS and SB are the same.

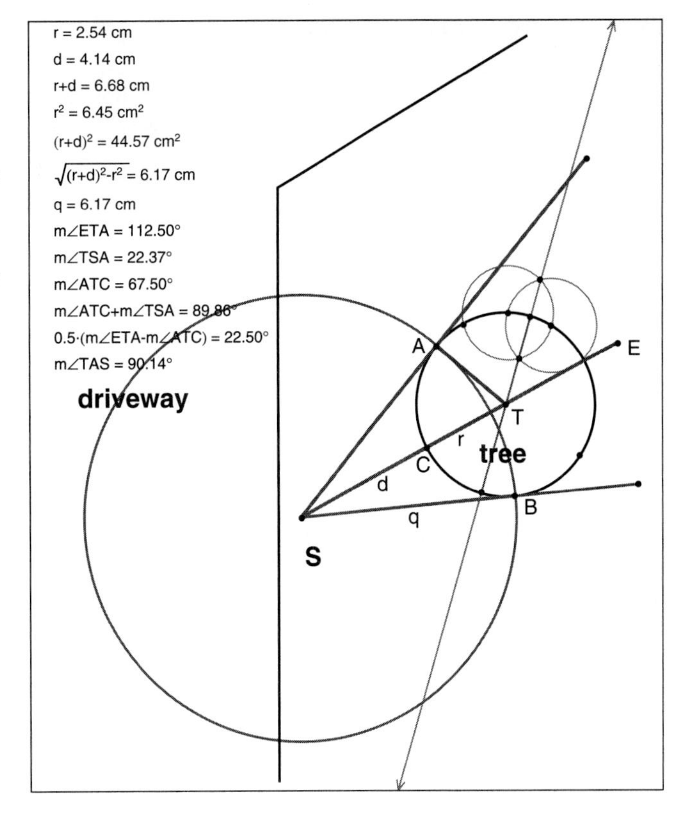

Fern: Because tangent lines are perpendicular to radii of circles, a right triangle is formed between the center of the tree, the point of tangency, and the sprinkler head. Using trig, I found the measure of the angle at S and rotated the ray ST the measure of the angle I found. My calculations are shown. The points of tangency must be the same distance from S (I rotated the same ray opposite directions by the same amount).

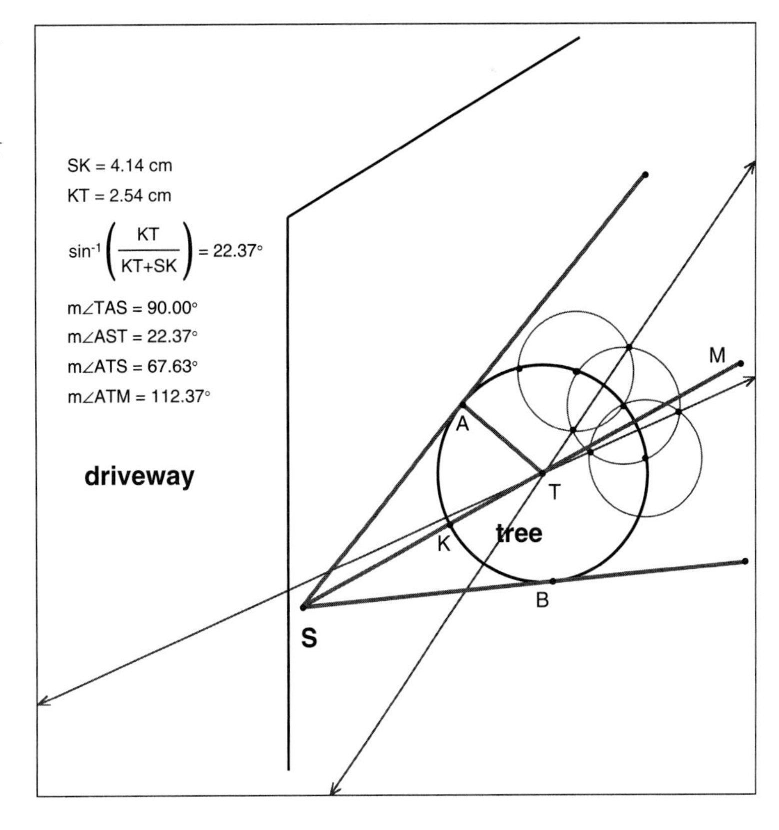

TEACHING AND LEARNING HIGH SCHOOL MATHEMATICS © 2010 John Wiley & Sons, Inc.

QRS Guide for Sprinkling around Obstacles

Questions or Learning Activities	Students' Responses (Expected or Observed)	Teacher Support (Using Stages of Questioning)
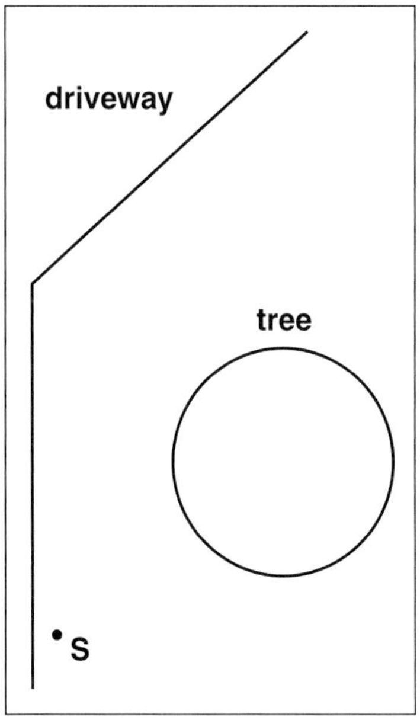 driveway tree S	*Misconceived*: Measure the diameter and find another diameter of the same length. Their intersection is the center of the circle.	• How do you measure the diameter? • How do you know the segment you have found is exactly the length of the diameter and not just a close approximation? • Have you constructed the center of the circle? Are the diameters "eye-balled" or constructed?
	Partial: I know I have to construct a secant to the circle and find its perpendicular bisector. I don't know where to go from here.	• Why do you think you need the perpendicular bisector of a secant? Student: "I remember doing this before." • How might the perpendicular bisector of a secant of the circle be useful? • How is the perpendicular bisector of a secant related to a circle?
	Satisfactory: I found the perpendicular bisector of a secant of the circle. This line coincides with the diameter of the circle. The part of this line whose endpoints are on the circle is the diameter. The midpoint of the diameter is the center of the circle.	
1. The circle represents the tree. Construct the center of the circle. Label the center T. Show and explain your process and describe how you know it works. **2.** Construct tangent lines to the circle from the sprinkler head. Label the points of tangency A and B, respectively. Show and explain your process for finding the points of tangency and describe how you know your process works.	*Misconceived*: Student eyeballs tangent lines.	• Have you constructed the tangent lines? • What is the difference between drawing and constructing tangent lines? • What relationships exist between tangent lines and circles? • Which of these relationships could help you construct a tangent line?
	Partial:	
	Satisfactory:	
3. Relative to the context of the problem, what do the tangent lines represent?	*Misconceived*:	
	Partial: Tangent lines are where the water hits the tree.	• Why do you think this?
	Satisfactory:	

TEACHING AND LEARNING HIGH SCHOOL MATHEMATICS © 2010 John Wiley & Sons, Inc.

Euclidean Geometry Postulates

P1. There is exactly one line through any two distinct points.

P2. The intersection of two lines is exactly one point.

P3. To every pair of distinct points, there corresponds a unique positive number. This number is called the distance between the two points.

P4. The points on a line can be placed in a correspondence with the real numbers such that:

- i. To every point of the line, there corresponds exactly one real number.
- ii. To every real number, there corresponds exactly one point on the line.
- iii. The distance between two distinct points is the absolute value of the difference of the corresponding real numbers.

P5. Through a point not on a given line, exactly one parallel line can be constructed to the given line (*Parallel Postulate*).

P6. To every angle, there corresponds a real number between 0 and 180.

P7. Let $\overrightarrow{AB}$ be a ray on the edge of the half-plane H. For every real number r between 0 and 180, there is exactly one ray $\overrightarrow{AP}$ with P in H such that $m\angle PAB = r$.

P8. If point B is on $\overline{AC}$ and between points A and C, then $AB + BC = AC$ (*Segment Addition Postulate*).

P9. If point D lies in the interior of ABC, then $m\angle ABD + m\angle DBC = m\angle ABC$ (*Angle Addition Postulate*).

P10. If two angles form a linear pair, then they are supplementary (*Linear Pair Postulate*).

P11. If two parallel lines are cut by a transversal, then the alternate interior angles are congruent. Conversely, if two lines are cut by a transversal forming congruent alternate interior angles, then the lines are parallel (*AIA Postulate*).

P12. If two sides and the included angle of one triangle are congruent to the corresponding parts of another triangle (or the first triangle), then the correspondence is a congruence (*SAS Postulate*).

References

Coxford, Arthur, Zalman Usiskin, and Daniel Hirschhorn. *The University of Chicago School Mathematics Project: Geometry.* Glenview, IL: ScottForesman, 1991.

Serra, Michael. *Discovering Geometry: An Inductive Approach* (Second Edition). Emeryville, CA: Key Curriculum Press, 1997.

© 2010 John Wiley & Sons, Inc.

Contents

- **Function Card Sort** (for use with the Unit Three Team Builder and Investigation)
- **Function Card Sort Answer Template**
- **Algebra Tiles**
- **Coordinate Grid**

All resources in this appendix are also on the course website.

Function Card Sort

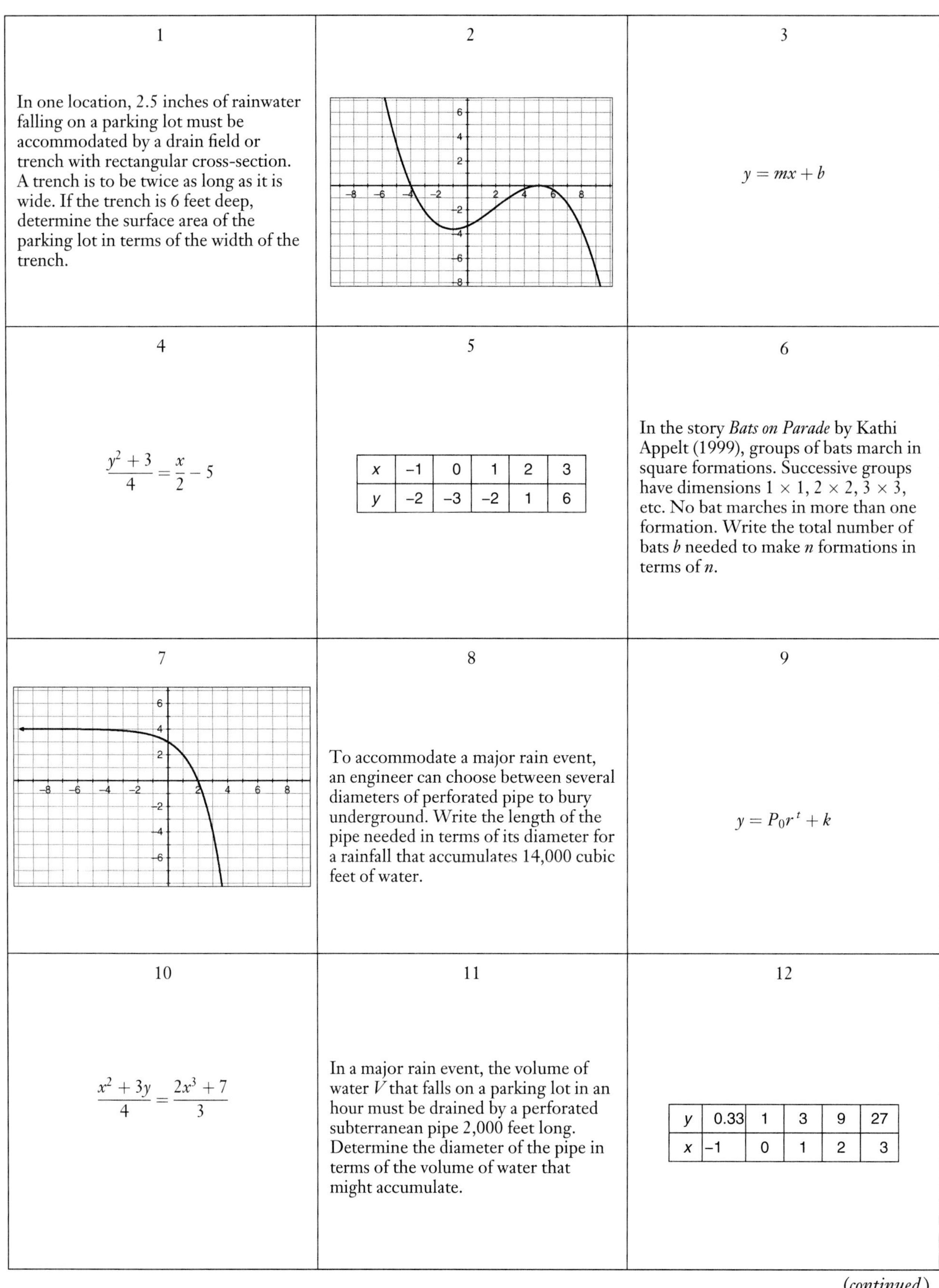

1

In one location, 2.5 inches of rainwater falling on a parking lot must be accommodated by a drain field or trench with rectangular cross-section. A trench is to be twice as long as it is wide. If the trench is 6 feet deep, determine the surface area of the parking lot in terms of the width of the trench.

2

3

$$y = mx + b$$

4

$$\frac{y^2 + 3}{4} = \frac{x}{2} - 5$$

5

x	−1	0	1	2	3
y	−2	−3	−2	1	6

6

In the story *Bats on Parade* by Kathi Appelt (1999), groups of bats march in square formations. Successive groups have dimensions 1×1, 2×2, 3×3, etc. No bat marches in more than one formation. Write the total number of bats b needed to make n formations in terms of n.

7

8

To accommodate a major rain event, an engineer can choose between several diameters of perforated pipe to bury underground. Write the length of the pipe needed in terms of its diameter for a rainfall that accumulates 14,000 cubic feet of water.

9

$$y = P_0 r^t + k$$

10

$$\frac{x^2 + 3y}{4} = \frac{2x^3 + 7}{3}$$

11

In a major rain event, the volume of water V that falls on a parking lot in an hour must be drained by a perforated subterranean pipe 2,000 feet long. Determine the diameter of the pipe in terms of the volume of water that might accumulate.

12

y	0.33	1	3	9	27
x	−1	0	1	2	3

(continued)

TEACHING AND LEARNING HIGH SCHOOL MATHEMATICS © 2010 John Wiley & Sons, Inc.

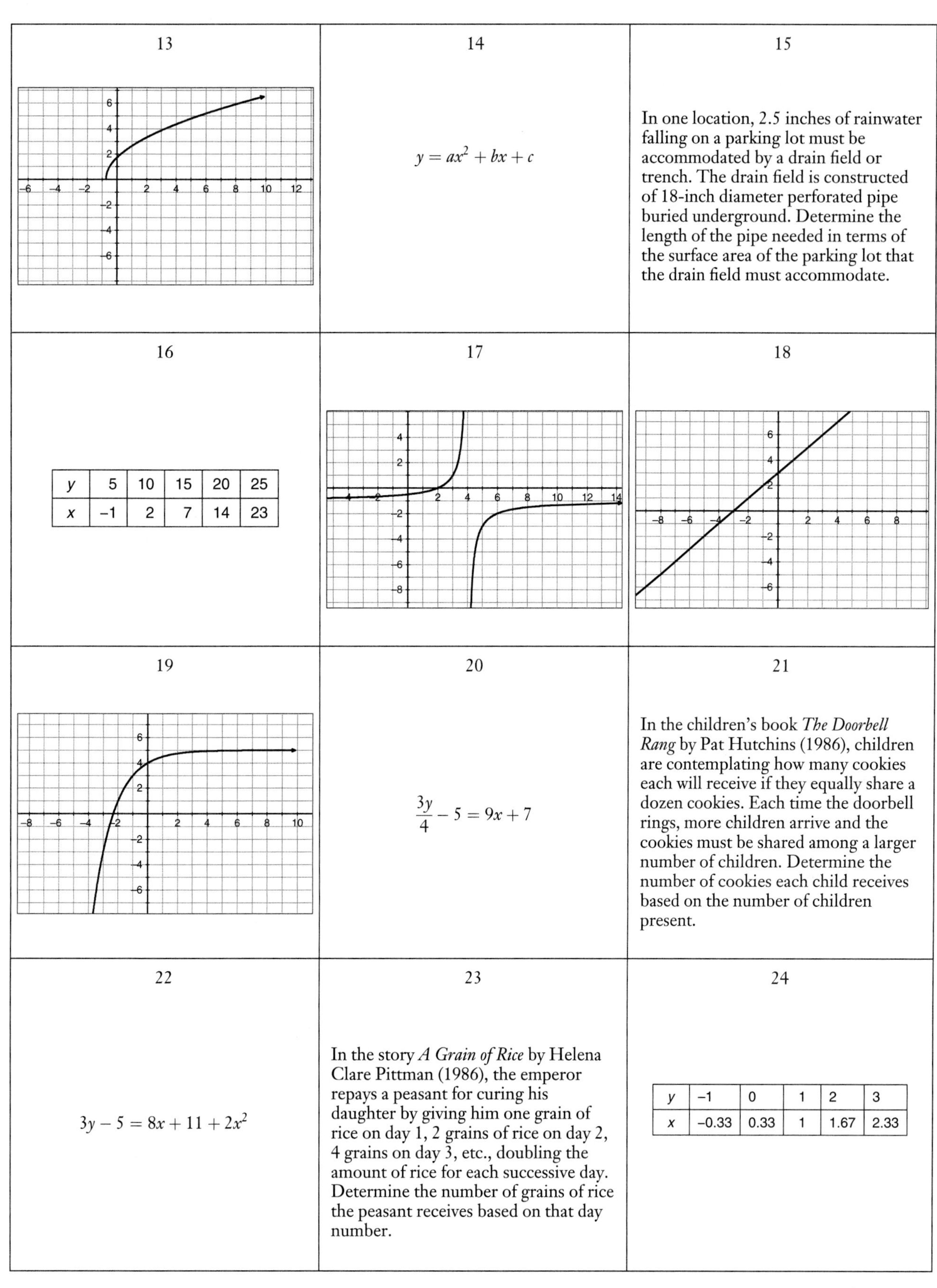

13

14

$y = ax^2 + bx + c$

15

In one location, 2.5 inches of rainwater falling on a parking lot must be accommodated by a drain field or trench. The drain field is constructed of 18-inch diameter perforated pipe buried underground. Determine the length of the pipe needed in terms of the surface area of the parking lot that the drain field must accommodate.

16

y	5	10	15	20	25
x	−1	2	7	14	23

17

18

19

20

$$\frac{3y}{4} - 5 = 9x + 7$$

21

In the children's book *The Doorbell Rang* by Pat Hutchins (1986), children are contemplating how many cookies each will receive if they equally share a dozen cookies. Each time the doorbell rings, more children arrive and the cookies must be shared among a larger number of children. Determine the number of cookies each child receives based on the number of children present.

22

$$3y - 5 = 8x + 11 + 2x^2$$

23

In the story *A Grain of Rice* by Helena Clare Pittman (1986), the emperor repays a peasant for curing his daughter by giving him one grain of rice on day 1, 2 grains of rice on day 2, 4 grains on day 3, etc., doubling the amount of rice for each successive day. Determine the number of grains of rice the peasant receives based on that day number.

24

y	−1	0	1	2	3
x	−0.33	0.33	1	1.67	2.33

(continued)

TEACHING AND LEARNING HIGH SCHOOL MATHEMATICS © 2010 John Wiley & Sons, Inc.

25

In the children's book *My Full Moon Is Square* by Elinor Pinczes (2002), fireflies arrange themselves in a square formation with f fireflies in the array. What is the length of a side, s, of the square in terms of the number of fireflies in the array?

26

x	–1	0	1	2	3
y	–6.5	–5	–3.5	–2	–0.5

27

$$y = \frac{A}{Bx + C} + D$$

28

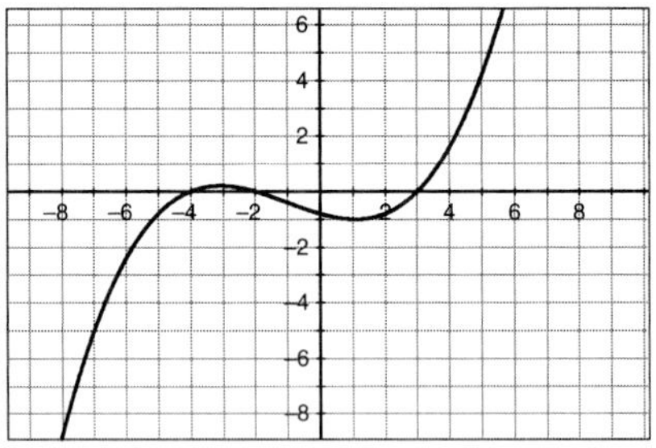

29

y	1	2	3	4	5
x	72	36	24	18	14.4

30

In the book *Counting Crocodiles* by Judy Sierra (1997), crocodiles appear one group per pair of pages. Each group of crocodiles contains one more crocodile than the previous group. Write the total number of crocodiles that have appeared in the book by the nth pair of pages as a function of n.

31

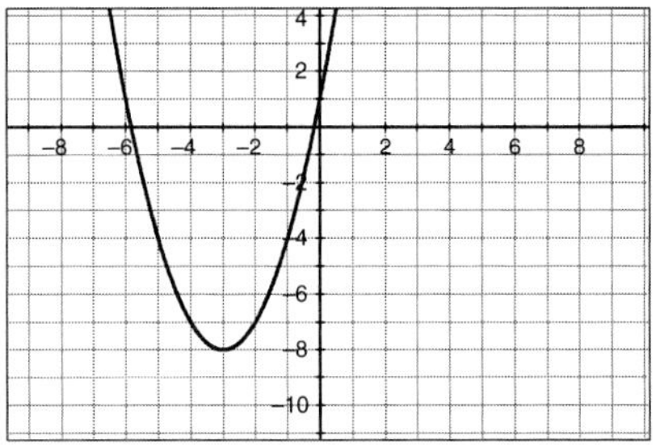

32

$$2y - 15 = 4^{x-3}$$

33

x	–3	–2	1	6	13
y	0	1	2	3	4

34

x	8	4	2	1	–1
y	3	6	12	24	–24

35

A septic tank whose diameter equals its height is buried underground. Determine the tank's volume in terms of its diameter.

36

x	–1	0	1	2	3
y	–2.6	–0.8	1.4	4	7

(continued)

TEACHING AND LEARNING HIGH SCHOOL MATHEMATICS © 2010 John Wiley & Sons, Inc.

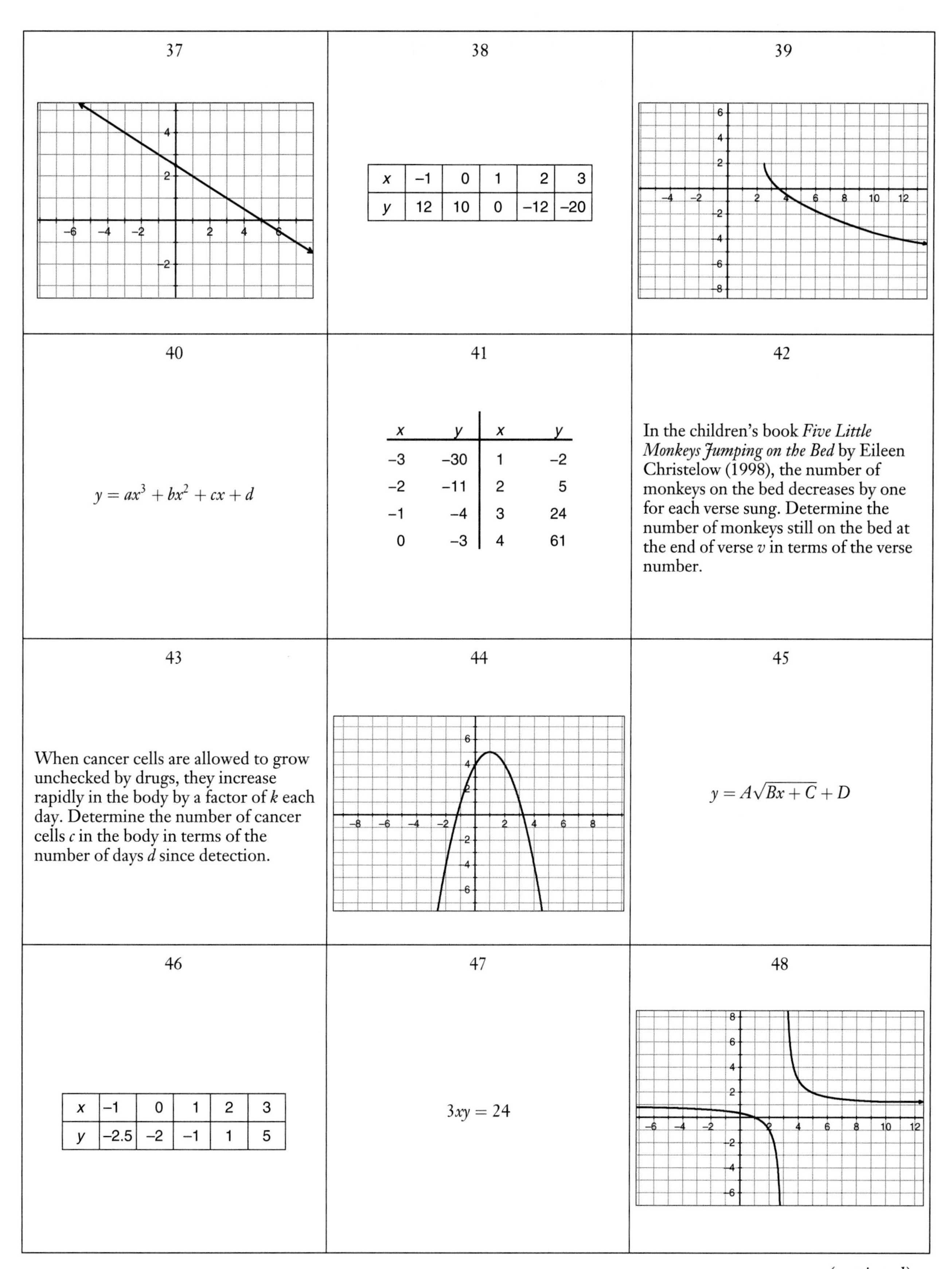

37

38

x	−1	0	1	2	3
y	12	10	0	−12	−20

39

40

$$y = ax^3 + bx^2 + cx + d$$

41

x	y	x	y
−3	−30	1	−2
−2	−11	2	5
−1	−4	3	24
0	−3	4	61

42

In the children's book *Five Little Monkeys Jumping on the Bed* by Eileen Christelow (1998), the number of monkeys on the bed decreases by one for each verse sung. Determine the number of monkeys still on the bed at the end of verse v in terms of the verse number.

43

When cancer cells are allowed to grow unchecked by drugs, they increase rapidly in the body by a factor of k each day. Determine the number of cancer cells c in the body in terms of the number of days d since detection.

44

45

$$y = A\sqrt{Bx + C} + D$$

46

x	−1	0	1	2	3
y	−2.5	−2	−1	1	5

47

$$3xy = 24$$

48

(continued)

TEACHING AND LEARNING HIGH SCHOOL MATHEMATICS © 2010 John Wiley & Sons, Inc.

Children's Literature References for Function Cards

Appelt, Kathi. *Bats on Parade.* New York: HarperCollins, 1999.
Christelow, Eileen. *Five Little Monkeys Jumping on the Bed.* New York: Clarion Books, 1998.
Hutchins, Pat. *The Doorbell Rang.* New York: Mulberry Books, 1986.
Pinczes, Elinor J. *My Full Moon Is Square*. Boston: Houghton Mifflin, 2002.
Pittman, Helena Clare. *A Grain of Rice.* New York: Bantam Skylark Books, 1986.
Sierra, Judy. *Counting Crocodiles.* Orlando, FL: Harcourt, 1997.

TEACHING AND LEARNING HIGH SCHOOL MATHEMATICS © 2010 John Wiley & Sons, Inc.

Function Card Sort Answer Template

Directions

Indicate the numbers of the **Function Card Sort** cards that fit in each cell.

Representation	Linear	Quadratic	Cubic	Exponential	Square Root	Rational Reciprocal
Equations						
Graphs						
Tables						
Children's Story						
Real Context						

TEACHING AND LEARNING HIGH SCHOOL MATHEMATICS © 2010 John Wiley & Sons, Inc.

Algebra Tiles—Units

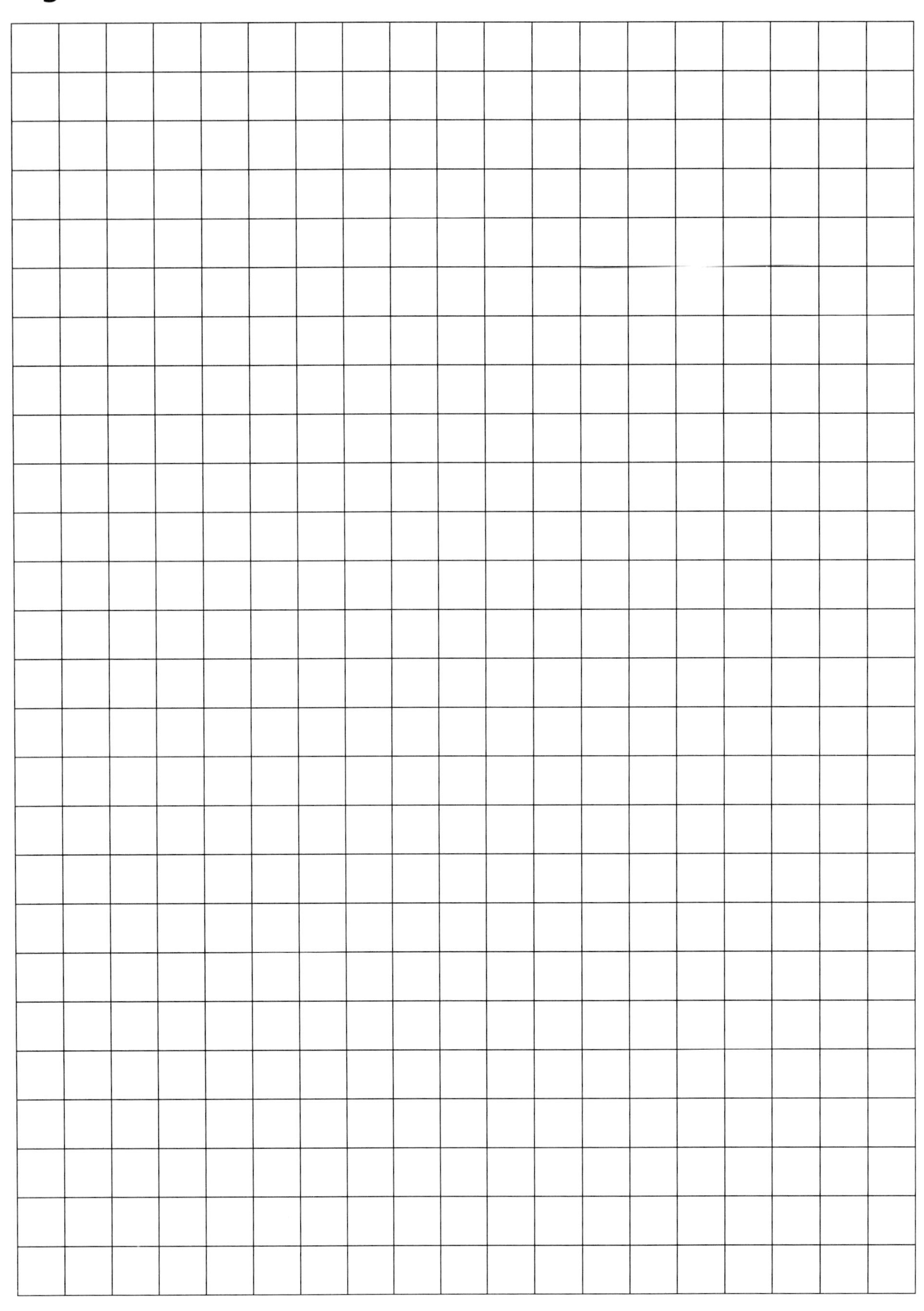

TEACHING AND LEARNING HIGH SCHOOL MATHEMATICS © 2010 John Wiley & Sons, Inc.

Algebra Tiles—1 by x

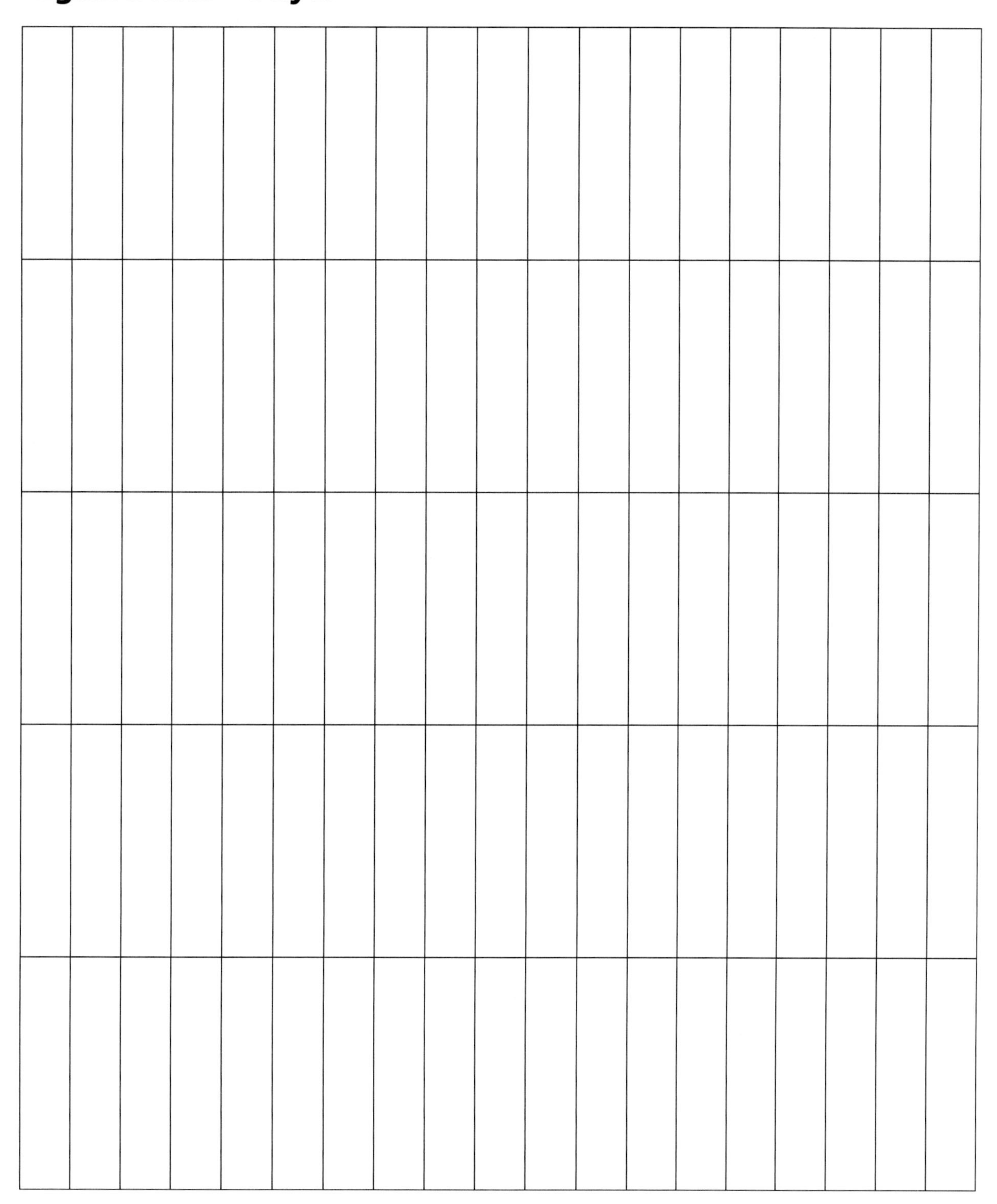

TEACHING AND LEARNING HIGH SCHOOL MATHEMATICS © 2010 John Wiley & Sons, Inc.

Algebra Tiles—x by x

TEACHING AND LEARNING HIGH SCHOOL MATHEMATICS © 2010 John Wiley & Sons, Inc.

Coordinate Grid

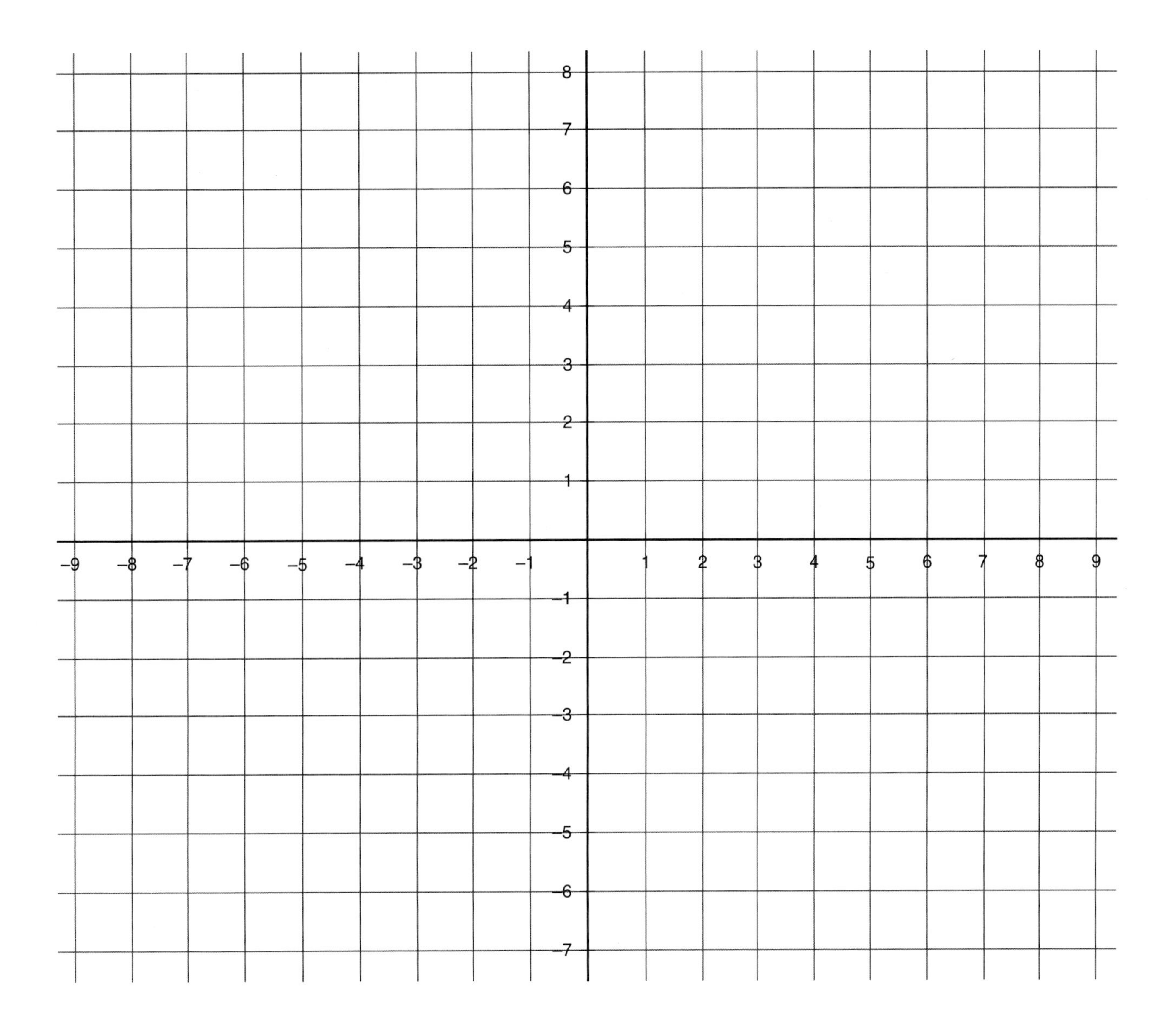

TEACHING AND LEARNING HIGH SCHOOL MATHEMATICS © 2010 John Wiley & Sons, Inc.

APPENDIX: UNIT FOUR

Contents

Except as noted, all resources in this appendix are also on the course website.

Lesson Planning Guide

Lesson Title: ____

Grade Level: ____________________ **Number of Class Periods:** ____________________

Text or Resource and Relevant Pages: ______________________________

Overview:

Preparation:

Continuity:

Previous Lessons	This Lesson	Next Lessons

Learning Outcomes:

Materials:

Use of Space:

Launch:

Explore: Include the following, as needed, in the **QRS Guide** at appropriate points.

Possible Segments of Direct Instruction:

Embedded Assessment:

Tone Setting:

Cautious Points:

Share and Summarize:

Homework or Extension:

Accommodations:

Assessments:

Attachments:

Reflection:

TEACHING AND LEARNING HIGH SCHOOL MATHEMATICS © 2010 John Wiley & Sons, Inc.

QRS Guide for ________________

Questions or Learning Activities	Students' Responses (Expected or Observed)	Teacher Support (Using Stages of Questioning)	Teacher Notes
Launch:			
Explore: Statement of Task 1	*Misconceived*		
	Partial		
	Satisfactory		
Explore: Statement of Task 2	*Misconceived*		
	Partial		
	Satisfactory		
Share and Summarize Questions			

TEACHING AND LEARNING HIGH SCHOOL MATHEMATICS © 2010 John Wiley & Sons, Inc.

Annotated Lesson Planning Guide

Items in italics are to be taken into consideration while planning, but not necessarily responded to individually in writing. Not all questions apply to every lesson.

Lesson Title: ______________________________

Grade Level: ______________ **Number of Class Periods:** ______________

Text or Resource and Relevant Pages: ______________________________

Overview: What will students do and learn? What are the mathematical points of the lesson?

Preparation: Use the **QRS Guide** provided ahead. (**QRS** for **Question**, **Response**, and **Support**)

- **a.** State the task as it will be posed for students; provide any sources used.
- **b.** Provide one clearly labeled satisfactory (correct) solution with full reasoning.
- **c.** Provide several other expected student solutions, including some that are misconceived, partial, or creative solutions using alternative reasoning. *Think about students who are "getting it," "struggling," or "ready for more challenge."*
- **d.** *If possible, have some students attempt the task and use their work to generate expected responses. Maintain a file of unusual or unexpected responses.*

Continuity:

Previous Lessons	This Lesson	Next Lessons
What did students learn previously related to this lesson?	What will students learn in this lesson?	What will students learn in the upcoming lessons related to this lesson?

Learning Outcomes:

- **a.** What specific mathematics will students develop, practice, or deepen during this lesson? (Key these to local, state, or national standards.)
- **b.** What ongoing learning outcomes (mathematical processes, attitudes, or collaborative skills or concepts) will students continue to build?

Materials: List all materials needed. *What special materials may be required (e.g., student pages, transparencies, physical materials, technology)? What supplies are students expected to have daily (e.g., graph paper, calculators, rulers, compasses, protractors)? What additional supplies do students need for this lesson?*

Use of Space: *What special space arrangements might be made (e.g., rearrangements of desks, use of computers, lab, hall space, library, or outdoor space)?*

Launch: Indicate how you will introduce the task. *How will you motivate students? What segue might be provided from earlier work? Is there a literature segment, everyday experience, visual aid, or other device that will spark students' interest or curiosity? What will you do to be sure students have the necessary prerequisites to understand the* **Explore** *task? What will you do to refrain from giving away too much? If there is a game, how will you help students understand the rules? How will you ensure that the Launch is brief?*

Explore: Continue to develop the **QRS Guide** with **Teacher Support** for each expected student response. Using Stages of Questioning, plan questions or responses that move students from where they are to a little higher level. Include rescaled tasks for students needing more support or ready for more challenge in the **QRS Guide** as appropriate.

The **QRS Guide** *is a tool to use during the lesson to collect further student responses and to help you plan for the* **Share and Summarize**. *When you see responses you had anticipated in your planning, use the* **Teacher Notes** *column to note which students used them. In addition, take notes on other unexpected responses that you observe or hear during the lesson and which student or team generated them. Make other*

TEACHING AND LEARNING HIGH SCHOOL MATHEMATICS © 2010 John Wiley & Sons, Inc.

notes as needed in the **Teacher Notes** *column while observing students. (These notes will also be helpful for planning similar lessons in the future.)*

Possible segments of direct instruction: *What brief segment of direct instruction might be appropriate during the* **Explore** *? Detail when you anticipate the need for an interruption and what you will say and do. If such segments do not include open questions for students, they can be written in paragraph form and inserted between* ***QRS*** *segments.*

Embedded Assessment: *What will you look and listen for during the* **Explore** *to know that students are developing the ideas/processes sought?*

Tone-setting: *How you will acknowledge to students, appropriately and in a timely manner, that the work is challenging and that they may be struggling with it? What will you say or do to help build students' confidence in their ability and their persistence to stay engaged in a difficult task?*

Cautious points: *What might you want to remember that requires particular attention in teaching this lesson? Add these where appropriate in the* **Teacher Notes** *column.*

Share and Summarize: *This is a critical phase when you want to support all students in clarifying the major ideas arising from the task. You need to think about the summary as you plan but also be prepared to continue to think about it as you observe students during the* **Explore**. *Use the lesson's learning outcomes to identify key ideas you want highlighted. If there is a pervasive misconception, include that response and be ready to ask questions to make certain the misconception is cleared up. If there are important relationships among different approaches, include at least one pair of solutions to have students compare and contrast. Most lessons include a strong, well developed solution as well.*

Identify in the **QRS Guide** column for **Teacher Notes** which samples of student work you want shared publicly and the sequence in which you anticipate having those responses presented. *Will you have a strong solution presented? If so, at what point? What about the solution will you highlight? Continue to keep notes as the lesson progresses of things you want to do or keep in mind.*

Extend the **QRS Guide** to pose a few major questions in the **Summary** to focus students on the major mathematical ideas to be derived from their work. Provide correct responses to those questions.

Things to think about in planning a summary:

- *How will you orchestrate the summary so students draw big ideas from the exploration or activity? What questions will elicit key concepts, representations, and processes?*
- *How will you help students integrate new learning with previous learning?*
- *How will you help students extend or generalize their understanding?*
- *What ideas in the lesson might you expect will not be fully developed and will need to be revisited in later instruction?*

Homework or Extension: Provide the homework or follow-up activity to be assigned.
Identify or design related learning activities, writing assignments, journal questions, or problems that students can do independently in class or at home to continue to develop, practice, apply, or extend their learning. Identify activities that set up a subsequent lesson.

Accommodations: *What considerations need to be made in light of students' different learning styles? What adaptations are needed for students with special needs (e.g., physical disabilities, emotional or behavior impairments)? Are there language issues or vocabulary that need attention for English language learners? Insert these, as needed, in appropriate places in the lesson plan or* **QRS Guide**.

Attachments: Include all materials needed for this lesson: e.g., relevant text or resource pages, presentation slides or transparencies, student pages, games with directions, or any other materials that clarify precisely what is planned.

Reflection: *After teaching the lesson, reflect on student progress and the effectiveness of teaching elements.*

a. *How are students progressing toward major mathematical concepts and processes?*

b. *How are lesson elements promoting learning (e.g., materials, tasks, directions, questions, guided discovery, open-ended explorations, brief segments of direct instruction, organization, management, uses of cooperative learning, assessment processes, other)?*

c. *How can I use what I have learned today to prepare more effectively for subsequent lessons?*

d. *What changes would make this plan more effective? Take the time after each use of the lesson to incorporate the changes.*

TEACHING AND LEARNING HIGH SCHOOL MATHEMATICS © 2010 John Wiley & Sons, Inc.

QRS Guide for ______________

[Insert lesson title or text section. Add additional sheets as needed.]

Materials:

Questions or Learning Activities	Students' Responses (Expected or Observed)	Teacher Support (Using Stages of Questioning)	Teacher Notes
Launch:			
Explore: Statement of Task 1	*Misconceived responses* (involves a misconception or notable mathematical error)		
	Partial responses (includes incomplete progress toward a solution)		
	One or more *satisfactory* (correct) *responses* with full reasoning.		
Explore: Statement of Task 2	*Misconceived*		
	Partial		
	Satisfactory		
Share and Summarize: Questions	Important responses sought		

TEACHING AND LEARNING HIGH SCHOOL MATHEMATICS © 2010 John Wiley & Sons, Inc.

Predicting Old Faithful Sample Student Work

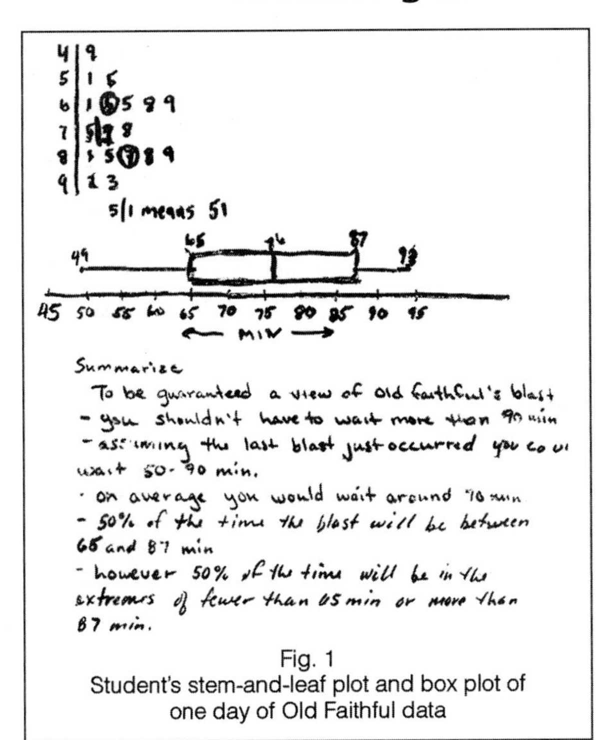

Fig. 1
Student's stem-and-leaf plot and box plot of one day of Old Faithful data

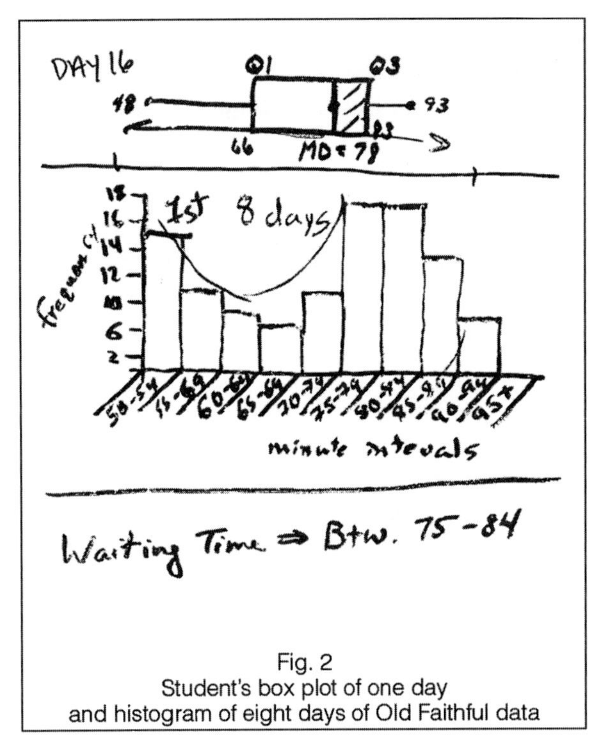

Fig. 2
Student's box plot of one day and histogram of eight days of Old Faithful data

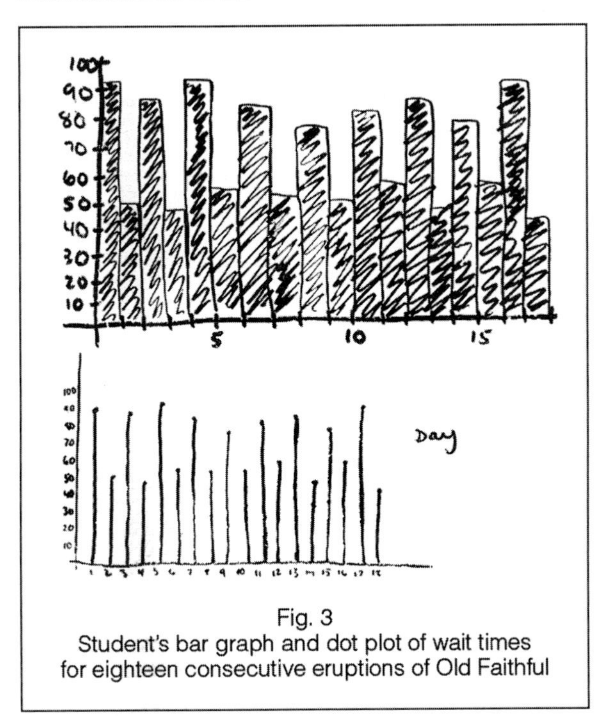

Fig. 3
Student's bar graph and dot plot of wait times for eighteen consecutive eruptions of Old Faithful

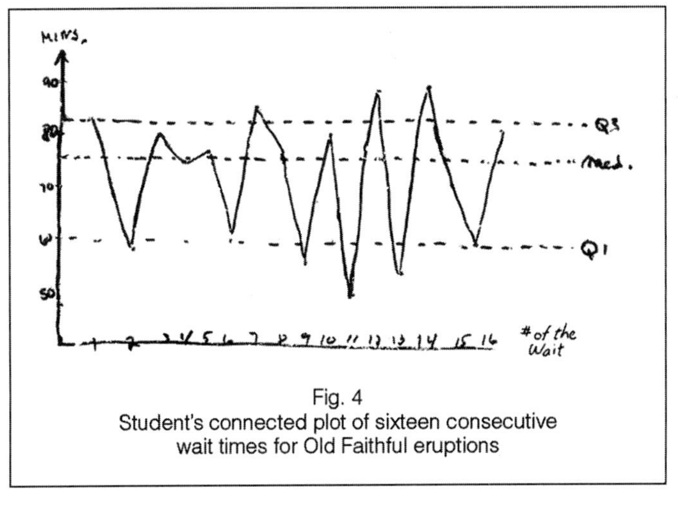

Fig. 4
Student's connected plot of sixteen consecutive wait times for Old Faithful eruptions

From " How Faithful Is Old Faithful? Statistical Thinking: A Story of Variation and Prediction" by Shaughnessy and Pfannkuch (*Mathematics Teacher*, 95, (April 2002): 252–259). Reprinted with permission of the *Mathematics Teacher* ©2002 by the National Council of Teachers of Mathematics.

Sample Lesson Plan for Predicting Old Faithful

Grade Level: 9 **Number of Class Periods:** 1

Text or Resource and Relevant Pages: *Predicting Old Faithful* student page, TLHSM, Lesson 4.1

Overview: Students examine data on times between eruptions of Old Faithful to predict how long a wait time might be from one eruption to the next. Though there are opportunities to review measures of center, measures of spread, and different forms of graphing (e.g., stem plot, boxplot, histogram), the main idea is to address the concept of variation: (1) predicting an interval is more meaningful than predicting a single value and (2) some displays reveal a distinct pattern of variation, alternating between longer and shorter wait times. Measures of center alone mask important information.

Preparation: See **QRS Guide**.

Continuity:

Previous Lessons	This Lesson	Next Lessons
Students have examined data, learned to display data in different forms, and worked with measures of center.	Students see the payoff in choosing displays judiciously and the idea of variability, both of which are important in working with data.	Students will work with sampling and measures of spread (e.g., range, interquartile range, standard deviation).

Learning Outcomes:

- Students will compare relative benefits of different displays of data.
- Students will recognize variability as a significant component of data analysis.

Materials:

- Copies of *Predicting Old Faithful* student page.
- Poster paper (grid paper if possible), markers, and tape for presented graphs and measures.

Use of Space: Students work at their tables in groups of three or four.

Launch: See **QRS Guide.**

Explore: See **QRS Guide.**

Share and Summarize: See **QRS Guide.**

Homework or Extension: Have students find mean monthly temperatures for two different cities over a few years (e.g., Rio de Janeiro and Fairbanks, Alaska). What patterns do they see in the data? What measures or displays are most revealing?

Accommodations: Encourage students who try to solve the problem numerically to attempt a visual solution. Support those who work only with graphs to analyze the data shown in the graphs. Visually handicapped students may need enlarged data sets and larger-scaled graph paper. Have all students make posters large and visible from the front to the back of the room. English Language Learners and others may need support in knowing meanings of *mean* and *median*. Remind students, as needed, about cooperative learning ground rules—for example, everyone needs to understand what the group is deciding, hold back if you participate a lot, encourage yourself if you are a reluctant participant.

Attachments: Student page *Predicting Old Faithful.*

Reflection: Reflect on the lesson after teaching.

TEACHING AND LEARNING HIGH SCHOOL MATHEMATICS © 2010 John Wiley & Sons, Inc.

QRS Guide for Predicting Old Faithful

Questions or Learning Activities	Students' Responses (Expected or Observed)	Teacher Support (Using Stages of Questioning)	Teacher Notes
Launch: Who has been to Yellowstone National Park? What are some of the thermal features?	Expect a few students to have been there and to mention hot springs, geysers like Old Faithful, acid lakes, and mudpots.		
Today, we are going to imagine that we are visiting Old Faithful, a geyser known for its regularity of erupting many times a day for many decades. Suppose your family arrived at the viewing area only to discover that the geyser has just erupted.			
Explore: How long would you expect to wait for the next eruption? What data would you like to have? How much data would you need? Discuss these questions with your partners.	Students speculate in small groups. Some might want to know how long the previous visitors waited for an eruption, some would like a day or week's worth of data, some will ask a ranger, and some will likely speculate that data vary by season.	**Tone setting:** Encourage students to find multiple ways to think about the problem. Acknowledge that they might have one way they feel comfortable with, but the goal is to find other ways to address the problem. Acknowledge that the work might be challenging, but you have high expectations for them.	**Cautious point:** Ask students who have visited Old Faithful to hold back what they might know about the wait times.
Distribute the student page and assign each team eruption data for one day. Tell them the data are the numbers of minutes between the end of one eruption to the beginning of the next eruption on a given day. Students are to analyze the data to get insights into addressing the question of the length of the wait until the next eruption. Remind students to use more than one means of analysis.	*Satisfactory:* (Expect this late in the **Explore.**) Students produce a connected scatterplot or bar graph showing alternating wait times. The independent variable is the eruption number (eruption 1 through eruption 18). The dependent variable is the length of time from the beginning of one eruption to the next in the order that the data are listed. Reasoning: The display shows how the wait times generally vary and alternate (or oscillate) between longer and shorter wait times.	How would you convince someone that your analysis is valid? What more might you investigate?	Share #4. **Cautious point:** If someone has an early "aha" by seeing the alternation of lengths of wait times, try to have them keep this quiet. Press this student or team to find other displays or insights.
	Partial response: Students may find an average and not be able to interpret it or to think of another way to analyze the data. (This is a partial response because averages mask a lot of information about what is happening, and the task required more than one analysis.) Example: mean for Set 1 = 72.5	How else might you get a sense of what is happening? What other ways are there to look at a set of numbers? Would drawing a graph or diagram help? What does the average tell you? How is it helpful?	Share #1.

(*continued*)

TEACHING AND LEARNING HIGH SCHOOL MATHEMATICS © 2010 John Wiley & Sons, Inc.

Questions or Learning Activities	Students' Responses (Expected or Observed)	Teacher Support (Using Stages of Questioning)	Teacher Notes
	Partial: Students may produce a graph but without analysis. Students do not see variability. Sample Stem Plot 5 | 0 4 6 7 6 | 0 0 7 | 1 3 5 7 7 8 | 0 1 3 6 6 9 9 | 0	What does your display help you see? How do you interpret this display? What important information does your display show? What does it tell you about Old Faithful?	Share #2
	Partial: Sample boxplot 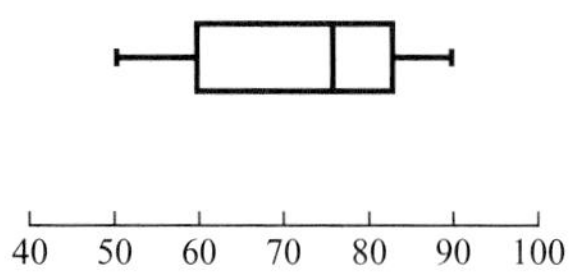 40 50 60 70 80 90 100 Students may produce bar graphs or histograms.	**Possible segment of direct instruction:** Students may need a reminder about the **five-point summary** that is needed for a boxplot. Students could be asked to explain this summary to one another. **Cautious point:** How accurately are students finding upper and lower quartiles?	Share #3 Compare with stem plot and with mean.
	Misconceived: Students may draw poorly constructed graphs.	Explain your graph to me. What are you trying to show? Can you make the graph clearer to someone who did not help you make it? Let's look closely at your scale (or other area needing attention).	Expect these to be self- or group-corrected during the **Explore**.
After students work for a short time with the data, ask: How does your prediction using data compare with your prediction before seeing the data?			
After groups have at least one display for their data, ask students to analyze a second day of data.		How do your predictions compare now after examining two data sets?	
Possible segment of direct instruction: Students may need a reminder about how to use the data features of a graphing calculator: `STAT, EDIT, Clear, STAT Plot, Window,` etc. To produce a connected scatterplot on the TI-83/84, make L2 the data and make L1 the			

(*continued*)

TEACHING AND LEARNING HIGH SCHOOL MATHEMATICS © 2010 John Wiley & Sons, Inc.

Questions or Learning Activities	Students' Responses (Expected or Observed)	Teacher Support (Using Stages of Questioning)	Teacher Notes
consecutive numbers, for example, 1–18. (Or students who figure this out may later explain it to others.)			
Share and Summarize: Questions	**Important Responses**		
Compare and contrast the different displays and measures. How are the different graphs related to one another?	They all show a wide range; times vary; stem and leaf plots could be rotated to be histograms; we would expect to wait around 50–80 min. Means mask the variations in times. Some teams may have found the bi-modal nature of the wait times without seeing how they alternate.		
What overall patterns do you see in the results?	Wait times alternate. The longer times seem to vary more than the shorter times. We could look more closely at the longer and shorter times separately. We could study the science more to learn more about what happens inside the volcano.		
Which displays are most effective in understanding the data?	Connected scatterplots showing eruption number versus wait time show more clearly the alternating lengths of wait times. Some histograms indicate the bimodal nature of the phenomenon.		
Why does it make more sense to predict a range of possible wait times than a single time?	You are more likely to be accurate when you predict a range of wait times. You can miss the specific time but be within the range. In this case, there might be two intervals, one following a longer wait and one following a shorter wait.		
What role does variability play in this analysis?	The wait times are not always the same. But there is a trend that longer times alternate with shorter times. Averaging loses information about the underlying pattern.		

TEACHING AND LEARNING HIGH SCHOOL MATHEMATICS © 2010 John Wiley & Sons, Inc.

Contents

All resources in this appendix are also on the course website.

Sample Unit Outline: Introducing Derivatives

Grade Levels: upper secondary or early college (precalculus or beginning calculus)

Time Frame

This unit is designed to take about two weeks, depending on the needs of students and the time available.

Prerequisite Knowledge

1. Read, use, and interpret function notation.
2. Model phenomena with linear, quadratic, exponential, and trigonometric functions.
3. Operate with functions (add/subtract, multiply/divide, compose, determine, and use inverse functions).

Learning Outcomes

1. Continue to build and extend meanings and contexts for rate of change.
2. Understand relationships between the graph of a distance/displacement function and secant lines between points on the graph as the distance between points decreases to zero.
3. Understand relationships between the graph of a distance/displacement function and its rate of change graph.
4. Be able to sketch a rate of change graph from a distance/displacement graph.

Overview and Learning Activities

This unit provides an informal introduction to concepts of derivatives and limits studied more formally in a calculus course.

1. *What Does Rate of Change Mean?*. In this lesson, students identify and generate several contexts where a rate of change is relevant (e.g., heart rate, birth rate, bowling average, oil spill expansion). The concept is symbolized by the quotient $\frac{\text{change in } y}{\text{change in } x}$ or $\frac{f(b) - f(a)}{b - a}$. Students apply the notation to a variety of contexts. This lesson takes one day.
2. *Representing Distance versus Time Graphically*. Students generate, then analyze several graphs of distance versus time using a calculator based ranger (CBR). Students match and justify why certain graphs represent certain sets of walking conditions (e.g., walk for 6 seconds, run for 6 seconds, then walk slowly back to your starting point). Students generate graphs for phenomena involving motion. This lesson takes one day.
3. *Distance versus Speed and Displacement versus Velocity*. Students are introduced to distinctions between distance (total distance traveled) and displacement (net distance from a starting point) and, comparably, speed (rate of change of distance with respect to time) and velocity (rate of change of displacement with respect to time). This lesson takes one day.
4. *Thinking More Carefully about Motion*. Students are introduced to smooth curves. They continue to interpret and connect displacement and velocity situations, including those with gradual changes. They also solve average velocity problems, and instantaneous velocity problems, making sense of the symbols $\frac{f(b) - f(a)}{b - a}$ and $\frac{f(c + h) - f(c)}{(c + h) - c} = \frac{f(c + h) - f(c)}{h}$ as h decreases to zero. They begin to consider the relationship between the graph of $y = f(x)$ and secant lines containing points $(c, f(c))$ and $(c + h, f(c + h))$ as h goes to zero.

   This lesson includes the student page **Rate of Change from Two Points of View**. This lesson takes one day.
5. *Rates of Change at Points and in Intervals*. Students examine graphs of curves under magnification to see that most approach a line. Within various motion contexts, they estimate and compare velocities visually. They sketch graphs of rates of change (velocity) from graphs of distance. They match graphs of distance versus time with graphs of their rates of change. This lesson takes one day.
6. *Secant versus Tangent Line and Definition of the Derivative*. Students examine rate of change functions locally by looking at a graph of $y = f(x)$, a fixed point of interest $P(c, f(c))$, and a movable point $Q(c + h, f(c + h))$, and lines

TEACHING AND LEARNING HIGH SCHOOL MATHEMATICS © 2010 John Wiley & Sons, Inc.

(secants to the graph) containing these two points. They use a computer program to slide Q along the graph to determine how the secant line appears when compared to the graph of $y = f(x)$ as Q approaches P. They determine that secant lines approach tangent lines as h approaches zero. Students summarize their work of the past several days to define the derivative at a point as $f'(c) = \lim_{h \to 0} \frac{f(c+h) - f(c)}{h}$. This lesson takes one day.

7. *Using the Definition of the Derivative.* Using the definition of the derivative, students estimate velocity-time graphs by estimating slopes of displacement-time graphs. They generate velocity-time graphs from displacement-time graphs and examine their relationships. They notice that for polynomials, rate of change graphs appear one degree lower than the functions from which they arise. They notice that derivatives of sine and cosine functions are closely related. This lesson takes one day.

Unit Project

Students generate a story (e.g., growth of a new business, population growth rates, running, mountain climbing) involving multiple changes in rates. They produce graphs for both the story and for its rates of change. They write about the relationships between the graphs and justify why they make sense. They produce a written report and present an oral report with visuals.

Materials

- calculator-based rangers (CBRs)
- graphing calculators with table and zoom capabilities

Assessments

1. Daily warm-up problems
2. Daily observations of students interacting in small groups
3. Daily homework
4. Quizzes or short writing assignments
5. Project described in Unit Project
6. Unit test

TEACHING AND LEARNING HIGH SCHOOL MATHEMATICS © 2010 John Wiley & Sons, Inc.

Template for Classifying Test Items

For each test item, classify the item using the categories as described in "Improving Classroom Tests as a Means of Improving Assessment" by Denisse R. Thompson, Charlene E. Beckmann, and Sharon L. Senk. *Mathematics Teacher*, 90 (January 1997): 58–64.

Item Number	**Skill**	**Level**	**Real Context**	**Open Ended**	**Representation**	**Reasoning**	**Technology**
	Yes (Y) **No (N)**	**Low (L)** **Other (O)**	**Yes (Y)** **No (N)**	**Yes (Y)** **No (N)**	**Interpret (I)** **Superfluous (S)** **Make (M)** **None (N)**	**Yes (Y)** **No (N)**	**Active (A)** **Neutral (N)** **Inactive (I)**
Totals	N:	O:	Y:	Y:	I : M :	Y:	A:
%							

TEACHING AND LEARNING HIGH SCHOOL MATHEMATICS © 2010 John Wiley & Sons, Inc.

Sample Confidential Group Evaluation Form

Project Topic: ____

Evaluator: ______________________ **Date:** ______________________

Directions:

Rate your contributions and those of each member of your group for work you completed together on the group project. This evaluation is confidential. For each group member:

- Write the person's name in the first row of the table below, one name per column.
- Use a rating scale from 0 to 4 for each category with the meaning:
  0: unacceptable performance
  1: inadequate and irregular performance
  2: adequate but irregular or last minute performance
  3: acceptable performance with minor follow-up required
  4: consistently strong performance
- Write the **Rating** in the labeled column. *This is not a rank ordering*. Several members may earn the same score. You may use decimals in your rating.
- Calculate **Rating** × **Weight** = Points. Enter this value in the **Points** column.
- Add the values in the **Points** column and write the total in the last row of the table.

For your project grade, individual scores will be computed as a percentage of the total points earned. The percentage earned is the average of the ratings provided by group members. For example, if a person is rated 80, 78, and 85 by self and other members of the group, the student will earn $((80 + 78 + 85) \div 3)\% = 81\%$ of the total project grade. The percentage amount will decrease by 5 for each part of this form that is not carefully completed. Be honest. Major discrepancies among group members' evaluations of each other will be investigated.

Person Evaluated		Yourself							
Category	Weight	Rating (0–4)	Points	Rating (0–4)	Points	Rating (0–4)	Points	Rating (0–4)	Points
Written Communications	2								
Spoken Communications	2								
Organization and Planning	2								
Initiative and Dependability	2								
Interpersonal Relations	2								
Quality of Research Completed Together or Independently	5								
Ability to Analyze and Summarize Information	10								
TOTAL POINTS									

Describe your contribution to the work completed toward the project. Be specific. How thorough, thoughtful, and accurate were your contributions?

TEACHING AND LEARNING HIGH SCHOOL MATHEMATICS © 2010 John Wiley & Sons, Inc.

Characterize the contribution of each member in your group. Indicate the group member's name in the space provided. Specify what each group member contributed to the project. How well did your group work together? Discuss each member's (including your own) level of preparedness for team meetings and the depth of thought, completeness, and thoroughness of each member's contribution. Explain in the space provided.

Name: ______________________________ Comments: ______________________________

Name: ______________________________ Comments: ______________________________

Name: ______________________________ Comments: ______________________________

Scheduling. Were there difficulties in scheduling times to meet? Explain.

What measures were taken by your group to ensure you had enough time to work together?

Other Comments/Concerns. Please comment on your level of satisfaction concerning the group's ability to get along with each other and to complete assigned work and activities to the satisfaction of each member of the group. Are you happy, unhappy . . . ? Keep in mind that this information is confidential.

TEACHING AND LEARNING HIGH SCHOOL MATHEMATICS © 2010 John Wiley & Sons, Inc.

Date: ___________ Evaluator: ___________

Sample Peer Evaluation for Oral Presentations

<table>
<tr>
<td>
Presenter ___________

Title ___________

Please circle the number that best represents your assessment of the presenter and information presented. Meanings for ratings of 0 and 4 are shown at ends of each number line.

The presenter was:
0-------------1-------------2-------------3-------------4
poorly prepared well prepared

The information presented was:
0-------------1-------------2-------------3-------------4
too incomplete complete, useful

The activity to help develop understanding was:
0-------------1-------------2-------------3-------------4
standard, routine interesting, engaging

Written materials for student activity were:
0-------------1-------------2-------------3-------------4
not useful well designed, impressive
</td>
<td>
A strength of the presentation was:

A suggestion for change is:

Overall comments:
</td>
</tr>
</table>

TEACHING AND LEARNING HIGH SCHOOL MATHEMATICS © 2010 John Wiley & Sons, Inc.

APPENDIX: UNIT SIX

Contents

The resource in this appendix is also on the course website.

Your Professional Portfolio

A professional portfolio includes evidence of your expertise related to your profession. It includes information about your views of your roles as a teacher and a professional. It provides evidence of your continued professional development. The portfolio will be a useful tool as you interview for positions and as you make the case for advancements in your career.

As you progress through your professional program and first several years of teaching, you will accumulate much more material to include in each section, but particularly for the sections on *Examples of Exemplary Teaching Materials* and *Examples of Assessments and Student Work*. Eventually, these sections will be removed from your Professional Portfolio to create a separate Teaching Portfolio. The distinction between these two is that the Teaching Portfolio provides evidence of your best work as a teacher. The Professional Portfolio provides evidence of work that you do outside of your classroom in addition to day-to-day teaching.

Include in your professional portfolio what is relevant to your situation from the following suggestions. The order of items is recommended but not rigid. Descriptions of each section are provided where needed. You are encouraged to update your portfolio regularly.

Mission Statement

The mission statement should be a brief, succinct statement about your beliefs about teaching and student learning. A mission statement indicates, in a straightforward manner, how you will assist students in their endeavors to learn. The statement should be about two sentences.

Professional Résumé

The résumé should include the following categories:

- Education
- Professional Work Experience (include tutoring for which you were paid, substitute teaching, teaching positions you have held, coaching for which you have been paid)
- Related Experience (tutoring, coaching, other volunteer experiences related to education, or work with young people)
- Grants and Awards
- Publications
- Presentations
- Professional Development Experiences (list conferences in which you participated, workshops/inservice experiences)
- Professional Affiliations (State Council of Teachers of Mathematics, NCTM)
- Hobbies and Interests

Personal Statement

Your personal statement should explain your interest in teaching mathematics and should give an indication of how your personal history led you to teaching. It needs to state why you want to work with students at the particular age groups for which you will be certified. Also, make evident your plans for staying current and for deepening your own understanding of mathematics and mathematics education.

Teaching Statement

Begin your teaching statement with your philosophy of teaching. Describe how you will teach and why. Make evident through your teaching statement what you have learned about how students learn mathematics. Reflect on the use of cooperative learning, technology, physical materials, a variety of assessment techniques, and so on. Through your teaching statement, display your familiarity with state and national standards for school mathematics, your familiarity with a variety of problem-solving methods, your use of multiple representations, and your interest in connecting mathematics to everyday situations.

TEACHING AND LEARNING HIGH SCHOOL MATHEMATICS © 2010 John Wiley & Sons, Inc.

Awards

Include copies of certificates or citations for awards you have received.

Letters of Support

Include letters from former teachers (particularly those who can vouch for your ability to teach mathematics), principals, teaching mentors, colleagues, and former employers who can speak about your work ethic, reliability, attitude, and the like.

Examples of Exemplary Teaching Materials

Include teaching materials, such as **QRS Guides**, lesson plans, or activities you created. Be sure to indicate in each lesson plan which state objectives and/or NCTM *Standards* are addressed by your lesson or unit. Also include supporting materials, such as review games or concept maps. Through this collection, make evident your skill in writing lesson plans and your ability to modify or design instructional materials.

Examples of Assessments and Student Work

Include assessment materials you have written. If possible, also include copies of student work at more than one level with names blacked out. Through this collection, make evident your familiarity with multiple roles and tools for assessment and your understanding of how some assessments might be more suitable for different purposes.

Pictures of You Working with Students

Include pictures of you interacting with students during classes or through volunteer experiences (in which you are not standing in front of the room, but instead are actively engaged with students). For each, write a brief description of what you and your students are doing and place the description with the picture. (Be sure to obtain appropriate permissions.)

Pictures of Exemplary Classroom Displays You Have Created

Include pictures of interesting bulletin boards or classroom displays that you and your students have created. Include a brief description of the display and how it is used in the classroom to help students learn or communicate their understanding.

Presentations

List professional presentations you have made, including the date, organization, and location. Include selected handouts you prepared.

Published Papers

Include papers you have authored or co-authored and their bibliographic information.

Grant Proposals

List grants you have led or co-led. Include dates, funding sources, and funding amounts.

The last three items will likely be added to your portfolio as your career advances.

TEACHING AND LEARNING HIGH SCHOOL MATHEMATICS © 2010 John Wiley & Sons, Inc.

REFERENCES

Albert, Lillie R. "Lessons Learned from the 'Five Men Crew': Teaching Culturally Relevant Mathematics." In Marilyn E. Strutchens, Martin L. Johnson, and William F. Tate (Eds.), *Changing the Faces of Mathematics: Perspectives on African Americans* (pp. 81–88). Reston, VA: National Council of Teachers of Mathematics, 2000.

Albrecht, Masha. "The Volume of a Pyramid: Low-Tech and High-Tech Approaches." *Mathematics Teacher*, 94 (January 2001): 58–61 plus lab pages.

Aliaga, Martha and Brenda Gunderson. *Interactive Statistics* (Second Edition). Upper Saddle River, NJ: Pearson Education, 2003.

American Association for the Advancement of Science. *Benchmarks for Science Literacy*. New York: Oxford University Press, 1993.

American Association for the Advancement of Science. "Instructional Criteria." (http://www.project2061.org/publications/textbook/algebra/summary/critdet.htm, accessed July 31, 2009.).

Anno, Masaichiro and Mitsumasa. *Anno's Mysterious Multiplying Jar*. New York: Philomel Books, 1983.

Anno, Mitsumasa. *Anno's Magic Seeds*. New York: Philomel Books, 1995.

Appelt, Kathi. *Bats on Parade*. New York: HarperCollins, 1999.

Ball, Deborah Loewenberg and Hyman Bass. "Interweaving Content and Pedagogy in Teaching and Learning to Teach: Knowing and Using Mathematics." In Jo Boaler (Ed.), *Multiple Perspectives on Teaching and Learning* (pp. 83–103). Westport, CT: Ablex Publishing, 2000.

Ball, Deborah Loewenberg, Imani Masters Gaffney, and Hyman Bass: "The Role of Mathematics Instruction in Building a Just and Diverse Democracy." *The Mathematics Educator*, 15 (2005): 2–6.

Baltus, Christopher. "A Truth Table on the Island of Truthtellers and Liars." *Mathematics Teacher*, 94 (December 2001): 730–732.

Barton, Mary Lee and Clare Heidema. *Teaching Reading in Mathematics* (Second Edition). Aurora, CO: Mid-continent Research for Education and Learning, 2002.

Becker, Jerry P. and Shigeru Shimada. *The Open-Ended Approach: A New Proposal for Teaching Mathematics*. Reston, VA: National Council of Teachers of Mathematics, 1997.

Beckmann, Charlene E. *Exploring Brief Calculus with a Graphing Calculator*. Reading, MA: Addison-Wesley, 1993.

Beckmann, Charlene E., Sharon L. Senk, and Denisse R. Thompson. "Assessing Students' Understanding of Functions in a Graphing Calculator Environment." *School Science and Mathematics*, 99 (December 1999): 451–456.

Beckmann, Charlene E. and Theodore A. Sundstrom. *Exploring Calculus with a Graphing Calculator*. Reading, MA: Addison-Wesley, 1992.

Birch, David. *The King's Chessboard*. New York: Puffin Pied Piper Books, 1988.

Bloom, Benjamin and David Krathwahl. *Taxonomy of Educational Objectives*. New York: Addison-Wesley, 1984.

Bogomolny, Alexander "Non-Euclidean Geometries: Drama of the Discoveries." *Interactive Mathematics Miscellany and Puzzles*. (http://www.cut-the-knot.org/triangle/pythpar/Drama.shtml, accessed May 8, 2009)

Bosangue, Martin V. and Gerald E. Gannon. "From Exploration to Generalization: An Introduction to Necessary and Sufficient Conditions." *Mathematics Teacher*, 96 (May 2003): 366–371.

Boyer, Carl B. and revised by Uta C. Merzbach. *A History of Mathematics* (Second Edition). New York: John Wiley & Sons, Inc., 1991.
Burke, Maurice J. and Ted R. Hodgson. "Using Technology to Optimize and Generalize: The Least-Squares Line." *Mathematics Teacher*, 101 (September 2007): 102–107.
Burrill, Gail, Christine A. Franklin, Landy Godbold, Linda J. Young, Johnny W. Lott, and Peggy A. House. *Navigating through Data Analysis in Grades 9–12*. Reston, VA: National Council of Teachers of Mathematics, 2003.
Canada, Dan, and Stephen Blair. "Intersections of a Circle and a Square: An Investigation." *Mathematics Teacher*, 100 (December 2006/January 2007): 324–328.
Carlson, Ronald J. and Mary Jean Winter. *Algebra Experiments II: Exploring Nonlinear Functions*. Menlo Park, CA: Addison-Wesley, 1993.
Christelow, Eileen. *Five Little Monkeys Jumping on the Bed*. New York: Clarion Books, 1998.
Chwast, Seymour. *The 12 Circus Rings*. San Diego, CA: Gulliver Books, 1993.
Cobb, Mary. *The Quilt-Block History of Pioneer Days*. Brookfield, CT: The Millbrook Press, 1995.
Cox, Rhonda L. "Using Conjectures to Teach Students the Role of Proof." *Mathematics Teacher*, 97 (January 2004): 48–52.
Coxford, Arthur, Zalman Usiskin, and Daniel Hirschhorn. *The University of Chicago School Mathematics Project: Geometry*. Glenview, IL: ScottForesman, 1991.
Craine, Timothy V. and Rheta N. Rubenstein. "A Quadrilateral Hierarchy to Facilitate Learning in Geometry." *Mathematics Teacher*, 86 (January 1993): 30–36.
Croom, Lucille. "Mathematics for All Students: Access, Excellence, and Equity." In Janet Trentacosta and Margaret J. Kenney (Eds.), *Multicultural and Gender Equity in the Mathematics Classroom: The Gift of Diversity* (pp. 1–9). Reston, VA: National Council of Teachers of Mathematics, 1997.
Crowley, Mary L. "The van Hiele Model of the Development of Geometric Thought." In Mary Montgomery Lindquist and Albert P. Shulte (Eds.), *Learning and Teaching Geometry, K–12* (pp. 1–16). Reston, VA: National Council of Teachers of Mathematics, 1987.
Dalton, LeRoy C. *Algebra in the Real World: 38 Enrichment Lessons for Algebra 2*. Palo Alto, CA: Dale Seymour Publications, 1983.
Davidson, Neil. *Cooperative Learning in Mathematics: A Handbook for Teachers*. Menlo Park, CA: Addison-Wesley, 1990.
Davis, Jon D. "Connecting Procedural and Conceptual Knowledge of Functions." *Mathematics Teacher*, 99 (August 2005): 36–39.
Dodge, Walter, Kathleen Goto, and Philip Mallinson. "I Would Consider the Following to Be a Proof . . . " *Mathematics Teacher*, 91 (November 1998): 652–653.
Donovan, M. Suzanne and John D. Bransford. "Introduction." In M. Suzanne Donovan and John D. Bransford (Eds.), *How Students Learn: Mathematics in the Classroom* (pp. 1–28). Washington, DC: National Academies Press, 2005. (http://books.nap.edu/openbook.php?record_id=11101&page=1, accessed August 4, 2009)
Edwards, Thomas G. and Kenneth R. Chelst. "Queuing Theory: A Rational Approach to the Problem of Waiting in Line." *Mathematics Teacher*, 95 (May 2002): 372–376.
Epp, Susanna S. "A Unified Framework for Proof and Disproof." *Mathematics Teacher*, 91 (November 1998): 708–713.
Franklin, Christine, Gary Kader, Denise Mewborn, Jerry Moreno, Roxy Peck, Mike Perry, and Richard Schaeffer. *Guidelines for Assessment and Instruction in Statistics Education (GAISE) Report*. Alexandria, VA: American Statistical Association, 2007.
Franklin, Christine A. and Madhuri S. Mulekar. "Is Central Park Warming?" *Mathematics Teacher*, 99 (May 2006): 600–605.
Frayer, D., Frederick, W. C., and Klausmeier, H. J. *A Schema for Testing the Level of Cognitive Mastery*. Madison, WI: Wisconsin Center for Education Research, (http://www.justreadnow.com/strategies/frayer.htm, accessed May 7, 2009)
Freedom Writers. Film. Produced by Danny DeVito, Michael Shamberg, and Stacey Sher. Directed by Richard LaGravenese. Paramount Pictures, 2006.
Garrison, Leslie and Jill Kerper Mora. "Adapting Mathematics Instruction for English-Language Learners: The Language-Concept Connection." In Luis Ortiz-Franco, Norma G. Hernandez, and Yolanda De La Cruz (Eds.), *Changing the Faces of Mathematics: Perspectives on Latinos* (pp. 35–47). Reston, VA: National Council of Teachers of Mathematics, 1999.
Gay, A. Susan. "Helping Teachers Connect Vocabulary and Conceptual Understanding." *Mathematics Teacher*, 102 (October 2008): 218–223.
Gilkey, Susan N. and Carla H. Hunt. *Teaching Mathematics in the Block*. Larchmont, NY: Eye on Education, 1998.

TEACHING AND LEARNING HIGH SCHOOL MATHEMATICS © 2010 John Wiley & Sons, Inc.

Goldin, Gerald A. "Observing Mathematical Problem Solving through Task-Based Interviews." In Anne R. Teppo (Ed.), *Qualitative Research Methods in Mathematics Education* (pp. 40–62). Reston, VA: National Council of Teachers of Mathematics, 1998.

Gordon, Sheldon P. "Placement Tests: The Shaky Bridge Connecting School and College Mathematics." *Mathematics Teacher*, 100 (October 2006): 174–178.

Gratzer, William and James E. Carpenter. "The Histogram-Area Connection." *Mathematics Teacher*, 102 (December 2008–January 2009): 336–340.

Groth, Randall E. "Linking Theory and Practice in Teaching Geometry." *Mathematics Teacher*, 99 (August 2005): 27–30.

Hale, Patricia. "Kinematics and Graphs: Students' Difficulties and CBLs." *Mathematics Teacher*, 93 (May 2000): 414–417.

Hart, Eric W. and W. Gary Martin. "Standards for High School Mathematics: Why, What, How?" *Mathematics Teacher*, 102 (December 2008–January 2009): 377–382.

Hiebert, James. "A Theory of Developing Competence with Written Mathematical Symbols." *Educational Studies in Mathematics*, 19 (August 1988): 333–355.

Hilberg, Ruth Soleste, R. William Doherty, Stephanie Stoll Dalton, Daniel Youpa, and Roland G. Tharp. "Standards for Effective Mathematics Education for American Indian Students." In Judith Elaine Hankes and Gerald R. Fast (Eds.), *Changing the Faces of Mathematics, Perspectives on Indigenous People of North America* (pp. 25–35). Reston, VA: National Council of Teachers of Mathematics, 2002.

Himmelberger, Kathleen S. and Daniel L. Schwartz. "It's a Home Run! Using Mathematical Discourse to Support the Learning of Statistics." *Mathematics Teacher*, 101 (November 2007): 250–256.

Hong, Lily Toy. *Two of Everything*. Morton Grove, IL: Albert Whitman & Company, 1993.

Hulme, Joy N. *Sea Squares*. New York: Hyperion Books, 1991.

Hutchins, Pat. *The Doorbell Rang*. New York: Mulberry Books, 1986.

Jenkins, Steve. *Actual Size*. Boston: Houghton Mifflin, 2004.

Jones, Joan Cohen. "Hmong Needlework and Mathematics." In Carol A. Edwards (Ed.), *Changing the Faces of Mathematics: Perspectives on Asian Americans and Pacific Islanders* (pp. 13–20). Reston, VA: National Council of Teachers of Mathematics, 1999.

Jones, Joan Cohen. "Promoting Equity in Mathematics Education through Effective Culturally Responsive Teaching." In Rheta N. Rubenstein and George W. Bright (Eds.), *Perspectives on the Teaching of Mathematics* (pp. 141–150). Reston, VA: National Council of Teachers of Mathematics, 2004.

Kagan, Spencer. *Cooperative Learning*. San Juan Capistrano, CA: Kagan Cooperative Learning Company, 1992.

Kalchman, Mindy and Kenneth R. Koedinger. "Teaching and Learning Functions." In M. Suzanne Donovan and John D. Bransford (Eds.), *How Students Learn: Mathematics in the Classroom* (pp. 351–393). Washington, DC: National Academies Press, 2005. (http://books.nap.edu/openbook.php?record_id=11101&page=163, accessed August 4, 2009)

Katz, Lawrence C. and Manning Rubin. *Keep Your Brain Alive*. New York: Workman Publishing Co., 1999.

Katz, Victor J. *A History of Mathematics: An Introduction* (Second Edition). Reading, MA: Addison-Wesley, 1998.

Kilpatrick, Jeremy and Edward A. Silver. "Unfinished Business: Challenges for Mathematics Educators in the Next Decade." In Maurice J. Burke and Frances R. Curcio (Eds.), *Learning Mathematics for a New Century* (pp. 223–235). Reston, VA: National Council of Teachers of Mathematics, 2000.

Knuth, Eric J. "Understanding Connections between Equations and Graphs." *Mathematics Teacher*, 93 (January 2000): 48–53.

Konold, Clifford. "Teaching Probability through Modeling Real Problems." *Mathematics Teacher*, 86 (April 1994): 232–235.

Lapp, Douglas A. and Vivian Flora Cyrus. "Using Data-Collection Devices to Enhance Students' Understanding." *Mathematics Teacher*, 93 (September 2000): 504–510.

Lappan, Glenda, James T. Fey, William M. Fitzgerald, Susan N. Friel, and Elizabeth Difanis Phillips. *Middle Grades Mathematics Project*. Menlo Park, CA: Dale Seymour Publications, 1998.

Lasky, Kathryn. *The Librarian Who Measured the Earth*. Boston: Little, Brown, and Company, 1994.

Leitze, Annette Ricks and Nancy A. Kitt. "Using Homemade Algebra Tiles to Develop Algebra and Prealgebra Concepts." *Mathematics Teacher*, 93 (September 2000): 462–466, 520.

Lipp, Alan. "The Angles of a Star." *Mathematics Teacher*, 93 (September 2000): 512–516.

TEACHING AND LEARNING HIGH SCHOOL MATHEMATICS © 2010 John Wiley & Sons, Inc.

Ma, Liping. *Knowing and Teaching Elementary Mathematics: Teachers' Understanding of Fundamental Mathematics in China and the United States*. Mahwah, NJ: Lawrence Erlbaum Associates, 1999.

Manouchehri, Azita and Douglas A. Lapp. "Unveiling Student Understanding: The Role of Questioning in Instruction." *Mathematics Teacher*, 96 (November 2003): 562–566.

Manouchehri, Azita and Dennis St. John. "From Classroom Discussions to Group Discourse." *Mathematics Teacher*, 99 (April 2006): 544–551.

Marshall, J. C., B. Horton, J. Smart, and D. Llewellyn. *EQUIP: Electronic Quality of Inquiry Protocol*. Clemson, SC: Clemson University's Inquiry in Motion Institute, 2008.

Martin, Tami S. (Ed.). *Mathematics Teaching Today*. Reston, VA: National Council of Teachers of Mathematics, 2007.

Mason, John, Leone Burton, and Kaye Stacey. *Thinking Mathematically* (Revised Edition). Harlow, UK: Prentice Hall, 1985.

McConnell, John. "Forging Links with Projects in Mathematics." In Peggy A. House and Arthur F. Coxford (Eds.), *Connecting Mathematics Across the Curriculum* (pp. 198–209). Reston, VA: National Council of Teachers of Mathematics, 1995.

McElhaney, Kevin W. "Demonstrating Boolean Logic Using Simple Electrical Circuits." *Mathematics Teacher*, 97 (February 2004): 126–134.

McGlone, Chris and Gary M. Nieberle. "Activities: Using Hooke's Law to Explore Linear Functions." *Mathematics Teacher*, 93 (May 2000): 391–398.

McTighe, Jay and Frank T. Lyman, Jr. "Cueing Thinking in the Classroom: The Promise of Theory-Embedded Tools." *Educational Leadership*, 45 (April 1988): 18–24.

Meserve, Dorothy T. and Bruce E. Meserve. "Sequences and Polynomials Part I: Guidelines for Finding a Next Term and a General Term for Any Given Finite Sequence." *Mathematics Teacher*, 100 (February 2007): 426–429.

Metz, James. "Seeing How Money Grows." *Mathematics Teacher*, 94 (April 2001): 278–280 plus lab sheets.

Middleton, James A. and Photini A. Spanias. "Findings from Research on Motivation in Mathematics Education: What Matters in Coming to Value Mathematics." In Judith Sowder and Bonnie Schappelle (Eds.), *Lessons Learned from Research* (pp. 9–15). Reston, VA: National Council of Teachers of Mathematics, 2002.

Mokros, Jan and Susan Jo Russell. "Children's Conceptions of Average and Representativeness." *Journal for Research in Mathematics Education*, 26 (January 1995): 20–39.

Moore, David S. *Statistics: Concepts and Controversies*. New York: W. H. Freeman and Co., 2001.

Moschkovich, Judit N. "Understanding the Needs of Latino Students in Reform-Oriented Mathematics Classrooms." In Luis Ortiz-Franco, Norma G. Hernandez, and Yolanda De La Cruz (Eds.), *Changing the Faces of Mathematics: Perspectives on Latinos* (pp. 5–12). Reston, VA: National Council of Teachers of Mathematics, 1999.

Murphy, Stuart J. *Divide and Ride*. New York: HarperCollins Publishers, 1997.

National Council of Teachers of Mathematics. "Are You Interested in Stretching Your Dollars?" *Student Math Notes*, November 2001.

National Council of Teachers of Mathematics. *Assessment Standards for School Mathematics*. Reston, VA, 1995.

National Council of Teachers of Mathematics. *Curriculum and Evaluation Standards for School Mathematics*. Reston, VA, 1989.

National Council of Teachers of Mathematics. "Equity in Mathematics Education." Position Statement of the National Council of Teachers of Mathematics, adopted January 2008. (http://www.nctm.org/about/content.aspx?id=13490, accessed August 4, 2009)

National Council of Teachers of Mathematics. *Principles and Standards for School Mathematics*. Reston, VA, 2000.

National Council of Teachers of Mathematics. *Professional Standards for Teaching Mathematics*. Reston, VA, 1991.

Natsoulas, Anthula. "Group Symmetries Connect Art and History with Mathematics." *Mathematics Teacher*, 93 (May 2000): 364–370.

O'Brien, Thomas C. "Parrot Math." *Phi Delta Kappan*, 80 (February 1999): 434–438.

O'Connor, J. J. and E. F. Robertson. "History Topic: Non-Euclidean Geometry." *MacTutor History of Mathematics* (February 1996). (http://www.gap-system.org/~history/HistTopics/Non-Euclidean_geometry.html, accessed May 8, 2009).

Paulos, John Allen. *Innumeracy: Mathematical Illiteracy and its Consequences*. New York: Hill and Wang, 1988.

TEACHING AND LEARNING HIGH SCHOOL MATHEMATICS © 2010 John Wiley & Sons, Inc.

Pay It Forward. Film. Produced by Peter Abrams, Robert Levy, and Steven Reuther. Directed by Mimi Leder. Screenplay by Leslie Dixon. Warner Brothers Pictures in association with Bellair Entertainment, a Tapestry Films production, 2000.

Perrin, John Robert and Robert J. Quinn. "The Power of Investigative Calculus Projects." *Mathematics Teacher*, 101 (May 2008): 640–646.

Philipp, Randolph A. "The Many Uses of Algebraic Variables." *Mathematics Teacher*, 85 (October 1992): 557–561.

Pinczes, Elinor J. *My Full Moon Is Square*. Boston: Houghton Mifflin, 2002.

Pinczes, Elinor J. *One Hundred Hungry Ants*. Boston: Houghton Mifflin, 1993.

Pitici, Mircea."Non-Euclidean Geometry Online: A Guide to Resources." June 2008. (http://www.math.cornell.edu/~mec/mircea.html, accessed May 8, 2009)

Pittman, Helena Clare. *A Grain of Rice*. New York: Bantam Skylark Books, 1986.

Pólya, George. *How to Solve It: A New Aspect of Mathematical Method* (Second Edition). Princeton, NJ: Princeton University Press, 1957.

Resnick, Lauren. Remarks at the University of Chicago School Mathematics Project's Third International Conference, Chicago, IL, 1991.

Romagnano, Lew. "The Myth of Objectivity in Mathematics Assessment." *Mathematics Teacher*, 94 (January 2001): 31–37.

Rousseau, Celia K. and Angiline Powell. "Understanding the Significance of Context: A Framework to Examine Equity and Reform in Secondary Mathematics." *The High School Journal*, 88 (April/May 2005): 19–31.

Rubel, Laurie H. "Good Things Always Come in Threes: Three Cards, Three Prisoners, and Three Doors." *Mathematics Teacher*, 99 (February 2006): 401–405.

Rubenstein, Rheta N. "Learning Mathematics: Perspectives from Researchers." Unpublished manuscript especially prepared for *Teaching and Learning High School Mathematics*, 2009.

Rubenstein, Rheta N., Charlene E. Beckmann, and Denisse R. Thompson. *Teaching and Learning Middle Grades Mathematics*. Hoboken, NJ: John Wiley & Sons, 2004.

Rubenstein, Rheta N. and Randy K. Schwartz. "Word Histories: Melding Mathematics and Meanings." *Mathematics Teacher*, 93 (November 2000): 664–669.

Scher, Daniel. "A Triangle Divided: Investigating Equal Areas." *Mathematics Teacher*, 93 (October 2000): 608–611.

Schultz, James E. and Rheta N. Rubenstein. "Integrating Statistics into a Course on Functions." *Mathematics Teacher*, 83 (November 1990): 612–617.

Scieska, Jon and Lane Smith. *Math Curse*. New York: Viking Books, 1995.

Senk, Sharon L. "How Well Do Students Write Geometry Proofs?" *Mathematics Teacher*, 78 (September 1985): 448–456.

Serra, Michael. *Discovering Geometry: An Inductive Approach* (Second Edition). Emeryville, CA: Key Curriculum Press, 1997.

Shaughnessy, J. Michael. "Research on Students' Understandings of Probability." In Jeremy Kilpatrick, W. Gary Martin, and Deborah Schifter (Eds.), *A Research Companion to Principles and Standards for School Mathematics* (pp. 216–226). Reston, VA: National Council of Teachers of Mathematics, 2003.

Shaughnessy, J. Michael and Thomas Dick. "Monty's Dilemma: Should You Stick or Switch?" *Mathematics Teacher*, 84 (April 1991): 252–256.

Shaughnessy, J. Michael and Maxine Pfannkuch. "How Faithful Is Old Faithful? Statistical Thinking: A Story of Variation and Prediction." *Mathematics Teacher*, 95 (April 2002): 252–259.

Shield, Mal and Kevan Swinson. "The Link Sheet: A Communication Aid for Clarifying and Developing Mathematical Ideas and Processes." In Portia C. Elliott and Margaret J. Kenney (Eds.), *Communication in Mathematics, K–12 and Beyond* (pp. 35–39). Reston, VA: National Council of Teachers of Mathematics, 1996.

Sierra, Judy. *Counting Crocodiles*. Orlando, FL: Harcourt, 1997.

Slavin, Robert E. "Research on Cooperative Learning and Achievement: What We Know, What We Need to Know." *Contemporary Educational Psychology*, 21 (January 1996): 43–69.

Smith, Margaret S., Victoria Bill, and Elizabeth K. Hughes. "Thinking through a Lesson: Successfully Implementing High-Level Tasks." *Mathematics Teaching in the Middle School*, 14 (October 2008): 132–138.

Smith, Margaret S., Elizabeth K. Hughes, Randi A. Engle, and Mary Kay Stein. "Orchestrating Discussions." *Mathematics Teaching in the Middle School*, 14 (May 2009): 549–556.

Smith, Margaret S. and Mary Kay Stein. "Selecting and Creating Mathematical Tasks: From Research to Practice." *Mathematics Teaching in the Middle School*, 3 (February 1998): 344–350.

TEACHING AND LEARNING HIGH SCHOOL MATHEMATICS © 2010 John Wiley & Sons, Inc.

Sowder, Larry and Guershon Harel. "Types of Justifications." *Mathematics Teacher*, 91 (November 1998): 670–675.

Stanton, Robert O. "Proofs that Students Can Do." *Mathematics Teacher*, 99 (March 2006): 478–482.

Steen, Lynn Arthur. "Facing Facts: Achieving Balance in High School Mathematics." *Mathematics Teacher*, 100 (Special Issue: *100 Years of Mathematics Teacher*, December 2006): 86–95.

Stein, Mary Kay and Suzanne Lane. "Instructional Tasks and the Development of Student Capacity to Think and Reason: An Analysis of the Relationship between Teaching and Learning in a Reform Mathematics Project." *Educational Research and Evaluation*, 2 (October 1996): 50–60.

Stevens, Jill. "Activities: Generating and Analyzing Data." *Mathematics Teacher*, 86 (September 1993): 475–478, 487–489.

Stiggins, Rick. "From Formative Assessment to Assessment FOR Learning: A Path to Success in Standards-Based Schools." *Phi Delta Kappan*, 87 (December 2005): 324–328.

Stutzman Rodney Y. and Kimberly H. Race. "EMRF: Everyday Rubric Grading." *Mathematics Teacher*, 97 (January 2004): 34–39.

Tappin, Linda. "Analyzing Data Relating to the *Challenger* Disaster." *Mathematics Teacher*, 87 (September 1994): 423–426.

Tarr, James E., Barbara J. Reys, David D. Barker, and Rick Billstein. "Selecting High-Quality Mathematics Textbooks." *Mathematics Teaching in the Middle School*, 12 (August 2006): 50–54.

Tennison, Alan D. "Promoting Equity in Mathematics: One Teacher's Journey." *Mathematics Teacher*, 101 (August 2007): 28–31.

Thompson, Denisse R., Charlene E. Beckmann, and Sharon L. Senk. "Improving Classroom Tests as a Means of Improving Assessment." *Mathematics Teacher*, 90 (January 1997): 58–64.

Thompson, Denisse R. and Rheta N. Rubenstein. "Learning Mathematics Vocabulary: Potential Pitfalls and Instructional Strategies." *Mathematics Teacher*, 93 (October 2000): 568–574.

Thompson, Denisse R., and Sharon L. Senk. "Using Rubrics in High School Mathematics Courses." *Mathematics Teacher*, 91 (December 1998): 786–793.

Tolkien, J. R. R. *The Lord of the Rings*. Boston: Houghton Mifflin, 1954.

Usiskin, Zalman. "The Case of the University of Chicago School Mathematics Project—Secondary Component." In Christian R. Hirsch (Ed.), *Perspectives on the Design and Development of School Mathematics Curricula* (pp. 173–182). Reston, VA: National Council of Teachers of Mathematics, 2007.

Usiskin, Zalman. "The Integration of the School Mathematics Curriculum in the United States: History and Meaning." In Sue Ann McGraw (Ed.), *Integrated Mathematics: Choices and Challenges* (pp. 13–32). Reston, VA: National Council of Teachers of Mathematics, 2003.

Usiskin, Zalman. *Van Hiele Levels and Achievement in Secondary School Geometry*. Final report of the Cognitive Development and Achievement in Secondary School Geometry Project. Chicago: University of Chicago, 1982. (ERIC Document Reproduction Services ED 220 288).

Usiskin, Zalman, Daniel Hirschhorn, Arthur Coxford, Virginia Highstone, Hester Lewellen, Nicholas Oppong, Richard DiBianca, and Merilee Maeir. *UCSMP Geometry* (Second Edition). Glenview, IL: Scott Foresman, Addison-Wesley, 1998.

Vazquez, Lorna Thomas. "A, E, I, O, U, and Always Y: A Simple Technique for Improving Communication and Assessment in the Mathematics Classroom." *Mathematics Teacher*, 102 (August 2008): 16–23.

Walker, Erica N. and Leah P. McCoy. "Students' Voices: African Americans and Mathematics." In Janet Trentacosta and Margaret J. Kenney (Eds.), *Multicultural and Gender Equity in the Mathematics Classroom: The Gift of Diversity* (pp. 71–80). Reston, VA: National Council of Teachers of Mathematics, 1997.

Wanko, Jeffrey J. "Giving Exponential Functions a Fair Shake." *Mathematics Teaching in the Middle School*, 11 (October 2005): 118–124.

Weisstein, Eric W. "Algebraic Function." *MathWorld*—A Wolfram Web Resource. (http://mathworld.wolfram.com/AlgebraicFunction.html, accessed July 31, 2009)

Weisstein, Eric W. "Transcendental Function." *MathWorld*—A Wolfram Web Resource. (http://mathworld.wolfram.com/TranscendentalFunction.html, accessed July 31, 2009)

Winter, Mary Jean and Ronald J. Carlson. *Algebra Experiments I: Exploring Linear Functions*. Menlo Park, CA: Addison-Wesley, 1993.

Wong, Michael. "The Human Body's Built-In Range Finder: The Thumb Method of Indirect Distance Measurement." *Mathematics Teacher*, 99 (May 2006): 622–626.

TEACHING AND LEARNING HIGH SCHOOL MATHEMATICS © 2010 John Wiley & Sons, Inc.

Woo, Elaine and Doug Smith. "California and the West: SAT Scores Rise, But Trouble Spots Remain." *Los Angeles Times*, August 27, 1993, p. A3.

Zbiek, Rose Mary. "Using Technology to Show Prospective Teachers the 'Power of Many Points.'" In William J. Masalski and Portia C. Elliot (Eds.), *Technology-Supported Mathematics Learning Environments* (pp. 291–301). Reston, VA: National Council of Teachers of Mathematics, 2005.

Zbiek, Rose Mary and Jeanne Shimizu. "Multiple Solutions: More Paths to an End or More Opportunities to Learn Mathematics." *Mathematics Teacher*, 99 (November 2005): 279–287.

TEACHING AND LEARNING HIGH SCHOOL MATHEMATICS © 2010 John Wiley & Sons, Inc.

INDEX

Note: *Titles of student pages are in italics.*

G

H

M

N

O

P

T

U